N-ACETYLASPARTATE

A Unique Neuronal Molecule
in the Central Nervous System

ADVANCES IN EXPERIMENTAL MEDICINE AND BIOLOGY

N-ACETYLASPARTATE

A Unique Neuronal Molecule
in the Central Nervous System

Edited by

John R. Moffett, Ph.D.
Uniformed Services University of the Health Sciences
Bethesda, Maryland

Suzannah B. Tieman, Ph.D.
University of Albany
Albany, New York

Daniel R. Weinberger, M.D.
National Institute of Mental Health
Bethesda, Maryland

Joseph T. Coyle, M.D.
McLean Hospital, Harvard University
Belmont, Massachusetts

and

Aryan M.A. Namboodiri, Ph.D.
Uniformed Services of the Health Sciences
Bethesda, Maryland

 Springer

John R. Moffett
Uniformed Services University of the
    Health Sciences
Bethesda, MD 20814
USA
jmoffett@usuhs.mil

Daniel R. Weinberger
National Institute of Mental Health
Bethesda, MD 20814
USA
weinberd@mail.nih.gov

Aryan M.A. Namboodiri
Uniformed Services University of
    the Health Sciences
Bethesda, MD 20814
USA
anamboodiri@usuhs.mil

Suzannah B. Tieman
University of Albany
Albany, NY 12222
USA
tieman@albany.edu

Joseph T. Coyle
McLean Hospital
Belmont, MA 02178
USA
joseph_coyle@hms.harvard.edu

Library of Congress Control Number: 2005938505

ISBN-10: 0-387-30171-2   e-ISBN 0-387-30172-0
ISBN-13: 978-0387-30171-6

Printed on acid-free paper.

9 8 7 6 5 4 3 2

springer.com

NAA, NAAG AND CANAVAN'S DISEASE: THE FIRST INTERNATIONAL SYPOSIUM ON *N*-ACETYLASPARTATE

Natcher Conference Center, NIH, Bethesda Maryland. September 13th and 14th, 2004. This volume chronicles the first comprehensive meeting held on *N*-acetylaspartate and Canavan's disease. The Symposium featured presentations by leading investigators and a poster session. The meeting was organized by the National Institute of Mental Health and was co-sponsored by the National Institute of Neurological Disorders and stroke, the National Institute of Child Health and Human Development, and the Office of Rare Diseases at the National Institute of Health, Bethesda, Maryland USA.

Organizing Committee:

Co-Chairs:
* Aryan Namboodiri Ph.D.
* Daniel Weinberger M.D.
* Joseph Coyle M.D.

Members:
* John R Moffett, Ph.D., Uniformed Services University, USA
* Reuben Matalon, M.D., Ph.D., University of Texas, USA
* Alessandro Burlina M.D., Ph.D, Universita de Padova, Italy
* Jeffrey Duyn, Ph.D., National Institutes of Health, USA
* Paola Leone, Ph.D., University of Medicine and Dentistry of New Jersey, USA
* Kishore Bhakoo Ph.D. Imperial College, London, UK
* P.M. Menon Ph.D., Jackson VAMC, USA
* Morris Baslow Ph.D. Nathan Kline Institute for Psychiatric Research, USA
* Thomas Sager, H. Lundbeck A/S, Denmark.

PREFACE: A BRIEF REVIEW OF
N-ACETYLASPARTATE

John R. Moffett and Aryan M.A. Namboodiri[*]

1. INTRODUCTION

The First International Symposium on N-acetylaspartate (NAA) was held on September 13th and 14th, 2004 in the Natcher Conference Center at the NIH, in Bethesda Maryland. The event showed clearly that NAA was no longer an obscure object of study by a few scattered researchers, and that instead it had become a subject of great interest to hundreds of scientists and clinicians from around the world. This newfound attention for one of the brain's most concentrated amino acid derivatives was a long time in coming, since the first report of NAA in the brains of cats and rats by Tallan and coworkers in 1956.[1] The Symposium also covered N-acetylaspartylglutamate (NAAG), a related dipeptide comprised of NAA and glutamate, which was first identified definitively as a brain-specific peptide by Miyamoto and colleagues in 1966.[2] Despite decades of research on the roles both these molecules play in the nervous system, their functions remain enigmatic, and controversial.

NAA in particular is a study in controversy, with virtually no consensus on its principle metabolic or neurochemical functions after nearly five decades of research. There is relative unanimity on one point; NAA is not thought to be a neurotransmitter or neuromodulator that is released synaptically upon neuronal depolarization. Beyond that, there is little consensus. At least four basic hypotheses have been offered for the principle role of NAA in the nervous system: 1) an organic osmolyte that counters the "anion deficit" in neurons, or a co-transport substrate for a proposed "molecular water pump" that removes metabolic water from neurons, 2) a source of acetate for myelin lipid synthesis in oligodendrocytes, 3) an energy source in neurons, and 4) a precursor for the biosynthesis of NAAG. These functions are not mutually exclusive, and indeed, NAA certainly serves multiple functions in the nervous system.

The two findings that catapulted NAA into mainstream scientific consciousness were the connection to the fatal hereditary genetic disorder known as Canavan disease, and the

[*] Uniformed Services University of the Health Sciences, 4301 Jones Bridge Rd, Bethesda MD, 20814 USA; email, jmoffett@usuhs.mil.

prominence of the NAA-signal in magnetic resonance spectroscopy (MRS). In the case of Canavan disease, it was found that a mutation in the gene for the enzyme aspartoacylase (ASPA) resulted in the inability to remove acetate from NAA, leading to a failure of CNS myelination, and preventing CNS maturation. In the case of the prominent NAA signal in MRS, it has been found that the levels of NAA in various parts of the brain correlated with neuronal health or dysfunction. Low levels of NAA as detected by MRS have been interpreted to indicate neuronal/axonal loss, or compromised neuronal metabolism. High levels of brain NAA were found in many Canavan patients, suggesting that excess NAA may have toxic effects in the CNS. The First International Symposium on NAA covered these, and many other issues involving the roles played by NAA in normal and pathological brain function. The following sections briefly outline several of the various hypotheses proposed by different research groups on the possible functions of NAA in the nervous system.

1.1. NAA and the Osmolyte Hypothesis

Perhaps the most controversial hypothesis regarding the possible functions of NAA in the nervous system is the neuronal osmolyte hypothesis. It is controversial because this hypothesis has relatively little definitive experimental support, but paradoxically is one of the most cited hypotheses in recent NAA literature. In the 1960s, several research groups proposed, based on the high concentration of NAA in the brain, that it might function to counter the so-called "anion deficit" in neurons.[3,4] In the 1990's, Taylor and colleagues proposed that NAA acts as a neuronal osmolyte involved in active neuronal volume regulation, or possibly in acid-base homeostasis.[5] Using microdialysis, they demonstrated that NAA concentrations in the extracellular fluid increased in response to hypoosmotic stress.[6] However, the changes in extracellular NAA concentrations were modest as compared with the substantially larger increases in extracellular taurine concentrations under the same conditions. Indeed, if NAA is involved significantly in neuronal osmoregulation, it is but one of many organic osmolytes that are responsible for maintaining water homeostasis in the brain. More recently, Davies et al. showed that under hypoosmotic conditions extracellular taurine levels increased almost 20 fold, whereas under the same conditions, extracellular NAA increased by only a few percent.[7] NAA efflux of that magnitude could occur by simple leakage along the concentration gradient when neurons are subjected to osmotic stress. As such, NAA may be a minor contributor to neuronal volume regulation, whereas taurine and other osmolytes provide the predominant regulation of water homeostasis in the brain.

More recently, Baslow and colleagues have proposed a modified version of the neuronal osmolyte hypothesis in which NAA acts as a cotransport substrate for an as yet undescribed molecular water pump which removes excess metabolic water from neurons.[8-10] The aspects of NAA that suggest that it could be involved in such a function include; 1) high concentration, 2) high intraneuronal to extracellular gradient, and 3) the fact that osmotic stress increases the extracellular NAA concentrations to a small degree (see chapter by Verbalis, this volume). Nonetheless, to date no proteins have been described in neurons that act to cotransport NAA and water out of neurons. However, it has been shown that the sodium-dependent NaDC3 transporter moves NAA into glial cells[11] (also see chapter by Ganapathy and Fujita, this volume).

1.2. NAA as Acetate Carrier for Myelin Lipid Synthesis

One hypothesis on the role of NAA in the nervous system that has mounting experimental support is the acetate hypothesis of myelin lipid synthesis in oligodendrocytes. In the 1960's, D'Adamo and coworkers showed that the acetate moiety of NAA was incorporated into brain lipids during CNS development, indicating that it was likely to be involved in the myelination of axons.[12,13] In the early 1990's Burri and colleagues showed that the acetate group from NAA was preferentially incorporated into brain lipids during brain development,[14] and in the mid 1990's Mehta and Namboodiri corroborated that radiolabeled NAA and acetate were incorporated into acetyl CoA and brain lipids.[15] In 2001, Ledeen and colleagues showed that radiolabeled NAA was transported down the optic nerve, and the acetate group was incorporated into the ensheathing myelin of the optic axons.[16] Recently, Namboodiri and coworkers demonstrated in the mouse model of Canavan disease (aspartoacylase knockout) that the rate of lipid synthesis was significantly reduced, and that free acetate levels in the brain were almost 5-fold lower than in wild-type mice (see Namboodiri et al., this volume).[17] Taken together, these data provide strong evidence for NAA as being a major source of free acetate and acetyl CoA for lipid synthesis in the brain during development, and that NAA-derived acetate is critical for proper CNS myelination.

It is important to state that the acetate-lipid synthesis hypothesis does not address the presence of the extraordinarily high levels of NAA in adult neurons. A theory that pertains directly to the high concentrations of NAA in the neurons of adult animals posits that NAA is intimately involved in neuronal energy metabolism.

1.3. NAA in Neuronal Energy Metabolism

The first report that NAA might be involved in brain metabolism was by Buniatian and coworkers in 1965.[18] The idea that NAA is involved in energy metabolism in the nervous system is based on a number of facts, including that 1) NAA is synthesized by aspartate N-acetyltransferase in neuronal mitochondria[19,20], 2) traumatic brain injury causes rapid and partially reversible decreases in NAA concentrations,[21-23] 3) inhibition of mitochondrial respiration results in the simultaneous decrease of NAA production, ATP production and oxygen consumption in isolated brain mitochondria,[24] and 4) NAA levels in the striatum of rats and primates were significantly decreased after the animals were treated with a mitochondrial toxin (3-ntiroproprionate).[25] Recently, Madhavarao and coworkers proposed a model whereby NAA is associated with neuronal energy production (see Namboodiri et al. and Madhavarao and Namboodiri, this volume). In this model, aspartate aminotransferase, the enzyme that synthesizes NAA, facilitates removal of excess aspartate from neuronal mitochondria via acetylation, thus favoring α-ketoglutarate formation from glutamate, and energy production via the citric acid cycle.[26]

1.4. NAAG Biosynthesis

NAAG biosynthesis has been studied for well over a decade, and yet to date, no NAAG biosynthetic enzyme has been isolated and characterized. The presence of a "NAAG synthetase" enzyme has been demonstrated indirectly in tissue culture and tissue explants,[27] but no data on the incorporation of radiolabeled precursors into NAAG have been reported in tissue homogenates. Protein synthesis inhibitors have no effect on incorporation of radiolabeled precursors into NAAG,[28] indicating that NAAG is not synthesized ribosomally as a portion of a protein, and then later cleaved to generate the active molecule. Because the ability of neurons to synthesize NAAG is lost when tissue explants are homogenized, it seems clear that "NAAG synthase" is either sequestered in a membrane compartment where optimal conditions are maintained for enzyme activity, or that the mixing of extracellular and intracellular constituents brings the enzyme in contact with salts, proteases or inhibitors that render it inactive.

1.5. NAA Toxicity in Canavan Disease

An additional controversy concerning NAA is that it is toxic to neurons when the concentration is substantially elevated in the brain, as in the case of many Canavan disease patients.[29] Canavan disease (CD) is a fatal, hereditary leukodystrophy that compromises myelination in the CNS. CD is caused by mutations in the gene for ASPA,[30,31] an enzyme that currently is thought to function exclusively to hydrolyze NAA into L-aspartate and free acetate. However, ASPA is strongly expressed in other tissues such as kidney, even though the only known substrate, NAA, is present predominantly in the nervous system.[32] Despite the established connection between mutations in the gene for ASPA in CD, and the lost capacity to deacetylate NAA, the specific connection between ASPA deficiency and the failure of proper CNS development is unclear.[33] It has been proposed that lack of deacetylase activity against NAA leads to toxic increases in the concentration of NAA in the brain.[34] Indeed, the level of extracellular NAA may be a critical factor in determining if it has toxic effects. For example, Pliss and colleagues reported that injection of 0.25 micromoles/ventricle of NAA into the lateral cerebral ventricles of rats did not induce any detectable neuronal death in the hippocampus,[35] whereas Akimitsu et al. reported that intraventricular injection of 8 micromoles of NAA into the lateral ventricle of rats induced strong seizures.[36] Seizures are one of the symptoms of CD, but it has not been conclusively demonstrated that elevated NAA levels are responsible for the seizure activity. The question of the toxicity of NAA has not been fully answered, but it is possible that exceptionally high levels of NAA in the brains of some CD patients may be involved in some of the pathogenesis, possibly by inducing seizure activity.

1.6. NAA and Magnetic Resonance Spectroscopy

Finally, it is important to note that there are two distinct NAA research communities; one group involved in basic research into the neurochemistry of NAA, and another group which employs MRS techniques for the non-invasive analysis of NAA levels in the brain with respect to clinical applications. NAA levels measured by MRS have been shown to be changed in a number of neurological disorders. These studies have mostly detected decreases in NAA, with the exception of Canavan disease which involves accumulation of NAA in the brain.[37] The diseases and disorders in which NAA levels are decreased include stroke,[38] Alzheimer's disease,[39] epilepsy,[40] brain cancer,[41] multiple sclerosis[42] and AIDS dementia complex.[43] Earlier, the decreases in NAA were interpreted to represent irreversible loss of neurons. However, more recent evidence indicates that these decreases also could represent reversible mitochondrial dysfunction.[23,24] In support of this view, some evidence has been presented showing that NAA levels are restored when patients recover clinically.[44] Two recent reports have shown that NAA levels in the brain are increased in rats after prolonged administration of haloperidol,[45] and in human children with sickle cell disease.[46]

The First International Symposium on NAA brought basic researchers and clinicians together, and offered these two research groups an opportunity to interact and discuss ideas, and while none of the controversies surrounding NAA neurochemistry were resolved, both groups benefited greatly from the exchange. The following proceedings from the first international NAA Symposium detail ongoing research in numerous laboratories, and present the interactions and exchanges between some of the top researchers in this field.

2. REFERENCES

1.   H. H. Tallan, S. Moore and W. H. Stein, N-Acetyl-L-aspartic acid in brain. *J. Biol. Chem.*, **219** (1), 257-264 (1956).
2.   E. Miyamoto, Y. Kakimoto and I. Sano, Identification of N-acetylaspartylglutamic acid in the bovine brain. *J. Neurochem.*, **13**, 999-1003 (1966).
3.   R. U. Margolis, S. S. Barkulis and A. Geiger, A comparison between the incorporation of 14C from glucose into N-acetyly-L-aspartic acid and aspartic acid in brain perfusion experiments. *J. Neurochem.*, **5**, 379-382 (1960).
4.   J. C. McIntosh, J. R. Cooper, Studies on the function of N-acetyl aspartic acid in brain. *J. Neurochem.*, **12** (9), 825-835 (1965).
5.   D. L. Taylor, S. E. Davies, T. P. Obrenovitch, J. Urenjak, D. A. Richards, J. B. Clark and L. Symon, Extracellular N-acetylaspartate in the rat brain: in vivo determination of basal levels and changes evoked by high K+. *J Neurochem.*, **62** (6), 2349-2355 (1994).
6.   D. L. Taylor, S. E. Davies, T. P. Obrenovitch, M. H. Doheny, P. N. Patsalos, J. B. Clark and L. Symon, Investigation into the role of N-acetylaspartate in cerebral osmoregulation. *J Neurochem.*, **65** (1), 275-281 (1995).
7.   S. E. Davies, M. Gotoh, D. A. Richards and T. P. Obrenovitch, Hypoosmolarity induces an increase of extracellular N-acetylaspartate concentration in the rat striatum. *Neurochem. Res.*, **23** (8), 1021-1025 (1998).
8.   M. H. Baslow, D. N. Guilfoyle, Effect of N-acetylaspartic acid on the diffusion coefficient of water: a proton magnetic resonance phantom method for measurement of osmolyte-obligated water. *Anal. Biochem.*, **311** (2), 133-138 (2002).
9.   M. H. Baslow, Evidence supporting a role for N-acetyl-L-aspartate as a molecular water pump in myelinated neurons in the central nervous system. An analytical review. *Neurochem. Int.*, **40** (4), 295-300 (2002).
10. M. H. Baslow, Brain N-acetylaspartate as a molecular water pump and its role in the etiology of canavan disease: a mechanistic explanation. *J. Mol. Neurosci.*, **21** (3), 185-190 (2003).
11. R. L. George, W. Huang, H. A. Naggar, S. B. Smith and V. Ganapathy, Transport of N-acetylaspartate via murine sodium/dicarboxylate cotransporter NaDC3 and expression of this transporter and aspartoacylase II in ocular tissues in mouse. *Biochim. Biophys. Acta*, **1690** (1), 63-69 (2004).
12. A. F. D'Adamo, L. I. Gidez and F. M. Yatsu, Acetyl transport mechanisms. Involvement of N-acetyl aspartic acid in de novo fatty acid biosynthesis in the developing rat brain. *Exp. Brain Res.*, **5**, 267-273 (1968).
13. A. F. D'Adamo, Jr., F. M. Yatsu, Acetate metabolism in the nervous system. N-acetyl-L-aspartic acid and the biosynthesis of brain lipids. *J. Neurochem.*, **13** (10), 961-965 (1966).
14. R. Burri, C. Steffen and N. Herschkowitz, N-acetyl-L-aspartate is a major source of acetyl groups for lipid synthesis during rat brain development. *Dev. Neurosci.*, **13**, 403-412 (1991).
15. V. Mehta, M. A. Namboodiri, N-acetylaspartate as an acetyl source in the nervous system. *Brain Res. Mol. Brain Res.*, **31** (1-2), 151-157 (1995).
16. G. Chakraborty, P. Mekala, D. Yahya, G. Wu and R. W. Ledeen, Intraneuronal N-acetylaspartate supplies acetyl groups for myelin lipid synthesis: evidence for myelin-associated aspartoacylase. *J. Neurochem.*, **78** (4), 736-745 (2001).
17. C. N. Madhavarao, P. Arun, J. R. Moffett, S. Suczs, S. Surendran, R. Matalon, J. Garbern, D. Hristova, A. Johnson, W. Jiang and M. A. Namboodiri, Defective N-acetylaspartate catabolism reduces brain acetate

levels and myelin lipid synthesis in Canavan's disease. *Proc. Natl. Acad. Sci. U. S. A*, **102** (14), 5221-5226 (5 A.D.).

18. H. C. Buniatian, V. S. Hovhannissian and G. V. Aprikian, The participation of N-acetyl-L-aspartic acid in brain metabolism. *J. Neurochem.*, **12** (8), 695-703 (1965).

19. C. N. Madhavarao, C. Chinopoulos, K. Chandrasekaran and M. A. Namboodiri, Characterization of the N-acetylaspartate biosynthetic enzyme from rat brain. *J. Neurochem.*, **86** (4), 824-835 (2003).

20. T.B. Patel and J. B.Clark, Synthesis of N-acetyl-L-aspartate by rat brain mitochondria and its involvement in mitochondrial/cytosolic carbon transport. *Biochem. J.* **184** (3), 539-46 (1979).

21. B. Tavazzi, S. Signoretti, G. Lazzarino, A. M. Amorini, R. Delfini, M. Cimatti, A. Marmarou and R. Vagnozzi, Cerebral oxidative stress and depression of energy metabolism correlate with severity of diffuse brain injury in rats. *Neurosurgery*, **56** (3), 582-589 (2005).

22. S. Signoretti, A. Marmarou, B. Tavazzi, G. Lazzarino, A. Beaumont and R. Vagnozzi, N-Acetylaspartate reduction as a measure of injury severity and mitochondrial dysfunction following diffuse traumatic brain injury. *J. Neurotrauma*, **18** (10), 977-991 (2001).

23. C. Gasparovic, N. Arfai, N. Smid and D. M. Feeney, Decrease and recovery of N-acetylaspartate/creatine in rat brain remote from focal injury. *J. Neurotrauma*, **18** (3), 241-246 (2001).

24. T. E. Bates, M. Strangward, J. Keelan, G. P. Davey, P. M. Munro and J. B. Clark, Inhibition of N-acetylaspartate production: implications for 1H MRS studies in vivo. *Neuroreport*, **7** (8), 1397-1400 (1996).

25. C. Dautry, F. Vaufrey, E. Brouillet, N. Bizat, P. G. Henry, F. Conde, G. Bloch and P. Hantraye, Early N-acetylaspartate depletion is a marker of neuronal dysfunction in rats and primates chronically treated with the mitochondrial toxin 3-nitropropionic acid. *J. Cereb. Blood Flow Metab*, **20** (5), 789-799 (2000).

26. C. N. Madhavarao, P. Arun, J. R. Moffett, S. Suczs, S. Surendran, R. Matalon, J. Garbern, D. Hristova, A. Johnson, W. Jiang and M. A. Namboodiri, Defective N-acetylaspartate catabolism reduces brain acetate levels and myelin lipid synthesis in Canavan's disease. *Proc. Natl. Acad. Sci. U. S. A*, **102** (14), 5221-5226 (5 A.D.).

27. L. M. Gehl, O. H. Saab, T. Bzdega, B. Wroblewska and J. H. Neale, Biosynthesis of NAAG by an enzyme-mediated process in rat central nervous system neurons and glia. *J. Neurochem.*, **90** (4), 989-997 (2004).

28. C. B. Cangro, M. A. Namboodiri, L. A. Sklar, A. Corigliano-Murphy and J. H. Neale, Immunohistochemistry and biosynthesis of N-acetylaspartylglutamate in spinal sensory ganglia. *J. Neurochem.*, **49**, 1579-1588 (1987).

29. P. Leone, C. G. Janson, S. J. McPhee and M. J. During, Global CNS gene transfer for a childhood neurogenetic enzyme deficiency: Canavan disease. *Curr. Opin. Mol. Ther.*, **1** (4), 487-492 (1999).

30. R. Kaul, G. P. Gao, K. Balamurugan and R. Matalon, Cloning of the human aspartoacylase cDNA and a common missense mutation in Canavan disease [see comments]. *Nat. Genet.*, **5**, 118-123 (1993).

31. B. J. Zeng, Z. H. Wang, L. A. Ribeiro, P. Leone, R. De Gasperi, S. J. Kim, S. Raghavan, E. Ong, G. M. Pastores and E. H. Kolodny, Identification and characterization of novel mutations of the aspartoacylase gene in non-Jewish patients with Canavan disease. *J. Inherit. Metab Dis.*, **25** (7), 557-570 (2002).

32. A. F. D'Adamo, Jr., J. C. Smith and C. Woiler, The occurrence of N-acetylaspartate amidohydrolase (aminoacylase II) in the developing rat. *J. Neurochem.*, **20** (4), 1275-1278 (1973).

33. R. Matalon, K. Michals and R. Kaul, Canavan disease: from spongy degeneration to molecular analysis. *J Pediatr.*, **127** (4), 511-517 (1995).

34. P. Leone, C. G. Janson, L. Bilaniuk, Z. Wang, F. Sorgi, L. Huang, R. Matalon, R. Kaul, Z. Zeng, A. Freese, S. W. McPhee, E. Mee, M. J. During and L. Bilianuk, Aspartoacylase gene transfer to the mammalian central nervous system with therapeutic implications for Canavan disease. *Ann. Neurol.*, **48** (1), 27-38 (2000).

35. L. Pliss, V. J. Balcar, V. Bubenikova, J. Pokorny, T. FitzGibbon and F. St'astny, Morphology and ultrastructure of rat hippocampal formation after i.c.v. administration of N-acetyl-L-aspartyl-L-glutamate. *Neurosci.*, **122** (1), 93-101 (2003).

36. T. Akimitsu, K. Kurisu, R. Hanaya, K. Iida, Y. Kiura, K. Arita, H. Matsubayashi, K. Ishihara, K. Kitada, T. Serikawa and M. Sasa, Epileptic seizures induced by N-acetyl-L-aspartate in rats: in vivo and in vitro studies. *Brain Res.*, **861** (1), 143-150 (2000).

37. H. J. Wittsack, H. Kugel, B. Roth and W. Heindel, Quantitative measurements with localized 1H MR spectroscopy in children with Canavan's disease. *J. Magn Reson. Imaging*, **6** (6), 889-893 (1996).

38. M. L. Lai, Y. I. Hsu, S. Ma and C. Y. Yu, Magnetic resonance spectroscopic findings in patients with subcortical ischemic stroke. *Chung. Hua. I. Hsueh. Tsa. Chih. (Taipei).*, **56** (1), 31-35 (1995).

39. W. E. Klunk, K. Panchalingam, J. Moossy, R. J. McClure and J. W. Pettegrew, N-acetyl-L-aspartate and other amino acid metabolites in Alzheimer's disease brain: a preliminary proton nuclear magnetic resonance study. *Neurology*, **42**, 1578-1585 (1992).
40. P. M. Matthews, F. Andermann and D. L. Arnold, A proton magnetic resonance spectroscopy study of focal epilepsy in humans. *Neurology*, **40**, 985-989 (1990).
41. S. S. Gill, D. G. Thomas, N. Van Bruggen, D. G. Gadian, C. J. Peden, J. D. Bell, I. J. Cox, D. K. Menon, R. A. Iles, D. J. Bryant and et al, Proton MR spectroscopy of intracranial tumours: in vivo and in vitro studies. *J. Comput. Assist. Tomogr.*, **14**, 497-504 (1990).
42. C. A. Davie, C. P. Hawkins, G. J. Barker, A. Brennan, P. S. Tofts, D. H. Miller and W. I. McDonald, Serial proton magnetic resonance spectroscopy in acute multiple sclerosis lesions. *Brain*, **117**, 49-58 (1994).
43. D. K. Menon, J. G. Ainsworth, I. J. Cox, R. C. Coker, J. Sargentoni, G. A. Coutts, C. J. Baudouin, A. E. Kocsis and J. R. Harris, Proton MR spectroscopy of the brain in AIDS dementia complex. *J. Comput. Assist. Tomogr.*, **16**, 538-542 (1992).
44. D. L. Arnold, P. M. Matthews, G. Francis and J. Antel, Proton magnetic resonance spectroscopy of human brain in vivo in the evaluation of multiple sclerosis: assessment of the load of disease. *Magn. Reson. Med.*, **14**, 154-159 (1990).
45. M. K. Harte, S. B. Bachus and G. P. Reynolds, Increased N-acetylaspartate in rat striatum following long-term administration of haloperidol. *Schizophr. Res.*, **75** (2-3), 303-308 (2005).
46. R. G. Steen, R. J. Ogg, Abnormally high levels of brain N-acetylaspartate in children with sickle cell disease. *AJNR Am. J. Neuroradiol.*, **26** (3), 463-468 (2005).

CONTENTS

A BRIEF OVERVIEW OF *N*-ACETYLASPARTATE AND *N*-ACETYLASPARTYLGLUTAMATE

Joseph T. Coyle[*]

1. INTRODUCTION

It has been 50 years since Tallan et al.[1] detected *N*-acetyl aspartate (NAA) in cat brain extracts, and 40 years since Curatolo et al.[2] reported the presence of *N*-acetyl aspartyl glutamate (NAAG) in horse brain. One of the major barriers to the discovery of these two moieties in brain was that they were *N*-blocked, thereby preventing the ready detection with ninhydrin reactivity, a common method for measuring amino acids. Tallan et al.[1] separated protein free extracts of cat brain over a Dowex ion exchange column and demonstrated that a late eluting ninhydrin negative fraction became positive after acid hydrolysis, thereby discovering N-acetyl aspartate. A decade later, Curatolo et al.[2] found that subsequent eluted fractions, also ninhydrin negative, contained aspartate and glutamate after acid hydrolysis, thereby disclosing the existence of NAAG. He also was able to demonstrate an uneven regional distribution of NAAG in brain that provided a hint of potential physiologic significance. These discoveries antedated by a decade or more the identification of trace neuropeptides in brain such as the enkephalins[3] and somatostatin.[4] Ironically, research on NAA and NAAG did not commence in earnest until the mid 1980s.[5-7]

It is indeed odd that the first symposium dedicated to these two moieties has been so long in taking place. The lack of attention to NAA and NAAG is reminiscent of Rodney Dangerfield's complaint: "I can't get no respect." In support of this contention, one might compare NAA and NAAG to somatostatin and enkephalin. Since its discovery in 1956, NAA, with a brain concentration of 3 mM, has been the topic 1759 publications. NAAG, discovered in 1965 with a concentration of 0.2 mM in rat cortex, undoubtedly the most abundant neuropeptide in cortex, is addressed in 282 publications. In contrast, enkephalin, discovered in 1975[3] with a 200 nM concentration in rat cortex, has received 16,448 publications; and somatostatin, discovered in 1972[4] with a 10 nM concentration in rat cortex, has been included in 21,866 publications. Thus, the universe of scientists involved in

---

[*] Department of Psychiatry and Neuroscience, Harvard Medical School, McLean Hospital, 115 Mill St, Belmont MA 02178-9106 USA, 617-855-2101, Fax: 617-855-2705, E-mail joseph_coyle@hms.harvard.edu

research on NAA and NAAG is rather small; and fortunately, most are participating in this conference.

2. *N*-ACETYL ASPARTATE (NAA)

Interest in NAA was stimulated by the fact that it represents the largest peak on proton nuclear magnetic resonance spectroscopy,[8] and alterations in its levels have been demonstrated in a wide range of disorders including Alzheimer's Disease, epilepsy and schizophrenia. Thus, it would be essential to know the precise cellular localization of NAA in order to interpret its role in disease processes. Simmons et al.[9] used NAA linked to bovine serum albumin by carbodiimide to develop a highly specific monoclonal antibody against with only 1% cross reactivity with NAAG-BSA. Immunocytochemical staining with avidin-biotin-peroxidase demonstrated that the NAA-like immunoreactivity was localized to neurons distributed throughout the brain. Immunoreactive neurons exhibited intense staining of the perikaraya, proximal dendrites and axons, although no consistent pattern of distribution of immunoreactivity was observed with regard to primary neurotransmitter characteristics of neurons. Neurons with long projections or extensive axonal arbors, such as the cortical pyramidal cells, the locus ceruleus, motor neurons, and Purkinje cells, stained much more intensely than local circuit neurons. These findings have been independently corroborated[10].

· Perikaryal specific excitotoxin lesions of the rat striatum substantially reduced NAA levels at the lesion site.[11] The NAA decrements were mirrored by a comparable reduction in the NAA signal as measured by magnetic resonance spectroscopy (MRS), thereby validating the inference that NAA measured by MRS was a reasonable surrogate for neuronal integrity. However, NAA reductions can not be simply equated with neuronal degeneration, because such reductions have been observed in conditions such as HIV encephalopathy, in which neuronal loss has not been documented.[12] Furthermore, Yurgelun-Todd et al. (in press) has found that the reductions in NAA levels in frontal cortex in schizophrenia may be an artifact of a change in relaxation time, which can be mimicked in vitro by adding increased concentrations of albumin to fixed concentrations of NAA. Thus, changes in intracellular environment associated with atrophy could also affect the apparent levels of NAA measured with MRS

Major questions concerning the role of NAA remain open and need to be addressed. First, the function of NAA within neurons remains unclear. Second, if NAA is involved in myelin synthesis how does its neuronal localization comport with oliogodendrogliocyte function? Third, how are the alterations in NAA signal in human MRS studies to be interpreted with regard to specific neurologic and psychiatric disorders. It is hoped that this conference will shed light on these important questions.

3. *N*-ACETYLASPARTYLGLUTAMATE (NAAG)

Several laboratories have studied the cellular localization of NAAG in the brains of a number of species using immunocytochemical methods. The immunogen has routinely been NAAG linked by to a carrier protein by carbodiimide. To minimize cross-reactivity with NAA, polyclonal antibodies purified by affinity chromatography,[13, 14] or a highly specific monoclonal antibody has been selected.[15] Consistent with the uneven regional distribution of

NAAG levels, NAAG expression is restricted to discrete subsets of neurons. For example, intense immunoreactivity is noted in the retinal ganglion cells and optic nerve, motor neurons and their axon and terminals, locus ceruleus noradrenergic neurons, GABAergic interneurons in the hippocampus and the subhuman primate and human cortico-limbic pyramidal cells[13, 14, 16-19] Ultra-structural studies demonstrate the localization of NAAG in storage vesicles,[20] consistent with its calcium-dependent evoked release upon stimulation of NAAG containing projections such as the retinal-tectal pathway.[21, 22]

The diverse co-localization of NAAG with different neurotransmitters including glutamate, GABA, acetylcholine and norepinephrine[16] is similar to that of other neuropeptides, which also co-localize with more than one neurotransmitter system. Nevertheless, this diverse co-localization, especially in motor neurons,[23] raises important but as yet unanswered questions about the role of NAAG in modulating neurotransmission.

The physiologic effects NAAG have been somewhat controversial. The initial finding of excitatory effects in the olfactory bulb[7] were likely the result of contaminating K+ and/or hydrolysis to free glutamate.[24] Other studies suggested a weak, agonist effect of NAAG at NMDA receptors.[25-27] Recent studies, however, indicate that NAAG antagonizes NMDA receptor responses in a glycine reversible fashion at the Shaffer collateral-CA1 pyramidal cell synapse.[28] In addition, Wroblewska et al.[29] screened the effects of NAAG in CHO cells expressing the different members of the mGLUR family and showed that NAAG is a full and selective agonist at mGluR3. This important discovery has been confirmed in the olfactory bulb with electrophysiological methods.[30]

Riveros and Orrego,[31] using partially purified synaptic membranes, were the first to report the hydrolysis of NAAG to yield glutamate. Blakely et al.[32] subsequently demonstrated that this hydrolysis was potently inhibited by quisqualic acid. Using this criterion, Slusher et al.[33] purified to homogeneity a 94 kda plasma membrane glycoprotein that possessed this quisqualate-sensitive peptidase activity, which glycoprotein is now known as Glutamate Carboxy Peptidase II (GCPII; EC 3.4.17.21). With antibodies raised against the homogeneously pure protein, Carter et al.[34] isolated a rat brain cDNA encoding GCPII. Sequence analysis indicates that it is a type II membrane protein in the M28 peptidase family with a short intracellular tail and a transmembrane domain for amino acids 20 through 43. Site-directed mutagenesis has clarified both the catalytic site of the enzyme and the location of the amino acids binding zinc.[35] Molecular studies demonstrate that GCPII is identical to folate hydrolase in the gut[36] and to prostate specific membrane antigen (PSMA),[34, 37] a cell surface protein highly expressed on metastatic prostate cancer cells.

Immunocytochemical as well as in situ hybridization studies demonstrate that GCPII is selectively expressed in astrocytes in the brain with an uneven regional distribution.[38, 39] Particularly high levels of expression are observed in the Bergmann glial cells. It is also expressed in non-myelating Schwann cells in the periphery, including those surrounding motor neuron terminals at the end plate.[40] In the periphery, GCPII is expressed in the testicles and in the tubules of the kidney. However, the exclusive role of GCPII in the hydrolysis of NAAG has been called into question by studies in mice homozygous for a null mutation of the GCPII gene. The brains of these mice exhibit a residual of 15% GCPII-like enzyme activity with no alterations observed in the levels of NAAG, NAA, glutamate or aspartate.[41] Recently, a second gene with high homology to GCPII has been cloned.[42, 43]

Many substantive unanswered questions remain with regard to the role of NAAG and its disposition in the nervous system. The mechanisms responsible for its synthesis remain obscure, although it is likely synthesized by an enzymatic and not an mRNA dependent process. Since GCPII is expressed in astrocytes, it is unclear what determines the levels of

NAAG in neurons. Given the absence of a behavioral phenotype with the GCPII knock-out mice, the role of NAAG as a co-transmitter in a variety of neuronal systems that do not appear to use glutamate as their primary neurotransmitter remains to be determined.

4. CONCLUSION

In closing, it is anticipated that this conference will not only provide a comprehensive review of the current state of knowledge of NAA and NAAG but will also define the many important and critical questions that need to be addressed in future research. It is hoped that the proceedings of the conference will attract a broader interest in these moieties, as they have been linked to serious disorders including schizophrenia, Alzheimer's Disease and pain. Given their remarkably high concentrations in the brain, the current level of research on NAAG and NAA seems to underestimate their potential significance.

5. REFERENCES

1. H. H. Tallan, S. Moore, and W. H. Stein. N-Acetyl-L-aspartic acid in brain. *J. Biol. Chem.* **219**, 257-264 (1956).
2. A. Curatolo, P. D'Arcangelo, and A. Lino. Distribution of *N*-acetyl-aspartic and *N*-acetyl-aspartyl-glutamic acids in nervous tissue. *J. Neurochem.* **12**, 339-342 (1965).
3. J. Hughes, T. Smith, B. Morgan, and L. Fothergill. Purification and properties of enkephalin - the possible endogenous ligand for the morphine receptor. *Life Sci.* **16**, 1753-1758 (1975).
4. W. Vale, P. Brazeau, G. Grant, A. Nussey, R. Burgus, J. Rivier, N. Ling, and R. Guillemin. Preliminary observations on the mechanism of action of somatostatin, a hypothalamic factor inhibiting the secretion of growth hormone. *C. R. Acad. Sci. Hebd. Seances Acad. Sci. D* **275**, 2913-2916 (1972).
5. K. J. Koller, R. Zaczek, and J. T. Coyle. *N*-acetyl-aspartyl-glutamate: regional levels in rat brain and the effects of brain lesions as determined by a new HPLC method. *J. Neurochem.* **43**, 1136-1142 (1984).
6. K. J. Koller and J. T. Coyle. Ontogenesis of N-acetyl-aspartate and N-acetyl-aspartyl- glutamate in rat brain. *Brain Res.* **317**, 137-140 (1984).
7. J. M. H. ffrench-Mullen, K. J. Koller, R. Zaczek, J. T. Coyle, N. Hori, and D. O. Carpenter. *N*-acetyl-aspartylglutamate: possible role as the neurotransmitter of the lateral olfactory tract. *Proc. Natl. Acad. Sci. USA* **82**, 3897-3900 (1985).
8. G. C. Tsai and J. T. Coyle. N-acetylaspartate in neuropsychiatric disorders. *Prog. Neurobiol.* **46**, 531-540 (1995).
9. M. L. Simmons, C. G. Frondoza, and J. T. Coyle. Immunocytochemical localization of *N*-acetyl-aspartate with monoclonal antibodies. *Neuroscience* **45**, 37-45 (1991).
10. J. R. Moffett, M. A. A. Namboodiri, C. B. Cangro, and J. H. Neale. Immunohistochemical localization of *N*-acetylaspartate in rat brain. *NeuroReport* **2**, 131-134 (1991).
11. A. R. Guimaraes, P. Schwartz, M. R. Prakash, C. A. Carr, U. V. Berger, B. G. Jenkins, J. T. Coyle, and R. G. Gonzalez. Quantitative in vivo ^{1}H nuclear magnetic resonance spectroscopic imaging of neuronal loss in rat brain. *Neuroscience* **69**, 1095-1101 (1995).
12. D. J. Meyerhoff, S. MacKay, L. Bachman, N. Poole, W. P. Dillon, M. W. Weiner, and G. Fein. Reduced brain *N*-acetylaspartate suggests neuronal loss in cognitively impaired human immunodeficiency virus-seropositive individuals: In vivo ^{1}H magnetic resonance spectroscopic imaging. *Neurology* **43**, 509-515 (1993).
13. J. R. Moffett, M. A. A. Namboodiri, and J. H. Neale. Enhanced carbodiimide fixation for immunohistochemistry: Application to the comparative distributions of *N*-acetylaspartylglutamate and *N*-acetylaspartate immunoreactivities in rat brain. *J. Histochem. Cytochem.* **41**, 559-570 (1993).
14. L. A. Passani, J. P. G. Vonsattel, and J. T. Coyle. Distribution of *N*-acetylaspartylglutamate immunoreactivity in human brain and its alteration in neurodegenerative disease. *Brain Res.* **772**, 9-22 (1997).

15. C. G. Frondoza, S. Logan, G. Forloni, and J. T. Coyle. Production and characterization of monoclonal antibodies to *N*-acetyl-aspartyl-glutamate. *J. Histochem. Cytochem.* **38**, 493-502 (1990).
16. G. Forloni, R. Grzanna, R. D. Blakely, and J. T. Coyle. Co-localization of *N*-acetylaspartylglutamate in central cholinergic, noradrenergic and serotonergic neurons. *Synapse* **1**, 455-460 (1987).
17. K. J. Anderson, M. A. Borja, C. W. Cotman, J. R. Moffett, M. A. A. Namboodiri, and J. H. Neale. *N*-acetylaspartylglutamate identified in the rat retinal ganglion cells and their projections in the brain. *Brain Res.* **411**, 172-177 (1987).
18. K. J. Anderson, D. T. Monaghan, C. B. Cangro, M. A. A. Namboodiri, J. H. Neale, and C. W. Cotman. Localization of *N*-acetylaspartylglutamate-like immunoreactivity in selected areas of the rat brain. *Neurosci. Lett.* **72**, 14-20 (1986).
19. S. B. Tieman, C. B. Cangro, and J. H. Neale. *N*-Acetylaspartylglutamate immunoreactivity in neurons of the cat's visual system. *Brain Res.* **420**, 188-193 (1987).
20. L. C. Williamson and J. H. Neale. Ultrastructural localization of *N*-acetylaspartylglutamate in vesicles of retinal neurons. *Brain Res.* **456**, 375-381 (1988).
21. G. Tsai, B. L. Stauch, J. J. Vornov, J. K. Deshpande, and J. T. Coyle. Selective release of *N*-acetylaspartylglutamate from rat optic nerve terminals in vivo. *Brain Res.* **518**, 313-316 (1990).
22. L. C. Williamson, D. A. Eagles, M. J. Brady, J. R. Moffett, M. A. A. Namboodiri, and J. H. Neale. Localization and synaptic release of *N*-acetylaspartylglutamate in the chick retina and optic tectum. *Eur. J. Neurosci.* **3**, 441-451 (1991).
23. U. V. Berger, R. E. Carter, and J. T. Coyle. The immunocytochemical localization of *N*-acetylaspartyl glutamate, its hydrolysing enzyme NAALADase, and the NMDAR- 1 receptor at a vertebrate neuro-muscular junction. *Neuroscience* **64**, 847-850 (1995).
24. E. R. Whittemore and J. F. Koerner. An explanation for the purported excitation of piriform cortical neurons by *N*-acetyl-L-aspartyl-L-glutamic acid (NAAG). *Proc. Natl. Acad. Sci. USA* **86**, 9602-9605 (1989).
25. G. L. Westbrook, M. L. Mayer, M. A. A. Namboodiri, and J. H. Neale. High concentrations of *N*-acetyl-aspartylglutamate (NAAG) selectively activate NMDA receptors on mouse spinal cord neurons in cell culture. *J. Neurosci.* **6**, 3385-3392 (1986).
26. P. Q. Trombley and G. L. Westbrook. Excitatory synaptic transmission in cultures of rat olfactory bulb. *J. Neurophysiol.* **64**, 598-606 (1990).
27. M. Sekiguchi, K. Wada, and R. J. Wenthold. *N*-Acetylaspartylglutamate acts as an agonist upon homomeric NMDA receptor (NMDAR1) expressed in *Xenopus* oocytes. *FEBS Lett.* **311**, 285-289 (1992).
28. R. Bergeron, J. T. Coyle, G. Tsai, and R. W. Greene. NAAG reduces NMDA receptor current in CA1 hippocampal pyramidal neurons of acute slices and dissociated neurons. *Neuropsychopharmacology* **30**, 7-16 (2005).
29. B. Wroblewska, J. T. Wroblewski, S. Pshenichkin, A. Surin, S. E. Sullivan, and J. H. Neale. *N*-acetylaspartylglutamate selectively activates mGluR3 receptors in transfected cells. *J. Neurochem.* **69**, 174-181 (1997).
30. J. Bischofberger and D. Schild. Glutamate and *N*-acetylaspartylglutamate block HVA calcium currents in frog olfactory bulb interneurons via an mGluR2/3- like receptor. *J. Neurophysiol.* **76**, 2089-2092 (1996).
31. N. Riveros and F. Orrego. A study of possible excitatory effects of *N*-acetylaspartylglutamate in different in vivo and in vitro preparations. *Brain Res.* **299**, 393-395 (1984).
32. R. D. Blakely, M. B. Robinson, R. C. Thompson, and J. T. Coyle. Hydrolysis of the brain dipeptide *N*-acetyl-L-aspartyl-L-glutamate: subcellular and regional distribution, ontogeny, and the effect of lesions on *N*-acetylated-α-linked acidic dipeptidase activity. *J. Neurochem.* **50**, 1200-1209 (1988).
33. B. S. Slusher, M. B. Robinson, G. Tsai, M. L. Simmons, S. S. Richards, and J. T. Coyle. Rat brain *N*-acetylated α-linked acidic dipeptidase activity. Purification and immunologic characterization. *J. Biol. Chem.* **265**, 21297-21301 (1990).
34. R. E. Carter, A. R. Feldman, and J. T. Coyle. Prostate-specific membrane antigen is a hydrolase with substrate and pharmacologic characteristics of a neuropeptidase. *Proc. Natl. Acad. Sci. USA* **93**, 749-753 (1996).
35. H. S. Speno, R. Luthi-Carter, W. L. Macias, S. L. Valentine, A. R. Joshi, and J. T. Coyle. Site-directed mutagenesis of predicted active site residues in glutamate carboxypeptidase II. *Mol. Pharmacol.* **55**, 179-185 (1999).
36. C. H. Halsted, E. Ling, R. Luthi-Carter, J. A. Villanueva, J. M. Gardner, and J. T. Coyle. Folylpoly-gamma -glutamate carboxypeptidase from pig jejunum. Molecular characterization and relation to glutamate carboxypeptidase II. *J. Biol. Chem.* **273**, 20417-20424 (1998).
37. C. W. Tiffany, R. G. Lapidus, A. Merion, D. C. Calvin, and B. S. Slusher. Characterization of the enzymatic activity of PSM: Comparison with brain NAALADase. *Prostate* **39**, 28-35 (1999).

38. U. V. Berger, R. Luthi-Carter, L. A. Passani, S. Elkabes, I. Black, C. Konradi, and J. T. Coyle. Glutamate carboxypeptidase II is expressed by astrocytes in the adult rat nervous system. *J. Comp. Neurol.* **415**, 52-64 (1999).
39. R. Luthi-Carter, U. V. Berger, A. K. Barczak, M. Enna, and J. T. Coyle. Isolation and expression of a rat brain cDNA encoding glutamate carboxypeptidase II. *Proc. Natl. Acad. Sci. USA* **95**, 3215-3220 (1998).
40. U. V. Berger, R. E. Carter, M. McKee, and J. T. Coyle. N-acetylated alpha-linked acidic dipeptidase is expressed by non-myelinating Schwann cells in the peripheral nervous system. *J. Neurocytol.* **24**, 99-109 (1995).
41. D. J. Bacich, E. Ramadan, D. S. O'Keefe, N. Bukhari, I. Wegorzewska, O. Ojeifo, R. Olszewski, C. C. Wrenn, T. Bzdega, B. Wroblewska, W. D. W. Heston, and J. H. Neale. Deletion of the glutamate carboxypeptidase II gene in mice reveals a second enzyme activity that hydrolyzes *N*-acetylaspartylglutamate. *J. Neurochem.* **83**, 20-29 (2002).
42. D. S. O'Keefe, S. L. Su, D. J. Bacich, Y. Horiguchi, Y. Luo, C. T. Powell, D. Zandvliet, P. J. Russell, P. L. Molloy, N. J. Nowak, T. B. Shows, C. Mullins *et al.* Mapping, genomic organization and promoter analysis of the human prostate-specific membrane antigen gene. *Biochim. Biophys. Acta* **1443**, 113-127 (1998).
43. T. Bzdega, S. L. Crowe, E. R. Ramadan, K. H. Sciarretta, R. T. Olszewski, O. A. Ojeifo, V. A. Rafalski, B. Wroblewska, and J. H. Neale. The cloning and characterization of a second brain enzyme with NAAG peptidase activity. *J. Neurochem.* **89**, 627-635 (2004).

EXPRESSION OF *N*-ACETYLASPARTATE AND *N*-ACETYLASPARTYLGLUTAMATE IN THE NERVOUS SYSTEM

John R. Moffett and Aryan M. A. Namboodiri[*]

1. INTRODUCTION

N-Acetylaspartate (NAA) and *N*-acetylaspartylglutamate (NAAG) are highly concentrated acetylated compounds found predominantly in the nervous system of vertebrates and invertebrates.[1-3] Their high concentrations make them good candidates for localization by immunohistochemistry. The two compounds are assumed to be related to one another in terms of biosynthesis. NAA is thought to be a direct precursor for NAAG biosynthesis,[4] despite numerous failures in several laboratories to isolate or characterize a NAAG synthase enzyme capable of coupling NAA to glutamate. NAAG and NAA are found primarily in neurons,[5-11] although much lower levels may be present in some glial cells,[12,13] and in somatic tissues.[2] Because of their predominant neuronal localization, and the fact that the two molecules provide strong acetate signals in water-suppressed proton magnetic resonance spectra, reductions in their acetate signals have been used as a non-invasive diagnostic marker for neuronal loss or dysfunction.[14-17] The N-terminal acetyl groups of NAAG and NAA make their localization by immunohistochemistry problematic, because both molecules lack an amine group, and thus lack a reactive group that would permit standard fixation coupling with glutaraldehyde. This problem has been examined in detail previously.[10]

The only chemically reactive groups available on NAAG and NAA for coupling to proteins are carboxyl groups; NAAG has three carboxyl groups, and NAA has two. The most effective reagents for inducing peptide bond formation between carboxyl-containing compounds and proteins are known as carbodiimides. Historically, carbodiimides were first used for tissue fixation for immunohistochemistry in the early 1970's.[18,19] We have used the water-soluble carbodiimide, EDAC (1-ethyl-3 [3-dimethylaminopropyl] carbodiimide hydrochloride), for the immunohistochemical localization of NAAG and

---
[*] Uniformed Services University of the Health Sciences, 4301 Jones Bridge Rd, Bethesda MD, 20814, USA, email; jmoffett@usuhs.mil.

NAA. EDAC-based fixation for immunohistochemistry is complicated by the fact that this coupling reagent is water soluble, but does not penetrate lipid-rich tissues, such as white matter in brain. To overcome this problem, we found that 5% DMSO could be used in the fixative solution to increase penetration of EDAC into white matter and other lipid-rich tissues. DMSO was found to be critical for uniform labeling of NAAG and NAA in nervous system white matter, but was also found to destroy the internal ultrastructure of neurons, making its use unsuitable for electron microscopy. Other improvements to carbodiimide fixations for light microscopy were achieved by increasing the temperature of the fixative solution to 37°C, and using the carbodiimide stabilizing agent; N-hydroxysuccinimide (1mM), to reduce non-productive hydrolysis of EDAC in solution.[10]

Another significant problem associated with small-molecule immunohistochemistry in general is cross-reactivity of the antibodies with non-specific epitopes in fixed tissue sections. The problems of cross-reactivity and high background staining in tissue sections were pronounced with affinity-purified NAAG and NAA antibodies. Solid phase immunoassays showed that affinity purified, polyclonal antibodies required an additional purification step to remove cross-reactive antibodies.[11] This was accomplished by the use of nitrocellulose-immobilized hapten-protein conjugates produced with EDAC. Structurally related compounds such as aspartate and aspartylglutamate were coupled to bovine serum albumin with EDAC, and then adsorbed to nitrocellulose strips. The nitrocellulose strips were then incubated first with the crude antisera to remove most of the cross-reactive antibodies. Then these partially purified anti-NAAG and NAA antibodies were affinity purified on a NAAG-coupled or NAA-coupled aminoalkyl-agarose gels respectively. Finally, the antibody solutions were incubated again with new nitrocellulose sheets containing related EDAC conjugates to eliminate remaining cross-reactivity. In the case of NAAG antibodies, NAA-BSA was the most cross-reactive conjugate, whereas in the case of NAA antibodies, NAAG-BSA was the most cross-reactive conjugate. This three-step purification process produced highly specific anti-NAAG and anti-NAA antibodies with less than 1% cross-reactivity to all related protein-hapten conjugates.

2. METHODS

Tissue fixation and immunohistochemistry protocols have been described in detail previously.[10,11,20] Briefly, carbodiimide fixations involved transcardial perfusion with 2 to 5 times the body volume of an aqueous solution of 6% EDAC and 5% DMSO containing 1mM N-hydroxysuccinimide. The fixative solutions were heated to 37°C before initiating the perfusions, because the coupling reaction proceeds significantly faster at higher fixative temperatures. Post-fixations are done in 4% freshly depolymerized paraformaldehyde, or 10% neutral buffered formalin for 24 to 48 hours. For invertebrates or other animals that are difficult to perfuse transcardially, tissues were fixed by removal and rapid immersion in the standard EDAC fixative containing 5% DMSO and 1mM N-hydroxysuccinimide for 20 minutes, followed by further fixation in buffered formalin for at least 24 hours.

Immunohistochemistry was done by the avidin-biotin complex method (Vectastain Elite; Vector Labs, Burlingame, CA). Primary rabbit antibodies were diluted appropriately in 2% normal goat serum (NGS) and incubated with tissue sections for 48 to 72 hours at room temperature with constant rotary agitation (with 0.1% sodium azide as preservative). The secondary antibody and HRP-labeled avidin-biotin complex solutions were incubated with sections for 60-70 minutes each, and after washing were developed with a nickel and cobalt enhanced diaminobenzidine chromogen system (Pierce Biotechnology, Rockford, IL).

3. RESULTS

NAAG and NAA immunoreactivities (NAAG-IR and NAA-IR) have been found to be distinct in many areas of the CNS in the species studied to date. The rat has been the most thoroughly studied species in terms of the extent of the coverage of NAA and NAAG localization in various CNS regions. As compared with other species studied, the rat has relatively high levels of NAA (approximately 8mM) and relatively moderate levels of NAAG (approximately 0.75mM).[2] We consider the rat to be a good representative species for the description of the distributions of NAA and NAAG in the CNS, despite having lower levels than carnivores and primates. In the rat, the staining patterns for NAAG and NAA were very distinct in forebrain and cerebellar cortex, but were more similar in areas of the brainstem and spinal cord, as will be shown below. The greatest disparities in localization are observed in the cerebral and cerebellar cortices of the rat, where NAA is present in most neurons, but NAAG is only present in subpopulations of neurons. For a thorough examination of the disparities between NAAG and NAA expression, see Moffett and Namboodiri.[11]

3.1. Comparative NAA and NAAG Expression in Neocortex

The comparative distributions of NAA-IR and NAAG-IR in rat neocortex are shown in Figure 1. Both immunoreactivities appeared punctate, which may represent localization in intracellular organelles or vesicles. NAAG-IR was also observed in putative extracellular NAAG-positive synaptic contacts (see insert, Figure 6C). NAA-IR was not observed in synaptic-like extracellular puncta, and was more diffuse in the cytoplasm of neurons. However, NAA-IR was often observed in large, organelle-sized intracellular inclusions in certain types of neurons, including cortical pyramidal cells and principle neurons of the hippocampus (Figure 1F inserts). These could possibly be DMSO-based artifacts associated with the fusion of internal membrane structures or organelles that contained high concentrations of NAA (e.g., neuronal mitochondria).

NAAG-IR was most prevalent in apparent interneurons in all cortical layers, and in the rat, was not observed significantly in pyramidal cells (Figure 1A, C, E). This is in contrast with carnivores and primates, where both interneurons and pyramidal cells were strongly immunoreactive for NAAG (see Figures 6 and 7 below). NAAG staining was present in the proximal dendrites of immunoreactive cells in the rat, but was not seen in distal dendrites. This also contrasts with carnivores and primates, where cortical pyramidal cells were immunoreactive for NAAG throughout their dendritic arborizations. NAA-IR was present in most or all neurons in all layers of neocortex, and was also observed in the apical and basal dendrites of pyramidal neurons (Figures 1B, D, F).

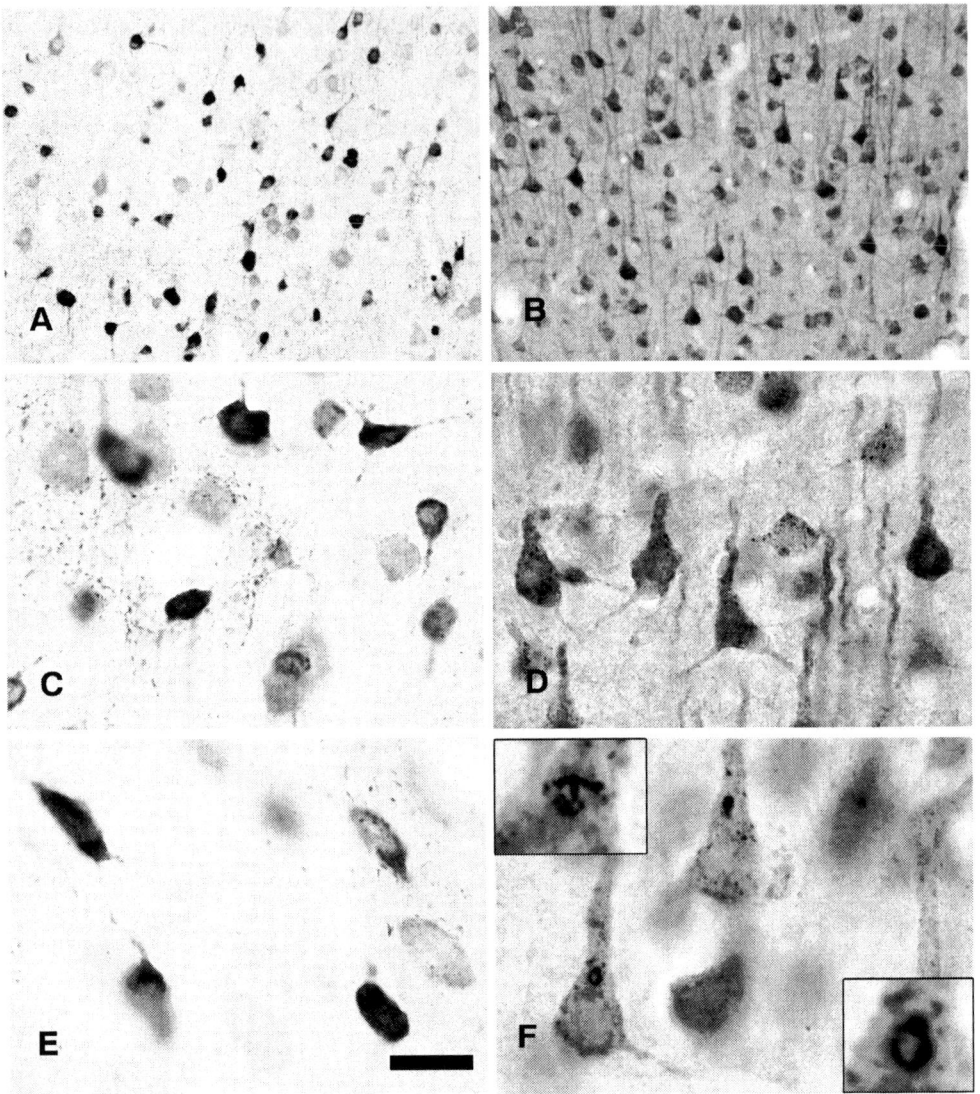

Figure 1. NAA and NAAG immunoreactivity in rat neocortex. NAAG-IR is relatively limited in rat neocortex, being present mostly in small interneurons (A), but NAA-IR is observed in most or all neurons, with high levels in cortical pyramidal cells (B). NAAG-IR is present in cell bodies, proximal dendrites, and probable synaptic contacts in the neuropil (C, E), whereas NAA-IR is present in cell bodies and throughout the dendritic arborizations of neurons (D). In cortical areas, many pyramidal neurons contained large NAA-IR inclusions, often located at the base of the apical dendrite (F). These ranged in shape from round to complex (inserts in F), and could represent DMSO-fused organelles, such as mitochondria, that contained high concentrations of NAA. Bar = 100μm A, B; 30μm C, D and 20μm E, F.

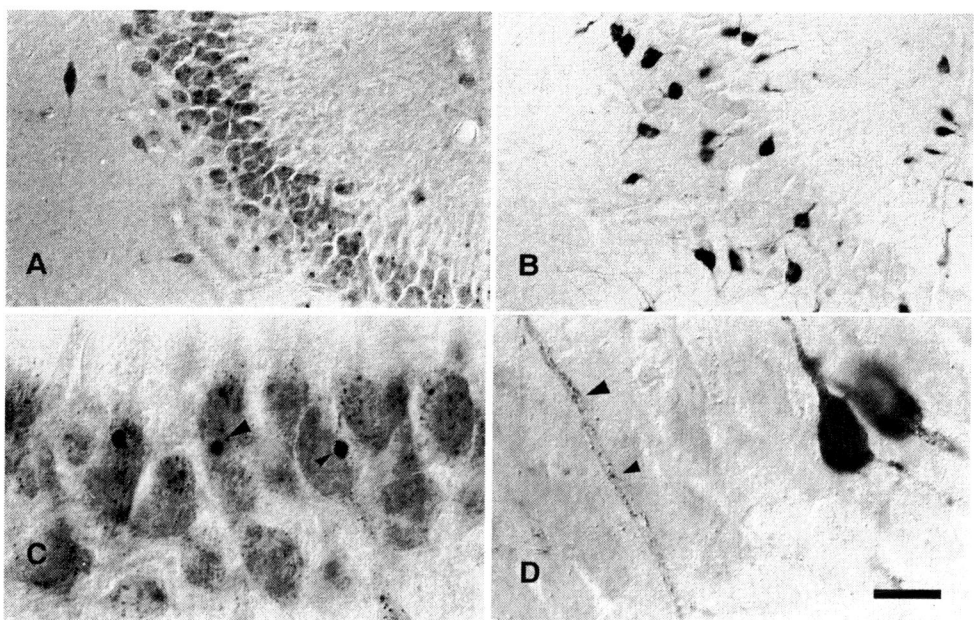

Figure 2. NAA-IR and NAAG-IR in rat hippocampus. NAA is present in most neurons in the rat hippocampus (A). Large intracellular NAA-stained elements were observed in many hippocampal pyramidal cells (arrowheads C). NAAG distribution is relatively restricted in the rat hippocampus, being expressed in scattered neurons in the pyramidal layer in all subdivisions (B), and in cells of the polymorph layer. Many unstained pyramidal cell dendritic shafts were covered with NAAG-stained puncta that appeared to be NAAG-containing synaptic contacts (arrowheads in D). Bar = 100μm A, B and 20μm C, D.

3.2. Comparative NAA and NAAG Expression in Hippocampus

NAA was much more widely distributed than NAAG in hippocampus, as was the case in neocortex (Figures 2A, B). NAAG staining was often opaque in the cytoplasm (Figure 2B), and punctate in processes such as dendrites and axons (Figure 2D, arrowheads). NAA appeared both diffusely in cytoplasm, and was also expressed in heavily-stained puncta within neurons (Figures 2A, C). Large, NAA-IR inclusions were also observed in many hippocampal pyramidal neurons (Figure 2C, arrowheads), as was the case in neocortical pyramidal neurons. NAAG-IR was restricted to scattered large neurons in the pyramidal cell layer, whereas NAA-IR was present in all pyramidal cells, and was moderate to strong in granule cells. No NAAG-IR was observed in the granule cell layers, but many cells in the polymorph layer expressed moderate to high levels of NAAG.

3.3. NAA, NAAG and GAD$_{67}$ Expression in Forebrain and Midbrain

Figure 3 shows the expression of NAAG, NAA and glutatmic acid decarboxylase (GAD$_{67}$) in rat neocortex and hippocampus as compared with midbrain structures. NAAG was distributed differentially, with midbrain and hindbrain having substantially higher levels than telencephalic cortical structures (Figure 3A). NAA distribution was relatively ubiquitous, and similar levels were observed in forebrain and hindbrain (Figure 3B).

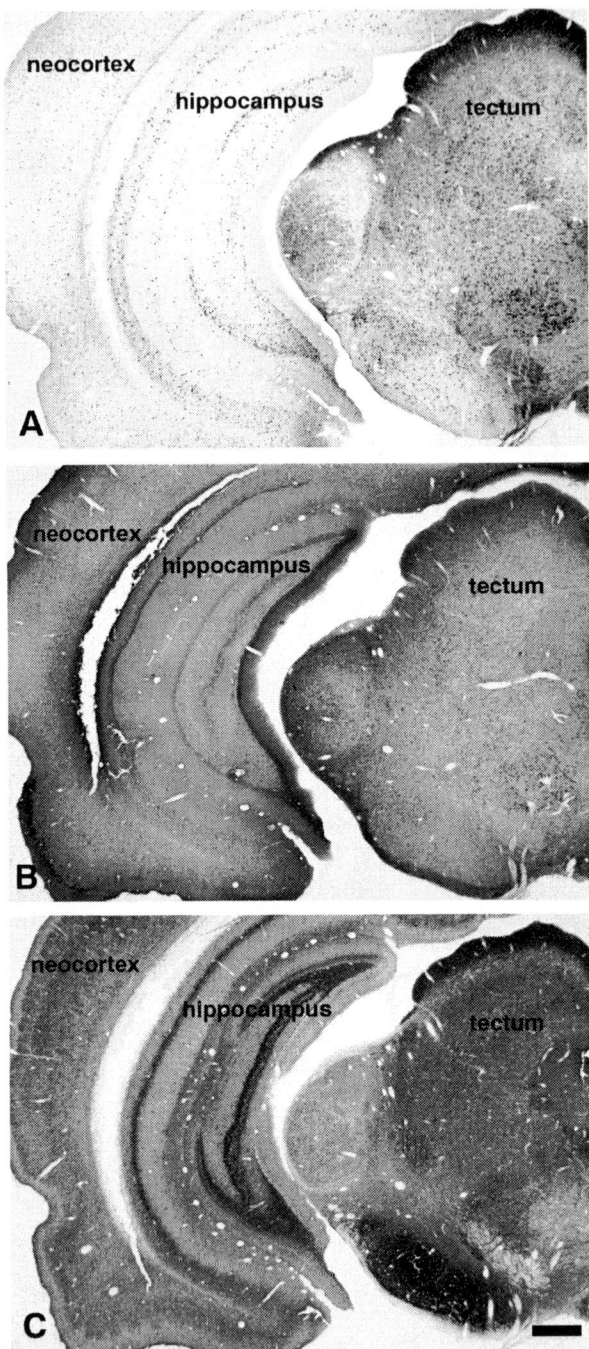

Figure 3. NAAG, NAA and GAD_{67} expression in rat forebrain and midbrain. NAAG expression is relatively low in neocortex and hippocampus as compared with the midbrain tectum, tegmentum and hindbrain (A). NAA immunoreactivity, in contrast, is expressed differentially in different neuronal groups, but in general, high levels are found throughout the CNS (B). GAD_{67} expression is extremely varied, as was the case with NAAG, with high levels in hippocampal granule cell and pyramidal cell layers, and very high expression in the reticular part of the substantia nigra and tectum (C). Bar = 400μm.

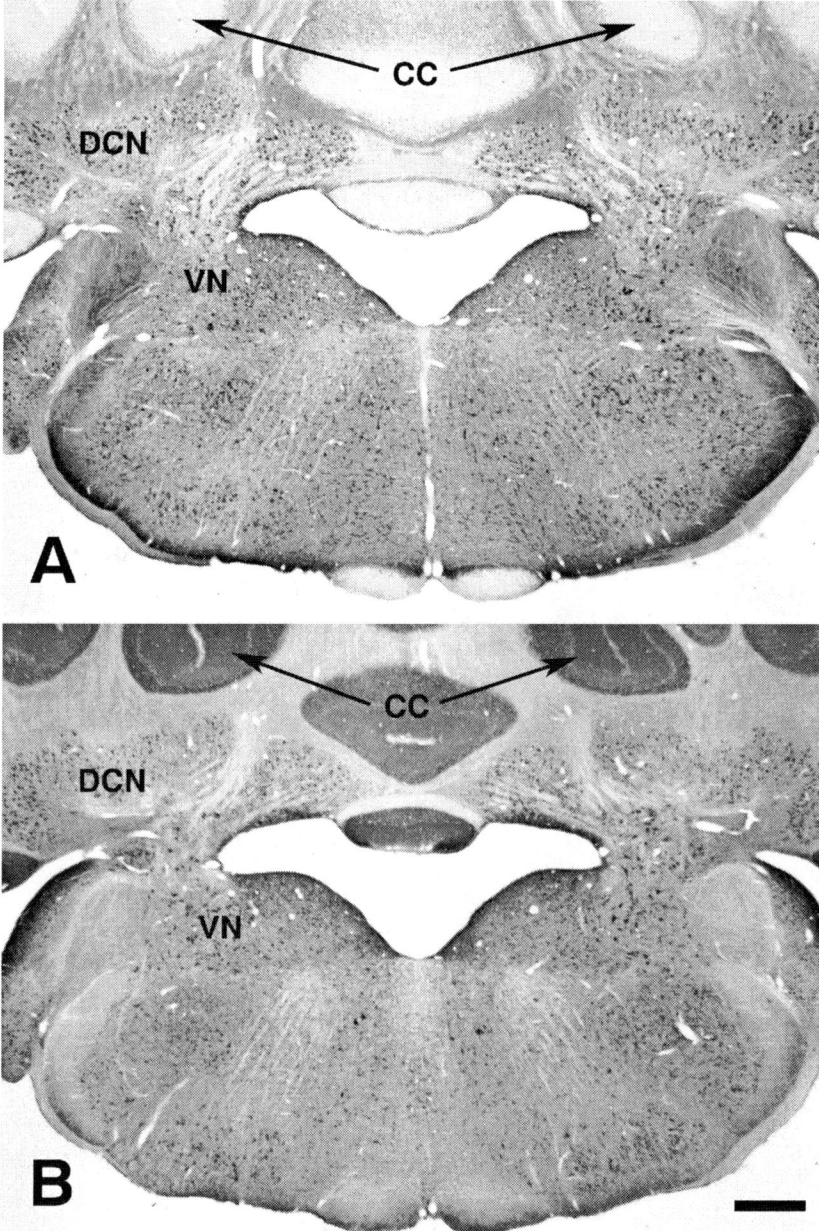

Figure 4. NAAG (A) and NAA (B) are both extensively distributed throughout the brainstem. NAAG expression is very high in the brainstem and deep cerebellar nuclei (DCN), but was much more limited in the cerebellar cortex (CC). In contrast, NAA expression was very high in the CC. Both NAAG and NAA were expressed in many cell groups throughout the brainstem, such as the vestibular nuclei (VN). Many NAAG and NAA-containing axons are present throughout the cerebellar white matter. Bar = 400μm.

GAD$_{67}$ expression was somewhat higher in many midbrain structures than in most cortical layers, but the increase along the rostro-caudal axis of the CNS was not as apparent as was the case with NAAG expression (Figure 3C).

3.4. NAA and NAAG Expression in the Cerebellum and Brainstem

NAAG and NAA were both extensively distributed throughout the cerebellum, brainstem and spinal cord (Figure 4). NAAG levels were low in cerebellar cortex, but were very high in the deep cerebellar nuclei and medulla (Figure 4A). NAA levels were high throughout cerebellar cortex, deep nuclei and medulla (Figure 4B).

3.5. NAA, NAAG and Aspartoacylase Expression in Corpus Callosum

Aspartoacylase is an enzyme that specifically de-acetylates NAA, and which was observed almost exclusively in oligodendrocytes (Figure 5A). In contrast, NAA and NAAG were expressed primarily in neurons (Figures 5B, C). Aspartoacylase-stained cells were arranged in characteristic rows between axon bundles in the corpus callosum. Low levels of NAA can also be seen in rows of oligodendrocytes. The level of NAA-IR in adult rat oligodendrocytes is substantially lower than the level in neurons. NAAG is present in scattered axons in the corpus callosum, whereas NAA is present in most or all axons in the corpus callosum, although at only a moderate level. It remains to be determined if the NAA present in oligodendrocytes is synthesized within the glial cells themselves, or if it is made in neurons, and passed from axon to glia.

3.6. NAAG Expression in Rhesus Monkey Motor Cortex

NAAG expression in Rhesus monkey motor cortex was more extensive than observed in the rat (Figure 6A). In particular, most cortical pyramidal cells in the Rhesus monkey were moderately to strongly stained, whereas in the rat, these cells were either unstained, or faintly stained.[11] NAAG-containing small neurons were relatively numerous in layer III of motor cortex, and the dendritic shafts of motor neurons could be seen coursing through this layer (Figure 6B). Medium and large motor neurons in layer V were moderately to strongly stained for NAAG (Figure 6 C). Many dendritic and axonal elements were immunoreactive for NAAG in layer V, including probable NAAG-containing synaptic contacts on apical motor neuron dendrites. Numerous NAAG-IR neurons and fibers were observed in layer 6b of motor cortex, including axons entering the corpus callosum (Figure 6D).

3.7. NAAG and GAD$_{67}$ Expression in Rhesus Monkey Motor Cortex

The distribution of small neurons stained for NAAG in neocortex was very similar to the distribution of GAD-positive interneurons (Figure 7). However, double labeling experiments showed that fewer than 50% of the NAAG-positive neurons in monkey

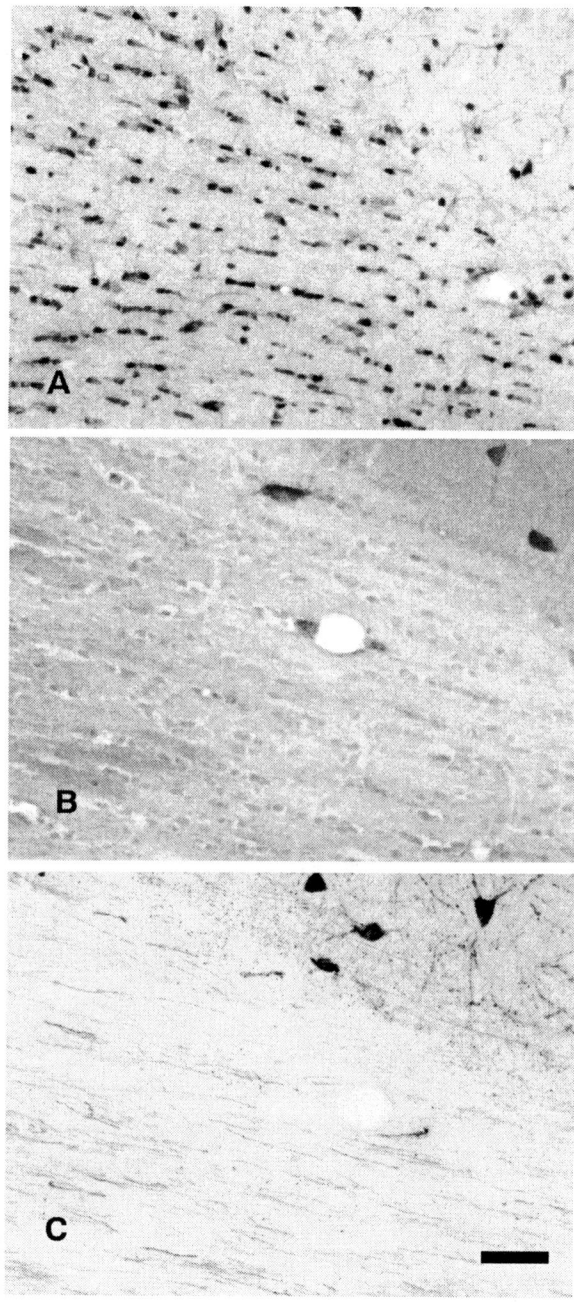

Figure 5. ASPA (A), NAA (B) and NAAG (C) in rat corpus callosum. Aspartoacylase (ASPA) is the only enzyme known in the brain that acts to deacetylate NAA. ASPA expression was predominantly observed in oligodendrocytes in white matter such as corpus callosum (A), as well as throughout the brain and spinal cord. NAA expression in oligodendrocytes was very low, possibly because it is broken down rapidly by ASPA (B). NAAG expression was relatively low in the corpus callosum of the rat, being present in a relatively small number of axons (C). In the cortical white matter from monkey, far more NAAG containing axons were observed, reflecting the greater expression of NAAG in pyramidal cells. Bar = 100μm.

neocortex were double-labeled for GAD_{67} (data not shown), suggesting that many small, NAAG-expressing neurons in cortex may be peptidergic rather than GABAergic.

3.8. NAAG and NAA in GABAergic Neuronal Groups

Unlike the partial colocalization of NAAG and GAD_{67} in neocortex, NAAG was expressed in virtually all GABAergic neurons in GABA projection areas such as the globus pallidus, thalamic reticular nucleus and lateral hypothalamus (Figure 8A, C, E). NAA-IR was also expressed at moderate to high levels in these cell groups (Figure 8B, D, F). One feature in common among many GABAergic projection systems containing NAAG was that NAAG was typically observed only in the cell body and basal dendrites, but not in the axons. This is in contrast to many glutamatergic projection systems containing NAAG, such as the visual projections, where NAAG is present throughout the axons and terminals.

3.9. NAAG in Groups with Known Neurotransmitters

NAAG is strongly expressed in several neuronal groups known to utilize specific neurotransmitters. Examples include the vertical limb of the diagonal band of Broca, which is known to use acetylcholine as the primary neurotransmitter (Figure 9A). The locus coeruleus is a noradrenergic nucleus in the brainstem which is strongly immunoreactive for NAAG (Figure 9B), and the substantia niagra pars compacta is a major dopaminergic nucleus in the midbrain which expresses high levels of NAAG-IR (Figure 9C). NAAG is also present in serotoninergic neurons of the raphe (Figure 9D).

3.10. Phylogeny

NAAG and NAA appear to be phylogenetically ancient molecules, being found in the nervous system of invertebrates,[3,21] and teleost vertebrates.[2] No data are available on elasmobranch vertebrates, but both compounds have been found in the brains of amphibians and reptiles. NAA and NAAG, in general, exhibit a concentration ratio in birds and mammals of approximately 10:1 respectively.[2] This ratio is reversed in amphibians such as the frog, where the ratio is approximately 1:10. In teleosts, such as the goldfish, the ratio is 100:1, with only trace amounts of NAAG [2] and only scattered NAAG-IR neurons present in the brain (Figure 10B, D and F). In contrast, NAA-IR was observed in most neurons throughout the goldfish brain (Figure 10A, C, and E).

4. DISCUSSION

NAAG and NAA have different distributions in the nervous system, and this disparity is most apparent in cortical areas of the CNS. NAA has a ubiquitous presence in most neurons in the nervous system, and yet NAA levels in different cell groups can vary

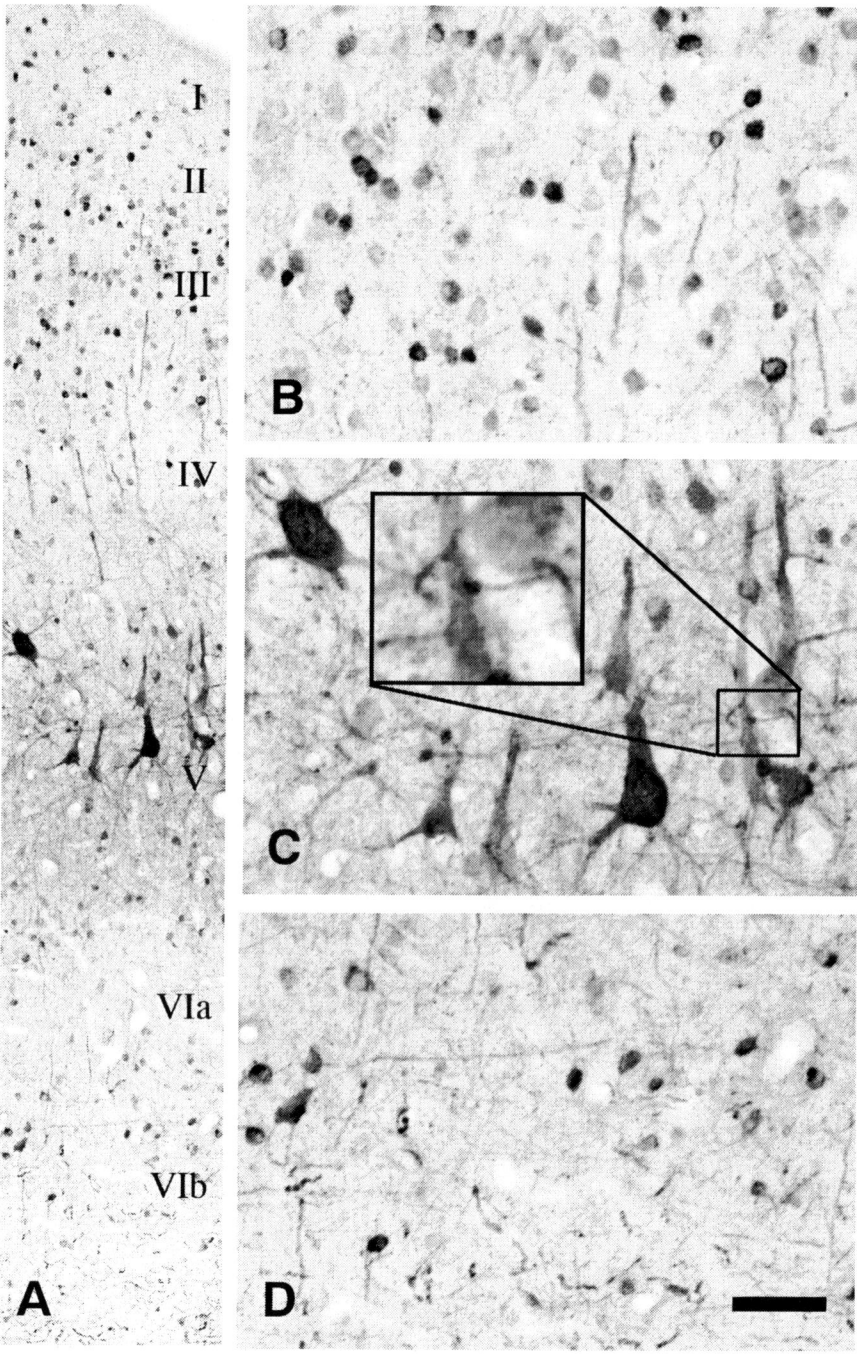

Figure 6. NAAG expression in rhesus monkey motor cortex. NAAG is expressed in all layers of neocortex in the monkey (A). In superficial cortical layers, NAAG was expressed in many small neurons, and in the apical dendrites of pyramidal cells (B). Unlike the rat, NAAG expression was highest in layer V, where it is present in pyramidal cells, and a dense plexus of fibers and synapses in the neuropil (C). In deeper layers, many small neurons expressed high levels of NAAG, as did axons entering the white matter and corpus callosum (D). Bar = 200μm A; 60μm C, D and E.

substantially (Figures 1B, 4B, 9B, D and F). NAAG has a more restricted distribution, and exhibits an increasing concentration gradient from the rostral to the caudal CNS (see Fig 3). NAAG-IR *in vivo* is only expressed in neurons, but it has been reported that oligodendrocytes and microglia can express low levels of NAAG *in vitro*.[12] In addition to being localized in most neurons, NAA was observed at low levels in oligodendrocytes (Figure 5B). Some brain capillary endothelial cells were also stained moderately to strongly for NAA, but only a small percentage of the total endothelial population of the CNS was NAA-positive (data not shown).

NAA has been shown to have a more ubiquitous distribution in the CNS as compared with NAAG, but both compounds are present in most or all regions of the brain and spinal cord of the rat.[8,11,20] Several broad generalizations can be made concerning their comparative localization, to which there are exceptions. First, both compounds are expressed at high levels primarily in neurons, although NAA is found in other cell types in the brain. Astrocytes and microglia did not stain for either compound in any species examined. NAA-IR is present in most neurons to a greater or lesser degree, and is particularly prominent in cortical pyramidal cells, and in granule cell layers in many brain regions, including hippocampus, retrosplenial cortex and cerebellar cortex. The distribution in NAAG in forebrain is also widespread, but only a relatively small percentage of neurons are NAAG-positive in many areas of the telencephalon. While cortical areas have relatively low numbers of NAAG immunoreactive neurons, NAAG is expressed in the majority of neurons in many brainstem and spinal cord regions. For example, virtually all the neurons of the deep cerebellar nuclei are strongly stained for NAAG, whereas a much lower density of stained neurons is seen throughout neocortex.

NAAG is colocalized extensively with GABA in GABAergic projection systems such as the globus pallidus, lateral hypothalamus, thalamic reticular nucleus (Figure 8) and the reticular region of the substantia nigra, but it is only partially colocalized with GABAergic interneurons in cortical brain areas. NAAG is known to be colocalized with other neurotransmitters, including acetylcholine, dopamine, norepinephrine and serotonin (see Figure 9). NAAG is also localized extensively in the retinal projections and retinorecipient terminals areas, which are known to be glutamatergic, perhaps to provide a means of preventing excitotoxicity associated with prolonged glutamate release.[22]

4.1. Comparative Phylogenetic Distribution

Miyake and colleagues were the first to show the concentrations of NAA and NAAG in various classes of vertebrates ranging from fish to mammals.[2] They demonstrated that teleost fish have very low levels of NAAG, but that amphibians, reptiles and mammals have relatively high NAAG levels in the CNS. Birds were found to have intermediate concentrations of NAAG in the brain. NAAG has also been reported in the nerve cord of a crustacean,[23] suggesting that it is present in the nervous system of most or all animals. Our immunohistochemical studies have confirmed that there are very few NAAG-positive neurons in the brains of teleosts (Figure 10), whereas birds have moderate numbers of NAAG positive neurons.[24] Among mammals, rats have lower levels of NAAG, and fewer NAAG-positive neurons (Figure 1) than primates (Figures 6 and 7). In

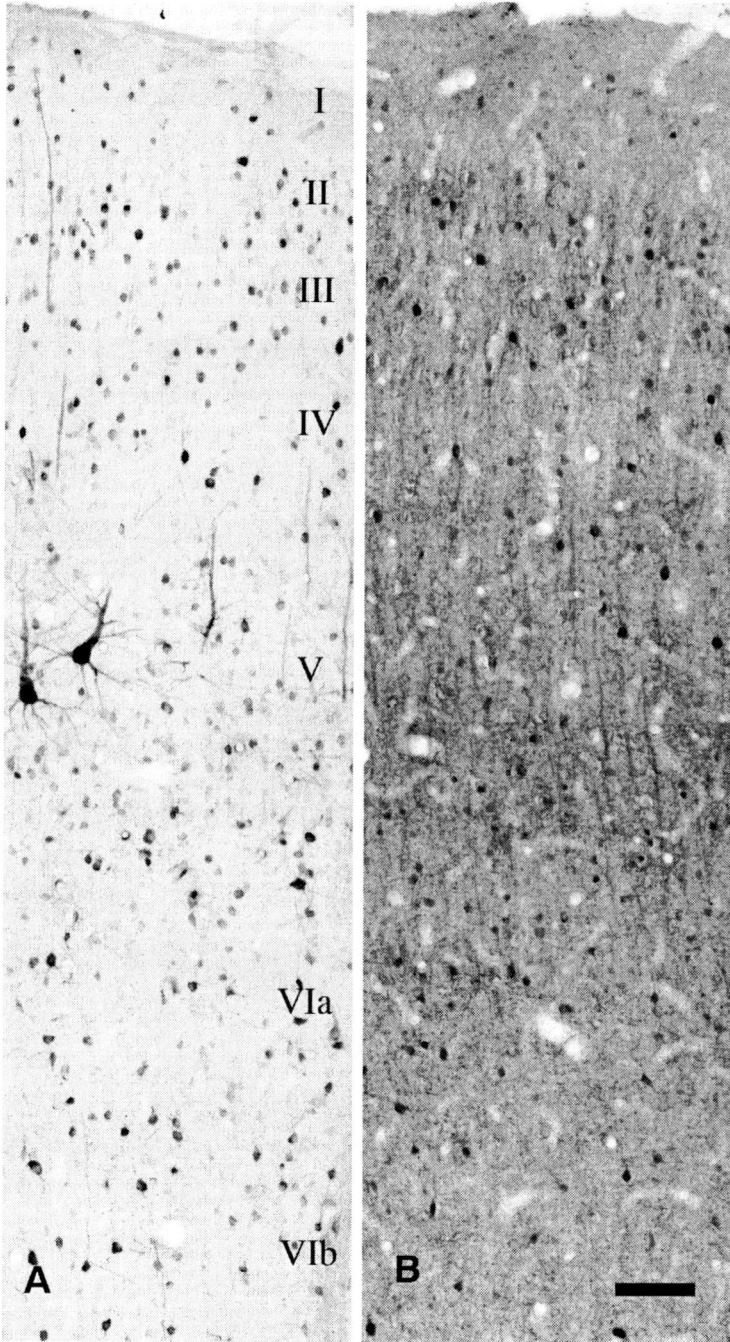

Figure 7. NAAG (A) and GAD$_{67}$ (B) in rhesus monkey cortex. NAAG and the 67 kilodalton form of glutatmic acid decarboxylase (GAD$_{67}$; the enzyme that synthesizes GABA) were present in small neurons throughout all layers of monkey neocortex (A, B). Bar = 200μm.

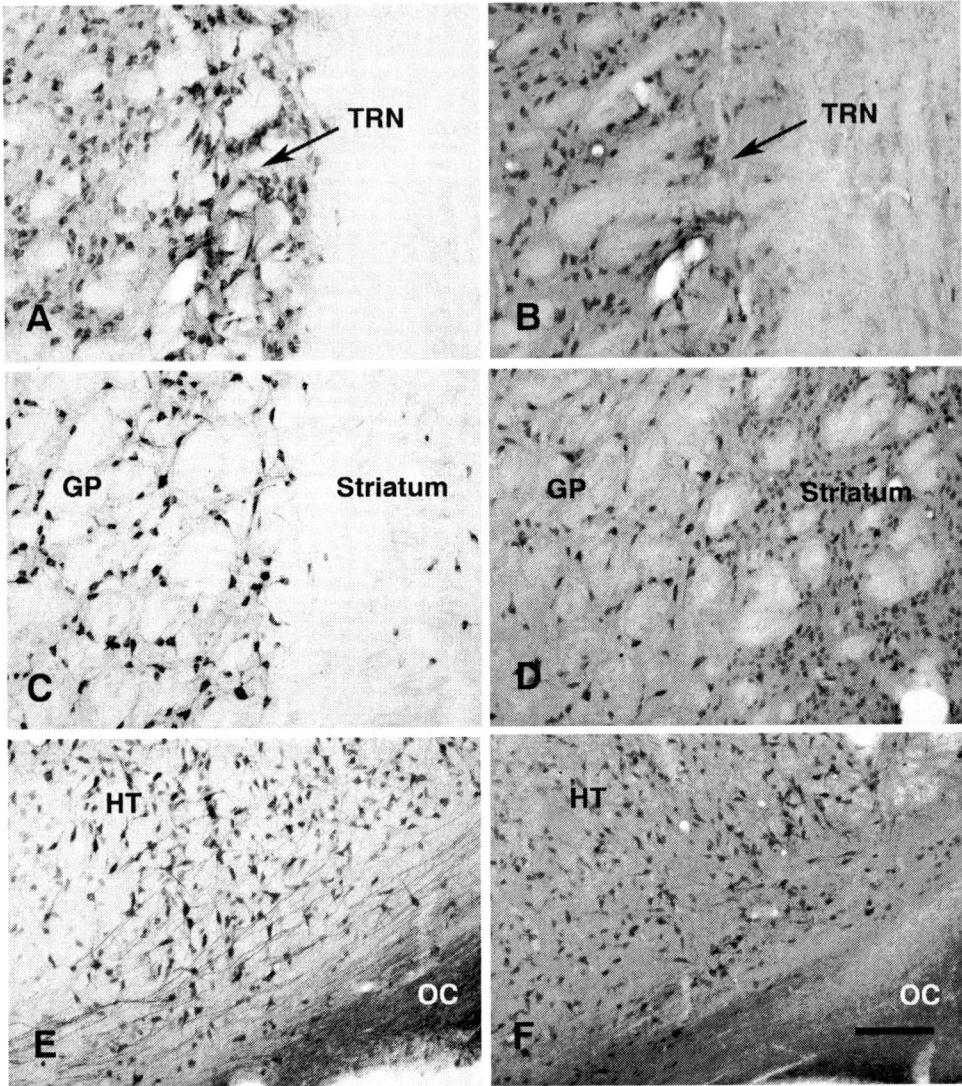

Figure 8. NAAG (A, C and E) and NAA (B, D and F) in GABAergic neuronal groups. NAAG and NAA are both expressed at relatively high levels in GABAergic projection groups, including the thalamic reticular nucleus (TRN), globus pallidus (GP) and the lateral hypothalamus (HT). Bar = 150μm.

contrast, NAA seems to be present in high concentrations in the CNS of most species of animals (with the notable exception of frogs), including invertebrates.

The expression level of NAAG in teleost cortex is extremely limited, with only scattered neurons stained for NAAG (Figure 10B). Substantially higher expression levels are seen in reptiles [25] and birds [24]. Among the mammalian species studied to date, the rat has the lowest levels of NAAG in cortical areas, whereas carnivores (cat) and primates (rhesus monkey) have higher levels. In the rat, very few cortical pyramidal cells

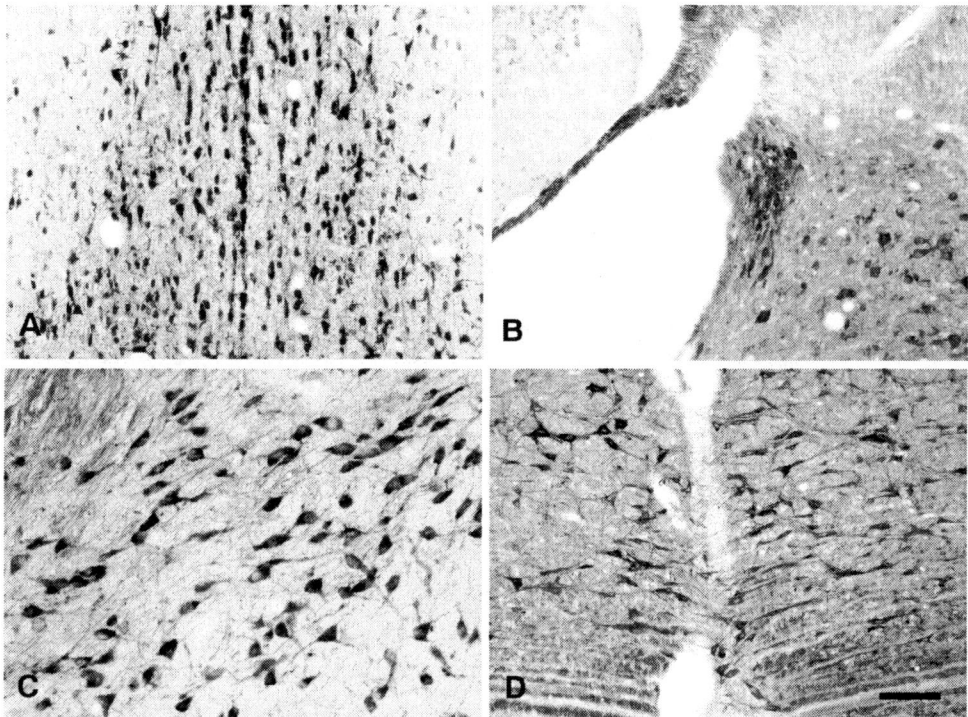

Figure 9. NAAG in groups with known neurotransmitters. NAAG is colocalized with virtually all major neurotransmitters including GABA (Figure 8), acetylcholine in the vertical limb of the diagonal band of Broca (A), norepinephrine in the locus coeruleus (B), dopamine in the compact part of the substantia nigra (C) and serotonin in the raphe (D). Bar = 100 μm.

contain NAAG, but in cat[26,27] and monkey, most layer V pyramidal cells express high levels of NAAG (see Figures 6 and 7).

4.2. Conclusions

Based on their cellular localization and known actions, several conclusions can be drawn concerning the roles played by these two compounds in brain function. In general, the distribution of NAA is consistent with it having a metabolic or housekeeping role in the nervous system, whereas the distribution of NAAG is more consistent with a role in neurotransmitter release modulation.

Because NAA is present at high concentrations in most neurons, and because it is not released from neurons in a calcium dependent manner after depolarization, it can not be acting as a classical neurotransmitter. The high concentration, lack of electro-physiological actions, and ubiquitous distribution all argue in favor of a metabolic role. These facts have led some researchers to propose that NAA acts to counter the "anion deficit" in neurons, [28] or acts as a "molecular water pump" to extrude metabolic water

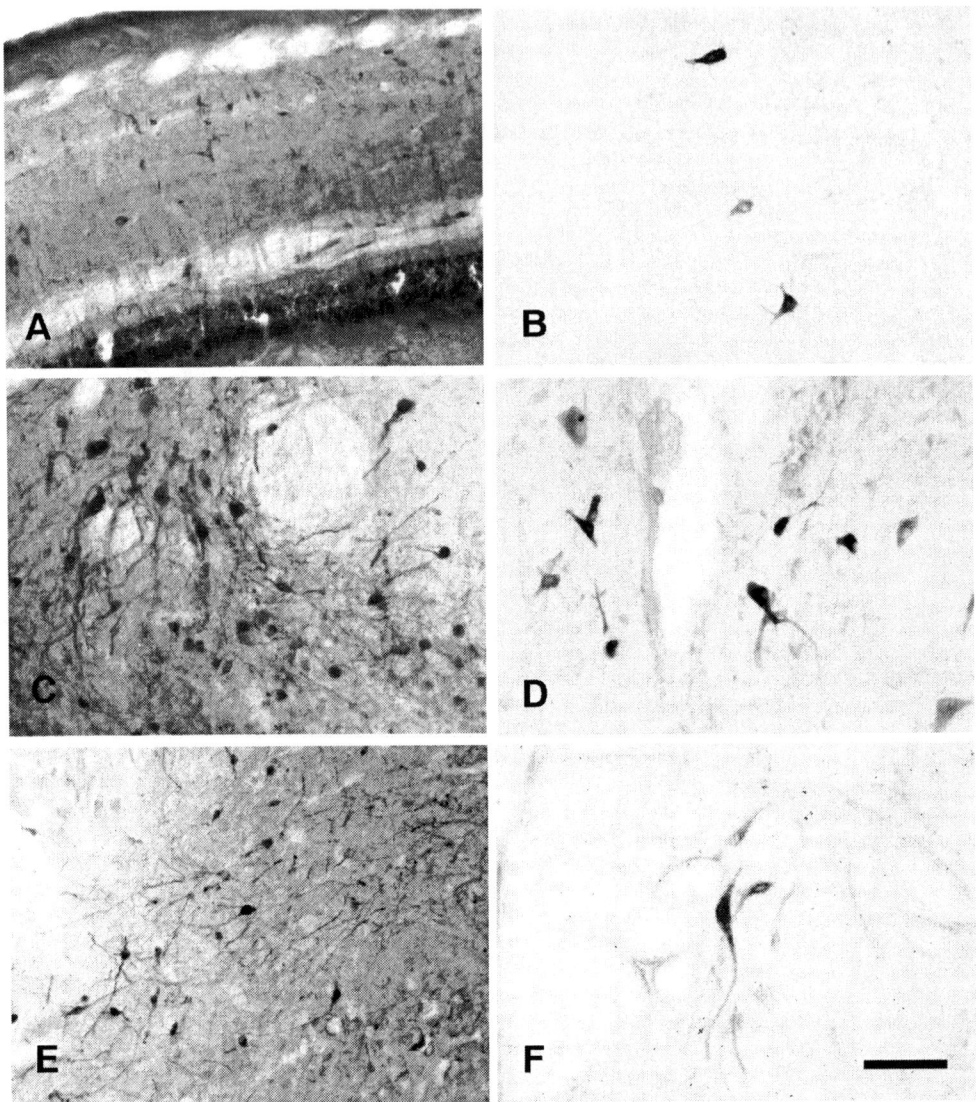

Figure 10. NAA (A, C and E) and NAAG (B, D and F) in goldfish brain. Both NAA and NAAG are phylogenetically conserved brain molecules found in the brains of teleost fish. NAA is expressed strongly throughout the goldfish brain, including cortex (A), hypothalamus (B) and hindbrain (C). NAAG, in contrast, has a very limited expression, with only scattered neurons being strongly stained in cortex (B, rotated 90 degrees compared with A), hypothalamus (D) and hindbrain (F). Bar = 100 μm

from neurons.[29,30] Recent findings clearly demonstrate that NAA provides a significant source of acetate for myelin lipid synthesis during CNS development,[31] and that NAA possibly plays a significant role in neuronal energy metabolism.[32]

   The evidence for NAAG acting as a neurotransmitter or neuromodulator is convincing. NAAG meets all the criteria of a classical neurotransmitter except for one. It is released in a calcium dependent manner after neuronal depolarization, it is broken down extracellularly by a specific enzyme (carboxypeptidase II) and it is known to act at postsynaptic NMDA receptors at high concentrations. However, NAAG application fails the test of eliciting the same postsynaptic actions as are elicited by stimulation of the afferent fibers to various brain regions receiving NAAG input (e.g., the lateral geniculate). Indeed, NAAG application often results in mixed actions on postsynaptic neurons. Electrophysiological studies have shown that different neurons in various brain regions could be depolarized, hyperpolarized, or not respond at all to exogenous NAAG application.[33,34]

The most compelling reason to think that NAAG is a peptide involved in neurotransmitter release modulation, rather than being a classical neurotransmitter, is that it is extensively colocalized with every major neurotransmitter known. Further, the specific and potent action at certain metabotropic glutamate receptors ($mGluR_3$) which are linked to neurotransmitter release modulation[35,36] argues strongly in favor of a modulatory role for NAAG. Finally, the extensive colocalization with both glutamatergic and GABAergic systems implicates NAAG in the regulation of the balance between excitatory and inhibitory neurotransmission in the nervous system.

5. ACKNOWLEDGEMENTS:

This work supported by the Samueli Institute for Information Biology, and by grants from the NIH (ROI, NS39387) and the National Eye Institute (EY 09085).

6. QUESTION AND ANSWER SESSION

PARTICIPANT: I was curious about your data with the monkey cortex.

DR. MOFFETT: Yes.

PARTICIPANT: Are you not finding NAAG in pyramidal neurons?

DR. MOFFETT: Yes, we do in the monkey. Those are all pyramidal cells in layer 5, these large ones. It was only in rat cortex where we did not see NAAG in pyramidal cells.

PARTICIPANT: You showed a monkey cortex, where you showed the comparison to GAD_{67}.

DR. MOFFETT: Yes. Here are some layer V pyramidal cells. And these pyramidal cells were chopped off, and are not visible in this particular 20-micron-thick section, but you can still clearly see that their apical dendrites are heavily labeled. So that's just a particular place where we only picked up two in that spot.

PARTICIPANT: You see it in pyramidal neurons in other layers, also?

DR. MOFFETT: Yes.

SESSION CO-CHAIR COYLE: That raises an interesting question in terms of animal models. We may potentially get misled about the salience of NAAG in cognitive processing because of the under representation of NAAG in mouse/rodent pyramidal neurons.

DR. MOFFETT: Yes. But in terms of non-cortical NAAG expression, the rodent brain is very similar to the situation in carnivores and primates. Even in rat cortex, NAAG immunoreactivity in non-pyramidal neurons is virtually identical to that in primates.

The one other question - we can't really answer too many functional questions with structural studies like this - but the one question that keeps coming back to me is that we have extremely high levels of NAA in virtually all neurons, especially in pyramidal neurons, but there is no deacetylase known that can metabolize NAA in neurons. So that's a question we still have to figure out. Why is there so much NAA in a cell type that cannot further metabolize it?

SESSION CO-CHAIR COYLE: I think you find that in terms of GABAergic neurons in particular, when you raise the question of what is colocalization all about; GABA is certainly an inhibitory neurotransmitter, plus NAAG activating mGluR3; Barbara Wroblewska has shown that, which further down-regulates excitatory neurotransmission, so that colocalization is coherent.

DR. MOFFETT: Yes.

Dr. WEINBERGER: It's also even beyond that level of coherence because mGluR3 is a heteroreceptor that is also found on all of these other neurotransmitter terminals. So there may actually be a local circuit of NAAG regulation of mGluR3 acting at various heteroreceptor terminals.

DR. MOFFETT: I did not focus on it, but you can quite clearly see examples of synaptic-like contacts on dendrites, where there are no obvious axons coming in. And to me, this suggests that these might be presynaptic on a dendrite. In other words, NAAG is being secreted from presynaptic sites on dendrites, which would also be unusual.

SESSION CO-CHAIR BURLINA: Other questions?

PARTICIPANT: Is there anything known about the localization of NALADase?

DR. MOFFETT: Primarily astrocyte, isn't it?

SESSION CO-CHAIR COYLE: Yes, it's in astrocytes, but it looks like it's on the end feet. When you make synaptosomal preparations you get NALADase is enriched in the synaptosomes, which could mislead you to think it is in nerve terminals, but those preparations also contain astrocyte end feet.

DR. MOFFETT: End feet, yes.

SESSION CO-CHAIR COYLE: Thank you very much.

DR. MOFFETT: Thank you.

7. REFERENCES

1. E. Miyamoto, Y. Kakimoto and I. Sano, Identification of N-acetylaspartylglutamic acid in the bovine brain. *J. Neurochem.*, **13**, 999-1003 (1966).
2. M. Miyake, Y. Kakimoto and M. Sorimachi, A gas chromatographic method for the determination of N-acetyl-L-aspartic acid, N-acetyl-aspartylglutamic acid and beta-citryl-L-glutamic acid and their distributions in the brain and other organs of various species of animals. *J. Neurochem.*, **36**, 804-810 (1981).
3. A. K. Urazaev, R. M. Grossfeld, P. L. Fletcher, H. Speno, B. S. Gafurov, J. G. Buttram and E. M. Lieberman, Synthesis and release of N-acetylaspartylglutamate (NAAG) by crayfish nerve fibers: implications for axon-glia signaling. *Neurosci.*, **106** (1), 237-247 (2001).
4. L. M. Gehl, O. H. Saab, T. Bzdega, B. Wroblewska and J. H. Neale, Biosynthesis of NAAG by an enzyme-mediated process in rat central nervous system neurons and glia. *J. Neurochem.*, **90** (4), 989-997 (2004).
5. C. B. Cangro, M. A. Namboodiri, L. A. Sklar, A. Corigliano-Murphy and J. H. Neale, Immunohistochemistry and biosynthesis of N-acetylaspartylglutamate in spinal sensory ganglia. *J. Neurochem.*, **49**, 1579-1588 (1987).

6. K. J. Anderson, M. A. Borja, C. W. Cotman, J. R. Moffett, M. A. Namboodiri and J. H. Neale, N-acetylaspartylglutamate identified in the rat retinal ganglion cells and their projections in the brain. *Brain Res.*, **411**, 172-177 (1987).
7. S. B. Tieman, K. Butler and J. H. Neale, N-acetylaspartylglutamate. A neuropeptide in the human visual system. *JAMA*, **259**, 2020 (1988).
8. J. R. Moffett, M. A. Namboodiri, C. B. Cangro and J. H. Neale, Immunohistochemical localization of N-acetylaspartate in rat brain. *Neuroreport*, **2**, 131-134 (1991).
9. S. B. Tieman, J. H. Neale and D. G. Tieman, N-acetylaspartylglutamate immunoreactivity in neurons of the monkey's visual pathway. *J. Comp. Neurol.*, **313**, 45-64 (1991).
10. J. R. Moffett, M. A. Namboodiri and J. H. Neale, Enhanced carbodiimide fixation for immunohistochemistry: Application to the comparative distributions of N-acetylaspartylglutamate and N-acetylaspartate immunoreactivities in rat brain. *J. Histochem. Cytochem.*, **41**, 559-570 (1993).
11. J. R. Moffett, M. A. Namboodiri, Differential distribution of N-acetylaspartylglutamate and N-acetylaspartate immunoreactivities in rat forebrain. *J. Neurocytol.*, **24**, 409-433 (1995).
12. L. Passani, S. Elkabes and J. T. Coyle, Evidence for the presence of N-acetylaspartylglutamate in cultured oligodendrocytes and LPS activated microglia. *Brain Res.*, **794** (1), 143-145 (1998).
13. A. K. Urazaev, J. G. Buttram, Jr., J. P. Deen, B. S. Gafurov, B. S. Slusher, R. M. Grossfeld and E. M. Lieberman, Mechanisms for clearance of released N-acetylaspartylglutamate in crayfish nerve fibers: implications for axon-glia signaling. *Neurosci.*, **107** (4), 697-703 (2001).
14. R. Burri, P. Bigler, P. Straehl, S. Posse, J. P. Colombo and N. Herschkowitz, Brain development: 1H magnetic resonance spectroscopy of rat brain extracts compared with chromatographic methods. *Neurochem. Res.*, **15**, 1009-1016 (1990).
15. S. H. Graham, D. J. Meyerhoff, L. Bayne, F. R. Sharp and M. W. Weiner, Magnetic resonance spectroscopy of N-acetylaspartate in hypoxic- ischemic encephalopathy. *Ann. Neurol.*, **35**, 490-494 (1994).
16. P. J. Pouwels, J. Frahm, Regional metabolite concentrations in human brain as determined by quantitative localized proton MRS. *Magn Reson. Med.*, **39** (1), 53-60 (1998).
17. S. M. Leary, C. A. Davie, G. J. Parker, V. L. Stevenson, L. Wang, G. J. Barker, D. H. Miller and A. J. Thompson, 1H magnetic resonance spectroscopy of normal appearing white matter in primary progressive multiple sclerosis. *J. Neurol.*, **246** (11), 1023-1026 (1999).
18. P. A. Kendall, J. M. Polak and A. G. Pearse, Carbodiimide fixation for immunohistochemistry: observations on the fixation of polypeptide hormones. *Experientia*, **27**, 1104-1106 (1971).
19. J. M. Polak, P. A. Kendall, C. M. Heath and A. G. Pearse, Carbodiimide fixation for electron microscopy and immunoelectron cytochemistry. *Experientia*, **28**, 368-370 (1972).
20. J. R. Moffett, M. Palkovits, M. A. Namboodiri and J. H. Neale, Comparative distribution of N-acetylaspartylglutamate and GAD_{67} immunoreactivities in the cerebellum and precerebellar nuclei of the rat utilizing enhanced carbodiimide fixation and immunohistochemistry. *J. Comp. Neurol.*, **347** (4), 598-618 (1994).
21. Tsukada Y, Uemura K, Hirano S and Nagata Y. Distribution of amino acids in different species. In: Richter D, ed. *Comparative Neurochemistry.* (Pergamon Press, Oxford, 1964), 79-183.
22. J. R. Moffett, Reductions in N-acetylaspartylglutamate and the 67 kDa form of glutamic acid decarboxylase immunoreactivities in the visual system of albino and pigmented rats after optic nerve transections. *J. Comp Neurol.*, **458** (3), 221-239 (2003).
23. B. Gafurov, A. K. Urazaev, R. M. Grossfeld and E. M. Lieberman, N-Acetylaspartylglutamate (NAAG) is the probable mediator of axon-to- glia signaling in the crayfish medial giant nerve fiber. *Neurosci.*, **106** (1), 227-235 (2001).
24. L. C. Williamson, D. A. Eagles, M. J. Brady, J. R. Moffett, M. A. Namboodiri and J. H. Neale, Localization and synaptic release of N-acetylaspartylglutamate in the chick retina and optic tectum. *Eur. J. Neurosci.*, **3**, 441-451 (1991).
25. M. M. Kowalski, M. Cassidy, M. A. Namboodiri and J. H. Neale, Cellular localization of N-acetylaspartylglutamate in amphibian retina and spinal sensory ganglia. *Brain Res.*, **406**, 397-401 (1987).
26. S. B. Tieman, C. B. Cangro and J. H. Neale, N-acetylaspartylglutamate immunoreactivity in neurons of the cat's visual system. *Brain Res.*, **420**, 188-193 (1987).
27. S. B. Tieman, J. R. Moffett and S. M. Irtenkauf, Effect of eye removal on N-acetylaspartylglutamate immunoreactivity in retinal targets of the cat. *Brain Res.*, **562**, 318-322 (1991).
28. Cooper JR, Bloom FE and Roth RH. Metabolism in the central nervous system. *The Biochemical Basis of Neuropharmacology.* (Oxford University Press, New York, 1991) , 75-87, .
29. M. H. Baslow, Evidence supporting a role for N-acetyl-L-aspartate as a molecular water pump in myelinated neurons in the central nervous system. An analytical review. *Neurochem. Int.*, **40** (4), 295-300 (2002).

30.   M. H. Baslow, Brain N-acetylaspartate as a molecular water pump and its role in the etiology of canavan disease: a mechanistic explanation. *J. Mol. Neurosci.*, **21** (3), 185-190 (2003).

31.   C. N. Madhavarao, P. Arun, J. R. Moffett, S. Suczs, S. Surendran, R. Matalon, J. Garbern, D. Hristova, A. Johnson, W. Jiang and M. A. Namboodiri, Defective N-acetylaspartate catabolism reduces brain acetate levels and myelin lipid synthesis in Canavan's disease. *Proc. Natl. Acad. Sci. U. S. A*, **102** (14), 5221-5226 (2005).

32.   B. Tavazzi, S. Signoretti, G. Lazzarino, A. M. Amorini, R. Delfini, M. Cimatti, A. Marmarou and R. Vagnozzi, Cerebral oxidative stress and depression of energy metabolism correlate with severity of diffuse brain injury in rats. *Neurosurgery*, **56** (3), 582-589 (2005).

33.   H. E. Jones, A. M. Sillito, The action of the putative neurotransmitters N- acetylaspartylglutamate and L-homocysteate in cat dorsal lateral geniculate nucleus. *J. Neurophysiol.*, **68**, 663-672 (1992).

34.   E. R. Whittemore, J. F. Koerner, An explanation for the purported excitation of piriform cortical neurons by N-acetyl-L-aspartyl-L-glutamic acid (NAAG). *Proc. Natl. Acad. Sci. USA*, **86**, 9602-9605 (1989).

35.   B. Wroblewska, J. T. Wroblewski, O. H. Saab and J. H. Neale, N-acetylaspartylglutamate inhibits forskolin-stimulated cyclic AMP levels via a metabotropic glutamate receptor in cultured cerebellar granule cells. *J. Neurochem.*, **61**, 943-948 (1993).

36.   B. Wroblewska, J. T. Wroblewski, S. Pshenichkin, A. Surin, S. E. Sullivan and J. H. Neale, N-Acetylaspartylglutamate selectively activates mGluR3 receptors in transfected cells. *J. Neurochem.*, **69**, 174-181 (1997).

N-ACETYLASPARTATE METABOLISM
IN NEURAL CELLS

Kishore K. Bhakoo, Timothy Craig and Daniel Pearce[*]

1. INTRODUCTION

N-Acetyl-Aspartate (NAA) is the second most abundant molecule (after glutamate) in the CNS, and its presence is used increasingly in clinical and experimental magnetic resonance spectroscopy (MRS) studies as a putative neuronal marker.

This acceptance originated from a body of work suggesting that NAA is localized primarily within the nervous system[1] and is synthesized by the synaptosomal mitochondria.[2, 3] Other evidence for the neuronal origin of NAA comes from studies of brain tumors of glial origin, which do not contain NAA, and those of neuronal origin, which do;[4] from effects of neuron-specific toxins which deplete NAA[5] and from immunohistochemical studies using antibodies to NAA, which were localized to a large number of neurons in the rat CNS, although many neurons remained unstained.[6, 7]

Subsequent studies using purified cell populations derived from the CNS and high-resolution ¹H-NMR spectroscopy confirmed earlier reports of the neuronal localization (cerebellar granule neurons) of NAA.[8, 9] Although this work also revealed its presence in the O-2A progenitor and immature oligodendrocyte, the metabolite was not detected in the mature oligodendrocyte. Thus NAA, which represents a major chemical constituent of the CNS, has been interpreted as a marker of functional neurons in the adult brain.

However, the function(s) of NAA remains an enigma and the proposed roles range from metabolism of specific brain fatty acids,[10] donor of acetyl groups for lipogenesis in myelination,[11, 12] involvement in ion balance[13] and neuromodulation.[14] There have also been suggestions that it may reflect brain mitochondrial energy metabolism.[3] It is widely

[*] MRC Clinical Sciences Centre, Imperial College London, Hammersmith Hospital Campus, Du Cane Road, London. W12 0NN. UK.

accepted that the NAA peak seen in the proton MR spectrum of brain represents viable neurons.

Moreover, in neurodegenerative pathologies as diverse as multiple sclerosis, Alzheimer's disease, mitochondrial encephalopathy with lactic acidosis and stroke-like episodes (MELAS)[15] and hypothyroidism,[16] loss or reduction of the NAA peak is related to neurological dysfunction. Additionally, decreases in NAA observed *in vivo* in several neurological conditions may be reversed - an observation that does not sit easily with the concept of NAA being a neuronal marker alone.

Hence, it is difficult to reconcile the paradox of the putatively neuron-specific location of NAA with the clinical evidence of myelin degeneration and the preservation of relatively normal gray matter in the early stages of the disease. It therefore raises the possibility of a contribution to the NAA signal from oligodendrocytes, the myelinating cells of the CNS. This reassessment is therefore central in the evaluation of NAA as a marker of neuronal pathology alone. The studies by Urenjak *et al.*[8, 9] have suggested that mature oligodendrocytes, derived from O-2A progenitors, do not express NAA. However, recent studies on oligodendrocyte biology[17, 18] have shown that these cells require CNTF for both maturation and survival in culture; information that was not available to Urenjak *et al.* at the time of their studies.

In light of these observations on oligodendrocyte development, the NAA synthesizing potential of cells derived from the O-2A progenitor lineage was re-examined.

Furthermore, in order to clarify the function of NAA, the distribution of its catabolic enzyme, aspartoacylase (amidohydrolase II) found in the mammalian brain, which specifically hydrolyses NAA into aspartate and acetate[19] was also investigated. Mutation of aspartoacylase leads to Canavan's Disease (CD[20], which is an autosomal recessive leukodystrophy, resulting in spongy degeneration of white matter tracts of the brain,[21] and the characteristic NAA aciduria).[22, 23]

As discussed the role of NAA hydrolysis in the adult brain is still uncertain, although there are several theories. One putative role of aspartoacylase may be to make the acetyl group available from NAA for lipid synthesis, i.e. N-acetyl-aspartate acts as an acetate shuttle from the mitochondrion to the cell membrane. Several lines of evidence support this model; intracerebrally injected radiolabeled NAA is incorporated into brain lipid fractions with increased efficiency with age.[11, 24] The highest proportion of radiolabeled lipids was found in myelin and mitochondrial fractions. Furthermore, immunostaining against aspartoacylase has shown that the enzyme is associated with myelin in the brain.[25] Uptake studies with radiolabeled NAA show that it is taken up exclusively by astrocytes,[26] whereas there is little NAA uptake by neurons.[25]

Canavan Disease is the only known genetic disorder, which is caused by a defect in the metabolism of a small metabolite, NAA; synthesized exclusively in the brain[27] in a cell specific manner.[6, 8] Since the initial discovery of NAA in brain,[1] its biological role has remained unclear.[10, 13, 28, 29] The stable level of NAA in the brain has made it a useful marker in ^{1}H-NMR spectroscopy of brain.[14] Significantly reduced levels of NAA have been reported in non-CD focal or generalized demyelinating disorders,[30] Huntington's disease,[31] HIV-seropositive individuals[32] and acute stroke.[33] While such decrease in NAA level has been proposed as a measure of neuronal loss, it is not specific for any particular pathology. The accumulation of NAA, and dystrophy of white matter, in brain of patients with CD is highly specific and elevated NAA level has been demonstrated both by biochemical as well as ^{1}H-NMR spectroscopy.[23, 30, 34]

Why should the failure to maintain normal levels of NAA in the brain produce the devastating effect on myelin, and why are they confined to white matter? Decreases in NAA are correlated with a number of disorders, including MS,[35] neurodegenerative disorders and stroke, as monitored using ^{1}H-NMR spectroscopy. NAA levels are also increased in the grey matter in patients with CD,[34] but this is spared any pathological damage. Although much about the onset of the disease remains an enigma, the identification of the gene defect to aspartoacylase[20] is now well established.

Previous studies have shown that aspartoacylase activity increases in the rat brain during development.[36] Furthermore, regional analysis of NAA in the developing rat brain demonstrated an increase in all areas of the brain,[37] the most noticeable being after 7 days. N-acetyl-aspartate levels are highest in gray matter,[37] whereas aspartoacylase activity is higher in white matter than in gray matter, with the largest increase seen between 10 and 30 days.[36, 38] More recently Baslow et al.[39] have demonstrated that the expression of aspartoacylase activity in cultured rat macroglial cells is limited to oligodendrocytes. In the present study the distribution of aspartoacylase activity was investigated in regional brain tissue from developing and mature rat brain, and primary cultures of purified glial and neuronal cells derived from the CNS. Some of this data presented in this study has been published previously.[40, 41]

2. METHODS

2.1. Preparation of Cell Cultures.

Cell cultures were maintained in Dulbecco's modified Eagle's medium (DMEM) containing 4.5g/L glucose (Life Technologies). The O-2A progenitor cells were cultured in DMEM-BS. DMEM-BS is a serum-free defined medium; a modification of the N2 medium described by Bottenstein and Sato.[42]

2.2. Preparation of O-2A Progenitor Cells.

Purification of O-2A progenitor cells from 7 day old rats was as described previously.[17] The O-2A progenitor cells were purified by immunopanning using antibody-coated dishes. Negative selection with Ran-2 antibody[43] was used to eliminate type-1 astrocytes, followed by anti-galactocerebroside (GalC) antibody[44] panning to remove oligodendrocytes. The O-2A progenitors were further purified from the remaining cell suspension by capture on a tissue culture dish coated with the A2B5 antibody.[45]

Platelet-derived growth factor (PDGF$_{AA}$, R&D Systems) and basic fibroblast growth factor (bFGF; Pepro Tech) were added daily, each at 10ng/ml, both to increase the number of progenitor cells and also to maintain them in an undifferentiated state.[46]

Increasing numbers of O-2A progenitors were obtained by repeated passage on to new plates. Immature and mature oligodendrocytes were derived from the same pool of cells as the O-2A progenitors following passage of O-2A progenitors on to new dishes. Immature oligodendrocytes were allowed to partially differentiate in the presence of bFGF, 10ng/ml added daily. O-2A progenitors were allowed to spontaneously differentiate into mature oligodendrocytes by withdrawal of growth factors (PDFG$_{AA}$ and bFGF), following passage on to new dishes. The mature oligodendrocytes were then cultured under two different conditions for at least five days prior to harvest. One set of cultures were supplemented with CNTF (cilliary neurotrophic factor - recombinant rat CNTF; Precision Research Biochemicals) at a concentration of 5ng/ml daily, while the other set of oligodendrocytes were cultured in DMEM-BS alone.[17]

2.3. Immunocytochemistry.

The maturation of oligodendrocyte was confirmed with antibody staining on parallel cultures grown on PLL coated glass coverslips. Antibody to GalC, a specific label of oligodendrocytes,[47] a cell surface antigen, was applied to fixed cells. To visualise cytoplasmic myelin basic protein (MBP), cells were permeabilized with methanol cooled to -70°C). Fluorescein and rhodamine-conjugated secondary layer antibodies were used to visualise the primary antibodies.

2.4. Oligodendrocyte Maturation and Survival Assay.

To examine the maturation and survival of oligodendrocyte in the presence and absence of CNTF, purified O-2A progenitor cells form 7-day-old rat pups were cultured at a density of 3000 cells/coverslip. CNTF was added daily and the coverslips stained after a total of 5 days in culture. The cells were stained with anti-GalC and anti-MBP antibodies. Dead cells were identified as GalC$^+$ 'ghosts' with no nuclei.

2.5. Preparation of Rat Cortical Astrocytes.

Enriched astrocytes were prepared by the method of Noble and Murray.[48] This procedure routinely produced astrocyte cultures of 95% purity as assessed by staining with a polyclonal serum against GFAP.

2.6. Preparation of Cortical Neurons.

Cortices from embryonic day 14-16 rats were used to harvest the neurons.[49] The cells were cultured in DMEM-BS supplemented with 10 ng/ml of basic fibroblast growth factor, and 50 ng/ml of nerve growth factor (Roche Diagnostics Ltd, Lewes, UK). The growth factors were added every second day to improve cell survival. Every second day, cells were pulsed with 20 mm cytosine arabinoside to prevent glial cell proliferation. This procedure routinely produced neuronal cultures of 95% purity as assessed by immunostaining with RT97.[50] Cells were harvested after 7 days of culture.

2.7. Preparation of Cerebellar Granule Neurons.

The procedure for the preparation of cerebellar granule neurons was the same as that of cortical neurons except that the source tissues were the cerebella of 7-day-old rat pups.

2.8. Preparation of Rat Meningeal Cells.

Meningeal sheaths from cortices were obtained from 7-day-old rat pups and cultivated as Bhakoo et al.[49] Cells were sequentially subcultured no more than 4 times prior to harvest (see below).

2.9. Cell Harvesting, Extraction and Preparation of Samples.

All cells were continuously cultured until sufficient material was available for high-resolution NMR (typically 10^7-10^8 cells for a sample) and enzyme assays. Cells were always harvested 24 hours after a final medium change for all cell types. Cells were washed 2 times with 25ml phosphate buffered saline while still attached to the culture plastics. The cells were then scraped off the plastic surface using silicon rubber cell scrapers. The cells were spun at 2000g in a microfuge for <1 min. The supernatant was removed and the pellet frozen immediately in liquid nitrogen (for further details see Bhakoo et al.[49])

2.10. ^{1}H-NMR Spectroscopy.

All spectra were obtained at 30°C on a Varian Unity-INOVA NMR spectrometer (Varian Associates Inc., NMR Instruments, Palo Alto, CA) operating at a proton frequency of 400 MHz. One-pulse fully relaxed spectra were acquired with 90° pulses applied every 14 sec, with a spectral width of 6000.6Hz and 64K resolution. For a satisfactory signal/noise ratio, 256 or 512 scans were accumulated and Fourier transformed.

2.11. Metabolite Identification and Quantification.

Metabolite peaks found in the one-dimensional spectra were identified by their chemical shift and coupling pattern as described in literature and by comparison with spectra of metabolites in known concentrations obtained at the same pH and spectroscopic conditions. The area of a signal in the proton spectrum is proportional to the amount of compound and number of protons contributing to the signal. NAA was quantified as an integral ratio with known amounts of TSP (1mM).

2.12. HPLC Method.

NAA was separated and quantified using the method of Koller et al.[51] Regular spike runs (of sample plus a known quantity of NAA standard) were performed in order to confirm the presence of NAA in a given cell extract. NAA concentrations were calculated from external standards. The NAA levels were normalized to sample protein content.

2.13. Tissue Harvest for Aspartoacylase Assays.

Tissues from various regions of the rat brain, at key ages during postnatal development and maturation, were analyzed for aspartoacylase activity. The regions examined were: cerebral cortex, corpus callosum, cerebellum, hippocampus, brain stem, olfactory bulb and optic nerve at ages from postnatal day 1 (P1) to adult. Adult samples were taken from rats aged over 6 months. Various regions of the brain were removed sequentially and placed into cryovials. The optic nerves were sectioned just behind the eye and the nerve tracts dissected anterior to the optic chiasm. Olfactory bulbs were dissected out. The brain stem and cerebellum were also removed, then the cerebrum was resected from the skull and placed in a Petri dish containing ice-cold nutrient medium (Leibovitz L-15; Life Technologies, Paisley, UK). The dissociation of the meninges from the cortex, the dissection of the hippocampus and the separation of the *corpus callosum* was performed using a dissecting microscope. Initially, the *corpus callosum* was dissected from the rest of the structure was to leave the cortical sample. In this way, cross contamination between *corpus callosum* and cortex was minimized. Cortex, hippocampus and *corpus callosum* were washed by immersion in ice-cold phosphate-buffered saline in order to remove all possible contaminants from the dissection medium. Samples were placed in cryovials and immediately frozen in liquid nitrogen after dissection, and stored at -85°C until assayed. In all studies, at least six separate litters were used to provide brain material.

2.14. Aspartoacylase Assay.

The aspartoacylase assay was based on previous method of Hagenfeldt *et al.*[22] The assay was optimized to yield the optimum aspartoacylase activity using the smallest amount of tissue. This was necessary due to the small quantities available for certain areas of the brain especially in younger animals and small quantities of cellular material available.

2.15. Calculation of Specific Activity.

Specific activity is defined as nmol of aspartate produced per mg of protein in 1 min (nmol/min/mg protein).

2.16. Protein Determinations.

The protein concentrations were determined by the method of Lowry *et al.*,[52] using BSA as standard reference protein.

2.17. Statistical Analysis.

Statistical analysis were carried out using one-way ANOVA followed by a least-square difference test, and significance was assumed at $p < 0.05$.

3. RESULTS

3.1. NAA identification and quantification.

Typical high-resolution spectra obtained from extracts of acid-soluble metabolites from O-2A progenitors, immature oligodendrocytes and mature oligodendrocyte (grown in the presence and absence of CNTF) are shown in Figures 1 to 4.

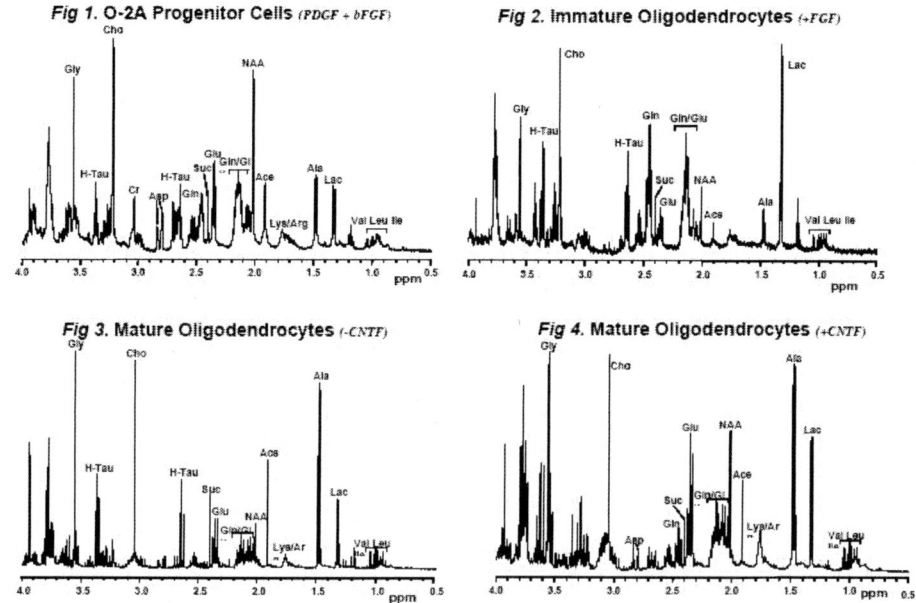

Figures 1-4. Typically $10^7 - 10^8$ cells were obtained for each extract. The protein content of samples ranged from 11.7 to 12.7 mg for O-2A progenitor cell extracts, from 3.4 to 14.5 mg in immature oligodendrocyte cell preparations, from 13.8 to 15.5 mg in CNTF treated mature oligodendrocytes, and from 8.8 to 14.1 mg for untreated mature oligodendrocytes. *Spectra of (1) O-2A progenitors, (2) immature oligodendrocyte, (3 and 4) Mature oligodendrocyte with and without CNTF respectively.*

These high field regions (0.5 - 4.5 ppm) contain signals from a variety of metabolites, including amino acids and related compounds, compounds involved in membrane biosynthesis (e.g. choline and choline-containing compounds) and compounds involved in intermediary metabolism (succinate and lactate). The metabolite of interest in the present study is the resonance assigned to NAA at 2.02 ppm, with reference to TSP. NAA was also identified and quantified using HPLC in the same samples. The O-2A progenitors expressed slightly lower levels of NAA then previously published data (15.1 $\pm$ 2.9 *vs* 21.1 $\pm$ 11.0 [mean $\pm$ SD] nmol/mg protein), whereas that for immature oligodendrocytes was similar (9.0 $\pm$ 4.1 *vs* 10.3 $\pm$ 5.4 [mean $\pm$ SD] nmol/mg protein).[8, 9] NAA was also seen in both preparations of mature oligodendrocytes, with higher levels in the CNTF treated population (9.5 $\pm$ 1.5 in the untreated cells and 15.5 $\pm$ 1.8 in the CNTF treated oligodendrocytes [mean $\pm$ SD] nmol/mg protein). The levels of NAA seen in mature oligodendrocytes were comparable to those seen in a variety of neuronal cultures; cerebellar granule neurons 12.3 $\pm$ 2.6;[9] cortical neurons 34.5 $\pm$ 3.7 and dorsal root ganglion neurons 15.6 $\pm$ 1.9 ([mean $\pm$ SD] nmol/mg protein).[49] It is however uncertain whether NAA levels seen in these different cellular preparations is found in similar proportions in the individual cell-types *in vivo*. Meningeal cells derived from the meninges surrounding the cortex were grown as a control population of cells. These cells were cultivated and harvested under similar conditions as the other cell types. As expected no NAA was seen in extracts of these cells, which is in agreement with previously published data.[8, 9, 53]

3.2. Aspartoacylase Assay.

Due the large dynamic range of tissue sizes required for analysis in the present study (average total protein 4.5gm for adult cortex to 30mg for P1 optic nerve, and 0.15mg for neurones to 0.5 gm for oligodendrocytes), it was important that the aspartoacylase assay was not only linear between different varieties of biological tissues but also consistent over the extended time course of the study. We therefore optimised the assay for the very small tissue size available for some samples such as optic nerves and cellular preparations. Assaying control samples on a regular basis enabled us to monitor the long-term consistency of the enzyme assay. Thus, all differences observed between samples represent valid differences in aspartoacylase activity.

3.3. Comparison of Tissue Aspartoacylase Activity and Developmental Evolution Pattern.

The level of aspartoacylase activity was measured in several brain regions in rats from newborn to adulthood. This study demonstrates that there was a differential expression of aspartoacylase activity, which not only varied with development but also showed regional variation in the induction of enzyme activity over time. Apart from the optic nerve, all areas of the brain had very low levels of activity at birth, with most tissues expressing enzyme levels at the limit of detection. The greatest developmental increase in expression of aspartoacylase was seen in white matter tracts such as the optic nerve with a five-fold increase from P1 to P21, followed by a significant reduction in adulthood. The largest rate of increase was seen after postnatal day 7. The brain stem and *corpus callosum* also showed similar trends, with accelerated expression seen after the seventh day of birth. Aspartoacylase activity in the cerebellum followed a similar profile but was much less marked. Gray matter such, as hippocampus, olfactory bulb and the cortex did not express much activity or show a developmental trend (see Figure 5). The largest differential in the expression of tissue aspartoacylase activity is seen between the optic nerve and the cortex at postnatal day 21 with an approximately 60-fold difference between the two regions. Aspartoacylase activity was determined in 'whole' cortex and compared to cortex 'stripped' of *corpus callosum*. There was a 18-fold difference between 'stripped' cortex and dissected *corpus callosum* (0.81 ± 0.05 μU), whereas there was only a 4-fold difference between 'whole' cortex (0.20 ± 0.05 μU) and 'stripped' cortex (0.05 ± 0.03 μU). Therefore unless the *corpus callosum* is dissected from the cortex, it leads to misleading levels of aspartoacylase activity. Moreover, the cortex, which shows the largest developmental increase in expression of NAA over a similar time-course[37] had the lowest levels of aspartoacylase activity, and was in most cases at the limit of detection. The opposite was true for the optic nerve, which had the highest levels of aspartoacylase activity, but the lowest levels of NAA (see Figure 5).

Fig 5. Relative tissue distribution of Aspartoacylase and NAA

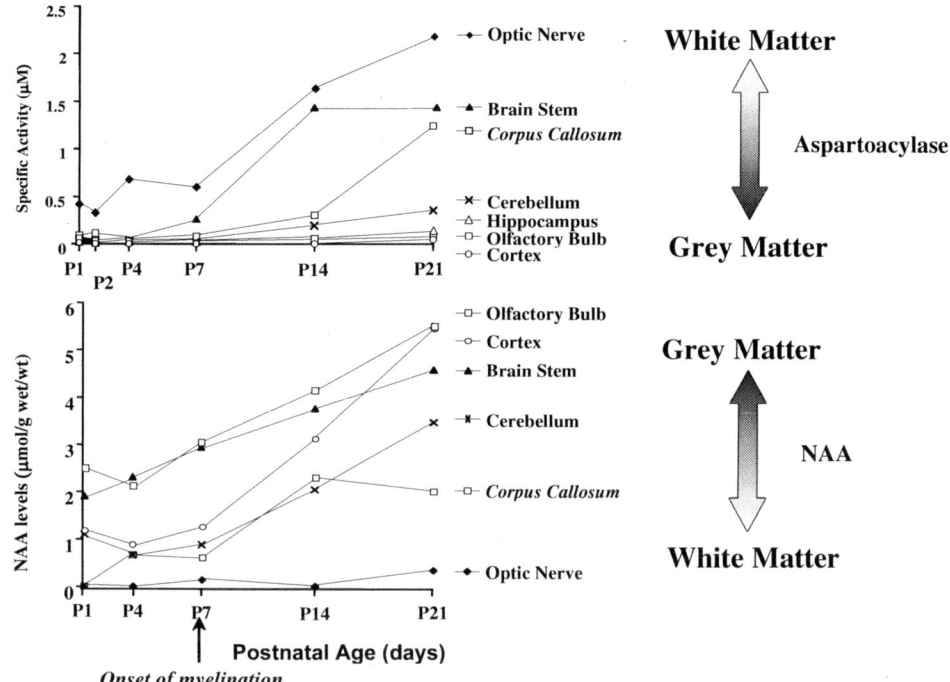

Figure 5. Compilation of aspartoacylase activities and NAA levels in various regions of the rat brain. Comparative distribution of brain tissue aspartoacylase activity (specific activity) and NAA levels (mmol/g wet weight) in different regions during postnatal development. For purposes of clarity the error bars have been omitted. NAA data is an adaptation of the numerical data from Florian et al. (1996).

3.4. Cellular Aspartoacylase Activity.

Purified cortical neurons and cerebellar granular neurones had no aspartoacylase activity. Primary O-2A progenitor cells had moderate activity (0.136 ± 0.073 nmol/min/mg protein) with approximately three fold higher activity in immature oligodendrocytes (0.341 ± 0.137 nmol/min/mg protein) and 13 fold higher levels in mature oligodendrocytes (1.749 ± 0.261 nmol/min/mg protein). However, the highest activity (20 fold) was seen in type-2 astrocytes (2.832 ± 0.785 nmol/min/mg protein) derived from the same source. Type-1 astrocytes had lower levels of aspartoacylase activity then those seen in mature oligodendrocytes (0.136 ± 0.073 nmol/min/mg protein).

Aspartoacylase activity was absent from freshly isolated astrocytes; however, there was an induction of enzyme activity in culture over time with significantly higher activity after 14 days in culture (see Figure 6).

Fig 6. Aspartoacylase activity in Astrocytes

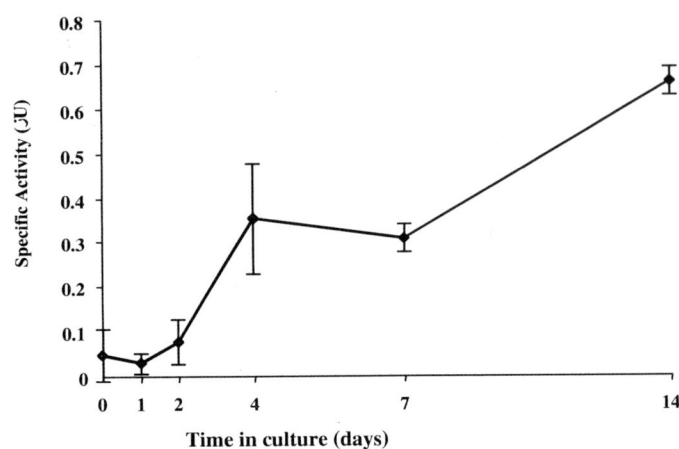

Figure 6. Development of aspartoacylase activity in cultured type-1 astrocytes over 14 days. Primary cultures of astrocytes were grown in parallel and four dishes were harvested at times indicated above. Aspartoacylase activity [specific activity (nmol/min/mg protein) - mean ± S.D.; n=4] was determined as outlined in Materials and Methods.

4. DISCUSSION

We have used cell culture techniques and whole tissues in combination with ^{1}H-NMR spectroscopy and HPLC analysis and biochemical enzyme assays, to compare the expression of NAA and aspartoacylase in cells derived from the O-2A progenitor lineage and expression of aspartoacylase levels brain regions during development.

This study casts new light on the metabolism of NAA at the cellular level and, furthermore, correlates with the expression of aspartoacylase activity with myelination.

Previous studies have demonstrated an absence of NAA in cell extracts of mature oligodendrocytes; the myelinating cells of the central nervous system. The data presented in this study demonstrate that mature oligodendrocytes can express NAA in quantities comparable to O-2A progenitors and neurons grown *in vitro* under similar conditions.[8, 9, 49] Meningeal cells derived from the cortical meninges, were cultivated under similar conditions and were used as a control cell population. NAA was undetectable in these cells.

In addition this study demonstrates the differential expression of aspartoacylase activity not only in different regions of the brain, during development, but also in purified cells derived from the CNS. The main findings are the marked difference in aspartoacylase activity between gray matter (cortex) compared to white matter (optic nerve; ~60 fold difference) after 21 days of postnatal development. Furthermore, the temporal rise in aspartoacylase correlated with the onset of myelination in the developing brain. Perhaps the most unexpected result was the total absence of activity in neurones, with enzyme activity being limited to the glial lineages alone.

One of the critical differences between the present studies and those of Urenjak *et al*, was the use of CNTF to enhance the maturation and survival of these cells in culture. During the development of the rat optic nerve, CNTF *mRNA* is first seen at birth, at a time when the first oligodendrocytes appear.[54] More recent studied have shown that the timing of oligodendrocyte differentiation, maturation and survival *in vitro* requires extrinsic factors.[17, 18, 55, 56] Of a range of small molecules and proteins, CNTF is the key protein with both survival and maturation activity on oligodendrocytes.

The localisation of NAA as a neuronal marker is based on previous immunohistochemical studies using antibodies to NAA, which were restricted to a large number of neurons in the rat CNS.[6, 7] It is however unclear, in the light of the present study, why the oligodendrocytes in these studies failed to show immunoreactivity to NAA, as did many neurons in all areas of the brain. Moreover, while the utility of NAA as a marker of neuronal viability is acknowledged, the fact that it may also be seen in other major cell-types in the brain in not wholly unexpected. More recent studies (Moffett et, personal communication) using newly developed antibodies to NAA, have shown limited immunoactivity in oligodendrocytes in tissue sections.

Using rat brain extracts, it has been shown that the concentration of NAA rises rapidly to adult levels between days 10 and 20 of life.[37, 57] This is also the period of active myelination in this animal. The rise in NAA has also been demonstrated by *in vivo* ¹H-NMR spectroscopy in developing rat brain.[58] The developmental changes in NAA have also been assessed in human foetal and child brains by high-resolution ¹H-NMR spectroscopy. NAA was detected in the cerebral cortex and white matter of foetuses as early as16 weeks' gestation. NAA increased gradually from 24 weeks' gestation with a marked rise from 40 weeks' gestation to 1 year of age. The developmental changes in tissue NAA of postnatal brains were found to be similar to those seen by clinical proton MRS.[59] Furthermore temporal studies using *in vivo* proton MR spectroscopy of healthy preterm, term, and infant brain also demonstrated an increase in NAA with development.[60] Since myelination is a period of rapid lipid synthesis, the simultaneous increase in NAA concentration suggests that NAA might function as an acetyl donor to CoA or to the acyl carrier protein involved in lipid synthesis. D'Adamo *et al.*[11, 61] have verified this hypothesis. Their work has demonstrated clearly that NAA provides acetyl groups for lipid synthesis in myelinating rat brain *in vivo*. All these data suggest that NAA has a role in myelin maturation. The earliest indication of non-neuronal contributions to the changes seen in NAA levels came from the longitudinal studies on MS, which involves demyelination in white matter tracts of the brain. MS is principally a disease involving the destruction of oligodendrocytes, thus leading to demyelination.[62-65] Using *in vivo* ¹H-NMR spectroscopy to follow the episodic changes in NAA during the relapse and remission phases of MS, Davie *et al*. demonstrated NAA, relative to creatine, was reduced in acute lesions, with a subsequent rise in the NAA/creatine ratio[66] This phenomenon had already been predicted earlier by Richards,[67] who hypothesised that serial *in vivo* ¹H-NMR spectroscopy could be used to monitor progression and evaluate therapy using changes in the NAA/creatine ratio in MS. Reversible changes in brain NAA to creatine ratio have been shown to correlate strongly with clinical disability.[15, 68] Despite the overwhelming support for NAA as a neuronal marker, some workers have however ventured as far as stating that a decrease in NAA levels does not imply axonal loss alone.[66, 69] Nevertheless the possibility of a reduction in NAA levels due to neuronal loss cannot be ruled out in MS lesions, especially in the chronic phase of the disease.[70, 71] Since we have shown that oligodendrocytes synthesise NAA in substantial quantities, it can therefore be surmised that a reduction in the NAA signal in MS lesions is not solely due to neuronal pathology.

The relationship between NAA and oligodendrocytes has been further augmented by the recent studies of Jagannathan et al.[16] on brains of hypothyroid patients by in vivo ^{1}H-NMR spectroscopy, before and after thyroxine treatment. The authors reported lower levels of NAA in hypothyroid subjects compared to control. Reversibility of NAA levels was seen even though thyroxine therapy was initiated at ages beyond which abnormalities in myelinogenesis are considered irreversible. Previously, proton NMR spectroscopic study on perchloric acid extract of neonatal hypothyroid rat brain also demonstrated a reduced level of NAA relative to control brain tissue.[72] Furthermore, experimental hypothyroidism has been documented to cause a variety of abnormalities in myelinogenesis;[73-77] and in an animal model of congenital hypothyroidism, the associated cerebral hypomyelination in neonatal mice was reversed only when thyroxine treatment was started before the 20th day of birth.[78] Thyroxine is known to regulate enzymes involved in myelinogenesis.[79-81] At a cellular level there is emerging evidence, which demonstrates that thyroid hormones accentuate O-2A progenitor differentiation into oligodendrocytes, both in vitro and in vivo.[82-84]

Recent studies by Bjartmar et al.,[85] used transected rat optic nerves to investigate NAA specificity for white matter axons, using HPLC and immunohistochemistry. In transected adult nerves, NAA and NAAG decreased in concordance with axonal degeneration and were undetectable 24 days post-transection. Non-proliferating oligodendrocyte progenitor cells, oligodendrocytes, and myelin were abundant in these axon-free nerves. The results also indicated that neuronal adaptation can increase NAA levels, and that 5 to 20% of the NAA in developing white matter is synthesized by proliferating oligodendrocyte progenitor cells.

In order to expand further our knowledge on NAA metabolism, we investigated the cellular and developmental expression of aspartoacylase in the brain. A developmental correlation on the distribution and expression of NAA and aspartoacylase activity has been demonstrated previously in the rat forebrain (gray matter[24]) and is in variance with the present study. However, the experiments on the differential distribution of aspartoacylase activity between whole cortex, cortex stripped of it corpus callosum and corpus callosum alone, has shown that the previous correlation must have been due to contaminating white matter in the rat forebrain.

The largest rate in the developmental induction of aspartoacylase activity was seen in the optic nerve. The developing optic nerve represents a simple model of brain tissue development. It has been used for developmental and cell lineage studies by the laboratories of Raff and Noble, who have demonstrated that the optic nerve of the rat is a 'gray matter tract' at birth.[86-88] On the day before birth (embryonic day 21) the optic nerve consists of axons (neuronal cell bodies lie outside the tract, either in the retina or the optic chiasm), type-1 astrocytes and O-2A progenitors. Oligodendrocytes are seen on the first day of birth but are immature and do not express myelin basic protein (MBP). Maturation of oligodendrocytes appears by the seventh day, when MBP is seen for the first time in the optic nerve. Myelination then occurs at an accelerated rate with maxima at day 21 of postnatal development. Hence the optic nerve in the rat brain develops from a 'gray matter tract' at birth to a 'white matter tract' over first three week of postnatal development, and therefore provides an ideal model to study developmental myelination. Furthermore, similar postnatal studies on maturation in mouse and rat CNS show similar temporal development of myelination in whole brains.[89] Our study on the developmental expression of aspartoacylase activity demonstrates that the induction of the enzyme is closely related to the myelination process in the brain as demonstrated in the developing optic nerve (see Figure 7).

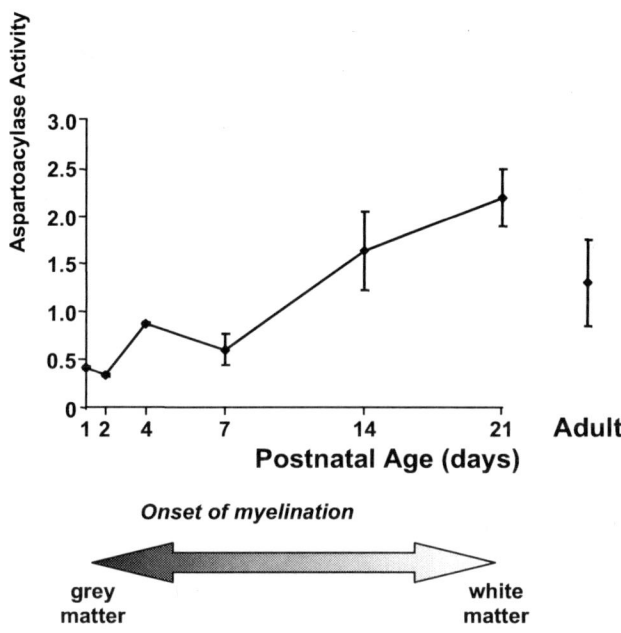

Figure 7. Evolution of aspartoacylase activity (specific activity, µM/ min/mg protein) - mean ± S.D.; n=4) in the optic nerve during postnatal development.

    A differential distribution of aspartoacylase activity between gray and white matter has been has been demonstrated in previous studies. The study by D'Adamo et al.[36] on the occurrence of aspartoacylase activity in the developing rat brain showed an approximately three-fold difference between white and gray matter in the adult cerebral hemispheres, whereas, in another study on adult bovine brain, the specific activity of aspartoacylase in the deepest tissue (white matter) was 15-fold higher than in tissue closest to the pial surface i.e. gray matter.[38]

    The experiments on the cellular distribution of aspartoacylase activity very much reflect the results of our findings on regional brain tissue studies. At the celluar level, the predominate expression of aspartoacylase activity is seen in the O-2A progenitor lineage that gives rise to the myelinating cells of the CNS. In contrast aspartoacylase activity in whole tissue is limited to white matter tracts of the brain, with the largest developmental increases seen in regions of greatest myelination. This is very much in contrast to the expression of its substrate, NAA, where the highest levels are found in gray matter and in all neuronal cell-types studied so far.[8, 9, 49] Thus the enzyme catalysing NAA catabolism appears to be located in a different cellular compartment from the principle NAA store. However, an interesting exception is the presence of both NAA and aspartoacylase activity in O-2A progenitor lineage with the highest levels of both found in mature oligodendrocytes. Type-2 astrocytes do not synthesise NAA (unpublished data) but do express the highest levels of aspartoacylase. The significance of this finding is discussed later. The development of aspartoacylase in cortical type-1 astrocytes is in contrast to the recent finding of Baslow et al.[39] where the expression of enzyme activity in cultured glial cells was limited to oligodendrocytes alone. In our present study freshly isolated type-1 astrocytes had activities at the lower end of the detection limit. However, there was an increased expression of aspartoacylase activity over time in culture. Another unexpected finding was the presence of the highest cellular aspartoacylase activity in type-2 astrocytes. It is generally accepted that type-1 astrocytes in culture represent cells that in vivo constitute reactive astrocytes, the predominant cell-type of glial scar tissue.[90] It may well be that type-2 astrocytes also represent a type of reactive astrocyte that does not appear in normal brain, but may develop in response to injury. These cells form scar tissue that limits the spread of damage,[91, 92] and express neurocan, a chondoitin sulphate proteoglycan, which is only seen on type-2 astrocytes in culture.[93] These cells may require aspartoacylase for either repair of damaged myelin, or hydrolysis of excess NAA released by cell death that, if remains unchecked, may result in the concomitant water imbalance.

Firstly, one could suggest a simple scavenging role for aspartoacylase, for example preventing potential neurotoxic effect on oligodendrocytes, such as that seen in Canavan's Disease where mutation of aspartoacylase results in accumulation of extracellular NAA in the brain, leading to spongy degeneration of the white matter tracts. However, there is no direct evidence that high levels of extracellular NAA are indeed toxic to oligodendrocytes. Work in our laboratories has shown no detrimental effects of NAA (concentrations of up to 5 mM for three days) on oligodendrocytes in culture (unpublished data). More recently Kitada et al.[94] have demonstrated that the absence-like seizure and spongiform degeneration in the CNS, exhibited in the tremor rat, are due to genomic deletion within a region in which the aspartoacylase gene is located. Accordingly, no aspartoacylase expression was detected in any of the tissues examined, and abnormal accumulation of NAA was shown in the mutant brain, in correlation with the severity of the vacuole formation. Interestingly, direct injection of NAA into normal rat cerebroventricle induced 4 to 10Hz polyspikes or spikewave-like complexes in cortical and hippocampal EEG. In addition NAA applied to a bath of normal rat brain slice preparations, rapidly induced a long-lasting depolarisation concomitantly with repetitive firings in hippocampal CA3 neurons.[95] These results suggested that accumulated NAA in the CNS would induce neuroexcitation and neurodegeneration directly or indirectly. A more plausible explanation may be that NAA utilisation is separate from its site of storage. In particular the majority of NAA could be released from the neurones to provide substrate for myelination by oligodendrocytes during development, whereas the NAA in oligodendrocytes is expressed or used in response to white matter damage when neuronal-oligodendrocyte interactions are unstable. Neuronal stores of NAA may be also be involved in the re-myelination process following injury. This would explain the reversible decrease in NAA sometimes observed in MS lesions.[96]

We note however that at the tissue level, gray matter contains high levels of NAA, but low levels of aspartoacylase. Therefore, a secondary role for NAA in modulating neuronal homeostasis throughout the CNS would not be inconsistent with our data.

Recent studies have confirmed the localization of aspartoacylase to white matter tracts,[97] where the authors used double-label immunohistochemistry for aspartoacylase and several cell-specific markers. The aspartoacylase was co-localized throughout the brain with CC1, a marker for oligodendrocytes. Many cells were labeled with aspartoacylase antibodies in white matter, including cells in the *corpus callosum* and cerebellar white matter. Moreover, only a few cells were labeled in gray matter. No astrocytes were labeled for ASPA. Neurons were unstained in the forebrain, although small numbers of large reticular and motor neurons were faintly to moderately stained in the brainstem and spinal cord.

The presence of aspartoacylase in the adult brain demonstrates that it may have a function after myelination is complete that involves repair of damaged myelin. It seems unlikely that such high activities would be required to maintain a repair function. This implies an additional function for aspartoacylase in the adult brain. The presence of aspartoacylase at low levels in the adult olfactory bulb and cortex, which are gray matter tracts, further suggests that its function is not solely for the repair of damaged myelin.

The inverse correlation seen between aspartoacylase and NAA in white and gray matter could be due to a higher turnover of NAA in the white matter during myelination, making the apparent concentration in white matter tracts lower. However, our study have shown that mature oligodendrocytes grown in culture express NAA to similar levels as neurones and may themselves contribute to the overall NAA pool observed *in vivo* by magnetic resonance.

In conclusion, this study demonstrates the presence of NAA and aspartoacylase in cultured mature oligodendrocytes and may thus present a clearer picture of NAA metabolism in the brain. The developmental and anatomical distribution of aspartoacylase correlates with the maturation of white matter tracts in the rat brain. Moreover, the cellular studies showed that the aspartoacylase activity is limited to glial cells, especially in the O-2A progenitor lineage, with no detectable enzyme activity in neurons. Despite the indications that NAA metabolism correlates closely with myelination, the function/s of NAA nevertheless, still remains obscure, and until this ambiguity is resolved, the utility of NAA as a diagnostic tool in certain brain disorders will require further work.

5. ACKNOWLEDGEMENTS

This work was funded by the Medical Research Council of Great Britain.

6. QUESTION AND ANSWER SESSION:

SESSION CO-CHAIR BURLINA: Are there any questions?

DR. ROSS: Very nice talk, but I just wanted to see how this would all play out when you look at your stem cell lines.

DR. BHAKOO: Right. We have dropped all of the NAA work due to lack of funding, almost feel like a one-man band across the ocean. John Clark doesn't work on NAA any more. Neither does Tim Bates. So, we tried to get some funding to carry on with the NAA work. There are major problems in Britain.

DR. LEONE: Hello. Very nice data. And I'm sorry to hear that you have lack of funding to continue this work, which is really interesting. I have two questions. One is, do you have any hypothesis on the mechanism of CNTF and cell survival? And, most importantly, did you look at the survival properties of CNTF on neuronal progenitors and perhaps some of the effect of CNTF on NAA in neurons in vitro, I mean, the same thing you have done on O2A cells?

DR. BHAKOO: The only thing I would say, that we didn't try the CNTF experiments on the neurons, but the neurons do need NGF for survival. NGF and CNTF share receptor populations. So we did not examine that, but that is a good question.

PARTICIPANT: Sorry. I'm a little out of my depth here, but I thought I had seen along the way that the O2A progenitor cells were so early in the developmental chain that they could also go on to develop into neurons. Is that correct or not correct?

DR. BHAKOO: No, no. There is a glial-restricted progenitor that is tri-potential that can only differentiate into astrocytes and oligodendrocytes and the type 1 astrocytes, but nobody has actually shown whether they go on to make neurons or not yet.

PARTICIPANT: This is a cell culture. I mean, this cell doesn't exist in vivo. It's strictly limited to cell culture. Let's see. I have activated Bruce. Maybe I'll let him make his comment here.

DR. TRAPP: Well, I think the majority of these O2A cells make oligodendrocytes. There is very little evidence they make astrocytes in vivo. I think occasionally they can.

By now I think there is increasing evidence that some of these cells make neurons, particularly the O2A or NG2 cells that are coming from the diencephalon migrating to the cortex for making GABAergic neurons. Several labs have provided evidence for that. So I think under these conditions, though, you are absolutely right.

They are making oligos under one condition of no serum, or they are making astrocytes with serum. So, I think the in vitro studies, Doug, are pretty clean for oligocentric studies.

D.R. BHAKOO: These cells were isolated from postnatal tissue, but the totipotential of the clear restrictive progenitors, you have to go back down to embryonic level. And I'm not sure what would happen if you go further back.

PARTICIPANT: I guess my concern is to what extent these kinds of observations reflect the particular cell culture environment that you create in order to have these things proliferate, and to what extent they reflect something that is relevant in vivo. What are your thoughts?

DR. BHAKOO: These results are not cast in stone. We can see that we can modulate the NAA levels in culture using different conditions. So, it may well be that you either don't see NAA in oligodendrocyte until this pathology, or that we only see NAA under these conditions. So I totally agree; nothing is cast in stone.

SESSION CO-CHAIR BURLINA: We'll take only two questions more.

PARTICIPANT: Just to add to the controversy about the translational aspects, maybe some MS experts can make some distinction that's better. There's a lot of controversy in multiple sclerosis regarding axonal pathology. So it is not clear that there is a lack of axonal neuropathy in these MS neurons.

And also thyroxin has multiple effects on neuronal function as well. So I don't know that those in vivo correlations necessarily provide very strong evidence for these.

DR. BHAKOO: Those neurons were grown in the presence, I thought, of thyroxin. We never grew neurons in the absence of thyroxin.

PARTICIPANT: I mean in other systems.
DR. BHAKOO: Yes.
PARTICIPANT: These are not neuronal in any sense.
DR. BHAKOO: No. The important thing was that for using these purified cultures, you don't need co-cultures for the expression of NAA either in neurons or in oligodendrocytes.
SESSION CO-CHAIR COYLE: Thank you very much.

7. REFERENCES

1. H.H. Tallan, S. Moore, and W.H. Stein, N-acetyl-L-aspartic acid in brain, *J. Biol. Chem.* **219**, 257-264 (1956).
2. T.B. Patel and J.B. Clark, Synthesis of N-acetyl-L-aspartate by rat brain mitochondria and its involvement in mitochondrial/cytosolic carbon transport, *Biochem. J.* **184**, 539-546 (1979).
3. T.E. Bates, M. Strangward, J. Keelan, G.P. Davey, P.M. Munro, and J.B. Clark, Inhibition of N-acetylaspartate production: implications for ¹H-MRS studies *in vivo*, *Neuroreport* 7, 1397-400 (1996).
4. J.V. Nadler and J.R. Cooper, N-Acetyl-L-aspartic acid content of human neural tumours and bovine periperal nervous tissues, *J. Neurochem.* **19**, 313-319 (1972).
5. K.J. Koller, R. Zaczek, and J.T. Coyle, N-acetyl-aspartyl-glutamate: regional levels in rat brain and the effects of brain lesions as determined by a new HPLC method, *J. Neurochem.* **43**, 1136-42 (1984).
6. J.R. Moffett, M.A. Namboodiri, C.B. Cangro, and J.H. Neale, Immunohistochemical localization of N-acetylaspartate in rat brain, *Neuroreport* **2**, 131-4 (1991).
7. M.L. Simmons, C.G. Frondoza, and J.T. Coyle, Immunocytochemical localization of N-acetyl-aspartate with monoclonal antibodies, *Neuroscience* **45**, 37-45 (1991).
8. J. Urenjak, S.R. Williams, D.G. Gadian, and M. Noble, Specific expression of *N*-acetylaspartate in neurons, oligodendrocyte-type-2 astrocyte progenitors, and immature oligodendrocytes *in vitro*, *J. Neurochem.* **59**, 55-61 (1992).
9. J. Urenjak, S.R. Williams, D.G. Gadian, and M. Noble, Proton nuclear magnetic resonance spectroscopy unambiguously identifies different neural cell types, *J. Neurosci.* **13**, 981-9 (1993).
10. H. Shigematsu, N. Okamura, H. Shimeno, Y. Kishimoto, L. Kan, and C. Fenselau, Purification and characterization of the heat-stable factors essential for the conversion of lignoceric acid to cerebronic acid and glutamic acid: identification of N-acetyl-L-aspartic acid, *J. Neurochem.* **40**, 814-820 (1983).
11. A.F.J. D'Adamo and F.M. Yatsu, Acetate metabolism in the nervous system.   N-acetyl-L-aspartic acid and the biosynthesis of brain lipids, *J. Neurochem.* **13**, 961-5 (1966).
12. A.F.J. D'Adamo, L.I. Gidez, and F.M.J. Yatsu, Acetyl transport mechanisms. Involvement of N-acetyl aspartic acid in *de novo* fatty acid biosynthesis in the developing rat brain, *Exp. Brain Res.* **5**, 267-73 (1968).
13. J.C. McIntosh and J.R. Cooper, Studies on the function of N-acetyl-L-aspartic acid in brain, *J. Neurochem.* **12**, 825-835 (1965).
14. D.L. Birken and W.H. Oldendorf, N-acetyl-L-aspartic acid: a literature review of a compound prominent in ¹H-NMR spectroscopic studies of brain, *Neurosci. Biobehav. Rev.* **13**, 23-31 (1989).
15. N. De Stefano, P.M. Matthews, and D.L. Arnold, Reversible decreases in N-acetyl-aspartate after acute brain injury, *Magn. Reson. Med.* **34**, 721-7 (1995).
16. N.R. Jagannathan, N. Tandon, P. Raghunathan, and N. Kochupillai, Reversal of abnormalities of myelination by thyroxine therapy in congenital hypothyroidism: localized *in vivo* proton magnetic resonance spectroscopy (MRS) study, *Dev. Brain Res.* **109**, 179-86 (1998).
17. M. Mayer, K. Bhakoo, and M. Noble, Ciliary neurotrophic factor and leukemia inhibitory factor promote the generation, maturation and survival of oligodendrocytes *in vitro*, *Development* **120**, 143-53 (1994).
18. B.A. Barres, J.F. Burne, B. Holtmann, H. Thoenen, M. Sendtner, and M.C. Raff, Ciliary neurotrophic factor enhances the rate of oligodendrocyte generation,   *Mol. Cell Neurosci.* **8**, 146-56 (1996).
19. S.M. Birnbaum, Amino acid acylases I and II from Hog Kidney, *Methods in Enzymology* **2**, 115-119 (1955).
20. R. Kaul, G.P. Gao, M. Aloya, K. Michals, and R. Matalon, Identification of mutations in human aspartoacylase (hasp) gene in Canavan disease, *Am. J. Hum. Genetics* **53**, 215-215 (1993).
21. M.M. Canavan, Schildler's encephalitis periaxialsis diffusa, *Arch. Neurol. Psychiatry* **25**, 299-308 (1931).
22. L. Hagenfeldt, I. Bollgren, and N. Venizelos, N-acetylaspartic aciduria due to aspartoacylase deficiency - a new aetiology of childhood leukodystrophy, *J. Inherit. Metab. Disease* **10**, 135-141 (1987).

23. R. Matalon, R. Kaul, J. Casanova, K. Michals, A. Johnson, I. Rapin, P. Gashkoff, and M. Deanching, Aspartoacylase deficiency: The enzyme defect in Canavan disease, *J. Inherit. Metab. Disease* **12**, 329-331 (1989).
24. R. Burri, C. Steffen, and N. Herschkowitz, N-acetyl-L-Aspartate is a major source of acetyl groups for lipid synthesis during rat brain development, *Dev. Neurosci.* **13**, 403-411 (1991).
25. M.H. Baslow and T.R. Resnik, Canavan disease. Analysis of the nature of the metabolic lesions responsible for development of the observed clinical symptoms, *J. Mol. Neurosci.* **9**, 109-25 (1997).
26. T.N. Sager, C. Thomsen, J.S. Valsborg, H. Laursen, and A.J. Hansen, Astroglia contain a specific transport mechanism for N-acetyl-L-aspartate, *J. Neurochem.* **73**, 807-11 (1999).
27. F.B. Goldstein, The enzymatic synthesis of N-acetyl-L-aspartic acid by subcellular preparations of rat brain, *J. Biol. Chem.* **244**, 4257-60 (1969).
28. C.B. Cangro, M.A. Namboodiri, L.A. Sklar, A. Corigliano Murphy, and J.H. Neale, Immunohistochemistry and biosynthesis of N-acetylaspartylglutamate in spinal sensory ganglia, *J. Neurochem.* **49**, 1579-88 (1987).
29. L. Ory Lavollee, R.D. Blakely, and J.T. Coyle, Neurochemical and immunocytochemical studies on the distribution of N-acetyl-aspartylglutamate and N-acetyl-aspartate in rat spinal cord and some peripheral nervous tissues,   *J. Neurochem.* **48**, 895-9 (1987).
30. W. Grodd, I. Krageloh Mann, D. Petersen, F.K. Trefz, and K. Harzer, *In vivo* assessment of N-acetylaspartate in brain in spongy degeneration (Canavan's disease) by proton spectroscopy [letter], *Lancet* **336**, 437-8 (1990).
31. D.S. Dunlop, D.M. Mchale, and A. Lajtha, Decreased brain N-Acetyl-aspartate in huntingtons disease, *Brain Research* **580**, 1-2 (1992).
32. D.J. Meyerhoff, S. MacKay, L. Bachman, N. Poole, W.P. Dillon, M.W. Weiner, and G. Fein, Reduced brain *N*-acetylaspartate suggests neuronal loss in cognitively impaired human immunodeficiency virus-seropositive individuals: *in vivo*[1]H magnetic resonance spectroscopic imaging, *Neurology* **43**, 509-15. (1993).
33. P. Gideon, O. Henriksen, B. Sperling, P. Christiansen, T.S. Olsen, H.S. Jorgensen, and P. Arliensoborg, Early time course of *N*-acetylaspartate, creatine and phosphocreatine, and compounds containing choline in the brain after acute stroke - a proton magnetic-resonance spectroscopy study, *Stroke* **23**, 1566-1572 (1992).
34. R. Kaul, K. Michals, J. Casanova, and R. Matalon, The role of N-acetylaspartic acid in brain metabolism and the pathogenesis in Canavan disease, *Intl. Pediat.* **6**, 40-43 (1991).
35. S. Confort-Gouny, J. Vion-Dury, F. Nicoli, P. Dano, A. Donnet, N. Grazziani, J.L. Gastaut, F. Grisoli, and P.J. Cozzone, A multiparametric data analysis showing the potential of localized proton MR spectroscopy of the brain in the metabolic characterization of neurological diseases, *J. Neurol. Sci.* **118**, 123-33 (1993).
36. A.F.J. D'Adamo, J.C. Smith, and C. Woiler, The occurrence of *N*-acetylaspartate amidohydrolase (aminoacylase II) in the developing rat, *J. Neurochem.* **20**, 1275-8 (1973).
37. C.L. Florian, S.R. Williams, K.K. Bhakoo, and M.D. Noble, Regional and developmental variations in metabolite concentration in the rat brain and eye: a study using ^{1}H-NMR spectroscopy and high performance liquid chromatography, *Neurochem. Res.* **21**, 1065-74 (1996).
38. R. Kaul, J. Casanova, A.B. Johnson, P. Tang, and R. Matalon, Purification, characterization, and localization of aspartoacylase from bovine brain, *J. Neurochem.* **56**, 129-35 (1991).
39. M.H. Baslow, R.F. Suckow, V. Sapirstein, and B.L. Hungund, Expression of aspartoacylase activity in cultured rat macroglial cells is limited to oligodendrocytes, *J. Mol. Neurosci.* **13**, 47-53 (1999).
40. K.K. Bhakoo and D. Pearce, *In vitro* expression of N-acetyl aspartate by oligodendrocytes: implications for proton magnetic resonance spectroscopy signal *in vivo*, *J. Neurochem.* **74**, 254-62 (2000).
41. K.K. Bhakoo, T.J. Craig, and P. Styles, Developmental and regional distribution of aspartoacylase in rat brain tissue, *J. Neurochem.* **79**, 211-20 (2001).
42. J.H. Bottenstein and G.H. Sato, Growth of a rat neuroblastoma cell line in serum-free supplemented medium, *Proc. Natl. Acad. Sci. U.S.A.* **76**, 514-517 (1979).
43. P.F. Bartlett, M.D. Noble, R.M. Pruss, M.C. Raff, S. Rattray, and C.A. Williams, Rat neural antigen-2 (RAN-2): a cell surface antigen on astrocytes, ependymal cells, Muller cells and lepto-meninges defined by a monoclonal antibody, *Brain Res.* **204**, 339-51 (1981).
44. B. Ranschl, P.A. Clapshaw, J. Price, M. Noble, and W. Seifert, Development of oligodendrocytes and Schwann cells studied with a monoclonal antibody against galactocerebroside, *Proc. Natl. Acad. Sci. U.S.A.* **79**, 2709-13 (1982).
45. G.S. Eisenbarth, F.S. Walsh, and M. Nirenberg, Monoclonal antibody to a plasma membrane antigen of neurons, *Proc. Natl. Acad. Sci. U.S.A.* **76**, 4913-7 (1979).

46. O. Bogler, D. Wren, S.C. Barnett, H. Land, and M. Noble, Cooperation between two growth factors promotes extended self-renewal and inhibits differentiation of oligodendrocyte-type-2 astrocyte (O-2A) progenitor cells, *Proc. Natl. Acad. Sci. U.S.A.* **87**, 6368-72 (1990).
47. M.C. Raff, R. Mirsky, K.L. Fields, R.P. Lisak, S.H. Dorfman, D.H. Silberberg, N.A. Gregson, S. Leibowitz, and M.C. Kennedy, Galactocerebroside is a specific cell-surface antigenic marker for oligodendrocytes in culture, *Nature* **274**, 813-816 (1978).
48. M. Noble and K. Murray, Purified astrocytes promote the *in vitro* division of a bipotential glial progenitor cell, *EMBO J.* **3**, 2243-7 (1984).
49. K.K. Bhakoo, I.T. Williams, S.R. Williams, D.G. Gadian, and M.D. Noble, Proton nuclear magnetic resonance spectroscopy of primary cells derived from nervous tissue, *J. Neurochem.* **66**, 1254-63 (1996).
50. J.N. Wood and B.H. Anderton, Monoclonal antibodies to mammalian neurofilaments, *Biosci. Rep.* **1**, 263-8 (1981).
51. K.J. Koller and J.T. Coyle, Ontogenesis of N-acetyl-aspartate and N-acetyl-aspartyl-glutamate in rat brain, *Brain Res.* **317**, 137-40 (1984).
52. O.H. Lowry, N.J. Rosenbrough, and A.L. Farr, Protein measurement with the Folin phenol regent, *J. Biol. Chem.* **193**, 265-275 (1951).
53. C.L. Florian, N.E. Preece, K.K. Bhakoo, S.R. Williams, and M. Noble, Characteristic metabolic profiles revealed by ^{1}H-NMR spectroscopy for three types of human brain and nervous system tumours, *NMR Biomed.* **8**, 253-64 (1995).
54. K.A. Stockli, L.E. Lillien, M. Naher Noe, G. Breitfeld, R.A. Hughes, M.C. Raff, H. Thoenen, and M. Sendtner, Regional distribution, developmental changes, and cellular localization of CNTF-mRNA and protein in the rat brain, *J. Cell Biol.* **115**, 447-59 (1991).
55. B.A. Barres, I.K. Hart, H.S. Coles, J.F. Burne, J.T. Voyvodic, W.D. Richardson, and M.C. Raff, Cell death and control of cell survival in the oligodendrocyte lineage, *Cell* **70**, 31-46 (1992).
56. B.A. Barres, R. Schmid, M. Sendnter, and M.C. Raff, Multiple extracellular signals are required for long-term oligodendrocyte survival, *Development* **118**, 283-95 (1993).
57. H.H. Tallan, Studies on the distribution of *N*-acetyl-L-aspartic acid in brain, *J. Biol. Chem.* **224**, 41-5 (1957).
58. K. Hida, [*In vivo* ^{1}H and ^{31}P NMR spectroscopy of the developing rat brain], *Hokkaido Igaku Zasshi* **67**, 272-80 (1992).
59. T. Kato, M. Nishina, K. Matsushita, E. Hori, T. Mito, and S. Takashima, Neuronal maturation and N-acetyl-L-aspartic acid development in human fetal and child brains, *Brain Dev.* **19**, 131-3 (1997).
60. P.B. Toft, Metabolite concentrations in the developing brain estimated with proton MR spectroscopy, *J. Magn. Reson. Imaging* **4**, 674-680 (1994).
61. A.F.J. D'Adamo and A.P. D'Adamo, Acetyl transport mechanisms in the nervous system. The oxoglutarate shunt and fatty acid synthesis in the developing rat brain, *J. Neurochem.* **15**, 315-23 (1968).
62. F.A. McMorris and R.D. McKinnon, Regulation of oligodendrocyte development and CNS myelination by growth factors: prospects for therapy of demyelinating disease, *Brain Pathol.* **6**, 313-29 (1996).
63. C.S. Raine, The Norton Lecture: a review of the oligodendrocyte in the multiple sclerosis lesion, *J. Neuroimmunol.* **77**, 135-52 (1997).
64. C.F. Lucchinetti, W. Brueck, M. Rodriguez, and H. Lassmann, Multiple sclerosis: lessons from neuropathology, *Semin. Neurol.* **18**, 337-49 (1998).
65. L. Zhou, B.D. Trapp, and R.H. Miller, Demyelination in the central nervous system mediated by an anti-oligodendrocyte antibody, *J. Neurosci. Res.* **54**, 158-68 (1998).
66. C.A. Davie, C.P. Hawkins, G.J. Barker, A. Brennan, P.S. Tofts, D.H. Miller, and W.I. McDonald, Serial proton magnetic resonance spectroscopy in acute multiple sclerosis lesions, *Brain* **117**, 49-58 (1994).
67. T.L. Richards, Proton MR spectroscopy in multiple-sclerosis - value in establishing diagnosis, monitoring progression, and evaluating therapy, *Am. J. Roentgenol.* **157**, 1073-1078 (1991).
68. N. De Stefano, P.M. Matthews, S. Narayanan, G.S. Francis, J.P. Antel, and D.L. Arnold, Axonal dysfunction and disability in a relapse of multiple sclerosis: longitudinal study of a patient, *Neurology* **49**, 1138-41 (1997).
69. P.A. Narayana, T.J. Doyle, D. Lai, and J.S. Wolinsky, Serial proton magnetic resonance spectroscopic imaging, contrast-enhanced magnetic resonance imaging, and quantitative lesion volumetry in multiple sclerosis, *Ann. Neurol.* **43**, 56-71 (1998).
70. P.M. Matthews, E. Pioro, S. Narayanan, N. De Stefano, L. Fu, G. Francis, J. Antel, C. Wolfson, and D.L. Arnold, Assessment of lesion pathology in multiple sclerosis using quantitative MRI morphometry and magnetic resonance spectroscopy, *Brain* **119**, 715-22 (1996).
71. L. Fu, P.M. Matthews, N. De Stefano, K.J. Worsley, S. Narayanan, G.S. Francis, J.P. Antel, C. Wolfson, and D.L. Arnold, Imaging axonal damage of normal-appearing white matter in multiple sclerosis, *Brain* **121**, 103-13 (1998).

72. H. Sugie, S. Tsurui, A. Ishikawa, F. Matsuda, Y. Sugie, Y. Igarashi, and Y. Fujise, [Effects of neonatal hypothyroidism on brain development: analysis of brain metabolites using high resolution phosphorus and proton magnetic resonance (NMR) spectroscopy]. [Article in Japanese], *No To Hattatsu.* **22**, 166-72 (1990).

73. M.J. Malone, N.P. Rosman, M. Szoke, and D. Davis, Myelination of brain in experimental hypothyroidism. An electron-microscopic and biochemical study of purified myelin isolates, *J. Neurol. Sci.* **26**, 1-11 (1975).

74. J.M. Matthieu, P.J. Reier, and J.A. Sawchak, Proteins of rat brain myelin in neonatal hypothyroidism, *Brain Res.* **84**, 443-51 (1975).

75. T. Valcana, E.R. Einstein, J. Csejtey, K.B. Dalal, and P.S. Timiras, Influence of thyroid hormones on myelin proteins in the developing rat brain, *J. Neurol. Sci.* **25**, 19-27 (1975).

76. R.A. Harris and H.H. Loh, Brain sulfatide and non-lipid sulfate metabolism in hypothyroid rats, *Res. Commun. Chem. Pathol. Pharmacol.* **24**, 169-179 (1979).

77. J.M. Pasquini, I.A. Faryna-de-Raveglia, N. Capitman, and E.F. Soto, Neonatal hypothyroidism and early undernutrition in the rat: defective maturation of structural membrane components in the central nervous system, *Neurochem. Res.* **6**, 979-91 (1981).

78. T. Noguchi, T. Sugisaki, I. Satoh, and M. Kudo, Partial restoration of cerebral myelination of the congenitally hypothyroid mouse by parenteral or breast milk administration of thyroxine, *J. Neurochem.* **45**, 1419-26 (1985).

79. G. Almazan, P. Honegger, and J.M. Matthieu, Triiodothyronine stimulation of oligodendroglial differentiation and myelination. A developmental study, *Dev. Neurosci.* **7**, 45-51 (1985).

80. G. Shanker, S.G. Amur, and R.A. Pieringer, Investigations on myelinogenesis *in vitro*: a study of the critical period at which thyroid hormone exerts its maximum regulatory effect on the developmental expression of two myelin associated markers in cultured brain cells from embryonic mice, *Neurochem. Res.* **10**, 617-25 (1985).

81. S.P. Porterfield and C.E. Hendrich, The role of thyroid hormones in prenatal and neonatal neurological development--current perspectives, *Endocr. Rev.* **14**, 94-106 (1993).

82. N. Ibarrola, M. Mayer-Proschel, A. Rodriguez-Pena, and M. Noble, Evidence for the existence of at least two timing mechanisms that contribute to oligodendrocyte generation *in vitro*, *Dev. Biol.* **180**, 1-21 (1996).

83. S.C. Ahlgren, H. Wallace, J. Bishop, C. Neophytou, and M.C. Raff, Effects of thyroid hormone on embryonic oligodendrocyte precursor cell development *in vivo* and *in vitro*, *Mol. Cell Neurosci.* **9**, 420-32 (1997).

84. F.B. Gao, J. Apperly, and M. Raff, Cell-intrinsic timers and thyroid hormone regulate the probability of cell-cycle withdrawal and differentiation of oligodendrocyte precursor cells, *Dev. Biol.* **197**, 54-66 (1998).

85. C. Bjartmar, J. Battistuta, N. Terada, E. Dupree, and B.D. Trapp, *N*-acetylaspartate is an axon-specific marker of mature white matter *in vivo*: a biochemical and immunohistochemical study on the rat optic nerve, *Ann. Neurol.* **51**, 51-8 (2002).

86. M. Noble and G. Wolswijk, Development and regeneration in the O-2A lineage: studies *in vitro* and *in vivo*, J. Neuroimmunol 40, 287-93 (1992).

87. M.C. Raff, Glial cell diversification in the rat optic nerve, *Science* **243**, 1450-5 (1989).

88. M.C. Raff, C. Ffrench Constant, and R.H. Miller, Glial cells in the rat optic nerve and some thoughts on remyelination in the mammalian CNS, *J. Exp Biol.* **132**, 35-41 (1987).

89. P.C. Brunjes, A comparative study of prenatal development in the olfactory bulb, neocortex and hippocampal region of the precocial mouse Acomys cahirinus and rat, *Dev. Brain Res.* **49**, 7-25 (1989).

90. A.K. Groves, A. Entwistle, P.S. Jat, and M. Noble, The characterization of astrocyte cell lines that display properties of glial scar tissue, *Dev. Biol.* **159**, 87-104 (1993).

91. C.A. Haas, U. Rauch, N. Thon, T. Merten, and T. Deller, Entorhinal cortex lesion in adult rats induces the expression of the neuronal chondroitin sulfate proteoglycan neurocan in reactive astrocytes, *J. Neurosci.* **19**, 9953-63 (1999).

92. R.J. McKeon, M.J. Jurynec, and C.R. Buck, The chondroitin sulfate proteoglycans neurocan and phosphacan are expressed by reactive astrocytes in the chronic CNS glial scar, *J. Neurosci.* **19**, 10778-88 (1999).

93. V. Gallo and A. Bertolotto, Extracellular matrix of cultured glial cells: selective expression of chondroitin 4-sulfate by type-2 astrocytes and their progenitors, *Exp. Cell Res.* **187**, 211-23 (1990).

94. K. Kitada, T. Akimitsu, Y. Shigematsu, A. Kondo, T. Maihara, N. Yokoi, T. Kuramoto, M. Sasa, and T. Serikawa, Accumulation of N-acetyl-L-aspartate in the brain of the tremor rat, a mutant exhibiting absence-like seizure and spongiform degeneration in the central nervous system, *J. Neurochem.* **74**, 2512-9 (2000).

95.   T. Akimitsu, K. Kurisu, R. Hanaya, K. Iida, Y. Kiura, K. Arita, H. Matsubayashi, K. Ishihara, K. Kitada, T. Serikawa, and M. Sasa, Epileptic seizures induced by N-acetyl-L-aspartate in rats: *in vivo* and *in vitro* studies, *Brain Res.* **861**, 143-50 (2000).

96.   N. De Stefano, P.M. Matthews, L. Fu, S. Narayanan, J. Stanley, G.S. Francis, J.P. Antel, and D.L. Arnold, Axonal damage correlates with disability in patients with relapsing-remitting multiple sclerosis. Results of a longitudinal magnetic resonance spectroscopy study, *Brain* **121**, 1469-77 (1998).

97.   C.N. Madhavarao, J.R. Moffett, R.A. Moore, R.E. Viola, M.A. Namboodiri, and D.M. Jacobowitz, Immunohistochemical localization of aspartoacylase in the rat central nervous system, *J. Comp. Neurol.* **472**, 318-29 (2004).

NAA SYNTHESIS AND FUNCTIONAL ROLES

Chikkathur N. Madhavarao and Aryan M.A. Namboodiri[*]

1. INTRODUCTION

Tallan[1] reported the occurrence of high concentrations (10-14 mM) of N-acetylaspartate (NAA) in cat brain almost 50 years ago. After a long gap, the past 20 years have seen tremendous progress with discoveries such as the NAA metabolic defect in Canavan disease[2,3] and the recognition of the value of NAA as an index of neuronal health using magnetic resonance spectroscopy.[4-7] However, the biochemical mechanisms of NAA synthesis remain unclear. The early reports on the NAA synthetic enzyme date back nearly 5 decades [8-10] and its subcellular localization to mitochondria was indicated in 1979.[11] The synthetic enzyme, L-Aspartate N-acetyltransferase (Asp-NAT; EC 2.3.1.17) was characterized only recently as a possible multi-subunit enzyme complex from rat brain.[12] However, further molecular characterization of this enzyme remains to be accomplished.

While a number of reports indicate that NAA is localized primarily in neurons in the central nervous system (CNS),[5,13-15] the NAA hydrolyzing enzyme aspartoacylase (amidohydrolase 2, ASPA; EC 3.5.1.15) has been localized primarily in oligo-dendrocytes, the myelinating cells in the CNS.[16-19] Unlike Asp-NAT, there has been a body of literature available on the NAA hydrolyzing enzyme, ASPA, after it was first reported in hog kidney.[20] ASPA was later detected in brain tissue[21] and it was subsequently purified from bovine brain.[22] This work led to the identification of the genetic sequence coding ASPA[23] and cloning and expression of human and mouse recombinant genes.[24,25] Mutations in the gene coding ASPA have been linked to the fatal genetic disorder, Canavan disease.[3,26,27]

[*] Department of Anatomy, Physiology and Genetics, Uniformed Services University of the Health Sciences, Bethesda, MD-20814; email, anamboodiri@usuhs.mil.

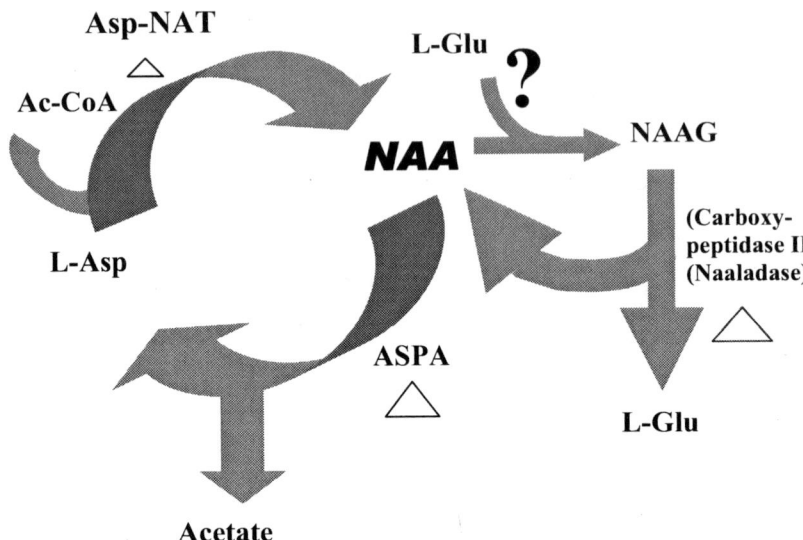

Figure 1. Synthesis of NAA and NAAG. Asp-NAT synthesizes NAA using acetyl CoA and L-aspartate in neuronal mitochondria. NAA possibly serves as a precursor of NAAG by peptide bond formation with L-glutamate, the enzyme catalyzing the reaction is yet to be described. NAAG is broken down by carboxy-peptidase II, localized to astrocytes, releasing NAA and L-glutamate. NAA is hydrolyzed by aspartoacylase localized to oligodendrocytes, releasing the acetate and L-aspartate.

The metabolism of NAA is summarized in Fig. 1. Asp-NAT synthesizes NAA using acetyl CoA and L-aspartate, most likely in neuronal mitochondria. NAA is proposed to serve as the precursor for the metabolite neuromodulator N-acetylaspartylglutamate (NAAG) synthesized by a yet to be described enzyme.[28] Because the hydrolysis of NAAG by carboxypeptidase II (NAALADase) yields the products NAA and L-glutamate[29,30], it has been hypothesized that the synthesis of NAAG could also occur by an enzyme catalyzed reaction, and preliminary observations are suggestive of an enzyme-mediated synthesis.[28,31,32] NAA is hydrolyzed by aspartoacylase, localized primarily in oligodendrocytes, producing acetate and aspartate.

2. PURIFICATION OF ASP-NAT

There are few reports on the cellular and sub-cellular localization of Asp-NAT *per se* other than those reported on its product molecule NAA, which is predominantly localized in neurons (see Moffett and Namboodiri, this volume).[13,14] An earlier report on the mitochondrial localization of Asp-NAT was based on efflux studies of NAA on isolated brain mitochondria.[11] Recently, it was convincingly shown to be localized to mito-chondria by demonstrating that Asp-NAT activity in rat brain was present in the highly purified mitochondrial subfractions, including the outer mitochondrial membrane, the contact sites of outer and inner mitochondrial membranes and the inner mitochondrial membrane + matrix of the mitochondria[12]. However, a recent report indicated detection of Asp-NAT activity in a significant amount in the microsomal fraction isolated from rat brain homogenates.[33]

The radiometric assay developed by modification of the assay for aspartoacylase[34] aided in the purification of Asp-NAT.[12] The purification procedure of Asp-NAT (summarized in Fig. 2) involved solubilization of Asp-NAT activity by treating crude mitochondrial pellets in a buffer containing 10 mM CHAPS, and subsequently purifying the enzyme activity in four steps: 1) anion exchange chromatography, 2) native gel electrophoresis/continuous elution (Model 491 Prep Cell, BioRad), 3) HPLC size exclusion chromatography and 4) HPLC size exclusion re-chromatography.

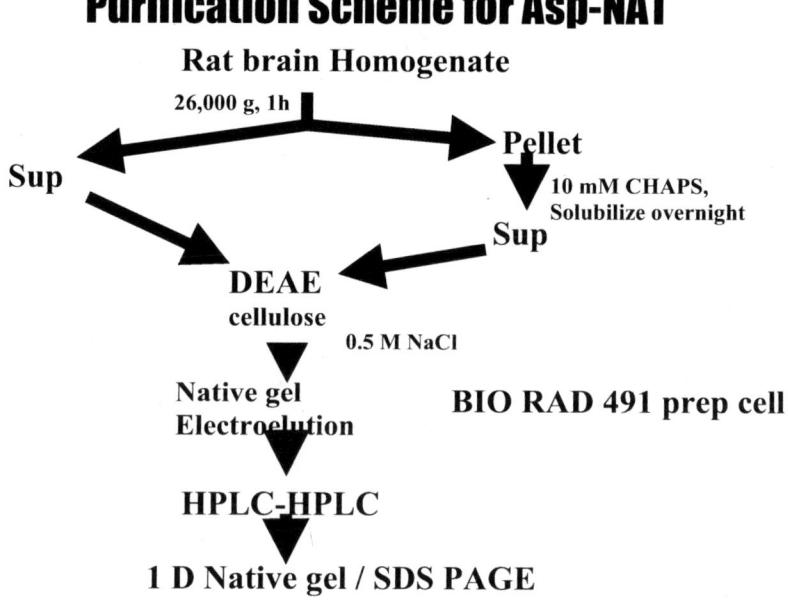

Figure 2. Purification procedures used for isolating Asp-NAT from rat brain.

2.1 Anion-exchange chromatography

Bio-Rad macro-prep DEAE cellulose resin (~100 ml in bed volume) was useful as an anion exchange matrix to separate the Asp-NAT from the bulk of the proteins. Asp-NAT was bound to the resin on mixing the resin with the solubilized enzyme, and chromatography was performed as follows. The column was washed sequentially in two steps: 1) with 10 column volumes of equilibrating buffer (EB) (0.15 M NaCl, 10 mM sodium phosphate buffer pH 7.1, 1 mM CHAPS, 1 mM DTT and 10 % glycerol), and 2) with 5 column volumes of EB with 0.25 M NaCl. The enzyme activity was eluted with the EB having 0.5 M NaCl in about 6 fractions of 45 ml each.

DEAE cellulose facilitated binding of ~80% of the enzyme activity and yielded ~90% of the bound enzyme activity on elution. This method enriched the enzyme activity by about 4-5 fold (activity increased from 5-6 nmol/h/mg protein to 24-26 nmol/h/mg protein). Fractions 2 and 3 containing the bulk of the enzyme activity were pooled, concentrated and applied to the native gel electrophoresis system.

2.2 Native gel electrophoresis

The Bio-Rad Model 491 Prep Cell was the apparatus of choice for gel electrophoresis purification of Asp-NAT after DEAE ion exchange chromatography, and was carried out by continuous flow electrophoresis under native conditions. Continuous buffer of Tris-Glycine-HCl of pH 8.0—8.5 was used in both the upper and lower chamber. Concentrations of Tris and Glycine were in the range of 6—25 mM and 60—192 mM, respectively. Asp-NAT activity was consistently detected in eluates after 21 hours when the electrophoresis buffer pH was 8.0. In order to retain the enzyme activity this buffer also consisted of 6% glycerol, 1 mM CHAPS and 1 mM DTT. Elution of Asp-NAT was carried out with a buffer consisting of PBS, 10% glycerol, 1 mM CHAPS, 1 mM DTT, pH adjusted to 7.2. Resolving gel (6%) of 3 cm height and stacking gel (3%) of 3 cm height were cast overnight according to the manufacturer's instructions using 30% acrylamide/Bis, 37.5:1 (2.6% C) stock solutions. Adding 10% glycerol, 1 mM CHAPS and Tris-Glycine-HCl buffer of pH 8.0 in the casting gel solution protected the enzyme activity. The entire electro-elution was carried out in a cold room (2-4 °C) at a constant power of 5W. About 2 hours of pre-electrophoresis was carried out before loading the protein sample, to remove the unreacted catalysts from the polyacrylamide gel. Sample protein (~6-8 ml) was dialysed against 2 liters of a medium containing 10 mM sodium phosphate pH 7.2, 10% glycerol, 1 mM CHAPS and 1 mM DTT before loading on to the native gel column, which retained the activity of the enzyme at the same level as opposed to dialysing in the electrophoresis buffer (Tris-HCl-Glycine, pH 8.0), which resulted in a loss of activity by ~ 20-25%. At the elution rate of 0.3 ml/min, about 250 fractions of 5 ml were collected. Every fourth fraction was assayed using the radiometric method ([^{14}C] L-Asp, 100 mCi/mmol) after concentrating 2 ml into ~0.25 ml (Centricon YM 10 concentrator).

Asp-NAT was also purified from the supernatant obtained after separating the crude mitochondrial pellet (26,000 x g, 1h). When a large number of rat brains (>100) are homogenized using domestic blenders there is an almost equal distribution of Asp-NAT activity in the supernatant and in the crude mitochondrial pellet. Interestingly, Asp-NAT in the soluble fraction (high-speed supernatant fraction) was found to be identical in substrate specificity and molecular weight to that obtained from solubilization of the crude mitochondrial pellet. A possible explanation would be that some of the Asp-NAT that is loosely held inside the mitochondrial membranes will be released into the soluble fraction on homogenization due to partial disruption of mitochondria. This explanation is consistent with the observation that most of the Asp-NAT activity is associated with the crude mitochondrial pellet on gentle homogenization that is used to prepare tightly coupled mitochondria.[12] However, a recent report suggests occurrence of Asp-NAT in the microsomal fraction as well.[33] Although, identical steps of purification were followed to purify Asp-NAT from the supernatant fraction and the CHAPS solubilized fraction of the crude mitochondrial pellet, some adjustments were made in the native gel electrophoresis to achieve maximal purity and recovery of enzyme activity. A comparison of the salient features of native gel electrophoresis is given in Table 1.

Table 1. Comparison of the purification of Asp-NAT from the crude mitochondrial pellet vs. the supernatant fraction after high-speed centrifugation.

Feature	Asp-NAT from pellet	Asp-NAT supernatant
(a) Source of Asp-NAT	Crude mitochondrial pellet (26,000 g), solubilized with 10 mM CHAPS	Supernatant after high-speed centrifugation (26,000 g) without CHAPS solubilization
(b) CHAPS	1mM CHAPS included in all buffers	CHAPS excluded
(c) DEAE anion exchange chromatography	Elutes with 0.5 M NaCl buffer	Elutes with 0.25 M NaCl and 0.5 M NaCl buffers
(d) Native gel electro-elution		
(i) Electrophoresis buffer	Tris-glycine (25mM-192mM), pH 8.0	Tris-glycine (6mM-60 mM), pH 8.5
(ii)Stacking and resolving gel	3 cm (3%) stacking and 3 cm (6%) resolving gel	2 cm (4%) stacking and 2 cm (6%) resolving gel
(iii)Dialysis medium	10 mM sodium phosphate buffer pH 7.0	Tris-Glycine (6mM-60mM), pH adjusted to 7.0
(iv) Elution rate and fraction size	0.3 ml/min; 5ml	0.6ml/min; 8 ml
(v) Elution time	24-48 h	6-9 h
(vi) Electrophoresis setting	5 watt constant; beginning Voltage ~280 and current ~17 mA with gradual decrease in V (up to 100 V) and increase in current (up to 47 mA).	5 watt constant; beginning Voltage ~490 and current ~6 mA with gradual decrease in V (up to 280 V) and increase in current (up to 20 mA)
(e) Size exclusion chromatography	Analytical grade with column volume of ~15 ml	Semi-preparative grade with total column volume of ~100 ml
(f) Peak activity	Single peak; ~670 kD	Single peak; ~670 kD

One important point of difference was that CHAPS was not required for activity or stability of the Asp-NAT preparation obtained from the supernatant fraction. The final enzyme preparation after HPLC (with the column exclusion limit of ~800 kD) showed a single peak on size exclusion chromatography and a single band on native gel electrophoresis (Fig. 3). However, multiple bands were detectable on SDS-PAGE as was the case with the mitochondria-solubilized enzyme preparation.[12] Although all these protein bands may not be part of Asp-NAT, the indication was that of an enzyme complex containing multiple enzymes or subunits, which is a characteristic of a variety of mitochondrial enzymes.

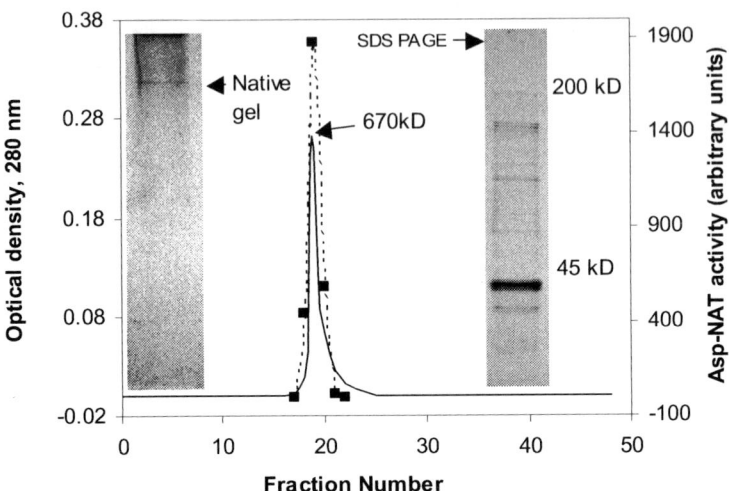

Figure 3. Analysis of the purity of Asp-NAT from HPLC size exclusion rechromatography. Detector tracing of elution profile of Asp-NAT in HPLC size exclusion re-chromatography (solid line, OD 280 nm) and the composition of proteins as detected by native-PAGE (Tris-acetate, 3-8%) and SDS-PAGE (Tris-glycine, 4-12%). A semi-preparative HPLC column of ~100 ml capacity (Shodex KW 2004) was used and elution was carried out at 1.0 ml/min collecting 2 ml/fraction. The single peak showed a single band in native gel but multiple bands on SDS-PAGE. Asp-NAT activity (broken line with symbols) corresponded to the peak of OD tracing.

3. MOLECULAR SIZE OF NATIVE ASP-NAT

Whether or not the large molecular size of native Asp-NAT was an artifact of the enzyme preparation methods was tested by using different homogenization and solubilization conditions. A buffer that is commonly used in mitochondrial protein purification (Tris-sucrose medium: Tris-HCl, 50 mM; 1 mM EDTA; 0.32 M sucrose; 1 mM DTT; protease inhibitor cock-tail added according to the quantity of processed tissue; pH adjusted to 7.4) was employed to obtain a crude mitochondrial pellet from rat brains. Various detergents were tested to solubilize the pellet: deoxycholate (DC) (negatively charged and one of the most commonly used detergents in mitochondrial studies), hexadecyl trimethyl ammonium bromide (positively charged, commonly called CTAB), Triton X-100 (non-ionic), laurylmaltoside (LM) (non-ionic, commonly used in mitochondrial protein reconstitution experiments) and CHAPS.

Crude mitochondrial pellets were obtained by homogenizing frozen rat brain tissue in the Tris-sucrose medium using a glass-teflon Potter-Elvehjem homogenizer, followed by two centrifugation steps: the first at low speed (1350g for 5 min) to separate the debris and the second at high speed (18,000g for 15 min) to obtain a crude mitochondrial pellet. The pellet was incubated in Tris-sucrose medium with various concentrations of detergents for 1h on ice with periodic agitation, followed by separation of the solubilized

Asp-NAT fraction by centrifugation (22,000g for 30 min). Asp-NAT activity was measured by the TLC-radiometric assay previously described.[12] The results are given in Figure 4. While CHAPS was the most effective detergent, Triton X-100 and laurylmaltoside were almost as effective. Deoxycholate was very low in effectiveness and CTAB was least effective.

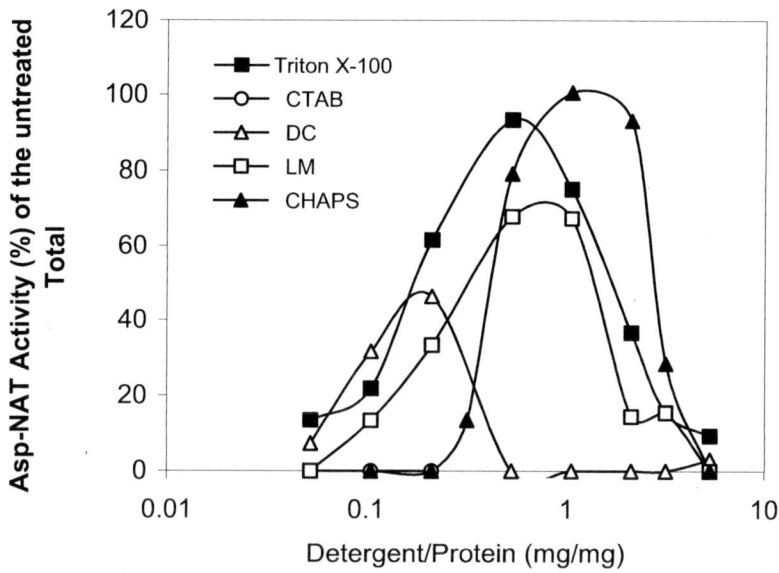

Figure 4. Effect of different detergents on solubilization of Asp-NAT activity. Measured activity in the supernatant after detergent treatment was normalized to the percentage of the Asp-NAT activity observed without treating with detergents. This is a normalized activity for comparison of the detergent suitability, and for choice of detergents, and does not reflect the exact percentage of the solubilization.

The molecular weight pattern of the Asp-NAT activity of the solubilized fractions was tested by HPLC (column with exclusion limit of ~800 kD). Deoxycholate and CTAB were not included because of their low efficiency in solubilization. The crude mitochondrial pellet solubilized overnight with an optimal concentration of each of the detergents, and the supernatant (22,000g, 30 min) was subjected to DEAE anion exchange chromatography as above, with the Tris-sucrose medium used for equilibration. Asp-NAT was eluted using Tris-sucrose medium containing 0.25 M NaCl. The peak activities were pooled, concentrated and subjected to size exclusion-HPLC to determine the molecular weight pattern (Fig. 5).

The molecular weight pattern is similar in all three preparations (Fig. 5A), although the mean molecular weight decreased slightly in the case of the Triton X-100 and laurylmaltoside solubilized preparations. A decrease in molecular weight of the CHAPS solubilized preparation was associated with considerable decreases in the Asp-NAT activity. The same effect was observed under the present conditions as evidenced by the lower specific activity of the Triton X-100 and laurylmaltoside solubilized preparations (Fig. 5B). These data suggest two points: (a) CHAPS is a preferable choice of detergent over Triton X-100 and laurylmaltoside for the purification of Asp-NAT and (b) the apparent size of ~670 kD of the Asp-NAT complex is not an artifact.

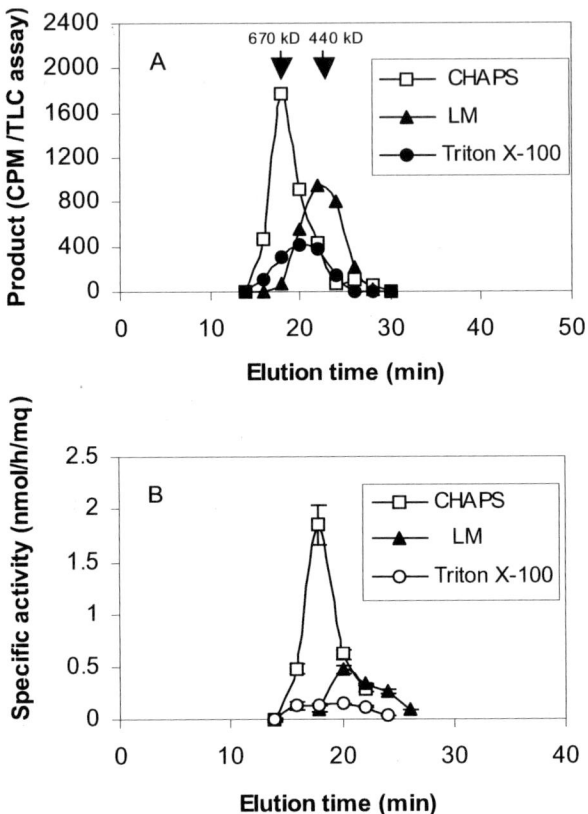

Figure 5. Decrease in size renders decreased activity for Asp-NAT. Nonionic detergents laurylmaltoside (LM) and triton X-100 both resulted in smaller Asp-NAT protein complex, after treatment and size exclusion HPLC (A). However, treatment with neutral detergent CHAPS resulted in larger sized Asp-NAT (elutes early) and higher specific activity (B).

4. STABILITY OF THE ASP-NAT COMPLEX

The large size and the multimeric nature of Asp-NAT called for an investigation on the stability characteristics of this complex. CHAPS and NaCl were used in increasing concentrations in the treatment medium to perturb the Asp-NAT complex. Enriched Asp-NAT preparations from a DEAE column were used in these studies in view of its higher enzyme activity. The results in Figure 6 show that CHAPS has a biphasic effect on Asp-NAT stability (Fig. 6A) as revealed by pre-treatment (0-4 °C, 1h). Asp-NAT activity decreased to 40% of the control at 12 mM CHAPS and 10% of the control at 20 mM CHAPS. Removal of excess CHAPS by dialysis had no restoring effect on the enzyme activity after pre-treatment at >12 mM CHAPS. Although NaCl decreased Asp-NAT activity at concentrations greater than 0.15 M, removal of excess NaCl by dialysis reversed the effect up to 1 M NaCl (Fig. 6B).

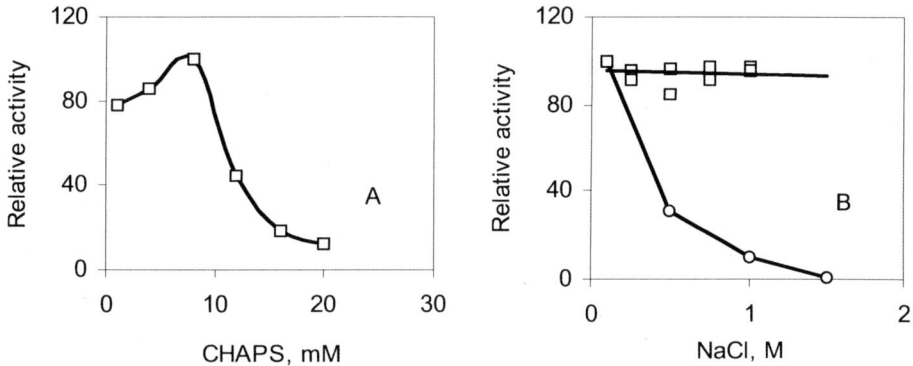

Figure 6. Asp-NAT activity in the presence of varying concentrations of CHAPS (A) and NaCl (B). Relative activity (%) with respect to 1 mM CHAPS in Fig. A, 0.15 M NaCl in Fig. B is represented in the y-axis. Enzyme activity was more or less fully restored by dialysis (squares) after treating in high concentrations of NaCl (Fig. B).

Pre-treatment for 1h in 0.15 M, 0.5 M and 2.0 M NaCl media and HPLC with a size exclusion column (~850 kD cut-off) gave unimodal distribution of Asp-NAT activity in the eluted fractions (Fig. 7A). Enzyme activity was somewhat decreased at 0.5 and 2.0 M NaCl, probably due to incomplete dialysis. However, pre-treatment with 10 mM CHAPS (as opposed to 1 mM CHAPS in the buffer) and HPLC size exclusion chromatography resulted in significant loss of activity, which was irreversible even after dialysis. The distribution pattern was also different with a right shift (Fig. 7B) and increased flatness, as compared to the control (1 mM CHAPS pre-treatment). These results are consistent with the possibility that some subunits that contribute or regulate the enzyme activity are separated from the complex during the chromatography, which results in the loss of enzyme activity. That these enzyme components are held primarily by hydrophobic interactions is indicated by the sensitivity to CHAPS, but not to increasing ionic strength.

The steady state kinetic parameters of Asp-NAT are presented in Table 2. Between the two substrates, Asp-NAT has an order of magnitude higher affinity (K_m) toward acetyl CoA than L-Asp. Both the products, NAA and CoA have inhibitory effect on Asp-NAT activity. This enzyme was specific toward L-Asp and showed ≤3% activity against L-Glu, L-Asn and L-Gln.

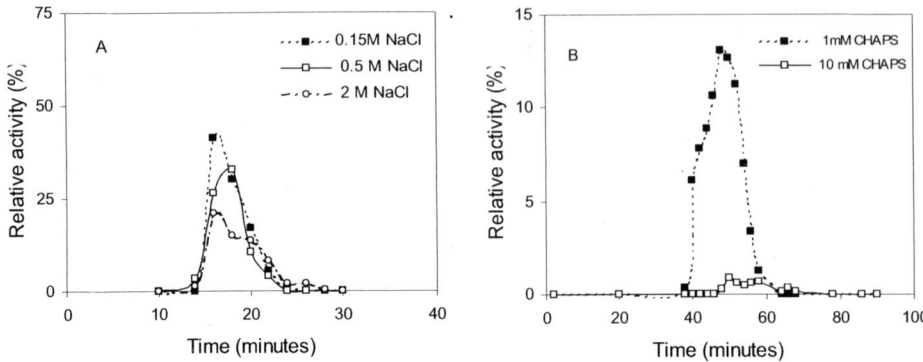

Figure 7. Structural perturbation of Asp-NAT by salt (A) and CHAPS (B) and its impact on enzyme complex size and activity. Salt concentration seems to decrease the Asp-NAT activity by interference (A) whereas detergent concentration decreases the activity by destroying the hydrophobic interactions within the Asp-NAT complex. The columns were pre-equilibrated and eluted with the respective media. The HPLC column size in Fig. A was 14 ml and fraction size was 1 ml each, and, in Fig. B, the column size was 100 ml and fraction size was 2 ml each. Asp-NAT activity was determined after dialyzing the sample against 0.15 M NaCl medium. The recovery of activity in Fig. A was 95%, 80% and 64% for 0.15M, 0.5 M and 2.0 M NaCl media, respectively. Asp-NAT activity was determined after dialyzing the sample against 1 mM CHAPS containing medium. About 85% of the activity was recovered in 1 mM CHAPS treated samples and only 5% of the activity was recovered in 10 mM CHAPS treated samples. The y-axis represents the activity recovered, relative to the input activity normalized to 100.

Table 2. Steady state kinetic parameters and inhibitory concentrations of NAA and CoA for Asp-NAT.

Substrate/ Product	K_m (µM)	V_{max} (nmol/h/mg protein)	IC-50% (µM)	I_{max} (%)[a]
L-Asp	580	27		
Acetyl CoA	58	23		
NAA			850	95
CoA			420	96

These data were generated from duplicate assays but analyzed using all data points by non-linear regression using Sigma Plot software (16 for L-Asp and 20 for acetyl CoA, for enzyme kinetics; 12 points for inhibitory studies). [a] I_{max} is the percent maximum inhibition exerted by the compound on Asp-NAT activity.

5. DOES NAA SYNTHESIS IN MITOCHONDRIA HAVE A BIOENERGETIC ROLE?

The presence of Asp-NAT in mitochondria and the apparently large multimeric structure observed in the rat brain enzyme suggest a possible involvement for this enzyme system in neuronal bioenergetics. NAA could be an integral player in mitochondrial energy metabolism in neurons, which is required to support their extraordinary energetic needs. A potential metabolic role for NAA is presented in schematic form in Figure 8.

The extra demand for ATP in neurons appears to be met in part by oxidation of glutamate via the aspartate aminotransferase (EC 2.6.1.1) pathway. The importance of this pathway in neuronal mitochondria was recognized by earlier investigators, and the name 'mini citric acid cycle' was coined to emphasize its role in neuronal energetics.[35,36] It is possible that neuronal mitochondria use this reaction instead of the glutamate dehydrogenase reaction to generate the key metabolite α-ketoglutarate, because this avoids the problem of ammonia toxicity associated with the glutamate dehydrogenase reaction. This metabolic distinction between neurons and somatic cells may be due to the absence of an effective urea cycle system in neurons. Thus, it seems likely that NAA synthesis is intimately associated with the proper functioning of the mini citric acid cycle by using the Asp-NAT reaction to remove excess aspartate from neuronal mitochondria. NAA synthesis serves two important functions. First, by converting the product aspartate into NAA, NAA synthesis helps to guide the aspartate aminotransferase reaction toward α-ketoglutarate facilitating energy production. Second, it substitutes for citrate as the acetate carrier to the cytoplasm and thus compensates for the lack of citrate production during the 'mini citric acid cycle'.

There are several findings in the literature that are consistent with or supportive of a bioenergetic role for NAA. First, inhibition of aspartate aminotransferase activity by β-methyleneaspartate decreased the oxygen consumption in state 3 respiration in the presence of glutamate and malate as substrates.[37] Second, efflux of NAA increased and that of L-Asp decreased with increasing concentration of pyruvate in the presence of glutamate and malate as substrates in neuronal mitochondria.[11] Third, ATP synthesis and NAA synthesis were directly correlated with oxygen consumption in studies involving respiratory chain inhibitors.[38] Fourth, concentrations of NAA and ATP were directly correlated in a study involving diffuse traumatic brain injury in rats.[39]

6. NAA AS A MARKER FOR NEURONAL HEALTH

The relatively high concentration of NAA permits its noninvasive detection by magnetic resonance spectroscopy (MRS), which combined with its predominant neuronal localization has enabled use of NAA as a diagnostic marker for neuronal dysfunction/loss in a variety of neurological disorders.[4,5,7] NAA levels measured by MRS have been reported to be altered in patients suffering from various neurological disorders[40-51] (Table 3). These studies have mostly detected decreases in NAA, but in case of Canavan disease (CD)[52] relative increases in NAA levels are common. The diseases in which NAA levels are decreased include Alzheimer's disease, epilepsy, schizophrenia, multiple sclerosis and AIDS dementia complex (Table 1). Earlier, the decreases in NAA were interpreted to represent irreversible loss of neurons. However, more recent evidence indicates that these decreases also could represent reversible neuronal mitochondrial dysfunction.[53]

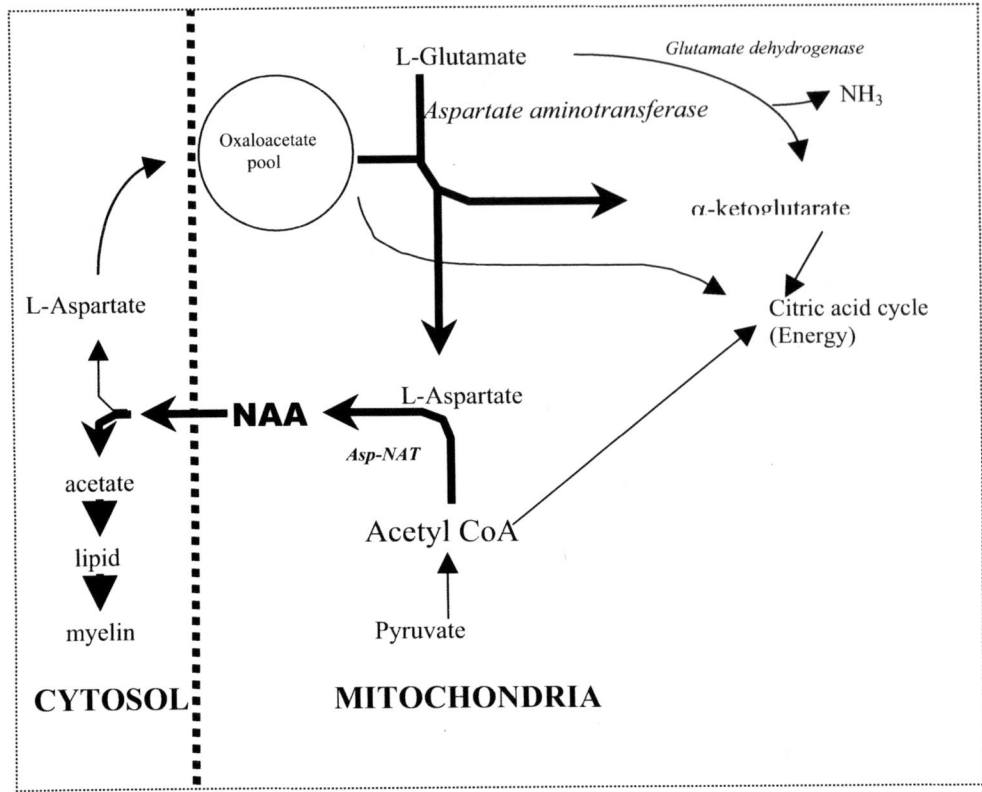

Figure 8. Schematic of the energetics model for NAA synthesis in neuronal mitochondria. The thick lines indicate the path of NAA synthesis and further its fate, the thin lines indicate other possible pathways the substrates can take.

In support of this view, some evidence has been presented showing that NAA levels are restored when patients recovered clinically[49]. These observations bring up a central question in NAA research. Does NAA play a part in the functional recovery, or is it just a marker for the recovery of mitochondrial energy metabolism?

The proposed bioenergetic model[12] described above suggests plausible pathways in mitochondria that are associated with the synthesis of NAA, and relies considerably on reported experimental evidence supporting the direction of reactions. There are reports of a positive correlation between neuronal ATP and NAA at the level of oxidative phosphorylation[38] and the responses observed following brain injury.[39,54] A subsequent investigation carried out on healthy individuals on the indices of mitochondrial/neuronal energetics did not, however, find a positive correlation between NAA levels and ATP.[55]

The lack of correlation between brain ATP and NAA levels was based entirely on *in vivo* MRS data. In addition, the determined concentration levels of free ADP and ATP varied by a factor of 1000. In terms of concentration however, NAA levels were comparable to that of ATP rather than ADP. On the other hand, *in vivo* studies on the application and withdrawal of the mitochondria toxin 3-nitropropionic acid (3-NP) were more striking. Rat and primate models were used in the study of progressive striatal degeneration induced by 3NP, and NAA levels were determined by proton MRS. While

Table 3. Changes in NAA levels in neurological disorders:

Disorder/Disease	Increase/Decrease in NAA level	Area of the Brain
Alzheimer's disease [40,43]	Decreases	Medial temporal lobe, parietal lobe, hippocampus
Schizophrenia [46,47]	Decreases	Thalamus, prefrontal cortex
Multiple sclerosis [45]	Decreases	Lesioned brain areas
Post traumatic stress disorder [41]	Decreases	Hippocampus
AIDS dementia complex [48]	Decreases	Medial frontal cortex
Stroke [51]	Decreases	Infarcted area
Huntington's disease [50]	Decreases	Striatum
Amyotrophic lateral sclerosis [49]	Decreases	Primary motor cortex
Gulf War Syndrome [44]	Decreases	Basal ganglia, brain stem
Canavan Disease [52]	Increases	White matter of brain

cell loss and dying cells were not detected, significant decreases in NAA concentrations were specifically restricted to the striatum and were associated with the appearance of motor symptoms by the 3rd day of treatment. The withdrawal of the toxin partially restored the NAA levels after 4 weeks.[53] These experimental data are consistent with the proposed bioenergtics model of NAA synthesis in neurons, and suggest a connection between NAA synthesis and ATP synthesis.

Although, a convincing link has been demonstrated between dysmyelination and the impaired catabolism of NAA in CD[56] (also see Chapter by Namboodiri et al., this volume), the exact link between the decreasing levels of NAA and the observed neurological impairments in other diseases remains to be established. Progress in the molecular characterization of Asp-NAT is expected to provide the necessary tools, such as antibodies and gene knockout models, to help unravel the neurobiology of NAA and its functional roles in the nervous system.

7. ACKNOWLEDGEMENTS

This work was supported by grants from NIH (R21, MH068341 and RO1, NS 39387) and an intramural research grant from the Uniformed Services University of the Health Sciences (C070PB). The authors thank Dr. John Moffett for critical reading of and suggestions on the manuscript.

8. QUESTION AND ANSWER SESSION

SESSION CO-CHAIR COYLE: Questions, discussion? Yes.

DR. BRUCE TRAPP: Could you clarify the purification procedure? Do you isolate mitochondria first? And do you find the enzymatic complex sequestered in mitochondria compared to other cellular compartments?

DR. MADHAVARAO: In fact, we did not take the route of purifying mitochondria, and then isolating the enzyme, because the yield would be very low. Rather than that, we thought that we would just homogenize the rat brain and then go about solubilizing this enzyme from the crude mitochondrial pellet.

That is to say, when we actually look for this enzyme activity with the different fractions, we found about 90 to 95 percent of the activity associated with crude mitochondrial pellet, which has myelin as well as the mitochondria. Therefore, we have to believe that it is probably synthesized in mitochondria and not elsewhere. Then we solubilized the crude mitochondrial pellet and we put it through the anion exchange. And then we further processed that fraction.

SESSION CO-CHAIR COYLE: Danny?

DR. WEINBERGER: Any evidence where ASPA is in oligodendrocytes? And is it a soluble enzyme?

DR. MADHAVARAO: Yes, ASPA is a soluble enzyme.

DR. WEINBERGER: Any idea where it is in oligos?

DR. MADHAVARAO: That is what we are actually focusing on by electron microscopy studies. There is a form that is in the cytosol. It is not associated with any membranes; maybe, we are not sure. It may be in the nucleus as well.

DR. WEINBERGER: Does the enzyme move out of the cell?

DR. MADHAVARAO: Does it leave the cell? I don't know.

DR. WEINBERGER: So that may leave open the remote possibility that perhaps the inconsistencies in NAA concentrations in oligos may have something to do with the packaging and handling of ASPA as much as it may have to do with where NAA really goes.

I mean, I guess it raises the possibility that this strange synthesis of NAA in neurons and transport very selectively into oligos for hydrolysis may well be what the fate of NAA is, but it sounds a bit strange. And it just raises the possibility that maybe this is a secretory enzyme and that, actually, the hydrolysis of NAA may take place in vivo, not in oligos.

DR. MADHAVARAO: Yes. Our hypothesis says that it has to take place in oligodendrocytes because they are synthesizing myelin. So there is transportation across neuronal/axonal membrane for NAA, actually, which goes into oligodendrocytes that probably the later speakers are going to focus on. There may be another hydrolyzing enzyme that we do not know of that may be there in neurons, therefore, which may facilitate the consumption of NAA as a fuel recycling back into the citric acid cycle that we do not know as yet.

SESSION CO-CHAIR COYLE: Thank you very much.

9. REFERENCES

1. Tallan HH, Moore S, Stein WH. N-Acetyl-L-aspartic acid in brain. *J Biol Chem* 1956;**219**:257-64.

2. Matalon R, Michals K, Sebesta D, Deanching M, Gashkoff P, Casanova J. Aspartoacylase deficiency and N-acetylaspartic aciduria in patients with Canavan disease. *Am J Med Genet* 1988;**29**:463-71.
3. Kaul R, Gao GP, Balamurugan K, Matalon R. Canavan disease: molecular basis of aspartoacylase deficiency. *J Inherit Metab Dis* 1994;**17**:295-7.
4. Birken DL, Oldendorf WH. N-acetyl-L-aspartic acid: a literature review of a compound prominent in 1H-NMR spectroscopic studies of brain. *Neurosci Biobehav Rev* 1989;13:23-31.
5. Tsai G, Coyle JT. N-acetylaspartate in neuropsychiatric disorders. *Prog Neurobiol* 1995;**46**:531-40.
6. Baslow MH. Functions of N-acetyl-L-aspartate and N-acetyl-L-aspartylglutamate in the vertebrate brain: role in glial cell-specific signaling. *J Neurochem* 2000;**75**:453-9.
7. Clark JB. N-acetyl aspartate: a marker for neuronal loss or mitochondrial dysfunction. *Dev Neurosci* 1998;**20**:271-6.
8. Goldstein FB. Biosynthesis of N-acetyl-L-aspartic acid. *J Biol Chem* 1959;**234**:2702-2706.
9. Knizley H, Jr. The enzymatic synthesis of N-acetyl-L-aspartic acid by a water-insoluble preparation of a cat brain acetone powder. *J Biol Chem* 1967;**242**:4619-22.
10. Goldstein FB. The enzymatic synthesis of N-acetyl-L-aspartic acid by subcellular preparations of rat brain. *J Biol Chem* 1969;**244**:4257-60.
11. Patel TB, Clark JB. Synthesis of N-acetyl-L-aspartate by rat brain mitochondria and its involvement in mitochondrial/cytosolic carbon transport. *Biochem J* 1979;**184**:539-46.
12. Madhavarao CN, Chinopoulos C, Chandrasekaran K, Namboodiri MA. Characterization of the N-acetylaspartate biosynthetic enzyme from rat brain. *J Neurochem* 2003;**86**:824-35.
13. Moffett JR, Namboodiri MA, Cangro CB, Neale JH. Immunohistochemical localization of N-acetylaspartate in rat brain. *Neuroreport* 1991;**2**:131-4.
14. Moffett JR, Namboodiri MA, Neale JH. Enhanced carbodiimide fixation for immunohistochemistry: application to the comparative distributions of N-acetylaspartylglutamate and N-acetylaspartate immunoreactivities in rat brain. *J Histochem Cytochem* 1993;**41**:559-70.
15. Tsai G, Cork LC, Slusher BS, Price D, Coyle JT. Abnormal acidic amino acids and N-acetylaspartylglutamate in hereditary canine motoneuron disease. *Brain Res* 1993;**629**:305-9.
16. Baslow MH, Suckow RF, Sapirstein V, Hungund BL. Expression of aspartoacylase activity in cultured rat macroglial cells is limited to oligodendrocytes. *J Mol Neurosci* 1999;**13**:47-53.
17. Kirmani BF, Jacobowitz DM, Kallarakal AT, Namboodiri MA. Aspartoacylase is restricted primarily to myelin synthesizing cells in the CNS: therapeutic implications for Canavan disease. *Brain Res Mol Brain Res* 2002;**107**:176-82.
18. Klugmann M, Symes CW, Klaussner BK, Leichtlein CB, Serikawa T, Young D, During MJ. Identification and distribution of aspartoacylase in the postnatal rat brain. *Neuroreport* 2003;**14**:1837-40.
19. Madhavarao CN, Moffett JR, Moore RA, Viola RE, Namboodiri MA, Jacobowitz DM. Immunohistochemical localization of aspartoacylase in the rat central nervous system. *J Comp Neurol* 2004;**472**:318-29.

20. Birnbaum SM. Amino acid acylases I and II from hog kidney. *Methods in Enzymology* 1955;**2**:115-119.
21. Goldstein FB. Amidohydrolases of brain; enzymatic hydrolysis of N-acetyl-L-aspartate and other N-acyl-L-amino acids. *J Neurochem* 1976;**26**:45-9.
22. Kaul R, Casanova J, Johnson AB, Tang P, Matalon R. Purification, characterization, and localization of aspartoacylase from bovine brain. *J Neurochem* 1991;**56**:129-35.
23. Kaul R, Balamurugan K, Gao GP, Matalon R. Canavan disease: genomic organization and localization of human ASPA to 17p13-ter and conservation of the ASPA gene during evolution. *Genomics* 1994;**21**:364-70.
24. Kaul R, Gao GP, Balamurugan K, Matalon R. Cloning of the human aspartoacylase cDNA and a common missense mutation in Canavan disease. *Nat Genet* 1993;**5**:118-23.
25. Namboodiri MA, Corigliano-Murphy A, Jiang G, Rollag M, Provencio I. Murine aspartoacylase: cloning, expression and comparison with the human enzyme. *Brain Res Mol Brain Res* 2000;**77**:285-9.
26. Kaul R, Gao GP, Aloya M, Balamurugan K, Petrosky A, Michals K, Matalon R. Canavan disease: mutations among Jewish and non-jewish patients. *Am J Hum Genet* 1994;**55**:34-41.
27. Zeng BJ, Wang ZH, Ribeiro LA, Leone P, De Gasperi R, Kim SJ, Raghavan S, Ong E, Pastores GM, Kolodny EH. Identification and characterization of novel mutations of the aspartoacylase gene in non-Jewish patients with Canavan disease. *J Inherit Metab Dis* 2002;**25**:557-70.
28. Gehl LM, Saab OH, Bzdega T, Wroblewska B, Neale JH. Biosynthesis of NAAG by an enzyme-mediated process in rat central nervous system neurons and glia. *J Neurochem* 2004;**90**:989-97.
29. Robinson MB, Blakely RD, Couto R, Coyle JT. Hydrolysis of the brain dipeptide N-acetyl-L-aspartyl-L-glutamate. Identification and characterization of a novel N-acetylated alpha-linked acidic dipeptidase activity from rat brain. *J Biol Chem* 1987;**262**:14498-506.
30. Slusher BS, Robinson MB, Tsai G, Simmons ML, Richards SS, Coyle JT. Rat brain N-acetylated alpha-linked acidic dipeptidase activity. Purification and immunologic characterization. *J Biol Chem* 1990;**265**:21297-301.
31. Urazaev AK, Grossfeld RM, Fletcher PL, Speno H, Gafurov BS, Buttram JG, Lieberman EM. Synthesis and release of N-acetylaspartylglutamate (NAAG) by crayfish nerve fibers: implications for axon-glia signaling. *Neuroscience* 2001;**106**:237-47.
32. Arun P, Madhavarao CN, Hershfield JR, Moffett JR, Namboodiri MA. SH-SY5Y neuroblastoma cells: a model system for studying biosynthesis of NAAG. *Neuroreport* 2004;**15**:1167-70.
33. Lu ZH, Chakraborty G, Ledeen RW, Yahya D, Wu G. N-Acetylaspartate synthase is bimodally expressed in microsomes and mitochondria of brain. *Brain Res Mol Brain Res* 2004;**122**:71-8.
34. Madhavarao CN, Hammer JA, Quarles RH, Namboodiri MA. A radiometric assay for aspartoacylase activity in cultured oligodendrocytes. *Anal Biochem* 2002;**308**:314-9.
35. Erecinska M, Zaleska MM, Nissim I, Nelson D, Dagani F, Yudkoff M. Glucose and synaptosomal glutamate metabolism: studies with [15N]glutamate. *J Neurochem* 1988;**51**:892-902.

36. Yudkoff M, Nelson D, Daikhin Y, Erecinska M. Tricarboxylic acid cycle in rat brain synaptosomes. Fluxes and interactions with aspartate aminotransferase and malate/aspartate shuttle. *J Biol Chem* 1994;**269**:27414-20.
37. Cheeseman AJ, Clark JB. Influence of the malate-aspartate shuttle on oxidative metabolism in synaptosomes. *J Neurochem* 1988;**50**:1559-65.
38. Bates TE, Strangward M, Keelan J, Davey GP, Munro PM, Clark JB. Inhibition of N-acetylaspartate production: implications for 1H MRS studies in vivo. *Neuroreport* 1996;**7**:1397-400.
39. Signoretti S, Marmarou A, Tavazzi B, Lazzarino G, Beaumont A, Vagnozzi R. N-Acetylaspartate reduction as a measure of injury severity and mitochondrial dysfunction following diffuse traumatic brain injury. *J Neurotrauma* 2001;**18**:977-91.
40. Block W, Jessen F, Traber F, Flacke S, Manka C, Lamerichs R, Keller E, Heun R, Schild H. Regional N-acetylaspartate reduction in the hippocampus detected with fast proton magnetic resonance spectroscopic imaging in patients with Alzheimer disease. *Arch Neurol* 2002;**59**:828-34.
41. Schuff N, Neylan TC, Lenoci MA, Du AT, Weiss DS, Marmar CR, Weiner MW. Decreased hippocampal N-acetylaspartate in the absence of atrophy in posttraumatic stress disorder. *Biol Psychiatry* 2001;**50**:952-9.
42. Martin E, Capone A, Schneider J, Hennig J, Thiel T. Absence of N-acetylaspartate in the human brain: impact on neurospectroscopy? *Ann Neurol* 2001;**49**:518-21.
43. Jessen F, Block W, Traber F, Keller E, Flacke S, Papassotiropoulos A, Lamerichs R, Heun R, Schild HH. Proton MR spectroscopy detects a relative decrease of N-acetylaspartate in the medial temporal lobe of patients with AD. *Neurology* 2000;**55**:684-8.
44. Haley RW, Marshall WW, McDonald GG, Daugherty MA, Petty F, Fleckenstein JL. Brain abnormalities in Gulf War syndrome: evaluation with 1H MR spectroscopy. *Radiology* 2000;**215**:807-17.
45. Gonen O, Catalaa I, Babb JS, Ge Y, Mannon LJ, Kolson DL, Grossman RI. Total brain N-acetylaspartate: a new measure of disease load in MS. *Neurology* 2000;**54**:15-9.
46. Deicken RF, Johnson C, Eliaz Y, Schuff N. Reduced concentrations of thalamic N-acetylaspartate in male patients with schizophrenia. *Am J Psychiatry* 2000;**157**:644-7.
47. Callicott JH, Bertolino A, Egan MF, Mattay VS, Langheim FJ, Weinberger DR. Selective relationship between prefrontal N-acetylaspartate measures and negative symptoms in schizophrenia. *Am J Psychiatry* 2000;**157**:1646-51.
48. Podell M, Maruyama K, Smith M, Hayes KA, Buck WR, Ruehlmann DS, Mathes LE. Frontal lobe neuronal injury correlates to altered function in FIV-infected cats. *J Acquir Immune Defic Syndr* 1999;**22**:10-8.
49. Kalra S, Cashman NR, Genge A, Arnold DL. Recovery of N-acetylaspartate in corticomotor neurons of patients with ALS after riluzole therapy. *Neuroreport* 1998;**9**:1757-61.
50. Jenkins BG, Rosas HD, Chen YC, Makabe T, Myers R, MacDonald M, Rosen BR, Beal MF, Koroshetz WJ. 1H NMR spectroscopy studies of Huntington's disease: correlations with CAG repeat numbers. *Neurology* 1998;**50**:1357-65.
51. Ford CC, Griffey RH, Matwiyoff NA, Rosenberg GA. Multivoxel 1H-MRS of stroke. *Neurology* 1992;**42**:1408-12.

52. Matalon R, Michals K, Kaul R. Canavan disease: from spongy degeneration to molecular analysis. *J Pediatr* 1995;**127**:511-7.

53. Dautry C, Vaufrey F, Brouillet E, Bizat N, Henry PG, Conde F, Bloch G, Hantraye P. Early N-acetylaspartate depletion is a marker of neuronal dysfunction in rats and primates chronically treated with the mitochondrial toxin 3-nitropropionic acid. *J Cereb Blood Flow Metab* 2000;**20**:789-99.

54. Signoretti S, Marmarou A, Tavazzi B, Dunbar J, Amorini AM, Lazzarino G, Vagnozzi R. The protective effect of cyclosporin A upon N-acetylaspartate and mitochondrial dysfunction following experimental diffuse traumatic brain injury. *J Neurotrauma* 2004;**21**:1154-67.

55. Pan JW, Takahashi K. Interdependence of N-acetyl aspartate and high-energy phosphates in healthy human brain. *Ann Neurol* 2005;**57**:92-7.

56. Madhavarao CN, Arun P, Moffett JR, Szucs S, Surendran S, Matalon R, Garbern J, Hristova D, Johnson A, Jiang W, Namboodiri MA. Defective N-acetylaspartate catabolism reduces brain acetate levels and myelin lipid synthesis in Canavan's disease. *Proc Natl Acad Sci U S A* 2005; **102** (14):5221-26.

IDENTITY OF THE HIGH-AFFINITY SODIUM/CARBOXYLATE COTRANSPORTER NaC3 AS THE N-ACETYL-L-ASPARTATE TRANSPORTER

Vadivel Ganapathy and Takuya Fujita[*]

1. INTRODUCTION

N-Acetyl-L-aspartate is synthesized primarily in neurons, continuously released into interstitial space in the brain, and taken up by glial cells for subsequent hydrolysis and use in myelin synthesis.[1-3] The intraneuronal concentration of N-acetyl-L-aspartate is ~15 mM while the extracellular concentration is ~100 μM.[4, 5] The synthesis of N-acetyl-L-aspartate in neurons and the hydrolysis of N-acetyl-L-aspartate in glial cells are consistent with the findings that neurons contain high levels of this compound whereas glial cells contain only low levels. N-Acetyl-L-aspartate is also present in retina where a similar situation may exist with respect to the differential functions of neurons and glia in terms of synthesis and degradation of this compound.[6] The presence of high concentrations of N-acetyl-L-aspartate in neurons is the basis for the clinical utility of measurements of the levels of this compound in the brain as an indicator of surviving neurons under pathological conditions.[3] The physiological function of N-acetyl-L-aspartate in neurons is not known, but the hydrolysis of this compound in glial cells with subsequent metabolic utilization of the released acetate is believed to be essential for optimal myelin synthesis.[1-3] The enzyme responsible for the degradation of N-acetyl-L-aspartate in glial cells is aspartoacylase II.[7] Genetic deficiency of aspartoacylase II leads to Canavan disease, a disorder associated with mental retardation, spongy degeneration of white matter in the brain, and optic neuropathy.[8] Aspartoacylase II is a cytoplasmic enzyme and therefore the extracellular N-acetyl-L-aspartate has to be first transported into glial cells

[*] Vadivel Ganapathy, Medical College of Georgia, Augusta, Georgia 30912, USA. Takuya Fujita, Department of Biochemical Pharmacology, Kyoto Pharmaceutical University, Yamashina, Kyoto 607-8414, Japan.

before the enzyme can have access to this compound. This raises the issue of the cellular processes that mediate the entry of this compound into glial cells. N-Acetyl-L-aspartate exists as a dicarboxylate anion at physiological pH and therefore non-specific diffusion is not likely to be a significant contributor to this process. Therefore, glial cells must possess a transport mechanism for the uptake of this compound. Here we review the evidence for the presence of an active transport system for N-acetyl-L-aspartate in glial cells and for molecular identity of this transport system with the high-affinity sodium/carboxylate cotransporter NaC3 (formerly known as NaDC3).

2. CHARACTERISTICS OF N-ACETYL-L-ASPARTATE TRANSPORT IN ASTROGLIAL CELLS

Sager et al[9] were the first to demonstrate the presence of a specific transport mechanism for N-acetyl-L-aspartate in astroglial cells. Using rat primary cultures of neurons and astroglia, these investigators showed that astroglial cells possess an active transport system for the uptake of N-acetyl-L-aspartate. Neuronal cultures do not have any detectable transport system for this compound that cannot be accounted for by astroglial contamination in these neuronal cultures. The transport system in astroglia is coupled to Na^+. The presence of Na^+ is obligatory for the transport function as removal of this ion from the uptake medium abolishes the transport function almost completely. The system also shows significant dependence on Cl^-, but this anion is not absolutely necessary for the transport function. There is still appreciable transport activity even in the total absence of Cl^-. It seems that this anion is not directly coupled to transport function and that its stimulatory effect may be indirect. The transport system is stereoselective. It prefers N-acetyl-L-aspartate over N-acetyl-D-aspartate as the substrate. Even though N-acetyl-L-aspartate is an amino acid derivative, the transport system does not interact with any of the naturally occurring amino acids. The amino acids tested in these studies as potential substrates for the transport system include taurine (a specific substrate for the β amino acid transport system), alanine, cysteine, threonine, tyrosine, valine, leucine, isoleucine, and histidine (substrates for various neutral amino acid transport systems), glutamate, aspartate, and L-homocysteate (substrates for anionic amino acid transport systems), arginine (a substrate for the various isoforms of cationic amino acid transport system), cystine (a substrate for the amino acid transport system known as $b^{0,+}$), γ-aminobutyrate (a specific substrate for the various isoforms of γ-aminobutyrate transport system), α-(methylamino)isobutyrate (a specific substrate for the amino acid transport system A), and 2-(-)-endoaminobicycloheptane-2-carboxylic acid (a substrate for the amino acid transport system L). Thus, none of the known amino acid transport systems appears to be responsible for the uptake of N-acetyl-L-aspartate in astroglial cells.

Brain also contains N-acetyl-L-aspartyl-L-glutamate (NAAG) at high concentrations (~0.3 mM).[10] NAAG may serve as a precursor for the excitatory amino acid glutamate in the brain. When NAAG is hydrolyzed by a specific peptidase (known as NAALADase for N-acetylated alpha-linked acidic dipeptidase or glutamate carboxypeptidase II or prostate-specific membrane antigen) in the brain, the resultant products are glutamate and N-acetyl-L-aspartate. The presence of a specific Na^+-dependent transport system for NAAG in the brain has been demonstrated.[11, 12] Available evidence indicates that the N-acetyl-L-aspartate transport system in astroglia is different from the transport system responsible for the uptake of NAAG. Competition studies have shown that NAAG inhibits N-acetyl-L-aspartate transport in astroglial cultures with an inhibition constant of

~400 µM.[9] In contrast, neuronal and glial cells take up NAAG via a transport system that shows very high affinity for NAAG with a Michaelis-Menten constant of ~2 µM.[12] Another difference is that while the N-acetyl-L-aspartate transport system is present almost exclusively in astroglia, the NAAG transport system appears to be present both in neurons and in glia.[12]

Taken collectively, the data from the literature show that there is a specific Na^+-coupled transport system for N-acetyl-L-aspartate in astroglia. The transport system is novel, does not interact with other naturally occurring amino acids, and exhibits comparatively low affinity for the structurally related compound NAAG.

3. TRANSPORT OF N-ACETYL-L-ASPARTATE VIA NaC3, A SODIUM-COUPLED HIGH-AFFINITY TRANSPORTER FOR DICARBOXYLATES

Though N-acetyl-L-aspartate is a derivative of the amino acid L-aspartate, it exists as a divalent anion at physiological pH. Since the amino group is acetylated, there is no positive charge on the molecule. Therefore, at the functional level, the novel N-acetyl-L-aspartate transport system identified in astroglial cells is actually a Na^+-coupled dicarboxylate transporter. Three different transporters that function as Na^+-coupled dicarboxylate transporters have been identified at the molecular level in mammals.[13, 14] As a group, these are called Na^+/carboxylate cotransporters (NaCs). This group consists of NaC1 (previously known as NaDC1), NaC2 (previously known as NaCT), and NaC3 (previously known as NaDC3). The corresponding human genes are designated $SLC13A2$, $SLC13A5$, and $SLC13A3$. NaC1 is a low-affinity transporter for dicarboxylates such as succinate, α-ketoglutarate, fumarate, and malate. It can also transport the tricarboxylate citrate but with very low affinity. NaC2 is a transporter with preference for the tricarboxylate citrate, but can transport succinate and other dicarboxylates. NaC3 is a high-affinity transporter for succinate and other dicarboxylates and has comparatively lower affinity for citrate. NaC1 is not expressed in the brain and therefore is unlikely to participate in N-acetyl-L-aspartate transport. In contrast, NaC2 and NaC3 are expressed in the brain.[15, 16] Therefore, these two transporters are candidates for the N-acetyl-L-aspartate transport system.

Detailed studies have been performed to investigate the ability of cloned rat, mouse, and human NaC3s to transport N-acetyl-L-aspartate.[16, 17] These studies have shown unequivocally, using two different heterologous expression systems, that NaC3 is indeed capable of mediating the uptake of N-acetyl-L-aspartate. In mammalian cells expressing the cloned NaC3s, N-acetyl-L-aspartate is able to compete with succinate for the transport process. The inhibition constants (i.e., the concentration of N-acetyl-L-aspartate necessary to cause 50% inhibition of succinate transport) are 59 ± 4 µM, 66 ± 10 µM, and 232 ± 20 µM for rat, mouse, and human NaC3s, respectively. The inhibition of NaC3-mediated succinate transport by N-acetyl-L-aspartate is strictly competitive as expected of a competing substrate for the same transport system.

Inhibition of NaC3-mediated succinate transport by N-acetyl-L-aspartate does not necessarily mean that the inhibitor is actually transported into cells via the transporter. It is possible that N-acetyl-L-aspartate binds to the substrate-binding site and blocks the transport of succinate without itself being translocated across the membrane. In other words, N-acetyl-L-aspartate may be a blocker rather than a transportable substrate. Subsequent studies with *Xenopus laevis* oocytes as the heterologous expression system provided direct evidence that N-acetyl-L-aspartate is indeed a transportable substrate for NaC3.[16] In oocytes, the transport function of the heterologously expressed NaC3 can be monitored directly by inward currents induced by transportable substrates using the two-microelectrode voltage-clamp technique. In this expression system, the interaction of the transporter with transportable substrates can be easily differentiated from interaction with blockers. Exposure of NaC3-expressing oocytes to transportable substrates will induce Na^+-dependent inward currents whereas blockers will not induce such currents. When oocytes expressing human NaC3 were used, exposure of the oocytes to N-acetyl-L-aspartate was found to induce marked inward currents, demonstrating directly that this compound is a transportable substrate for human NaC3.[16] The currents induced by N-acetyl-L-aspartate were obligatorily dependent on the presence of Na^+. The substrate-induced currents were saturable with a Michaelis-Menten constant of 300 ± 40 μM. There was no effect on N-acetyl-L-aspartate-induced currents when the studies were performed in the absence of Cl^-, indicating that the transport process is not dependent on this anion. The induced currents increased in magnitude as the concentration of Na^+ was increased in the medium. The Na^+-activation kinetics showed a sigmoidal relationship between the concentration of Na^+ and the magnitude of currents. The Hill coefficient for this process was approximately 3. Thus, the Na^+ : N-acetyl-L-aspartate stoichiometry for the transport process was 3:1. Such a stoichiometry would predict transfer of one net positive charge into the oocytes because of the dianionic nature of the substrate. This provides the basis for the electrogenic nature of the transport process. The Michaelis-Menten constant for Na^+ was 32 ± 3 mM. The affinities of the transporter for N-acetyl-L-aspartate and Na^+ were found to be influenced significantly by membrane potential. Hyperpolarization of the membrane increased the affinity for both of the cotransported substrates whereas depolarization decreased the affinity. Similarly, the maximal velocity of the transport process as assessed by the inward currents induced by saturating concentrations of N-acetyl-L-aspartate also increased with hyperpolarization and decreased with depolarization. The coupling ratio however remained fairly independent of membrane potential.

A comparison of the features of N-acetyl-L-aspartate transport via cloned NaC3s with those of N-acetyl-L-aspartate transport in cultured astroglial cells highlights significant similarities. Transport of N-acetyl-L-aspartate mediated by NaC3 is coupled to Na^+ as is the transport in astroglial cells. The Michaelis constant for N-acetyl-L-aspartate transport in rat astroglial cell cultures is ~80 μM. This value is similar to the corresponding value for transport via rat NaC3 (~60 μM). The affinity of N-acetyl-L-aspartate for human NaC3 is 4- to 5-fold less than that for rat NaC3. Whether this is the case in astroglial cells is not known because there is no information in the literature regarding the affinity of N-acetyl-L-aspartate for transport in human astroglial cells. The astroglial transport system does not interact with any of the naturally occurring amino acids. Likewise, NaC3 also does not interact with compounds other than dicarboxylates and tricarboxylates. There appears to be some discrepancy in terms of Cl^- dependence between the transport of N-acetyl-L-aspartate via NaC3 and the transport in astroglial cell cultures. Transport in astroglial cells is significantly compromised when Cl^- is removed from the extracellular medium. In contrast, there is no evidence of Cl^- involvement in NaC3-mediated transport. However, this discrepancy may not be real for the following reasons. The dependence of transport on Cl^- in astroglial cells is only partial. This suggests that the transport process is not tightly coupled to this anion. The dependence on Cl^- seen in astroglial cells may be due to some indirect effect. For example, the presence of Cl^- in the uptake medium during measurements in astroglial cells may lead to hyperpolarization caused by the Cl^- diffusion potential and such an effect may stimulate the transport of N-acetyl-L-aspartate. The transport process is certainly electrogenic as evidenced in oocytes. Whether or not the transport of N-acetyl-L-aspartate in astroglial cells is coupled to membrane potential is not known because appropriate studies have not been done to address this issue. In oocytes, the transport of N-acetyl-L-aspartate mediated by NaC3 is absolutely independent of Cl⁻. This is understandable because the transport function in oocytes is measured under voltage-clamped conditions where there are no changes in membrane potential with or without Cl⁻ in the extracellular medium. It appears reasonable to conclude from these findings that NaC3 is most likely responsible for the transport of N-acetyl-L-aspartate in astroglial cells.

4. DIRECT EVIDENCE FOR THE INVOLVEMENT OF NaC3 IN N-ACETYL-L-ASPARTATE TRANSPORT IN GLIAL CELLS

Studies described in the previous section have demonstrated unequivocally that cloned NaC3s are capable of mediating the transport of N-acetyl-L-aspartate in a Na⁺-coupled manner. The functional characteristics of NaC3-mediated N-acetyl-L-aspartate transport show marked similarities to those of N-acetyl-L-aspartate transport in astroglial cells. However, to implicate NaC3 definitively as the transporter responsible of N-acetyl-L-aspartate transport in astroglial cells, the following questions need to be addressed. Is NaC3 expressed functionally in glial cells? Is there a direct interaction between N-acetyl-L-aspartate and the known substrates of NaC3 for uptake in glial cells? What is the role, if any, of NaC2, also a Na⁺-coupled transporter for dicarboxylates and tricarboxylates, in N-acetyl-L-aspartate transport in glial cells?

We have sought to answer these questions in a systematic manner using primary cultures of astrocytes from rat brain.[18] These studies have shown that astrocytes in culture take up succinate by a Na⁺-dependent process. The uptake process is saturable with a Michaelis constant of ~35 μM. Succinate uptake in these cells is inhibitable by N-acetyl-L-aspartate with an inhibition constant of ~140 μM. Citrate is also capable of inhibiting succinate uptake but with an inhibition constant of ~900 μM. The relative affinities of succinate and citrate for the uptake process suggest that NaC3 is likely to be responsible for the process. Reverse transcriptase-polymerase chain reaction (RT-PCR) has confirmed that these cells do indeed express NaC3 mRNA. We then used these cells for characterizing the uptake of N-acetyl-L-aspartate. The cells are able to take up radiolabeled N-acetyl-L-aspartate avidly. The uptake process is saturable with a Michaelis constant of ~110 μM. Succinate is able to inhibit the uptake in a strictly competitive manner. The dependence of N-acetyl-L-aspartate uptake on Na⁺ shows a sigmoidal relatiolship with a Hill coefficient of 3. Thus, the Na⁺ : N-acetyl-L-aspartate stoichiometry is 3 : 1. Furthermore, the uptake is inhibitable by α-ketoglutarate and fumarate. NaC2 is not expressed in these cells as assessed by RT-PCR and by the lack of high-affinity citrate uptake. These studies provide direct answers to the questions raised above and demonstrate unequivocally that NaC3 is indeed responsible for N-acetyl-L-aspartate uptake in astrocytes.

5. CONCLUSIONS

The active transport of *N*-acetyl-L-aspartate in astrocytes is mediated by NaC3, a Na$^+$-coupled transporter belonging to the *SLC13* gene family. Since the uptake of *N*-acetyl-L-aspartate into glial cells is a prerequisite for intracellular hydrolysis of this compound by aspartoacyalse II, a process involved in the utilization of *N*-acetyl-L-aspartate in myelination, we speculate that functional defects in NaC3 may lead to defective myelination. The *SLC13A3* gene coding for human NaC3 is located on chromosome 20q12-13.1 and the structural organization of the gene has been elucidated.[19] The molecular identification of the *N*-acetyl-L-aspartate transport system in glial cells as NaC3 raises the possibility that genetic defects in this gene may lead to inheritable forms of disorders associated with defective myelination.

6. QUESTION AND ANSWER SESSION

SESSION CO-CHAIR COYLE: Thank you. This presentation is open for discussion.

DR. MADHAVARAO: You said that this transporter is found only is astrocytes, but the hydrolyzing enzyme is in oligodendrocytes, and also there is no mention of identification of NAA in astrocytes. So what is the significance of this being in astrocytes?

DR. GANAPATHY: This astrocyte culture that we do, as I have been listening to the talks, we are not characterizing these astrocytes in detail, like type I, type II, oligodendrocytes. All we did was to follow the published procedure for astrocyte cultures where the uptake of NAA had been demonstrated. And then we got these results from such cultures. Other investigators have already shown that there is a transport system for the N-acetylaspartate in this type of astrocyte cultures. So all I can say is that this is a crude culture of astrocytes, and they contain an N-acetylaspartate transport system. Therefore, in future studies, if you can do it in a differential way with the various types of astrocytes, i.e., oligodendrocytes, type I astrocytes, and type II astrocytes, we'll be able to resolve the issue of what is the role of the transport system in these different cell types.

PARTICIPANT: Do you think NAAG has an affinity for this transport system?

DR. GANAPATHY: Other investigators have shown that NAAG has a very low affinity for this transport system, as has been shown in the Journal of Neurochemistry paper.[5]

PARTICIPANT: Okay. Did you try NAAG?

DR. GANAPATHY: That is what we are planning to do now. NAAG is a tricarboxylate. NaCT is actually a tricarboxylate transporter and, therefore, it would be of interest to know whether this transporter can transport NAAG. We are doing those studies now, thinking that NaDC3 is the N-acetylaspartate transport system and NaCT is the NAAG transport system.

DR. ROSS: I have just one question in relation to succinate and citrate, both of which are transported. One thinks that these molecules on the whole are intramitochondrial and are never going anywhere. So what is the role of this transporter in the astrocyte?

DR. GANAPATHY: You will be surprised to know the plasma level of succinate and citrate. Succinate levels in plasma are about 30 micromolar. Citrate levels in the plasma are about 160 micromolar. These concentrations are quite significant.

In the brain, for example, if you take succinate with the 30 micromolar concentration with the N-acetylaspartate 100 micromolar, I would say that the primary function of this NaDC3 in the brain is really to transport the N-acetylaspartate, because you will see very little competition from succinate under in vivo conditions. However, NaDC3 is also expressed in other tissues, such as the placenta and the liver, where N-acetylaspartate may not be the physiological substrate. So, I am not suggesting that with NaDC3, the only function is the transport of N-acetylaspartate. Probably that is what it does in the brain, but it has other functions in the rest of the body, in other peripheral tissues.

PARTICIPANT: I guess this is the only transport talk on this. So could I ask a question about the other essential component of this? Is there any idea of how NAA may be transported out of the axon, whether that is an active process, whether it occurs along the entire axon or at specific locations along the axon?

DR. GANAPATHY: No. I have no idea. Here people were talking about the N-acetyl-aspartate synthesis in the mitochondria. So this transporter has nothing to do with the release of N-acetylaspartate from the mitochondria into the cytoplasm. This transporter is present on the plasma membrane of the cells.

Therefore, we are looking at the transport into cells, rather than the release from the cells. Therefore, the release of N-acetylaspartate from the neuron should be using different mechanisms. It will have nothing to do with NaDC3.

DR. NAMBOODIRI: One question. Your *in situ* studies indicate that the transport is present more in oligodendrocytes and astrocytes. Now, is it possible that with respect to oligodendrocytes, it is transporting into cells, and with respect to astrocytes, it is transporting out of the cells? Do you have any reason to think that this may be what is happening?

DR. GANAPATHY: I don't know. I know very little about the biological functions of N-acetylaspartate because I came to this area because of my interest in transport.

So you all have to help me in trying to understand why this N-acetylaspartate transport system is present in astrocytes as well as in the oligodendrocytes and why the hydrolyzing enzyme is present only in the oligodendrocytes. And, what could be the other roles of N-acetylaspartate in astrocytes?

PARTICIPANT: Your talk I think raises a question of whether the transport mechanism in astrocytes is a means by which NAA actually reaches the circulation. The endfeet of the astrocyte connect with the endothelial cells in the brain capillaries. And, therefore, one needs to transport NAA on the node to the astrocyte.

There are patients who are now receiving succinate who have Canavan's disease. I am wondering now whether succinate is really helpful in that it is inhibiting the transport of NAA out of the nervous system.

PARTICIPANT: Could I just add to Ed's question, too? I'm wondering if you looked at the transporter distribution on other cell types, like ependymal cells and endothelial cells.

DR. GANAPATHY: No. I am not telling you that the transport system is responsible for the release of N-acetylaspartate. It is actually for the uptake into the cells. I say this mainly because it is sodium-coupled, with three sodium ions coupled to the transport of one NAA. It is electrogenic.

It would take a very high concentration of N-acetylaspartate in the cytoplasm in order to reverse this transport function in the opposite direction. Thermodynamically, we can predict that N-acetylaspartate transport via NaDC3 has the ability to concentrate N-acetylaspartate inside the cell more than 1,000-fold, mainly because of the transmembrane sodium gradient and the membrane potential.

Therefore, if you have the N-acetylaspartate concentration inside the cell, say about 1,000 times much higher than what is outside, then you can change the direction of the transport system such that NaDC3 can function in the reverse direction. Only under these conditions, you can release the N-acetylaspartate from the cell to outside via NaDC3.

Now, with succinate therapy, you can argue that, actually, the succinate is taken up into the cell via NaDC3. And then succinate can feed into the citric acid cycle to make oxaloacetate and then aspartate. So, this pathway might feed into the synthesis of the N-acetylaspartate. And that might be the therapeutic beneficial use of the succinate therapy.

I don't know whether you all know that recently a G-protein-coupled receptor for succinate has been identified. When you are treating the patient with succinate, how do we know that the beneficial effect of succinate is actually due to just its transport into the cell? It may be acting on the G-protein-coupled receptor and doing something actually to the biology of the cell, because there is this specific G-protein-coupled receptor for succinate?

So I have been wondering whether the G-protein-coupled receptor for succinate can also interact with N-acetylaspartate. In the brain, for example, the G-protein-coupled receptor might use N-acetylaspartate as a high-affinity ligand. So this is something that the investigators interested in N-acetylaspartate should think about.

SESSION CO-CHAIR COYLE: I think we can open up this discussion to a general discussion for a few minutes.

DR. GANAPATHY: Sure.

SESSION CO-CHAIR COYLE: Do you have time for it?

DR. GANAPATHY: Yes.

DR. MOFFETT: Basically astrocytes make end-feet contacts with capillaries in the brain. Is it possible that extracellular NAA just has to be cleared from the system and astrocytes take it up and transport it to capillaries? Because it is known that NAA is excreted in the urine, in particular, in Canavan disease.

DR. GANAPATHY: In the kidney, the NaDC1 is expressed at a very high level. The NaDC1 has a low affinity for succinate as well as N-acetylaspartate. We lose very little succinate and citrate in the urine. NaDC3 in the kidney is present in the basolateral membrane where it plays a role in the entry of succinate and may be N-acetylaspartate into the tubular cells. NaDC1 in the kidney, on the other hand, is present in the brush border membrane and is responsible for reclamation of the succinate from the tubular filtrate.

Therefore, NaDC1 might be responsible for the reabsorption of N-acetylaspartate in the kidney. Once N-acetylaspartate that is present in the circulation is filtered at the glomerulus, it can be recaptured very effectively by NaDC1. So that would explain, actually, why we don't lose a lot of N-acetylaspartate in the urine. But then, we are talking about the release of the N-acetylaspartate from the astrocytes. I do not know which transport mechanism is responsible for this.

PARTICIPANT: Well, is the system you described reversible?

DR. GANAPATHY: It theoretically is. If N-acetylaspartate inside the cell is very high, for example, or the cells are depolarized thus reducing the driving force the function of the transport system as an entry mechanism, then you can reverse the direction of NaDC3, facilitating the release of N-acetylaspartate from the cells. This is possible because all these transport systems are bidirectional. And the direction is completely determined by the driving forces.

Therefore, if we can change the driving force, for example, for the glutamate transport systems such as GLT1, which is the sodium-coupled, you can change the direction of glutamate flux. Normally, glutamate is transported into the neurons or into the glial cells. But you can reverse it with appropriate changes in driving forces. There are certain pathological conditions in which glutamate transport occurs in the reverse mode, i.e., glutamate is released from the neurons, thus increasing excitotoxicity.

Therefore, NaDC3 is not a unidirectional transport system. It is bidirectional, and the direction is purely determined by the driving forces.

PARTICIPANT: Considering that the operative transporter in neurons is NaCT, what would be the effect of succinate on that?

DR. GANAPATHY: If you use the murine neuronal model system, I would expect the NaCT to interact with the citrate as well as the succinate, because rodent NaCT has as high an affinity for citrate as it has for succinate. In the case of human NaCT, the affinity for citrate is much higher than that for succinate. Therefore in human neuronal cultures, you would expect a much higher affinity for tricarboxylate compounds, like citrate, as the preferred substrate.

DR. NAMBOODIRI: I have one question. Is this transporter exclusively on the plasma membrane?

DR. GANAPATHY: Yes.

DR. NAMBOODIRI: Or is it also on other membranes?

DR. GANAPATHY: No. NaDC3 is present only in the plasma membrane because you don't find any sodium-coupled transport system anywhere else, for example, in the mitochondria. In the mitochondria, the inner mitochondrial membrane has a huge number of transport systems. All of them are driven by the proton gradient, which is ideal in the *in vivo* situation because you have a proton-motive force across the innermitochondrial membrane. But, you don't find any sodium-coupled transport system in the innermitochondrial membrane. So the moment you see a transport system which is sodium-coupled, you can think about only the plasma membrane as the most likely location of the transport system.

But, you can find a sodium-coupled transport system that is inside the cell but not functional. But these transport proteins are sitting there only for trafficking purposes because you can modulate that the transporter going into the intracellular store and then into the plasma membrane. Thus, you can modulate the density of the transporter in the plasma membrane acutely without the participation of de novo synthesis of new transporter protein. But you don't find any sodium-coupled transport system that is functional inside the cytoplasm or inside the cell.

DR. ROSS: To change the topic, this is a general discussion, right? When I read the title of your paper, I thought you were going to educate me on NAA transport. My understanding of that title was not the same as yours. NAA transport along the axon is something we have all been taught about. Is somebody going to illuminate me on that subject?

DR. GANAPATHY: You are talking about the transport of the N-acetylaspartate along the axon?

DR. ROSS: Yes.

DR. GANAPATHY: Oh. I have no clue.

SESSION CO-CHAIR COYLE: On that note, I would like to thank the speakers for their very timely presentations and the audience for a vigorous discussion and my colleague Alessandro Burlina, who is the co-chair.

7.   REFERENCES

1.   G. Tasai and J. T. Coyle, N-Acetylaspartate in neuropsychiatric disorders, *Prog. Neurobiol.* **46,** 531-540 (1995).
2.   M. H. Baslow, A review of phylogenetic and metabolic relationships between the acylamino acids, N-acetyl-L-aspartic acid and N-acetyl-L-histidine, in the vertebrate nervous system, *J. Neurochem.* **68,** 1335-1344 (1997).
3.   J. B. Clark, N-Acetyl aspartate: A marker for neuronal loss or mitochondrial dysfunction. *Dev. Neurosci.* **20,** 271-276 (1998).
4.   D. L. Taylor, S. E. C. Davies, T. P. Obrenovitch, J. Urenjak, D. A. Richards, J. B. Clark, and L. Symon, Extracellular N-acetylaspartate in the rat brain: In vivo determination of basal levels and changes evoked by high K$^+$, *J. Neurochem.* **62,** 2349-2355 (1994).
5.   T. N. Sager, A. Fink-Jensen, and A. J. Hansen, Transient elevation of interstitial N-acetylaspartate in reversible global brain ischemia. *J. Neurochem.* **68,** 675-682 (1997).
6.   M. H. Baslow and S. Yamada, Identification of N-acetylaspartate in the lens of the vertebrate eye: a new model for the investigation of the function of N-acetylated amino acids in vertebrates. *Exp. Eye. Res.* **64,** 283-286 (1997).
7.   F. B. Goldstein, Amidohydrolases of brain: enzymatic hydrolysis of N-acetyl-L-aspartate and other N-acyl-L-amino acids. *J. Neurochem.* **26,** 45-49 (1976).
8.   R. Matalon and K. Michals-Matalon, Biochemistry and molecular biology of Canavan disease, *Neurochem. Res.* **24,** 507-513 (1999).
9.   T. N. Sager, C. Thomsen, J. S. Valsborg, H. Laursen, and A. J. Hansen, Astroglia contain a specific transport mechanism for N-acetyl-L-aspartate, *J. Neurochem.* **73,** 807-811 (1999).
10. K. L. Reichelt and E. Kvamme, Acetylated and peptide bound glutamate and aspartate in brain, *J. Neurochem.* **14,** 987-996 (1967).
11. L. C. Williamson and J. H. Neale, Uptake, metabolism, and release of N-[^{3}H]acetylaspartylglutamate by the avian retina, *J. Neurochem.* **58,** 2191-2199 (1992).
12. M. Cassidy and J. H. Neale, Localization and transport of N-acetylaspartyglutamate in cells of whole murine brain in primary culture, *J. Neurochem.* **60,** 1631-1638 (1993).
13. A. M. Pajor, Molecular properties of sodium/dicarboxylate cotransporters, *J.   Membr. Biol.* **175,** 1-8 (2000).
14. D. Markovich and H. Murer, The *SLC13* gene family of sodium sulphate/carboxylate cotransporters, *Pflugers Arch.-Eur. J. Physiol.* **447,** 594-602 (2004).
15. K. Inoue, L. Zhuang, D. M. Maddox, S. B. Smith, and V. Ganapathy, Structure, function, and expression pattern of a novel sodium-coupled citrate transporter (NaCT) cloned from mammalian brain, *J. Biol. Chem.* **277,** 39469-39476 (2002).
16. W. Huang, H. Wang, R. Kekuda, Y. J. Fei, A. Friedrich, J. Wang, S. J. Conway, R. S. Cameron, F. H. Leibach, and V. Ganapathy, Transport of N-acetylaspartate by the Na$^+$-dependent high-affinity dicarboxylate transporter NaDC3 and its relevance to the expression of the transporter in the brain, *J. Pharmacol. Exp. Ther.* **295,** 392-403 (2000).
17. R. L. George, W. Huang, H. A. Naggar, S. B. Smith, and V. Ganapathy, Transport of N-acetylaspartate via murine sodium/dicarboxylate cotransporter NaDC3 and expression of this transporter and aspartoacylase II in ocular tissues in mouse, *Biochim. Biophys. Acta,* **1690,** 63-69 (2004).
18. T. Fujita, H. Katsukawa, E. Yodoya, M. Wada, A. Shimada, A. Yamamoto, and V. Ganapathy, Transport characteristics of N-acetyl-L-aspartate in rat astrocytes: Involvement of sodium-coupled high affinity carboxylate transporter NaC3/NaDC3-mediated transport system. J. Neurochem. **93,** 706-714 (2005).
19. H. Wang, Y. J. Fei, R. Kekuda, T. L. Yang-Feng, L. D. Devoe, F. H. Leibach, P. D. Prasad, and V. Ganapathy, Structure, function, and genomic organization of human Na$^+$-dependent high-affinity dicarboxylate transporter, *Am. J. Physiol.* **278,** C1019-C1030 (2000).

CANAVAN DISEASE: STUDIES ON THE KNOCKOUT MOUSE

Reuben Matalon, Kimberlee Michals-Matalon, Sankar Surendran and Stephen K. Tyring[*]

1. ABSTRACT

Canavan disease (CD) is an autosomal recessive disorder, characterized by spongy degeneration of the brain. Patients with CD have aspartoacylase (ASPA) deficiency, which results accumulation of N-acetylaspartic acid (NAA) in the brain and elevated excretion of urinary NAA. Clinically, patients with CD have macrocephaly, mental retardation and hypotonia. A knockout mouse for CD which was engineered, also has ASPA deficiency and elevated NAA. Molecular studies of the mouse brain showed abnormal expression of multiple genes in addition to ASPA deficiency. Adenoassociated virus mediated gene transfer and stem cell therapy in the knockout mouse are the latest attempts to alter pathophysiology in the CD mouse.

2. INTRODUCTION

Canavan disease (CD) is an autosomal recessive leukodystrophy characterized by spongy degeneration of the white matter of the brain (1,2). Deficiency of the enzyme Aspartoacylase (ASPA) was found to be the basic defect in Canavan disease (3). Aspartoacylase hydrolyzes *N*-acetyl aspartate (NAA) to aspartate and acetate (4). The deficiency of ASPA leads to accumulation of NAA in the brain and other tissues of patients with Canavan disease (5). The clinical features of the disease include

[*] Department of Pediatrics, University of Texas Medical Branch, Galveston, TX 77555; Stephen K. Tyring Department of Dermatology, University of Texas Health Science Center, Houston, TX 77030. Corresponding author: Dr. Reuben Matalon, Pediatrics, Childrens Hospital, UTMB, Galveston, TX 77555-0359. Tel: 409 772 3453, Fax: 409 772 9595, e-mail: rmatalon@utmb.edu.

psychomotor retardation, megalencephaly and hypotonia (2). Abnormalities in the brain of patients with CD include spongy degeneration with swollen astrocytes and elongated mitochondria (1,2,6-8). Aspartoacylase gene has been cloned and mapped to the 17p13-ter region (9). Although CD is observed in various ethnic groups, the disease is most prevalent in Ashkenazi Jewish population (10-19).

Treatment for CD is symptomatic at this time. Treatment with acetazolamide for 5 months reduced the intracranial pressure and did not affect the water concentration and NAA level (20). Aspartoacylase gene transfer into the brain of two children with CD reduced NAA accumulation, but long term follow up is lacking (21).

3. ASPARTOACYLASE GENE

Aspartoacylase gene was cloned and localized on the short arm of chromosome 17 (9,22). The human aspartoacylase gene spans 30 kb, contains five introns and six exons coding for 313 aminoacids, an enzyme with a molecular weight of approximately 36 kDa.

There are two mutations, E285A and Y231X that account for over 98% of the mutations among Ashkenazi Jewish population (14,18,23-25). Carrier frequency for CD among Ashkenazi Jewish populations has ranged from 1/37 to 1/40 (26,27). This high frequency of carriers indicates that routine preventive measure using DNA analysis for the common Jewish mutations is recommended for Ashkenazi Jews. Currently, carrier detection for CD is included in a panel of several diseases with high frequency in the Ashkenazi Jewish population.

The most common mutation in non-Jewish patients is A305E (14). There have been over 40 mutations identified in various ethnic groups (14,17,18). Some mutations in exon 6 of the gene shows polymorphism that does not change the enzyme activity (28,29). Therefore expression studies are important to understand functional significance of mutations (30).

4. ASPARTOACYLASE DEFICIENCY IN CD

Aspartoacylase is one of two amino acylases, aminoacylase I hydrolyses acetate from other aminoacids. Aspartoacylase (aminoacylase II or ASPA) is specific for *N*-acetylaspartic acid (NAA), which is hydrolyzed to aspartate and acetate (4). Aspartoacylase is localized in the oligodendrocytes, with little or no enzyme activity in astrocytes for the oligodendrocyte participation in the formation of a myelin sheath (31). Deficiency of the enzyme results in accumulation of NAA in the brain, causing spongy degeneration. Aspartoacylase deficiency can also be determined in peripheral organs such as lungs and kidneys (14). Aspartoacylase activity in the brain increases with age as mylination progresses (32).

The substrate for ASPA, NAA, is synthesized from acetyl-coenzyme A and aspartate by L-aspartate N-acetyltransferase (33). *N*-Acetylaspartic acid may be involved in osmotic regulation, neuromodulation and lipogenesis during myelination (34). The rapid increase of the NAA in the brain during the active myelination, and the localization of ASPA in oligodendrocytes support a roll for NAA in the myelination process (35,36). Deficiency of ASPA with severe myelin disease further indicates the importance of NAA in promoting intake myelin metabolism.

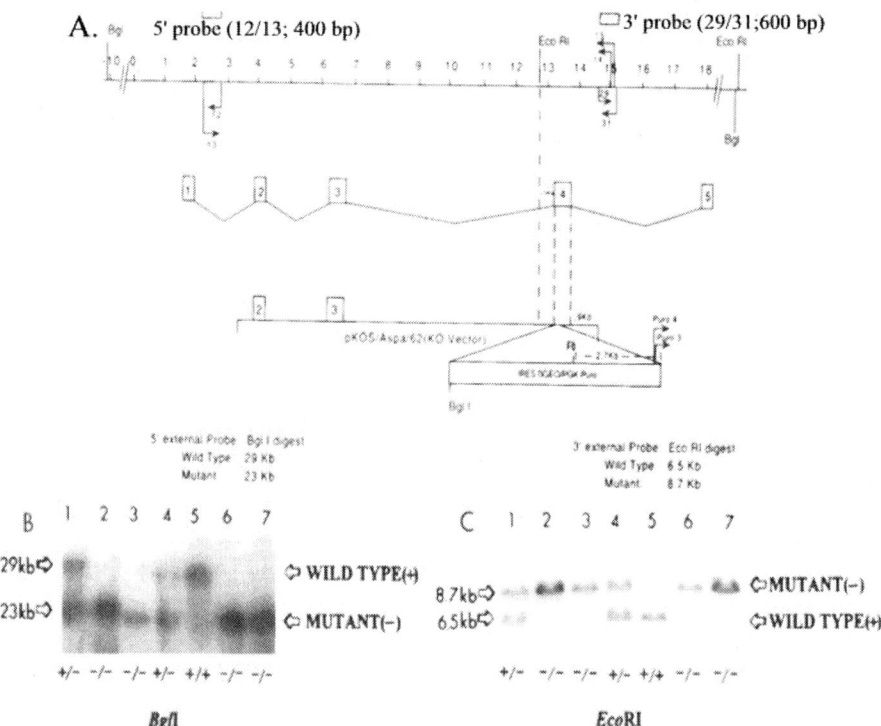

Figure 1. Targeted disruption of the murine aspartoacylase gene. (A). Map of the murine ASPA gene, the targeting vector (pKOS/ASPA/62) restriction enzyme cutting sites. The position of a 400 bp sized probe 5' from the targeting vector and the position of 600 bp sized probe 3' from the targeting vector are shown. The probes were generated by PCR using 12/13 and 29/21 primer pairs. The numbered boxes denote exons, however exon six is not shown in the figure. A targeting vector for deleting 10 bp from exon 4 was engineered. B. Southern blot analysis of BglI digested genomic DNA from F1 intercross progeny using the PCR generated 5' probe. C. Southern blot analysis of EcoRI digested genomic DNA from F1 intercross progeny using the PCR generated 3' probe. The homozygous knockout mouse (Fig 2) showed pathophysiology similar observed in patients with CD including ASPA deficiency, NAA accumulation in the brain, excessive urinary NAA excretion (Table 1) and spongy degeneration of the brain.

The central nervous system contains high level of NAA (36). The concentration of NAA in the human fetus brain in utero is approximately 2.5μmol/g (36), while in the normal adult human brain the level varies 8-10 μmol/g (36-39). The accumulation in the brain of NAA due to ASPA deficiency, and the efflux of this compound form the brain, leads to elevated urinary NAA. Thus, in Canavan disease, patients had urine NAA 1440.5 ± 873.3 μmol/mmol Cr in contrast to 23.5 ± 16.1 μmol/mmol Cr in normal subjects (3,14). Therefore, urinary NAA is one of the markers to determine CD. However, mild elevation of NAA with macrocephaly must not be predicted as Canavan disease, unless spongy degeneration of the brain is confirmed (30).

Wild type   Knockout

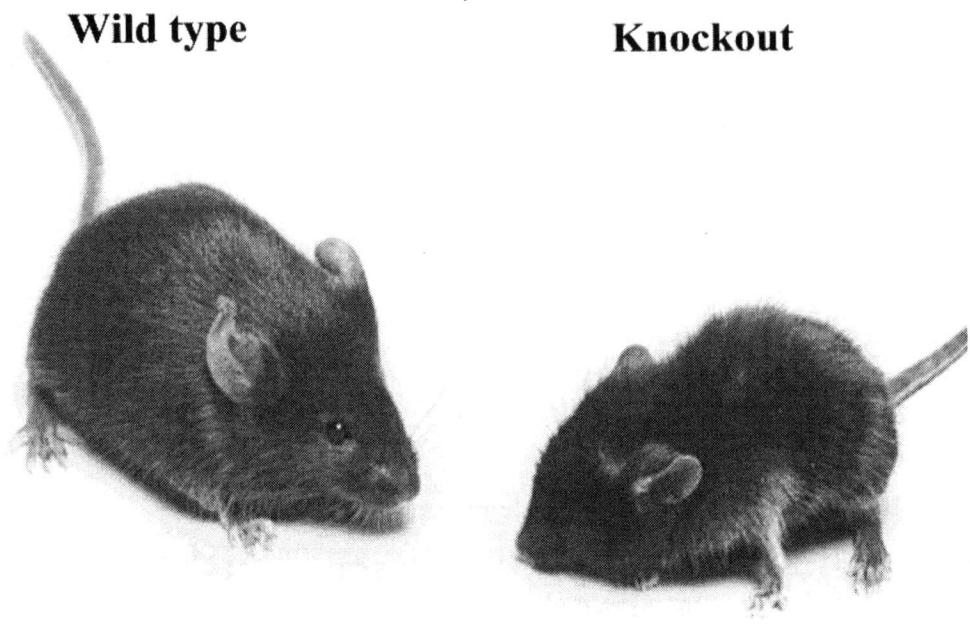

Figure 2. The knockout mouse reflects Canavan's phenotype

Table 1. Aspartoacylase activity and NAA levels in the homozygous knockout mouse and wild type.

	ASPA activity (mU/mg protein)		NAA (ng/mg protein)
	Brain	Kidney	urine
Knockout mouse	0.004 ± 0.004	0.016 ± 0.004	1541± 146
Wild type	0.260 ± 0.015	0.505 ± 0.059	170 ± 24

5. ASPA GENE KNOCKOUT IN THE MOUSE

In order to study the pathophysiology of CD, a knockout mouse for CD has been generated. The knockout mouse was created by deleting 10 base pair of exon four of the murine ASPA gene (Fig. 1) (40). The homozygous knockout mouse (Fig 2) showed pathophysiology similar observed in patients with CD including ASPA deficiency, NAA accumulation in the brain, excessive urinary NAA excretion (Table 1) and spongy degeneration of the brain.

5.1. NAAG Levels in CD Mouse

N-acetylaspartylglutamate is an NAA-glutamate adduct that is synthesized primarily in neurons, is affected in patients with CD. The enzyme, glutamate carboxypeptidase II that cleaves NAAG into NAA and glutamate. In neurons, NAA and glutamate are converted into NAAG in the presence of *N*-acetylaspartate-L-glutamate ligase (41,42). Patients with CD showed approximately twenty-one fold higher NAAG level in the urinary excretion (43,44). However, the knockout mouse brain for CD did not show any difference in the levels of NAAG or the hydrolyzing enzyme, NAALADase (Fig. 3). The ratio of NAAG/Cr. In the knockout mouse brain was 0.032 ± 0.003 compared to that in the wild type, 0.028 ± 0.004 (45). These studies in the knockout mouse brain suggest that ASPA deficiency does not affect NAAG. Therefore it is likely that NAAG seen in the urine of patients with CD may be due to some other factors and not by ASPA deficiency (CD). The activity of NAALADase has also been described in renal tissue (41,46). However, NAAG is located only in the central nervous system suggesting that NAALDase in peripheral tissue may function as house keeping enzyme (47).

5.2. Low Level of Glutamate in the CD Brain and Gene Expression Studies

In the central nervous system, the termination of chemical neurotransmission involves the rapid removal of neurotransmitter, such as glutamate, γ-aminobutyric acid (GABA) from synapse by specific transport systems. Glutamate is generated from the hydrolysis of N-acetylaspartylglutamate (NAAG) by the enzyme, glutamate carboxy-peptidase II (NAALADase). The levels of NAAG and glutamate carboxy-peptidase II were found to be normal in the knockout mouse brain (45). Glutamate transporter, EAAT4 may have combined function of transporter and inhibitor glutamate receptor (48), is mainly located in the cerebellum and its concentrations increase with development (48,49) suggesting an important role of EAAT4 in the developing brain. Abnormal gene expression in the CD mouse brain is shown in Table 2. There is down-regulation of glutamate transporter, EAAT4, which may be associated with the cause for low glutamate level in the CD mouse brain. There is up-regulation of genes involved in glial activation, apoptosis, and inflammatory response and cell death (50).

5.3. Low GABA Levels in CD

Transport of glutamine from astrocytes to neurons is important for the synthesis of γ-aminobutyric acid (GABA). The converted glutamate from glutamine, serves as a precursor for GABA (50). The glutamate product, GABA is lower in the CD mouse brain (40). GABA is a major inhibitory neurotransmitter. The low level of GABA seen in the CD mouse brain is possibly associated with the down-regulation of GABA receptor, GABRA6 (Table 1) to result abnormal cortical excitation. The enzyme associated with the glutamate-GABA pathway, succinyl semiladehyde dehydrogenase (SSADH) was reported with the mental retardation, hypotonia and seizures (52), symptoms also seen in patients with CD. However, the knockout mouse brain did not show any difference from the wild type (Table 3) suggesting that SSADH is not involved in the phenotype seen in CD (53).

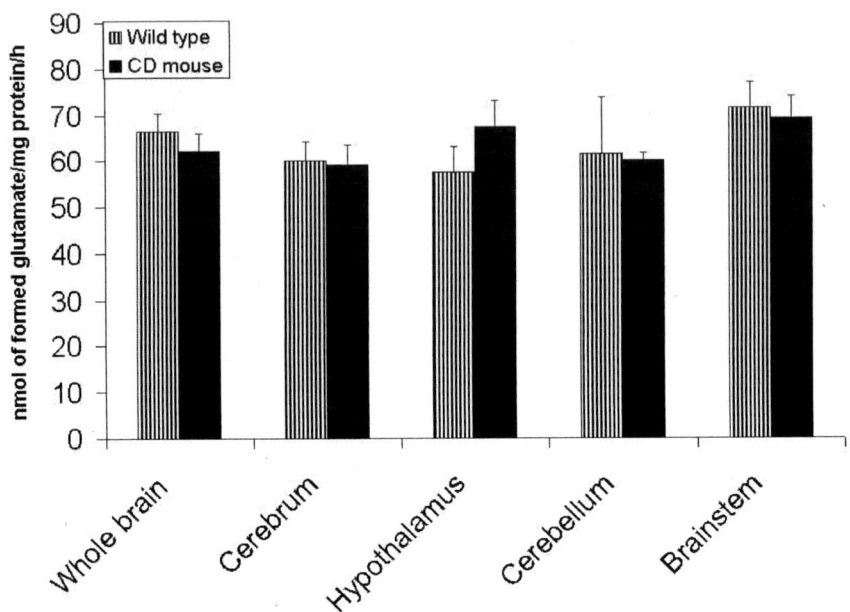

Figure 3. Measurement of glutamate carboxypeptidase II (NAALADase) activity in the brain of knockout mouse for CD. Glutamate carboxypeptidase II activity in the whole brain, cerebrum, hypothalamus, cerebellum and brainstem of knockout mouse did not have significant difference compared to that in the wild type.

Table 2: Aspartoacylase gene knockout leading to abnormal gene expression in the brain of Canavan mouse.

Gene Bank accession no.	Genes	Expression ratio	
Downregulation			
D83262	Glutamate transporter-EAAT4	9.7	▼
AJ222970	Gamma-aminobytyric acid A receptor, subunit alpha 6	119.1	▼▼
AF071562	G-substrate mRNA	16.4	▼▼
Y11682	Mitochondrial ribosomal protein S12	10.7	▼
Upregulation			
D10210	Dao-1d mRNA for D-amino acid oxidase	16.0	▲
X51547	Lzp-s mRNA for lysozyme p	25.4	▲
Y13089	Caspase-11 mRNA	4.4	▲
L28095	Interleukin 1-beta converting enzyme	3.8	▲
M64086	Serine proteinase inhibitor 2	29.8	▲

Table 3. Levels of succinic semialdehyde dehydrogenase and glutamate dehydrogenase in the brain of CD and wild type mice.

Tissue source	NAD$^+$ dependent succinic semialdehyde dehydrogenase (SSADH) activity (mU/mg protein)	Glutamate dehydrogenase (GDH) activity (mU/mg protein)
Wild type:		
Cerebrum	0.27 ± 0.02	3.81± 0.64
Hypothalamus	0.64 ± 0.02	8.29 ± 1.04
Cerebellum	0.47 ± 0.05	8.94 ± 0.74
Brainstem	0.38 ± 0.02	6.9 ± 0.45
CD mouse:		
Cerebrum	0.27 ± 0.01	2.7 ± 0.45
Hypothalamus	0.76 ± 0.04	7.25 ± 1.67
Cerebellum	0.36 ± 0.03	3.9 ± 0.42
Brainstem	0.30 ± 0.03	3.53 ± 0.31

5.4. Altered Expression of Cell Death Influencing Genes

Serine proteinase inhibitor 2 (SPI2), Caspase-11 mRNA and interleukin 1 beta (IL-1β) converting enzyme (Caspase I) expressions (Table 1) are higher in the brain of CD mouse (54). The Spi2 induces apoptosis, gliosis and neuronal death and vacuolization (52,53). Caspase-11 is another important enzyme that induces cell death, up-regulation of this gene expression is seen in the mouse model for amyotrophic lateral sclerosis (54). Caspase-11 mediation of cell death has also been reported in the animal model for multiple sclerosis (55) and rat model of focal cerebral ischemia (56,57). These studies suggest that the high expression of Caspase-11 in the CD mouse brain is likely to induce cell death in the central nervous system. Caspase-11, which leads to the synthesis of the functional form of IL-1 β, which leads to the accumulation of the inactive pro of IL-1 β. Thus, high expression of Caspase-11 seen in the CD mouse brain may be to increase the level of active IL-1 β. Interlukin-1 β belongs to a family of proinflammatory cytokines (58). During neuronal damage or apoptosis, high level of IL-1 β is expressed (58). Thus, abnormal expression of these cell death inducing genes are at least partly, may lead to the sponginess seen in CD (59).

6. THERAPEUTIC APPROACHES FOR CD

So far therapy for CD is symptomatic. Study on therapy for CD is limited. The development of CD model mouse gives the opportunity to evaluate therapeutic approaches for CD (60,61).

6.1. ASPA Gene Transfer to CD Mouse

Recombinant adenoassociated virus (rAAV) is an interesting gene transfer vehicle, because it is nonpathogenic wild type human parvovirus, all viral genes are removed and replaced by the transgene in the replication-incompetent recombinant version, and it has been shown to be almost completely non toxic after CNS delivery in mammals. Therefore, efficacy of AAV-ASPA injection to the CD mouse was examined (62).

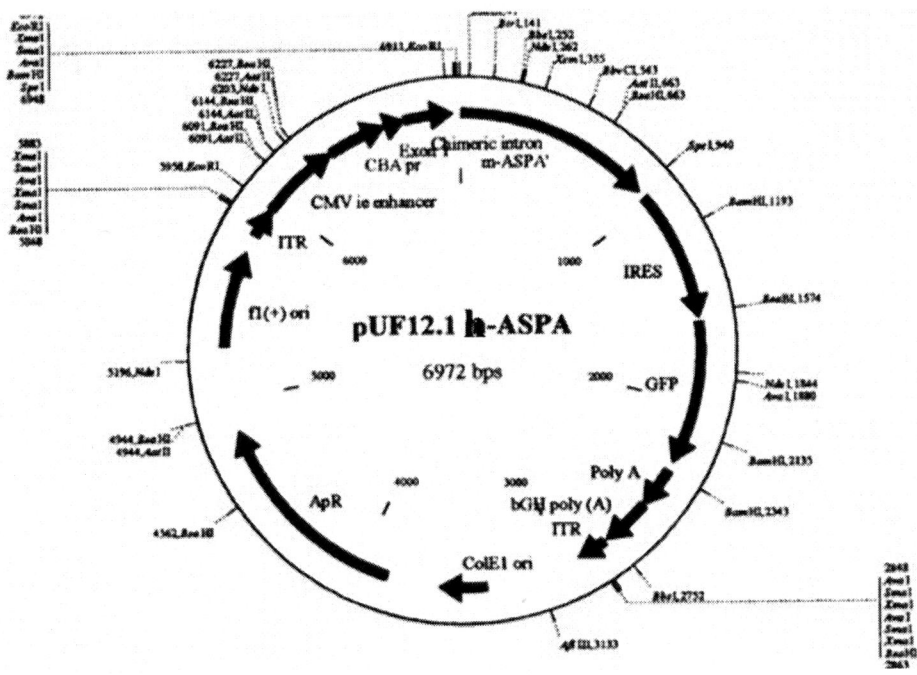

Figure 4. Map of human ASPA containing adeno-associated virus vector.

Recombinant adenoassociated viral construct was prepared at the UFL Gene Vector Core (UFLGTCVC, Gainesville, FL) (Fig.4). Plasmid CBA is a CMV-based promoter with a chicken β-actin intervening sequence. The vector was gel purified and a NotI-digested human aspartoacylase or GFP PCR product was cloned. A double cotransfection procedure was used to introduce a rAAV vector plasmid together with pDG-AAV helper plasmid carrying the AAV rep and cap genes, as well as Ad helper genes required for rAAV replication and packaging at a 1:1 molar ratio. The final preparation was having 4.2 x 1011 infectious units/ml (62).

Two μl of rAAV2-ASPA was stereotaxically injected into the striatum or thalamus of the knockout mouse brain. The gene transfer in the CD mouse brain improved ASPA activity, normalized the NAA level (Fig. 5) and improved spongy degeneration without any side effect even five months after treatment, the maximum period studied. The mean value of ASPA activity in three rAAV-ASPA-GFP injected brains showed 0.116 mU/mg protein, while the uninjected brains had no ASPA activity.

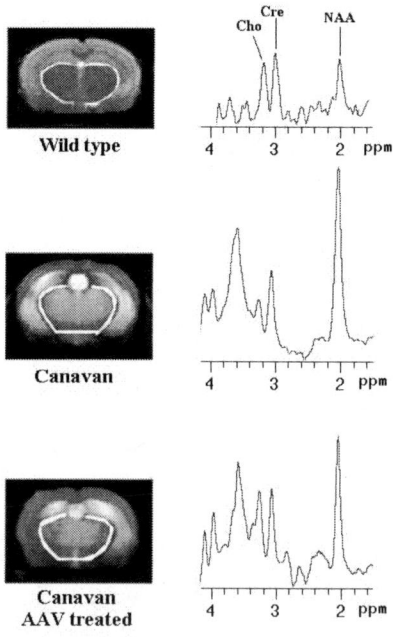

Figure 5. MRI/MRS study of rAAV-hASPA-GFP injected mouse brain. Treatment with AAV-hASPA-GFP normalizes the hyperintensity in the subcortical region, which includes thalamus and subthalamus, in contrast to high signal intensity in the CD mouse. The level of NAA was lower in rAAV-hASPA-GFP treated mouse compared to the CD mouse brain. The reading 2.018 is the location of NAA.

Histopathological study of rAAV-ASPA-GFP injected brain revealed a remarkable decrease in vacuolation in the thalamus AAV, which is probably the structure most affected by the vacuolar spongiform neuropathology in the CD mouse brain (Fig 6A-D). The reduction in vacuolation was seen in 3 as well as 6 months aged mice after rAAV injection. Neuropathology distal to the vector injection sites, such as vacuolation in the hippocampus ot the enlarged ventricles was not affected by the gene transfer (Fig 6 E-H).

6.2. Stem Cell Therapy in the Knockout Mouse

Stem cell therapy is another promising approach for the treatment of neuro-degenerative diseases (63). The source of stem cells may be from various sources including bone marrow, amniotic epithelial cells, embryonic layers and brain tissue (64-66). Of particular interest are neural stem cells (NSCs) or neural progenitor cells (NPCs) that can be readily isolated from the central nervous system of fetal or postnatal animals (67,68). NSCs are self renewing and multipotent whose differentiative capacity is limited to cells of the central nervous system – neurons, astrocytes and oligodendrocytes (69).

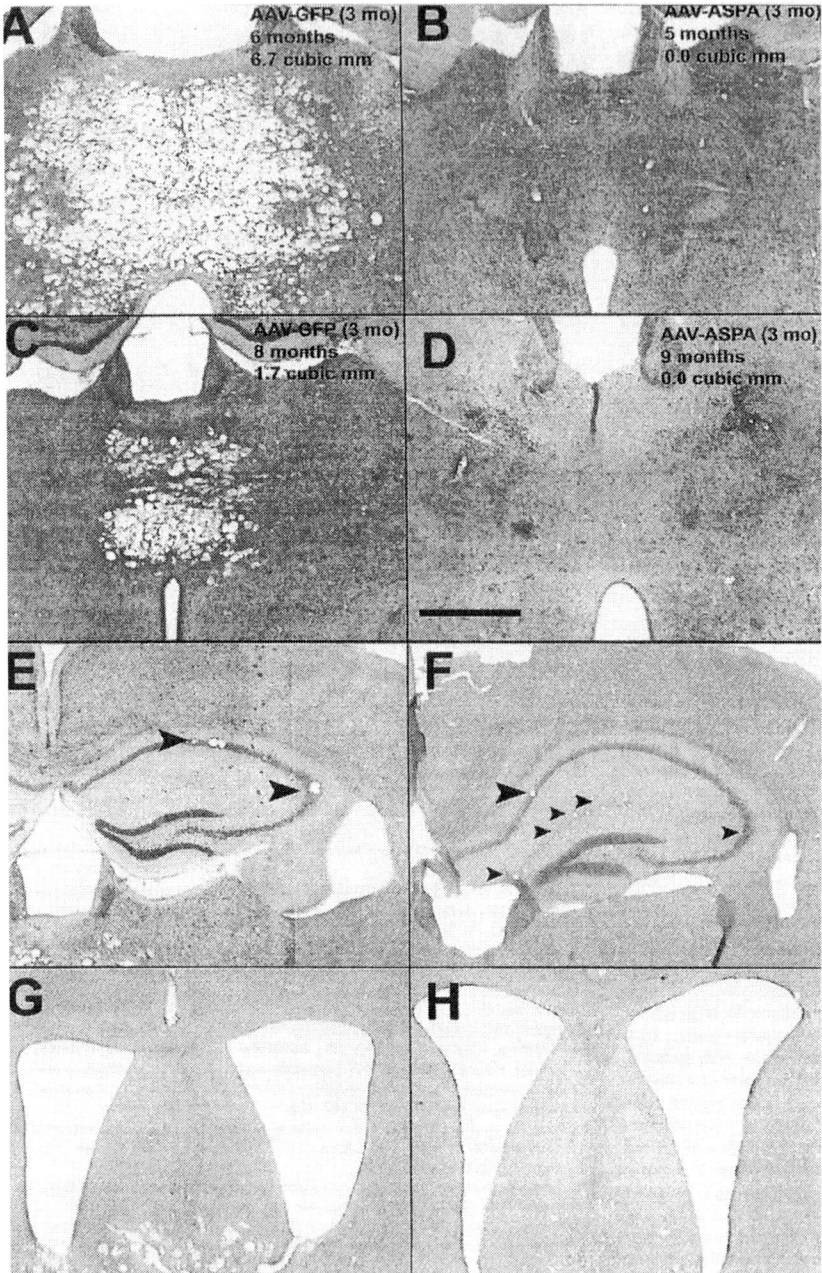

Figure 6. Neuropathology of mouse brain using H&E stain. A. Thalamic vacuolization in an rAAV-GFP-injected 3 months old CD mouse survived until 6 months of age, B. survived until 5 months of age, C. survived until 8 months of age. D.   Thalamus of rAAV-hASPA-GFP injected 3 months old mouse survived until 9 months of age. E. Hippocampus of same mouse shown in A. F. Hippocampus of same mouse shown in B. G. Enlarged lateral ventricle in an rAAV-hASPA-GFP injected mouse shown in A and E.   H. Enlarged lateral ventricle in an rAAV-hASPA-GFP injected mouse shown in B and F.   These studies suggest that ASPA mediated gene transfer can be one of the safest method to treat abnormal ASPA and NAA levels and sponginess seen in CD. Whether ASPA gene transfer recovers secondary effects seen in CD is yet to be studied.

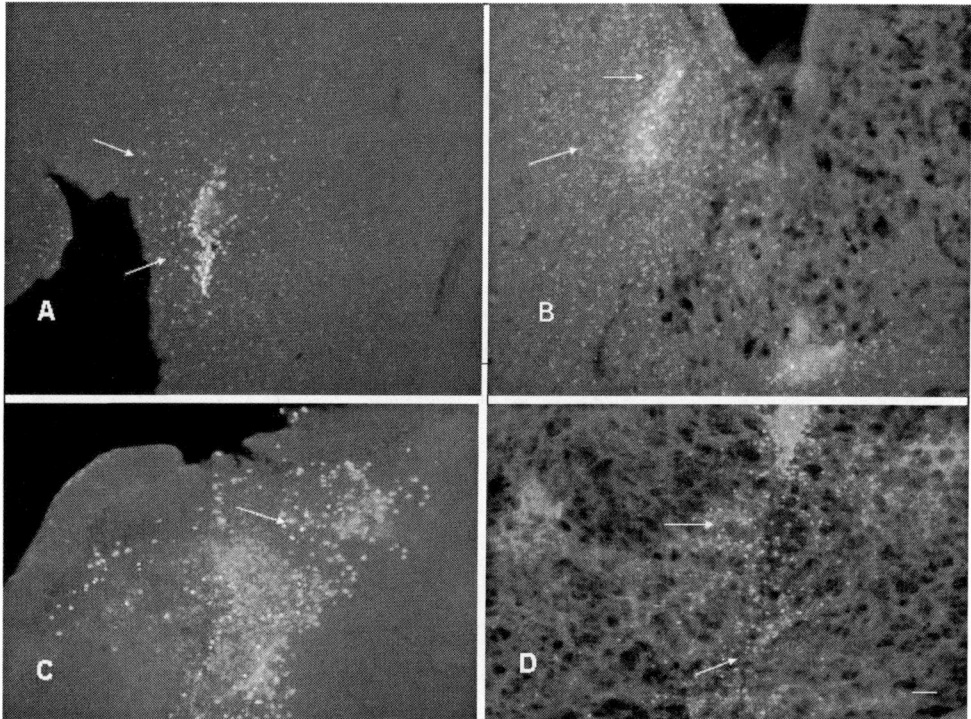

Figure 7. Neural stem cells transplantation to the knockout mouse brain. The fluorescent cells are BrdU positive implanted cells in the (A) wild type forebrain, (B) knockout mouse forebrain, (C) wild type cerebellum and (D) knockout mouse cerebellum mouse brain. Following implantation, neural stem cells survive and migrate widely thoughout the brain and differentiate into neuronal and glial cells in an appropriate site-specific manner (Snyder et al., 1997; Na et al., 1998; Brüstle et al., 1999). Our study showed that transplantation of mouse NSCs into the knockout mouse brain survived for one month, the maximum period studied (Fig. 7) (Surendran et al., 2004b). These studies suggest that stem cell therapy is an important method to replace the lost cells in CD.

Since white matter cells are lost in CD, NSC transplantation were performed in order to test the survival and therapeutic effect of such cells.

Following implantation, neural stem cells survive and migrate widely thoughout the brain and differentiate into neuronal and glial cells in an appropriate site-specific manner (64,66,70). Our study showed that transplantation of mouse NSCs into the knockout mouse brain survived for one month, the maximum period studied (Fig. 7) (71). These studies suggest that stem cell therapy is an important method to replace the lost cells in CD.

7. CONCLUSIONS

Canavan disease is caused by ASPA deficiency, which results in accumulation of NAA in the brain. Aspartoacylase gene knockout in the mouse alters other genes expression without affecting NAAG. Treatment with rAAV2-ASPA resulted in increase ASPA activity over 3 months period. The levels of NAA in brain decreased and the spongy degeneration improved. Stem cell therapy was tried, but did not show a marked effect, most likely because of the small number of cells received. Gene expression studies may be of help in the follow up of therapeutic experiments.

8. QUESTION AND ANSWER SESSION

SESSION CO-CHAIR KOLODNY: Questions for Dr. Matalon? Paolo?

DR. LEONE: Thank you, Reuben. Paola Leone, Robert Wood Johnson Medical School. Thank you, Reuben, for summarizing so many years of work in this presentation. So I have a lot of questions, but I will just make one or two questions of the handouts that I have. Perhaps later I will have more one on one.

The first question is I noticed immediately in the construct, you have an IRS sequence with the GFP expressing gene that is very useful to track down expression of ASPA. But, as you and the Mendel group, the Florida group, know, because of the system that you have, this may backfire; one, in the concentration of the AAV that you have, the final concentration, because the gene that you have is pretty big, having two genes, the ASPA and the GFP. Also, you might have even less expression.

So there is a possibility -- and I will get to the question -- that using just aspartoacylase gene might have even emphasized increase the positive results that you had on NAA and perhaps in the diffusion of the vector itself to further areas.

So did you look at any of the subgroups just using the aspartoacylase gene or did you use always in all experimental groups the ASPA IRS GFP?

DR. MATALON: No.

DR. LEONE: You didn't? Okay.

DR. MATALON: We did just the ASPA.

DR. LEONE: And presumably it was an AAV II. Perhaps the Florida group may also help suggest the point or later on to use different serotypes to look at different effusion of the gene.

As you know, there are several reports showing that different AAVs or hybrids might have better diffusion since you got to the point of what vector to use. And the serotype may make a difference.

DR. MATALON: I don't know.

DR. LEONE: You don't know?

DR. MATALON: What we are waiting for now, is the use of another vector. We are working on a different system to get better widespread expression. We are currently trying methods to spread the vector better. When we get those done, then we can plan to show, hopefully, that may lead to better therapy. Not just ASPA itself, but the other parameters, not just showing ASPA in certain areas of the brain. Those other parameters are important to be used for follow-up, like glutamate, GABA, other genes, and so forth. This is our plan at this moment.

DR. LEONE: And the last question that I have is just a comment about the neural stem cell status that you have. I just flew back from the myelin projects meeting last night. The focus of that meeting is really stem cell therapy for muscular dystrophies and multiple sclerosis. There were a lot of talks that looked at just engraftment, any differentiation of embryonic cells, the stem cells, neural stem cells, and the data that you show are pretty good, but I think if you look at, for instance, in autologous cells, rat versus rat, instead of human versus rat, or even looking at the possibility of immunosupressing the animals.

I'm saying you could also in plane with several of the variables emphasize the engraftment, perhaps the efficacy of the stem cells in the neural mouse model.

Thank you.

DR. MATALON: Thank you. I think neural stem cell is what I summarized here with one slide. It's a little bit of a different bird, it depends on the origin we use. For example, I didn't tell you neural stem cell that we received from Harvard, from Dr. Snyder. There was no activity at all.

Those with some activity which I showed, were done in collaboration with the group from Genzyme, Greg Stuart and others. And those stem cells show ASPA activity and differentiate to oligodendrocytes. So I wouldn't close the book on such experiments, but one has to try with different types, different origins of stem cells.

PARTICIPANT: I just want to add to that last question. Could you be more specific in terms of the stem cells that you did use, not the Harvard cells but the Genzyme cells, the source and also what, if anything, you did to overexpress ASPA with those cells or if they are just near the stem cells?

Could you describe the etiology of the stem cells that you used? What were the genotype cells that you used? What source.

DR. MATALON: I really don't have the detail. The stem cells we got from Genzyme were adult mouse brain stem cells. The ones we got from Harvard were, I believe were embryonal rat stem cells. But I don't know more than that.

SESSION CO-CHAIR KOLODNY: Dr. Philippart?

DR. PHILIPPART: Philippart, Los Angeles.

Reuben, it's a beautiful, convincing document, the improvement of the sponginess in your treated mice, but how about some other document, like behavioral testing or NAA level in the brain?

DR. MATALON: NAA went down. I mentioned that to you. We haven't checked other parameters, like glutamate didn't go up. We tried to get glutamate up, even in patients with Canavan. We did some of the work with you and Brian Ross. We did not check behavioral changes, because the experiment was for short time.

The reason we didn't go into the gene expression at the DNA level or the mRNA level. I felt at least I need better spread-out therapy in order to go into that area.

SESSION CO-CHAIR KOLODNY: We'll take one last question. Then we need to go on.

SESSION CO-CHAIR NAMBOODIRI: I have a question. Now, where do you see the expression? In the neuron? Oligodendrocytes? Astrocytes? Where is it expressed? Is it expressed in the neurons?

DR. MATALON: When we measure N-acetylaspartate we do it in large areas of the brain and not specific cells.

SESSION CO-CHAIR KOLODNY: Thank you very much, Dr. Matalon.

9. REFERENCES

1. Canavan, M.M., 1931, Schilder's encephalitis periaxialis diffusa. *Arch. Neurol. Psychiatry* 25, 299-308.
2. van Bogaert, L., and Bertrand, I., 1949, Su une idiotie familiale avec degenerescence spongieuse de neuraxe (note preliminaire), *Acta. Neurol. Belg.* 49, 572-587.
3. Matalon, R., Michals, K., Sebesta, D., Deanching, M., Gashkoff, P., and Casanova, J., 1988, Aspartoacylase deficiency and N-acetyl-aspartic aciduria in patients with Canavan disease, *Am. J. Med. Genet.* 29, 463-471.
4. Birnbaum, S.M., Levinton, L., Kingsley, R.B., and Greenstei, J.P., 1952, Specificity of aminoacid acylases, *J. Biol. Chem.* 194, 455-462.
5. Matalon, R.M., Michals-Matalon-K., 2000, Spongy degeneration of the brain, Canavan disease : biochemical and molecular findings, *Front Biosci,* Mar 1 :5 :D307-11.
6. Globus, J.H. and Strauss, I., 1928, Progressive degenerative subcortical encephalopathy (Schilder's disease), *Arch. Neurol. Psychiatry* 20, 1190-1228.
7. Adachi, M., Torii, J., Schneck, L., and Volk, B.W., 1972, Electron microscopic and enzyme histochemical studies of the cerebellum in spongy degeneration (van Bogaert and Bertrand type), *Acta Neuropathol.* (Berl) 20, 22-31.
8. Adornato, B.T., O'Brien, J.S., Lampert, P.W., Roe, T.F., and Neustein, H.B., 1972, Cerebral spongy degeneration of infancy: a biochemical and ultrastructural study of affected twins, *Neurology* 22, 202-210.
9. Kaul, R., Balamurugan, K., Gao, G.P., and Matalon, R., 1994a, Canavan disease: genomic organization and localization of human ASPA to 17p13-ter: conservation of the ASPA gene during evolution, *Genomics* 21, 364-370.
10. Banker, B.Q.,. and Victor, H., 1979, Spongy degeneration of infancy, in: Genetic diseases among Ashkenazi Jews, R.M. Goodman., and A.G. Motulsky., Eds., Raven Press, New York, pp. 201-217.
11. Matalon, R., and Michals-matalon, K., 2000, Spongy degeneration of the brain, Canavan disease: biochemical and molecular findings, *Front. Biosci.* D, 307-311.
12. Matalon, R., Michals-Matalon, K, 1998, Molecular: basis of Canavan disease, *Eur J Paediatr Neurol.* 2(2):69-76.
13. Ozand, P.T.; Gascon, G., and Dhalla, M., 1990, Aspartoacylase deficiency and Canavan disease in Saudi Arabia, *Am. J. Med. Genet.* 35, 266-268.
14. Matalon, R., Michals-Matalon, K., 1999, Recent advances in Canavan disease, Adv. Pediatr. 46, 493-506.
15. Matalon, R., Matalon, K.M., 2002, Canavan disease prenatal diagnosis and genetic counseling, *Obstet Gynecoi Clin* North Am. Jun:29(2):297-304.
16. Matalon, R., Kaul, R., Casanova, J., Michals, K., Johnson, A., Rapin, I., Gashkoff, P., and Deanching, M., 1989, Aspartoacylase deficiency: the enzyme defect in Canavan disease, *J. Inherit. Metab. Dis.* 12, 329-331.
17. Zeng,B.J., Wang,Z.H., Ribeiro,L.A., Leone,P., De Gasperi, R., Kim, S.J., Raghavan,S., Ong,E., Pastores,G.M., Kohodny,E.H., 2002, Identification and characterization of novel mutations of the Aspartoacylase gene in non-Jewish patients with Canavan disease, *J Inherit.Metab. Dis.,* Nov:25(7):557-570.
18. Shaag, A., Anikster, Y., Christensen, E., Glustein, J.Z., Fois, A., Michelakakis, H., Nigro, F., Pronicka, E., Ribes, A., Zabot, M.T., Elpeleg, O.N., 1995, The molecular basis of Canavan (aspartoacylase deficiency) disease in European non-Jewish patients, *Am. J. Hum. Genet.* 57, 572-580.
19. Olsen, T.R., Tranebjaerg, L., Kvittingen, E.A., Hagenfeldt, L., Moller, C., and Nilssen, O., 2002, Two novel aspartoacylase (ASPA) gene missense missense mutations specific to Norweigian and Swedish patient with Canavan disease, *J. Med. Genet.* 39, e55.
20. Bluml, S., Seymour, K., Philippart, M., Matalon, R., and Ross, B., 1998, Elevated brain water in Canavan disease: impact of a diuretic therapy, Proceeding of the 6th International Society for Magnetic Resonance in Medicine, P 171.
21. Leone, P., Janson, CG., Bilaniuk, L, Wang, Z., Sorgi, F., Huang, L., Matalon, R., Kaul, R., Seng, Z., Freese, A., McPhee, S.W., Mee, E., During, MJ.,2000, Aspartoacylase gene transfer to the mammalian central nervous system with therapeutic implications for Canavan disease. *Ann: Neurol.* Jul;48(1):27-38.
22. Kaul, R., Gao, G.P., Balamurugan, K. and Matalon, R., 1993, Human aspartoacylase cDNA and mis-sense mutation in Canavan disease, *Nat. Genet.* 5, 118-123.
23. Kaul, R., Gao, G.P., Aloya, M., Balamurugan, K., Petrosky, A., Michals, K., and Matalon, R., 1994b, Canavan disease: Mutations among non-Jewish patients, *Am. J. Hum. Genet.* 55, 34-41.

24. Kaul, R., Gao, G.P., Michals, K., Whelan, D.T., Levin, S. and Matalon, R., 1995, Novel (cys 125 arg) missense mutation in an Arab patient with Canavan disease, *Hum. Mutat.* 5, 269-271.

25. Kaul, R., Gao, G.P., Matalon, R., Aloya, M., Su, M., Jin, M., Johnson, A.B., Shutgens, R.B. and Clarke, J.T., 1996, Identification and expression of eight novel mutations among non-Jewish patients with Canavan disease, *Am. J. Hum. Genet.* 59, 95-102.

26. Matalon, R., Kaul, R. and Michals, K, 1994, Carrier rate of Canavan disease among Ashkenazi Jewish individuals, *Am. J. Hum. Genet.* 55, A157.

27. Kronn, D., Oddoux, C., Phillips, J. and Ostrer, H., 1995, Prevalence of Canavan disease heterozygous in the New York metropolitan Ashkenazi Jewish individuals, *Am. J. Hum. Genet.* 57, 1250-1252.

28. Alford, R.L., DeMarchi, J.M., and Richards, C.S., 1998, Frequency of a DNA polymorphism at position Y231 in the aspartoacylase gene and its impact on DNA-based carrier testing for Canavan disease in Ashkenazi Jewish population, *Hum. Mutat.* (suppl. 1), S161-162.

29   Propheta, O., Magal, N., Shohat, M., Eyal, N., Navot, N. and Horowitz, M., 1998, A benign polymorphism in the aspartoacylase gene may cause misrepresentation of Canavan gene testing, *Eur. J. Hum. Genet.* 6, 635-637.

30. Surendran, S., Bamforth, F.J., Chan, A.,Tyring, S.K., Goodman, S.I., and Mataon, R.,2003a, Mild elevation of N-acetyl aspartic acid, macrocephaly: diagnostic problem for Canavan disease, *J. Child Neurol.* 18, 809-812..

31. Baslow, M.H., Suckow, R., Sapirstein, V., and Hungund, B.L., 1999, Expression of aspartoacylase activity in cultured rat macroglial cells is limited to oligodendrocytes, *J .Mol. Neurosci.* 13, 47-53.

32. Surendran, S., Matalon, K.M., Szucs, S., Tyring, S.K., Matalon, R., 2003 Metabolic changes in the knockout muse for Canavan's disease implications for patients with Canavan's disease, *J Child Neurol,* Sep: 18(9):611-5.

33. Goldstein, F.B., 1969, The enzymatic synthesis of N-cetyl-L-aspartic acid by subcellular preparations of the rat brain, *J. Biol. Chem.* 244, 4257-4260.

34. Birken, D.L., and Olendroff, W.H., 1989, N-acetyl-Laspartic acid: a literature review of a compound prominent in 1H-NMR spectroscopic studies of brain, *Neurosci. Behav. Rev.* 13, 23-31.

35. Jacobson, K.B., 1957, Studies on the role of N-acetylaspartic acid in mammalian brain, *J. Gene Physiol.* 43, 323-333.

36. Tallan, H.H., Moore, S., and Stein, W.H., 1956, N-acetyl-L-aspartic acid in brain, *J. Biol. Chem.* 219, 257-264.

37. Kato, T., Nishina, M., Matsushita, K., Hori, E., Mito, T., and Takashima, S., 1997, Neuronal maturation and N-acetyl-L-aspartic acid development in human fetal and child brains, Brain Dev. 19, 131-133.

38. Frahm, J., Bruhn, H., Gyngell, M.L., Merboldt, K.-D., Hanicke, W., and Sauter, R., 1989, Localized proton NMR spectroscopy in different regions of the human brain in vivo. Relaxation times and concentrations of cerebral metabolites, *Magn. Reson. Med.* 11, 47-63.

39. Kreis, R., Ernst, T., and Ross, B.D., 1993, Absolute quantitation of water and metabolites in the human brain. Part II: Metabolite concentrations, *J. Magn. Reson.* 102, 9-19.

40. Matalon, R., Rady, P.L., Platt, K.A., Skinner, H.B., Quast, M.J., Campbell, G.A., Matalon, K., Ceci, J.D., Tyring, S.K., Nehls, M., Surendran, S., Wei, J., Ezell, E.L., and Szucs, S., 2000, Knock-out mouse for canavn disease: a model for gene transfer to the central nervous system, *J. Gene Med.* 2, 165-175.

41. Slusher, B.S., Robinson, M.B., Tsai, G., Simmons, M.L., Richards, S.S., and Coyle, J.T. ,1990, Rat brain N-acetylated α linke acidic dipeptidase activity: purification and immunologic characterization, J. Biol.Chem. 265, 21297-21301.

42. Tyson, R.L., and Sutherland, G.R., 1998, Labeling of N-acetylaspartate and N-acetylaspartylglutamate in rat neocortex, hippocampus and cerebellum from [1-13C]glucose, *Neurosci. Lett.* 251, 181-184.

43. Burlina, A.P., Corazza, A., Ferrari, V., Erhard, P., Kunnecke, B., Seelig, J., and Burlina, A.B., 1994, Detection of increased urinary N-acetylaspartylglutamate in Canavan disease, Eur. J. Pediatr. 153, 538-539.

44. Krawczyk, H., and Gradowska, W., 2003, Characterization of the 1H and 13C NMR spectra of N-acetylaspartylglutamate and its detection in urine from patients with Canavan disease, *J. Pharm. Biomed. Anal.* 31, 455-463.

45. Surendran, S., Ezell, E., Quast,MJ, Wei, J., Tyring SK., Michals-Matalon, and Matalon, R., 2004c, Aspartoacylase deficiency does not affect N-acetylaspartylglutamate level or glutamate carboxypeptidase II activity in the knockout mouse brain, *Brain Res.* 1016, 268-271.

46. Blakeley, R.D., Robinson, M.B., Thompson, R.C., and Coyle, J.T., 1988, Hydrolysis of the brain dipeptide N-acetyl-L-aspartyl-glutamate: subcellular and regional distribution, ontogeny and the effect of lesions on N-acetylated alpha-linked acidic dipeptidase activity, *J. Neurochem.* 50, 1200-1209.

47. Stauch, B.L., Robinson, M.B., Forloni, G., Tsai, G., and Coyle, J.T., 1989, The effects of N-acetylated alpha-linked acidic dipeptidase (NAALADase) inhibitors on [3H]NAAG catabolism in vivo, *Neurosci. Lett.* 100, 295-300.
48. Dehnes, Y., Chaudhry, F.A., Ullensvang, K., Lehre, K.P., Storm-Mathisen, J., and Danbolt, N.C., 1998, The glutamate transporter EAAT4 in rat cerebellar purkinje cells: a glutamate-gated chloride channel concentrated near the synapse in parts of the dendritic membrane facing astroglia, *J. Neurosci.* 18, 3606-3619.
49. Furuta, A., Rothstein, J.D., and Martin, L.J., 1997, Glutamate transporter protein subtypes are expressed differentially during rat CNS development, *J. Neurosci.* 17, 8363-8375.
50. Surendran, S., Rady, P.L., Matalon, K., Quast, M.J., Rassin, D.K., Campbell, G.A.,Ezell, E.L., Wei, J., Tyring, S.K., Szucs, S., and Matalon, R., 2003b, Expression ofglutamate transporter, GABRA6, serine priteinase inhibitor 2 and low levels of glutamate and GABA in the brain of knock-out mouse for Canavan disease, *Brain Res. Bull.* 61, 427-435.
51. Petroff, O.A., Hyder, F., Mattson, R.H., and Rothman, D.L., 1999, Topiramate increases brain GABA, homocarnosin and pyrolidone in patients with epilepsy, *Neurology* 52, 473-478.
52. Gibson, KM., Hoffman GF., Hodson, AK., Bottiglieri, T., Jacobs, C., 1988, 4-Hydroxybutyric acid and the clinical phenotype of succinic semialdehyde dehydrogenase deficiency, an inborn error of GABA metabolism, *J. Inherit. Metab. Dis.* 16, 704-715.
53. Surendran, S., Ezell, EL., Quast, MJ., Wei, J., Tyring, SK., Michals-Matalon, K. and Matalon, R (2004d) Mental retardation and hypotonia seen in the knockout mouse for Canavan disease is not due to succinate semialdehyde deficiency, *Neurosci. Lett.* 358, 29-32.
54. Kang, S.J., Sanchez, I., Jing, N., Yuan, J., 2003, Dossociation between neurodegeneration and caspase-11-mediated activation of caspase-1 and caspase-3 in a mouse model for amyotrophic lateral sclerosis, *J. Neurosci.* 23, 5455-5460.
55. Hisahara, S., Yuan, J., Momoi, T., Okano, H., and Miura, M., 2001, Caspase-11 mediates oligodendrocyte cell death and pathogenesis of autoimmune mediated demyelination, *J.Exp.Med.* 193, 111-122.
56. Harrison, D.C., Davis, R.P., Bond, B.C., Campbell, C.A., Jmes, M.F., Parsons, A.A., and Philpott, K.L., 2001, Caspase mRNA expression in a rat model of focal cerebral ischemia, *Brain Res. Mol. Brain Res.* 89, 133-146.
57. Thiemmara, V., Pays, L., Danty, E., Jourdan, F., Moyse, E., and Mehlen, P., 2002, Serine protease inhibitor 2 Spi2 mediated apoptosis of olfactory neurons, *Cell Death Differ.* 9, 1343-1351.
58. Dinarello, C.A., 1998, Interlekin-1 receptors and interleukin-1 receptor antagonist. Int. Rev. Immunol. 16, 457-499.
59. Dandoy-Dron, F., Benboudjema, L., Guillo, A., Jaegly, A., Jasmin, C., Dormont, D., Tovey, M.G., and Dron, M., 2000, Enhanced levels of scrapie responsive gene mRNA in BSE-infected mouse brain, *Brain Res. Mol. Brain Res.* 76, 173-179.
60. Surendran, S., Tyring, S.K., Michals-Matalon, K and Matalon, R., 2004a, Therapeutic options in prevention and treatment of aspartoacylase gene mutation resulting abnormalities in Canavan disease, *Current pharmacogenomics* 2, 13-20.
61. Rothwell, N.J., and Luheshi, G.N., 2000, Interleukin 1 in the brain: biology, pathology, and therapeutic target, *Trends Neurosci.* 23, 618-625.
62. Matalon, R., Surendran, S., Rady, P., Quast, J.J., Campbell, G.A., Matalon, K.M., Tyring,S.K., Wei, J., Peden, C.S., Ezell, E.L., Muzyczka, N., and Mandel, R.J., 2003, Adeno-associated virus-mediated Aspartoacylase gene transfer to the brain of knockout mouse for Canavan disease, *Eur.J.Pediatr.* 153, 538-539.
63. Snyder, E.Y., Deitcher, D.L., Walsh, C., Arnold-Aldea, S., Hartwieg, E.A., and Cepko, C.L., 1992, Multipotent neural cell lines can engraft and participate in development of mouse cerebellum, *Cell* 68, 33-51.
64. Brüstle, O., Maskos, U., and McKay, R.D.G.,1995, Host-guided migration allows targeted introduction of neurons into the embryonic brain, *Neuron* 15, 1275-1285.
65. Sun, S., Guo, Z., Xiao, X., Liu, B., Liu, X., Tang, P.H., and Mao, N., 2003, Isolation of mouse marrow mesenchymal progenitors by a novel and reliable method, *Stem cells* 21, 527-535.
66. Turnpenny, L., Brickwood, S., Spalluto, C.M., Piper, K., Cameron, I.T., Wilson, D.I.,and Hanley, N.A., 2003, Derivation of human embryonic germ cells: an alternative source of pluripotent stem cells, *Stem Cells* 21, 598-609.
67. Rosario, C.M., Yandava, B.D., Kosaras, B., Zurakowski, R., Sidman, L., and Snyder, E.Y., 1997, Differentiation of engrafted multipotent neural progenitors towards replacement of missing granule neurons in meander tail cerebellum may help determine the locus of mutant gene action, *Development* 124, 4213-4224.

68. Snyder, E.Y., Yoon, C., Flax, J.K., and Macklis, J.D., 1997, Multipotent neural precursors can differentiate toward replacement of neurons undergoing targeted apoptotic degeneration in adult mouse neocortex, *Proc. Natl. Acad. Sci.* USA 94, 11663-11668.
69. Brundin, L., Brismar, H., Danilov, A.I., Olsson, O., and Johansson, C.B., 2003, Neural stem cells: a potential source for remyelination in neuroinflammatory disease, *Brain Pathol.* 13, 322-328.
70. Na, R., McCarthy, M., Neyt, C., Lai, E., and Fishell, G., 1998, Telencephalic progenitors maintain anteroposterior identities cell autonomously, *Curr. Biol.* 8, 987-990
71. Surendran, S., Shihabuddin, L.S., Clarke, J., Taksir, T.V, Stewart, G.R., Parsons, G., Yang, W., Tyring SK., Michals-Matalon K., and Matalon, R., 2004b, Mouse neural progenitor cells differentiate into oligodendrocytes in the brain of a knockout mouse model of Canavan Disease, *Dev. Brain Res.* 2004 Oct 15:153(1):19-27.

FUNCTIONS OF *N*-ACETYLASPARTATE AND *N*-ACETYLASPARTYL-GLUTAMATE IN BRAIN

Evidence of a Role in Maintenance of Higher Brain Integrative Activities of Information Processing and Cognition

Morris H. Baslow and David N. Guilfoyle*

1. INTRODUCTION

1.1. Brain function

The brain is a complex information-processing organ that cannot see, smell, hear, taste or feel. For these things it relies on meaningful encoded electrophysiological signals that originate in a variety of environmental neuronal sensors and without which input, normal cognitive abilities cannot be sustained. The brain's metabolic lifeline to the environment is the vascular system, upon which it relies for a continuous supply of nutrients, and for removal of waste products. Thus, the well being of both the brain and the whole organism depend on continuous interactions between it, its environmental sensors, and the vascular system. Superimposed on basic individual neuronal requirements for energy and waste product removal, is a very complex neuronal network that receives, interprets and responds to brain messages of both external and internal origin. At the most integrated level of neural networking is the realm of higher cognitive functions.

1.2. Brain information coding

The method used by a neuron to transmit information is by generation a series of wave-like "spikes" or depolarization's of the plasma membrane. These spikes are interspersed with relative refractory periods during which time the membrane is re-

---

* Nathan S. Kline Institute for Psychiatric Research, 140 Old Orangeburg Road, Orangeburg, New York 10962 (USA). Phone: 845-398-5530, Fax: 845-398-5531, E-mail, Baslow@nki.rfmh.org

polarized using energy derived from ATP. The encoded information transmitted is in the form of these spike-refractory period sequences, and at neuron-neuron synaptic interfaces the information is translated into chemical neurotransmitters for network processing. The spike-generation process is also metabolically costly and requires that ATP supplies be constantly replenished or the timing of the spike-refractory periods will be altered, and meaningful encoded information lost.

1.3. Brain- vascular system interactions, a neuron control mechanism hypothesis

There are two relatively simple substances synthesized by, and present in great abundance in neurons in the brains of vertebrates, whose possible functions have been the subject of research efforts over a period of many decades. One is an N-acetylated derivative of L-aspartic acid (Asp), *N*-acetyl-L-aspartic acid (NAA), and the other, a dipeptide derivative of NAA, *N*-acetylaspartylglutamic acid (NAAG), in which L-glutamic acid (Glu) is joined to the Asp moiety via a peptide bond.

In this review, evidence is presented, based on results of studies in a number of different scientific disciplines, which provides new and unique perspectives from which to evaluate the physiological roles of NAA and NAAG, and allows development of insights into their importance in brain function. This evidence leads to the conclusion that NAA and NAAG operate as a linked metabolic system, which functions as a homeostatic neuronal control mechanism that interacts with the vascular system to remove metabolic water and supply energy in order to maintain the ability of neurons to receive and transmit meaningful encoded information. In addition, it has become clear that any attempt to understand the roles of NAA and NAAG separately, or to try to deduce their linked roles based on analysis of results obtained in only a few disciplines, would probably not be successful.

2. DISCUSSION

In this section, the elements involved in developing the neuron-vascular control mechanism hypothesis are presented in a logical sequence, with each step evaluated by considering what purpose is served by the particular process or metabolic step involved. Indeed, what follows is more of a description of events than a hypothesis, a description that leads to the conclusions that form the basis of a possible understanding the roles of NAA and NAAG in the brain.

2.1. Neuron requirements for energy and for waste metabolic product removal

The basic function of a neuron is to communicate, and they do this by generating intracellular electrophysiological signals in the form of action potentials or spikes. The spikes are then translated into intercellular neurochemical signals transmitted to other neurons at synapses, and subsequently interpreted at some level in the CNS neural network (Clifford and Ibbotson, 2000). These energy-dependent spike trains are both ephemeral and transient in nature, and a neuron must be able to quickly indicate its needs for increased energy supplies, and for waste metabolic product removal in order to sustain this spiking activity. Signal transmission is also a metabolically expensive process with energy demands tightly coupled to encoding of information (Smith et al., 2002).

The timeframe in which normal information processing activities must operate is short, and is bounded by the time between neuronal action potentials measured in ms (Clifford and Ibbotson, 2000; Mackel and Brink, 2003) and changes in rate of energy supply and waste removal in seconds (Vanzetta and Grinvald, 1999). Within this timeframe, dynamic metabolic feedback interactions between neurons and the vascular system, changes in rates of neuronal energy supply and ATP production, and energy-dependent neuronal re-polarization all occur. Since the ATP energy required by neurons increases directly with spike frequency, use of stores of ATP to generate an increase in spiking must be replaced by an equivalent increase in Glc oxidation and ATP production. Without a timely replenishment of ATP supplies, it can be anticipated that neuron re-polarization and absolute refractory periods would increase, reducing the maximum spike frequency and thus altering information processing. On the global level, this could be associated with a loss of cognitive ability. The waste metabolic products of Glc oxidation are CO_2 which can exit neurons down its gradient, and water, which must be removed against a water gradient.

2.2. Measures of neuron-vascular system interactions

As neurostimulation is a metabolically expensive process, it requires the continuous supply of energy in the form of Glc and oxygen from the vascular compartment. One method used to evaluate neurostimulated areas of the brain is by measuring changes in vascular oxygen levels. This can be measured locally using magnetic resonance (MR), based on magnetic properties of blood, which are in turn dependent on the oxygenation state of hemoglobin (Thomas et al., 2000). Blood oxygen exists in two states, a dissolved but freely diffusible gas, and a bound form associated with hemoglobin in red blood cells. As oxygen gas diffuses down its gradient out of the vascular system, additional bound oxygen in the vascular compartment is dissociated from hemoglobin.

Since deoxygenated hemoglobin is more paramagnetic than oxygenated hemoglobin, it can then act as an intravascular paramagnetic contrast agent. Thus, hemoglobin deoxygenation results in an increased magnetic susceptibility difference resulting in a signal loss in vascular water, caused by changes in the apparent proton transverse relaxation time (T_2*). It has also been observed that in addition to measuring a vascular water T_2* signal loss, these MR changes in intravascular magnetic susceptibility can also be calibrated to obtain values for focal changes in the cerebral metabolic rate of O_2 consumption (Smith et al., 2002; Hyder et al., 2002).

Because the changes in blood oxygenation levels affect the water signal, and are also associated with neurostimulation, this MR imaging technique (MRI) is used to visualize areas of the brain that appear to be stimulated as a function of brain activation tasks. The acronym for these functional MRI (fMRI) measurements is "BOLD" imaging, which stands for blood oxygen level dependent-imaging. The basis of the fMRI induced BOLD signal has been recently reviewed (Logothetis, 2003). An initial BOLD effect phase in 1-2 s, which is a measure of the change in oxygenation level of hemoglobin in the vascular system, is separate from a second BOLD phase within 3-12 s, which is linked to hyperemia due to vascular expansion (Logothetis, 2003). However, since both are associated with neurostimulation, the interrelationship and balance between these two parameters are incorporated into most BOLD measurements (Zarahn, 2001).

Using another technique, based on temporal data obtained using an oxygen-dependent phosphorescence quenching method of an exogenous indicator, it has also been observed that the first event after sensory stimulation was a localized increase in O_2 consumption, and this was followed by hyperemia and a more regional increase in blood flow (Vanzetta and Grinvald, 1999). Therefore, it has been suggested that fMRI focused on the earliest initial phase after stimulation would better co-localize with the actual site of neurostimulation.

To summarize, the BOLD effect in response to focal neurostimulation, while complex in origin, reflects two basic elements; an initial increase in deoxyhemoglobin as a result of increased O_2 demand, and a subsequent increase in blood flow to meet this demand. This focal neurostimulation-vascular system interaction serves two purposes for neurons. It increases the vascular sink capacity for waste metabolic water and, at the same time, increases the available supply of energy in areas where it is required by increased neuronal activity, both factors which are important to maintenance of neuron function.

2.3. Synthesis of NAA and NAAG by neurons

2.3.1. NAA

NAA is an amino acid that is present in the vertebrate brain, and in human brain, at about 10 mM, its concentration is among the highest of all free amino acids. Although NAA is synthesized and stored primarily in neurons, it cannot be hydrolyzed in these cells. However, neuronal NAA is dynamic in that it turns over more than once each day by virtue of its continuous efflux down a steep gradient, in a regulated inter-compartmental cycling via extracellular fluids (ECF), between neurons and a second compartment, primarily in oligodendrocytes (Madhavarao et al., 2004), where it is rapidly deacetylated.

The neuronal membrane transport mechanism for NAA into ECF and its oligodendrocyte docking mechanism are presently unknown. In addition, the specific neuronal sites of NAA efflux are unclear. While the NAA synthetic enzyme has only been partially characterized (Madhavarao et al., 2003), the gene for its hydrolytic enzyme, amidohydrolase II (aspartoacylase), in oligodendrocytes has been cloned. The compartmental metabolism of NAA, between its anabolic compartment in neurons and its catabolic compartment in oligodendrocytes, and its possible physiological role in the brain has been reviewed (Birken and Oldendorf, 1989; Baslow, 1997, 2000, 2003).

2.3.2. NAAG

NAAG, a dipeptide derivative of NAA and Glu, is also present in abundance in neurons as well as being present in oligodendrocytes and microglia, and at about 1 mM, it is one of the most abundant dipeptides in the vertebrate brain. NAAG is metabolically unusual in that there are three cell types involved in its metabolism. NAAG that is synthesized in neurons is first exported to astrocytes, via an as yet unknown transport mechanism, where it docks with a metabotropic Glu receptor, and then an astrocyte-specific enzyme located on the astrocyte surface, NAAG peptidase (GCP II,

NAALADase, NAAG peptidase I), hydrolyzes the Glu moiety, which is then taken up by the astrocytes and converted into glutamine (Gln), which is then transported back to neurons. The residual NAA metabolic product then diffuses to oligodendrocytes, where the cell-specific enzyme, amidohydrolase II, removes the acetate (Ac) moiety, which is then taken up by the oligodendrocytes. Finally, the ECF-liberated Asp diffuses back to neurons where it is taken up and subsequently used for recycling into NAA and then into NAAG, completing the cycle.

The nature of the NAAG synthase is presently unknown, but the NAAG peptidase (GCP II) has been cloned (Kozikowski et al., 2004), as has been a second membrane-bound brain astrocyte NAAG peptidase, NAAG peptidase II (GCP III) that has about 6% of the activity of NAAG peptidase I on NAAG (Bacich et al., 2002; Bzdega et al., 2004). The distribution and unusual metabolism of NAAG in brain has also been reviewed (Birken and Oldendorf, 1989; Coyle, 1997; Neale et al., 2000; Baslow, 2000; Karelson et al., 2003).

From these observations, it is clear that neurons synthesize and release NAA and NAAG for the purpose of interacting with oligodendrocytes and astrocytes respectively, and that these metabolic interactions are highly complex and cyclical in nature, with Asp continuously being recycled to neurons. What remains to be clarified are the reasons for this intercellular traffic.

2.4. NAA is structurally and metabolically coupled to Glc metabolism

Using MRS involving combined ^{13} C MRS and [1-^{13} C] glucose (Glc) infusion, the rate of NAA synthesis in the human brain has been measured *in vivo*, and it has been demonstrated that NAA synthesis is both structurally and metabolically coupled to Glc metabolism. The NAA carbon 6 (Ac moiety) is derived from acetyl-coenzyme A (AcCoA), which is derived in turn from [1-^{13} C] Glc metabolism (Moreno et al, 2001). This study connects both the structure, and the rate of synthesis of NAA directly with the rate of Glc energy metabolism in the human brain. Similar results have also been reported in rat brain (Choi and Gruetter, 2001; Henry et al., 2003; Karelson et al., 2003). The synthesis of NAA and oxidation of Glc are also physically connected in that their metabolism occurs in the same organelle, the mitochondrion. The NAA metabolic sequence starting from neuron Glc is unidirectional in nature and is only completed when Asp is regenerated by the action of oligodendrocyte amidohydrolase II.

There are two human inborn errors associated with the Glc-NAA metabolic sequence. One is Canavan disease (CD), a rare inborn error in which amidohydrolase II activity is lacking, and as a consequence, there is a buildup of NAA (hyperacetylaspartia) in brain (Baslow, 2003a). The second is a singular human case of hypoacetylaspartia, where brain NAA may not be synthesized at all (Boltshauser et al., 2004). In both cases the outcomes are profound. In CD, there is extensive neuron-oligodendrocyte and astrocyte pathology, along with severe cognitive dysfunction. In hypoacetylaspartia, the NAA metabolic deficit is also closely associated with cognitive dysfunction, but without any other striking cellular, electrophysiological, or brain macro-structural pathology. Thus, in this inborn error, neuron viability and spiking activity appear to be intact, but without meaningful information being processed. Similarly, in cultured neurons and organotypic brain slices of rats, which represent two levels of brain deconstruction, NAA levels are also significantly reduced or absent, but without evidence of loss of neuron viability or function (Baslow et al., 2003).

2.5. NAA and NAAG are structurally, metabolically and dynamically coupled

NAA and NAAG are also metabolically and structurally connected since Glc metabolism is the source of Ac in both molecules. In rat brain, NAAG is synthesized from NAA and Glu (Tyson and Sutherland, 1998) at a rate of about 0.06 μmol/g/h or about 1 molecule of NAAG for every 10 molecules of NAA synthesized, and under steady-state conditions they are maintained at this ratio. Based on its rate of synthesis, and a brain NAAG content of about 1 mM, the turnover time of NAAG at the calculated rate of 6.0 %/h is 16.7 h, a value very similar to the turnover rate of NAA (14.2 h) in the rat brain. Thus, even though brain NAAG content is lower than that of NAA, their turnover rates are similar, suggesting that there is also a dynamic connection. The NAAG metabolic sequence starting with neuron Glc is also unidirectional, and is also completed when Asp is regenerated from NAA by the action of amidohydrolase II, following the hydrolysis of NAAG to form NAA by astrocyte NAAG peptidase. The relationships between rates of NAA and NAAG synthesis, and rates of efflux and hydrolysis in rat brain are shown in Table 1.

Table 1. Rat brain NAA and NAAG dynamic values [a]

Symbol	Function	Units	Measured/Derived	
			NAA [b]	NAAG [c]
(S)	Rate of synthesis	μmol/g/h	0.60	0.06
(T)	Turnover [K_3/S]	h	14.2	16.7
		%/h	7.1	6.0
(E)	Rate of efflux [=S]	μmol/g/h	0.60	0.06
(H)	Rate of hydrolysis [=S]	μmol/g/h	0.60	0.06
(K_3)	Brain content	μmol/g	8.50	1.0

[a] Based on the NAA metabolic model (Baslow, 2002)
[b] Rate of synthesis in whole brain (Choi and Grutter, 2001). Rate of synthesis of NAA has also been reported in brain homogenates of 30-60 day-old rats, exclusive of cerebella, and in two subcellular components, isolated mitochondrial and microsomal fractions (Lu et al., 2004). In these fractions, S was 3.5, 6.3 and 29.0 nmol/mg protein/h respectively. Assuming a nominal value of about 10 % protein in whole brain homogenates, S in the portion of brain assayed in this study would be about 0.35 μmol/g/h.
[c] Value for NAAG synthesis. Based on NAAG ^{13}C Glu enrichment of 20.0 % (6.0 %/h) by metabolism of [1-^{13}C] Glc, infused over a period of 200 min (Tyson and Sutherland, 1998), and an average brain NAAG content of about 1 mM for rat and other homeothermic vertebrate species (Robinson et al., 1987).

Based on these structural relationships, it is evident that in neurons, Glc, NAA and NAAG form a linked metabolic sequence in which each molecule provides a part of the carbon skeleton used for the formation of the next molecule. Ac is the carbon portion supplied by Glc, which in turn is formed into NAA, and then only used to form NAAG since there is no other anabolic pathway known for NAA. Therefore, in this sequence, not only are their structures connected, but also the metabolic rates of formation of NAA and NAAG are directly coupled to the rate of Glc utilization and energy production. Considering the intercellular catabolic metabolism of NAA and NAAG by oligodendrocytes and astrocytes, and the control of rates of their synthesis in neurons, it is apparent that neurons, via changes in the rate of Glc metabolism, directly control the intercellular trafficking of NAA and NAAG. These metabolic relationships, based on whole brain values, are shown in the following summary equation. However, based on neurons comprising about 50 % of whole brain volume, and having a similar metabolic rate as other brain cells, it is estimated that 200 mols of Glc would be metabolized in neurons, producing 1200 mols of water.

$$400 \text{ Glc} + 2400 \text{ O}_2 \longrightarrow 2400 \text{ CO}_2 + 2400 \text{ H}_2\text{O} + 14400 \text{ ATP} + 10 \text{ NAA} + 1 \text{ NAAG}$$

2.6. Functions of NAA and NAAG

2.6.1. NAA as a water transporting mechanism

The ability of a substance to diffuse down its gradient across a plasma membrane while transporting water against its gradient has been described as a molecular water pump (MWP) (Baslow, 2002). Under normal conditions, Asp present in neurons is acetylated by an NAA synthase in mitochondria (Madhavarao et al., 2003). Since there is no acylase for hydrolysis of NAA in neurons, its concentration builds, reaching a cytosolic equilibrium level of about 20 mM. As each new NAA molecule is then formed from Glc metabolism at a ratio of approximately 1 NAA for each 40 Glc metabolized in human brain (Baslow, 2002; 2003b), a hydrophilic NAA molecule is also released into ECF.

In ECF, NAA diffuses down its gradient to oligodendrocytes where it is rapidly deacetylated to form Asp, most of which is again taken up by neurons from ECF and recycled into NAA (Karelson et al., 2003). The half-life of NAA in ECF, and that of formed Asp is short due to the presence of high levels of amidohydrolase II activity in oligodendrocytes in both white and gray matter (Klugmann et al., 2003), and a rapid uptake mechanism for Asp in neurons. The enzyme liberated Ac is primarily taken up by glial cells where it becomes part of the Ac pool.

The net result of the intercellular NAA cycle is that neuronal water, including metabolic water derived from complete Glc oxidation at a ratio of 6 molecules of water to 1 molecule of Glc, is transported against a water gradient to ECF as each molecule of NAA with a minimum of 32 molecules of obligated water (Baslow and Guilfoyle, 2002) is transported down its gradient. From ECF, after hydrolysis of NAA and recycling of Asp, the released water can then pass into the vascular sink compartment, via astrocyte aquaporin-4 channels and the vascular capillary epithelium, for its eventual excretion (Saadoun et al., 2002). Since removal of metabolic water from a neuron is vital to maintenance of its structural, metabolic, and functional integrity, the MWP function of NAA appears to be an integral component in this process, directly linked to Glc metabolism, the source of most of the neuron's metabolic water.

2.6.2. NAAG as a water transporting mechanism

Evidence from the dynamics of NAAG synthesis indicates that under normal conditions, NAAG made in neurons is liberated to ECF simultaneously with NAA, where they function similarly as MWP's. NAA is specifically targeted to oligodendrocytes, where upon its hydrolysis it liberates at least 32 molecules of water to ECF. NAAG in ECF is specifically targeted to astrocytes where Glu is first removed, releasing about 21 molecules of bound water per molecule of NAAG to ECF. The secondary hydrolysis of NAAG-liberated NAA by oligodendrocytes then releases an additional 32 molecules of bound water thus completing the NAA-NAAG spatial and temporal MWP cycles. As a MWP, NAAG transports about 14 % of metabolic water exported to ECF. The oligodendrocyte enzyme, amidohydrolase II, is obviously of great importance to this process in the brain in that it is the key component, and an irreplaceable mechanism, for the successful operation of both the NAA and NAAG inter-compartmental cycles.

2.6.3. NAAG as an intercellular signal molecule

An additional role in which the astrocyte-targeted NAAG participates is to stimulate one type of metabotropic Glu receptor (mGluR) on the astrocyte surface. NAAG is a selective Group II, mGluR3 agonist (Schoepp et al., 1999), one of the several mGluR's that trigger Ca^{2+} oscillations in neuron-associated astrocytic processes which then spread to astrocytic endfeet in contact with arterioles, where release of vasoactive agents induce vascular expansion and increased local blood flow (Zonta et al., 2003). Thus, the rate of release of NAAG can serve to continuously signal astrocytes about the state of focal neurostimulation, the neurons requirements for vascular Glc or astrocyte lactate (Tsacopoulos, 2002), oxygen, and for increased vascular sink capacity for metabolic water removal. The unidirectional metabolic linkage of NAAG to Glc oxidation via NAA insures that its signal function with astrocytes always indicates the current rate of neuron Glc utilization and the neuron's requirements for replenishment of energy supplies.

2.7. NAA and NAAG operate as a linked metabolic system

The synthesis and simultaneous release of both NAA and NAAG from neurons appears as a continuous, compensatory and coordinated process in brain osmotic and metabolic function. First, metabolic water produced from oxidation of Glc is transported to ECF against a water gradient by both of these substances, and second, the local blood flow is affected by the NAAG-astrocyte mGluR3 signal, a signal resulting in vasodilation with increased availability of Glc and O_2, and an increased ability of the vascular system to function as a sink for exported water and CO_2. The NAAG signal function appears to be a homeostatic mechanism that can quickly respond to any changes in the rate of neuron stimulation, and to resulting changes in local, regional or global circulatory requirements. This is because the rate of synthesis and efflux of NAAG is directly and dynamically coupled to changes in the rate of neuron stimulation, Glc utilization and NAA formation.

2.8. The NAA-NAAG system as a neuronal mechanism used to maintain the ability of the brain to receive and transmit meaningful frequency-coded information

2.8.1. The NAA-NAAG system maintains neurons at peak efficiency

When viewed from an overall brain perspective, it is evident that the NAA-NAAG system functions as a unit, and that these substances play important and complimentary homeostatic and regulatory roles in support of the ever-changing local and global requirements of brain metabolism. Both participate in MWP osmoregulatory activity, and in addition, NAAG has a role in altering brain microcirculation in order to regulate rates of energy supply and waste removal. Thus, the system appears to play a role in maintaining neurons at a level of their peak efficiency and to ensure that energy resources remain adequate at any level of demand.

2.8.2. The NAA- NAAG system interacts with the rate of neuron re-polarization

Attentiveness and responsive performance are key elements in animal survival, and therefore, specific and rapid changes in energy supply to activated areas of the brain are important to both the processing of input information, and to performance-related output responses. In this regard, both NAA and NAAG appear to be integral components in the maintenance of normal brain function, in that they are involved in sustaining the ability of neurons to function at maximum efficiency, and thus to receive, interpret, and transmit meaningful time-coded information. In Table 2, the posited time course and proposed interplay of the NAA-NAAG system with the three phases of neuron-neuron interactions are shown.

In the event of delays in astrocytic responses to chemical signals elicited during neurostimulation, it can be anticipated that the vascular hyperemic response would also be delayed, and the focal energy supply and vascular sink capacity would not be adequately increased. As a result, the vascular system would become less responsive to a neuron's immediate needs, and the lack of sufficient energy for re-synthesis of ATP, the primary energy source for ongoing neuron metabolism (Raichle and Gusnard, 2002), would alter the neuron's ability to send and receive messages. This would be reflected in an increased re-polarization time, an increase in the absolute refractive period, and a decrease in the neuron's maximal rate of firing.

To the extent that NAAG interacts with astrocyte mGluR3 receptors and with astrocytic control of focal hyperemia, it would therefore be directly involved in the maintenance of the maximal rate of neuron firing and as a consequence, with the basic function of neurons which is the transmission of inter-neuronal frequency-encoded communications.

A summary of observed metabolic relationships between NAA and NAAG linked syntheses, and astrocytic mGluR3 signals affecting vascular microcirculation is illustrated in Figure 1, and the likely interaction of the NAA-NAAG system with neuron re-polarization rate and information processing in Figure 2.

Table 2. Functional and temporal association of NAA and NAAG inter-compartmental metabolism with three phases of a neuron-neuron transmitted impulse.

Phase I. Steady-state (a dynamic homeostatic resting state)

    1. Preformed NAA and NAAG are maintained at high cellular-ECF gradients
    2. Cell membrane is polarized (resting potential)
    3. Plasma membrane surface pump-leak systems are operating
    4. Energy producing systems are operating
    5. Preformed ATP is available for use
    6. Preformed transmitter is stored at pre-synaptic efferent membrane

Phase II. Stimulation and signal transmission

    1. Neurotransmitter arrives at post-synaptic afferent receptor
    2. Depolarization is induced and a plasma membrane surface wave (Na+, K+) is carried along the axon
    3. Simultaneous efflux and diffusion of preformed NAAG to juxtaposed astrocyte mGluR3
    4. Pre-synaptic neurotransmitter released to post-synaptic efferent receptor resulting in synaptic signal transmission

Phase III. Recovery (including neuron refractive sub-period)

    1. Neurotransmitter removed from post-synaptic afferent membrane receptor
    2. ATP depletion is coupled with Na/K membrane re-polarization
    3. Increased rate of Glc and O_2 use, and of supply via NAAG-astrocyte induced hyperemia
    4. Glc oxidation and ATP re-synthesis from ADP
    5. Glc oxidation products, water and CO_2 formed at increased rate
    6. Efflux of preformed NAA with its obligated water to ECF and diffusion to oligodendrocytes
    7. Glc coupled NAA re-synthesis from oligodendrocyte amidohydrolase II recycled Asp
    8. NAA coupled NAAG re-synthesis from Glu regenerated from astrocyte surface NAAG-peptidase and astrocyte formed Gln
    9. Pre-synaptic efferent transmitter synthesized and stored
  10. Dynamic steady-state conditions resumed

2.8.3. The NAA-NAAG system and inborn errors in metabolism

In light of the observed osmoregulatory and signal functions of NAA and NAAG, the effects of inborn errors in their metabolism can be assessed. In CD, in the absence of sufficient amidohydrolase II activity, the MWP activities of both NAA and NAAG, and microcirculatory activity of NAAG are compromised. With the global buildup of NAAG in ECF due to a product inhibition effect of NAA on NAAG peptidase activity, the astrocyte mGluR3 receptors are continuously exposed to NAAG signal molecules that cannot be readily hydrolyzed and removed. Thus, in addition to extensive neuron-oligodendrocyte pathology and brain edema in this disease, it is highly likely that any

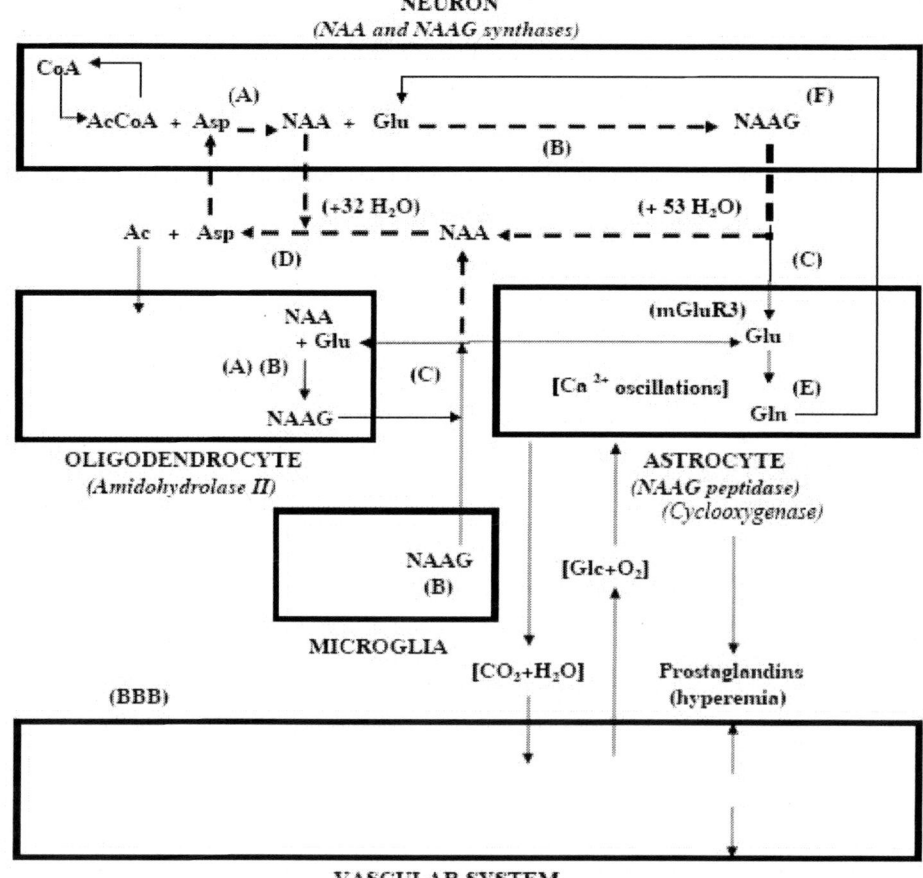

Figure 1. Coupling of NAA, NAAG and Glu metabolism with brain hyperemia, energy availability, and metabolite flux. Enzymes involved in metabolic sequences; (A) NAA synthase, (B) NAAG synthase, (C) NAAG peptidase, (D) Amidohydrolase II, (E) Gln synthase, (F) Glutaminase. From Baslow (2004)

ability of NAAG to control focal or regional hyperemia, and resultant feedback regulation of energy availability and metabolite sink capacity, is severely affected.

In hypoacetylaspartia, the singular known case of apparent lack of brain NAA synthase activity, NAAG is also not present, and the absence of both NAA and NAAG precludes any of their osmoregulatory or microcirculatory functions. Since there are a variety of cellular osmoregulatory mechanisms available, some may be upregulated to make up for the lack of the NAA inter-compartmental system. However, a lack of

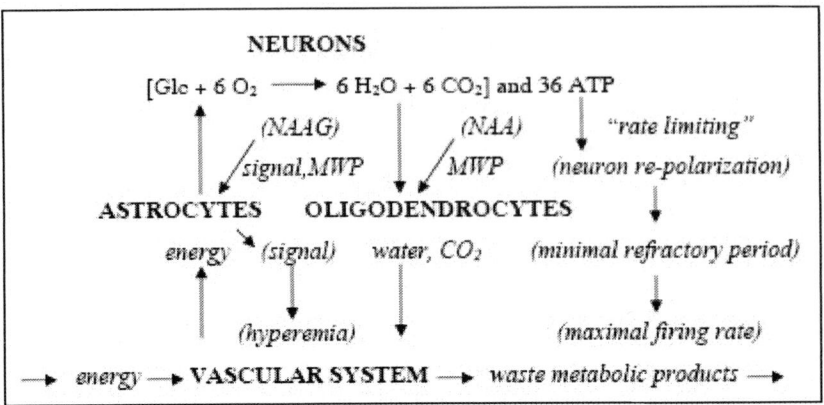

Figure 2. Dynamic interactions of the NAA-NAAG system with neuron energy flux, neuron energy dependent minimal re-polarization and refractory periods, and neuron maximal firing rate [1,2,3,4]

[1] Energy flux used in this example is based on complete oxidation of Glc. However, the neuron energy source may be lactate, which has a lower ATP value. Also, Glc is subject to "proton leak in mitochondria" which would reduce the Glc/ATP ratio to 31 (Attwell and Laughlin, 2001).
[2] Neurons transmit information in the form of single spikes or bursts of spikes, and in either case, the relative refractoriness is related to interspike intervals, and therefore to information processing (Chacron et al., 2001).
[3] Any reduction in rate of energy supply that increases the minimal refractory period reduces the maximum firing rate, thereby truncating a neuron's signaling repertoire at it's highest rates. However, that portion of information transmitted at less than the maximal firing rate would not be affected (Berry and Meister, 1998).
[4] From Baslow, 2004

NAAG, and its specific signal effect on the mGluR3 astrocyte receptor, suggests that an important neuron-microcirculatory control mechanism is inoperative in this case. This could explain why it is in this case, that neurons survive (brain structure is normal), are myelinated (unremarkable MRI), and signal one another (normal EEG), but cannot process information well (Boltshauser et al., 2004).

2.9. Supporting evidence based on enzyme inhibition

There are two lines of enzyme inhibition evidence in support of the MWP and signal hypotheses presented in this review.

2.9.1. Inhibition of amidohydrolase II

It has been proposed that a primary function of NAA is to transport neuron metabolic water to ECF against a water gradient, and that release of the NAA-obligated water in ECF is a function of amidohydrolase II activity. Based on this hypothesis, a predictable outcome of amidohydrolase II inhibition would be a buildup of NAA-water and hydrostatic pressure in ECF. In support of a role for NAA and NAAG as MWP's, are the outcomes of the many natural experiments in which amidohydrolase II (enzyme "D" in figure 1) is inactivated by inborn errors that alter the structure of the enzyme. These result in human CD, a genetic osmotic disease characterized by the buildup of NAA and NAAG in brain, brain cell and ECF edema, and an osmotic-hydrostatic imbalance resulting in increased hydrostatic pressure (with megalocephaly) on the brain side of brain-barrier membranes (Baslow, 2003a). In CD, the primary osmotic defect is that only half of the NAA metabolic sequence (catabolic) is inactivated, which results in a buildup of NAA-water in ECF. In the singular case of hypoacetylaspartia, in which the entire NAA metabolic sequence appears to be absent, there is no evident osmotic component associated with the inborn error.

2.9.2. Inhibition of NAAG peptidase

It has been proposed that a primary function of NAAG is to initiate focal hyperemic responses in the brain in order to provide neurons with required energy components and remove products associated with generation of ATP, and that NAAG peptidase is a key component in this process. If true, a predictable outcome of NAAG peptidase inhibition would be a change in cerebral blood flow. In support of an important neuron-NAAG microcirculatory role via the astrocyte mGluR3 receptor are the results of a murine MRI experiment in which NAAG peptidase (enzyme "C" in figure 1) was inhibited by 2-(phosphonomethyl) pentanedioic acid (2-PMPA). In this study, inhibition of NAAG peptidase resulted in a rapid and sustained global attenuation of T_2* related BOLD signals, an indication of elevated deoxyhemoglobin that is associated with decreased cerebral blood flow (Baslow et al., 2005). These 2-PMPA induced decreases in BOLD signals appear to indicate that blood deoxyhemoglobin is elevated when endogenous NAAG cannot be hydrolyzed, thus linking the efflux of NAAG from neurons and its hydrolysis by astrocytes, to hyperemic oxygenation responses in brain.

3. CONCLUSIONS

As described in this review, there is a complex chain of electrophysiological, metabolic and diffusion events, related to the degree of neuronal activity in the brain, that interact with homeostatic feedback mechanisms to provide neurons with sufficient energy to continuously maintain their full repertoire of signaling capabilities. Maintenance of this signaling ability is the key element in information processing, without which input information cannot be properly interpreted nor appropriate output responses provided.

Neuronal activity requires use of energy in the form of ATP for re-polarization, which is replenished primarily by the oxidation of Glc. It is proposed that the NAA-NAAG system is a neuronal control mechanism that plays an integral role in this process by ensuring both the timely delivery of adequate energy components and removal of waste products. NAA is released to ECF where it diffuses to oligodendrocytes and functions primarily as a MWP. NAAG is also released to ECF, where it diffuses to an astrocyte surface mGluR3 receptor and initiates a Ca^{2+} wave that results in a secondary astrocytic messenger sent to the vascular system that results in focal hyperemia, increasing both the availability of energy and of sink capacity for neuron waste metabolites. It is posited, that as a result of the integrated interactions of neurons, the NAA-NAAG system and the vascular system, that ATP supplies are continuously replenished in neurons, thus maintaining the neurons minimum re-polarization times and absolute refractory periods, and thereby retaining their maximal spike frequency-coding capabilities. This hypothesis has been developed in greater detail (Baslow 2005).

Based on the evidence: (A) that NAA and NAAG are not required for individual neuron survival or viability; (B) that NAA and NAAG participate in the inter-compartmental transport of water; (C) that NAAG is a specific mGluR3 signal molecule associated with focal hyperemia, and; (D) of a unique tri-cellular distribution of NAA and NAAG synthetic and hydrolytic enzymes, the following is a description of how the NAA-NAAG system appears to work, and its apparent role and importance in brain function.

1. NAA and NAAG are structurally and metabolically linked, and their functions involve dynamic cyclic metabolic processes between neurons, astrocytes and oligodendrocytes. It is a unique metabolic system in which the metabolic components are partitioned between various brain cell types, and therefore is only operational in the intact brain.

2. The primary role of NAA is osmoregulatory in that it functions as a MWP to continuously remove neuronal metabolic water to ECF against a water gradient. The cells involved in its inter-compartmental metabolic cycle are neurons and oligodendrocytes.

3. The primary role of NAAG is communication whereby it continuously signals astrocytes about the state of neurostimulation, and of the neurons changing requirements for vascular energy supplies and metabolic waste removal. The cells involved in this signaling sequence are neurons, astrocytes, vascular epithelium and vascular smooth muscle cells, and the product of the stimulatory process is the continuous control of degree of focal hyperemia.

4. The functions of NAA and NAAG are linked and compensatory in nature. Metabolic water, continuously produced as a result of neuron activity and Glc oxidation to renew ATP energy supplies for re-polarization, is rapidly transported to ECF by both substances for removal via the vascular sink. Concurrently, the state of neuronal activity is also continuously signaled to astrocytes by NAAG, which in turn influences the availability of energy supplies and vascular sink capacity by interacting with cells of the vascular microcirculation. The system is dynamic, homeostatic and responsive on focal, regional or global levels in order to serve the ever-changing and ongoing requirements of the brain.

5. Cognitive and other information processing aspects of brain function require a high degree of reliability of complex signal pathways and of temporal signal sequences at specific synapses. These information-coded ephemeral signal sequences are in turn dependent on a stable synaptic architecture and availability of energy. The NAA-NAAG system provides responsive energy and waste removal mechanisms to maintain these conditions. Thus, the operation of the NAA-NAAG system appears to be essential for normal brain integrative activities and higher brain functions in that it supports ongoing and meaningful communication between neurons.

6. The function of the NAA-NAAG system is both subtle and powerful. While it is not required for individual neuron survival, it can, by control of access to energy supplies, affect rates and outcomes of neural information processing and retrieval, which are limiting and vital elements in brain function. Therefore, modest differences in dynamic aspects of the NAA-NAAG system may be an as yet unrecognized, but important factor

underlying elements of brain-motor coordination, the ability to learn, and other normal cognitive functions.

4. ACKNOWLEDGMENTS

Supported in part by a NIH grant from the NIBIB (1 R21 EB004727-01).

5. QUESTION AND ANSWER SESSION

SESSION CO-CHAIR KOLODNY: Thank you, Dr. Baslow.

PARTICIPANT: Thank you, Dr. Baslow. Very elegant presentation, but how do you explain the differential distribution of NAAG and NAA that was presented so elegantly earlier this morning?

DR. BASLOW: We don't know exactly where the NAA comes out of a neuron. We also don't know where the NAAG comes out of a neuron. In addition, we don't know the sites where these substances interact with any of the other cells. However, all around neuron cell bodies and axons you also have oligodendrocytes and astrocytes. In myelinated white matter far from cell bodies there are also unmyelinated nodes along the axon approximately every millimeter, ten for every centimeter, and at each node, you have an oligodendrocyte that can interact with the neuron as well as an end foot of an astrocyte that can also interact with the neuron. Therefore, even though these substances are differentially distributed in gray and white matter and in different areas of the brain, both NAA and NAAG liberated by neurons can interact with oligodendrocytes and astrocytes.

I can't say where these interactions take place at present and this is a question that remains to be resolved. So there are still many, many unknowns.

SESSION CO-CHAIR KOLODNY: Dr. Ledeen?

DR. LEDEEN: The 32 molecules of water that leave with the NAA are then hydrolyzed by ASPA, aspartoacylase, which is part of the oligodendrocyte and then the enzyme hydrolyzes the NAA and the aspartate is recycled back into the neuron. I think that's the model you presented.

Wouldn't this require aspartoacylase to be on the external membrane of the oligodendrocyte; whereas, most of the evidence we have is that it's inside the cell?

DR. BASLOW: Yes. I don't know. I've seen a lot of data on the immunochemistry of the myelin sheaths. And this enzyme seems to be associated with the myelin sheath. Well, the myelin in the compact myelin sheath is really an external feature because there is very little cellular water, and consequently there is very little cytoplasm left in those portions where we see both the myelin and high level of aspartoacylase.

It may be that the enzyme is not only present inside the cell, but also in the membrane or at least in the myelin-associated membrane. In the brain there is another acylase, amidohydrolase I. However, in the eye, in the ocular fluid the acylase is free. It's in the extracellular fluid. So this is an example of another brain enzyme sometimes associated with a cell, sometimes not.

SESSION CO-CHAIR KOLODNY: Dr. Parhu?

PARTICIPANT: Dr. Baslow, one of the things that happens when you stimulate at fairly high frequency either on myelinated or non-myelinated nerves, is that the astrocytes and the oligodendrocytes, or the Schwann cells in the peripheral nervous system, swell as though they were taking up excess potassium that is coming out of neurons. This same thing happens in non-myelinated nerve as well. The glial cells surrounding it swell. Would you expect that or would you comment on the possibility that NAA/NAAG may also be involved in osmoregulation, volume regulation of these cells as well as dealing with metabolic sequences?

DR. BASLOW: I can't say exactly how NAA/NAAG may interact with glial cell volume regulation. I don't know. But being placed in extracellular fluid, they may increase the osmotic pressure in the extracellular fluid, and this may cause some movement of water out of the glial cells. It's a question I can't answer at this point.

DR NAMBOODIRI: Normally when you put cells under osmotic stress, water will flow through aquaporin channels controlled by the osmotic pressure. What is it about neurons that prevents them from osmoregulating that way? Do neurons have aquaporins?

DR. BASLOW: Okay. Neurons do not have aquaporins. That's primarily a property of astrocytes which have aquaporin channels for free movement of water, but neurons have no aquaporins for this purpose. The neuron has a second and compounding problem. It's the most active cell in the brain, contributing to the small one and one half pound brain burning 20 percent of the body's energy. Therefore, neurons are producing a tremendous amount of metabolic water, at the same time being surrounded in large parts by a hydrophobic myelin sheath and without aquaporin water channels. Neurons have to get rid of this metabolic water, and I believe that the NAA/NAAG system is a neuronal mechanism to pump metabolic and other water into surrounding extracellular fluids.

Now, once the water is in extracellular fluid from where it is then liberated from NAA and NAAG by oligodendrocyte and astrocyte enzymes, it can flow down its gradient into the vascular system which is a water sink, having less water than extracellular fluid and with an osmotic pressure differential of about one and a half milliosmoles.

Once liberated in extracellular fluid, the free water can either flow directly across the vascular epithelium into the vascular system or it can flow through astrocytes surrounding the vascular system because the astrocytes do have aquaporin water channels. In this way movement of metabolic water from neurons out of the brain can be accounted for.

SESSION CO-CHAIR NAMBOODIRI: Just one more question up in the back.

SESSION CO-CHAIR KOLODNY: Dr. Leone?

DR. LEONE: Well, I would like to make a comment rather than a question, which is very important in this setting where we are scientists, clinicians, and parents of children affected by Canavan's disease; which is that your data very elegantly show a very close fitness with our *in vivo* data in patients, collected through MRI studies and MR spectroscopy studies for about eight years. We have MRI's, and have the T1 quantitative measures usually associated with the brain and myelin development.

DR. BASLOW: Yes.

DR. LEONE: What we found is that specifically we have very elevated T1 in these children.

DR. BASLOW: Yes.

DR. LEONE: Which exactly fits with your idea that elevated brain NAA would create edema.

DR. BASLOW: Okay. And then probably T2 is also elevated. Here in this particular case, we are measuring T2*, which is used for BOLD-signal measurements as an indicator of changes in cerebral blood flow. My co-author on the paper will be here tomorrow if anybody wants to talk to him about the methodology, a technique which is rather unique to do in a mouse brain, as I understand.

6. REFERENCES

Attwell, D. and Laughlin, S.B., 2001, An energy budget for signaling in the grey matter of the brain. J. Cerebral *Blood Flow and Metabol.* **21**: 1133-1145

Bacich, D.J., Ramadan, E., O'Keefe, D.S., Bukhari, N., Wegorzewska, I., Ojeifo, O., Olszewski, R., Wrenn, C.C., Bzdega, T., Wroblewska, B., Heston, W.D.W. and Neale, J.H., 2002, Deletion of the glutamate carboxypeptidase II gene in mice reveals a second enzyme activity that hydrolyzes N-acetylaspartylglutamate. *J. Neurochem.* **83**: 20-29

Baslow, M.H., 1997, A review of phylogenetic and metabolic relationships between the N-acylamino acids, N-acetyl-L-aspartic acid and N-acetyl-L-histidine in the vertebrate nervous system. *J. Neurochem.* **68** (4): 1335-1344.

Baslow, M.H., 2000, Functions of N-acetyl-L-aspartate and N-acetyl-L-aspartylglutamate in the vertebrate brain. Role in glial cell-specific signaling. *J. Neurochem.* **75**: 453-459

Baslow, M.H., 2002, Evidence supporting a role for N-acetyl-L-aspartate as a molecular water pump in myelinated neurons in the central nervous system. An analytical review. *Neurochem. Int.* **40** (4): 295-300

Baslow, M.H., 2003a, Brain N-acetylaspartate as a molecular water pump and its role in the etiology of Canavan disease: A mechanistic explanation. *J. Mol. Neurosci.* **21**: 185-189

Baslow, M.H., 2003b, N-acetylaspartate in the vertebrate brain: Metabolism and function. *Neurochemical Res.* **28** (6): 941-953

Baslow, M.H., 2004, N-acetylaspartate and N-acetylaspartylglutamate. In: *Handbook of Neurochemistry and Molecular Neurobiology*, 3 rd Ed., Abel Lajtha, Editor-in-Chief. Vol. 6, Chapter 17, Kluwer Academic/Plenum Publishers, New York, N.Y., USA (*in process*)

Baslow, M.H. and Guilfoyle,D.N., 2002, Effect of N-acetylaspartic acid on the diffusion coefficient of water: A proton magnetic resonance phantom method for measurement of osmolyte-obligated water. *Analyt. Biochem.* **311** (2): 133-138

Baslow, M.H., Suckow, R. F., Gaynor, K., Bhakoo, K.K., Marks, N., Saito, M., Saito, Mitsuo, Duff, K., Matsuoka, Y. and Berg, M.J., 2003, Brain damage results in downregulation of N-acetylaspartate as a neuronal osmolyte. *NeuroMol. Med.* **3** (2): 95-103

Baslow, M.H., Dyakin, V.V., Nowak, K., Hungund, B. and Guilfoyle, D.N., 2005, 2-PMPA, a NAAG peptidase inhibitor, attenuates magnetic resonance BOLD signals in brain of anesthetized mice. Evidence of a link between neuron NAAG release and hyperemia. *J. Mol. Neurosci.* **26**, (3); 1-16

Berry II, M.J. and Meister, M., 1998, Refractoriness and neural precision. *J. Neurosci.* **18**: 2200-2211

Birken, D.L. and Oldendorf, W.H., 1989, N-acetyl-L-aspartic acid: a literature review of a compound prominent in H-NMR spectroscopic studies of brain. *Neurosci. Biobehav. Rev.* **13**: 23-31

Boltshauser, E., Schmitt, B., Wevers, R.A., Engelke, U., Burlina, A.B. and Burlina, A.P., 2004, Follow-up of a child with hypoacetylaspartia. *Neuropediatrics* **35**: 255-258

Bzdega, T., Crowe, S.L., Ramadan, E.R., Sciarretta, K.H., Olszewski, R.T., Ojeifo, O.A., Rafalski, V.A.., Wroblewska, B. and Neale, J.H., 2004, The cloning and characterization of a second brain enzyme with NAAG peptidase activity. *J. Neurochem.* **89**: 627-635

Chacron, M.J., Longtin, A. and Maler, L., 2001, Negative interspike interval correlations increase the neuronal capacity for encoding time-dependent stimuli. *J. Neurosci.* **21**: 5328-5343

Choi, I-Y. and Gruetter, R., 2001, In vivo [13] C NMR measurement of total brain glycogen concentrations in the conscious rat. *Proc. Intl. Soc. Mag. Reson. Med.* **9**: 210

Clifford, C.W. and Ibbotson, M.R., 2000, Response variability and information transfer in directional neurons of the mammalian horizontal optokinetic system. *Visual Neurosci.* **17**: 207-215

Coyle, J.T., 1997, The nagging question of the function of N-acetylaspartylglutamate. *Neurobiol. of Dis.* **4**:231-238

Henry, P.-G., Tkac, I. And Gruetter, R., 2003, ¹H-localized broadband [13] C NMR spectroscopy of the rat brain in vivo at 9.4 T. *Mag. Reson. Med.* **50**: 684-692

Hyder, F., Rothman, D.L. and Shulman, R.G., 2002, Total neuroenergetics support localized brain activity: implications for the interpretation of fMRI. *Proc. Nat. Acad. Sci.* **99**:10771-10776

Karelson, G., Ziegler, A., Kunnecke, B. and Seelig, J., 2003, Feeding versus infusion: a novel approach to study the NAA metabolism in rat brain. *NMR in Biomed.* **16**: 413- 423

Klugmann, M., Symes, C.W., Klaussner, B.K., Leichtlein, C.B., Serikawa, T., Young, D. and During, M.J., 2003, Identification and distribution of aspartoacylase in the postnatal rat brain. *Neuroreport* **14**:1837-1840

Kozikowski, A.P., Zhang, K., Nan, F., Petukov, P.A., Grajkowska, E., Wroblewski, J.T., Yamamoto, T., Bzdega, T., Wroblewska, B., and Neale, J.H., 2004, Synthesis of urea-based inhibitors as active site probes of glutamate carboxypeptidase II: efficacy as analgesic agents. *J. Medicinal Chem.* **47**: 1729-1738

Logothetis, N.K., 2003, The underpinnings of the BOLD functional magnetic resonance imaging signal. *J. Neurosci.* **23**: 3963-3971

Lu, Z.H., Chakraborty, G., Ledeen, R.W., Yahya, D. and Wu, G., 2004, N-acetylaspartate synthase is bimodally expressed in microsomes and mitochondria in brain. *Brain Res. Mol. Brain Res.* **122**: 71-78

Mackel, R. and Brink, E., 2003, Conduction of neural impulses in diabetic neuropathy. *Clin. Neurophysiol.* **114**: 248-255

Madhavarao, C.N., Chinopoulos, C., Chandrasekaran, K. and Namboodiri, M.A.A., 2003, Characterization of the N-acetylasparate biosynthetic enzyme from rat brain. *J. Neurochem.* **86**: 824-835

Madhavarao, C.N., Moffett, J.R., Moore, R.A., Viola, R.E., Namboodiri, M.A. and Jacobowitz, D.M., 2004, Immunohistochemical localization of aspartoacylase in the rat central nervous system. *J. Comp. Neurol.* **472**: 318-329

Moreno, A., Ross, B.D. and Bluml, S., 2001, Direct determination of the N-acetyl-L-aspartate synthesis rate in the human brain by ^{13} C MRS and [1-^{13} C] glucose infusion. *J. Neurochem.* **77**: 347-350

Neale, J.H., Bzdega, T. and Wroblewska, B., 2000, N-acetylaspartylglutamate: The most abundant peptide neurotransmitter in the mammalian central nervous system. *J. Neurochem.* **75**: 443-452

Raichle, M. E. and Gusnard, D. A. 2002, Appraising the brain's energy budget, *Proc. Natl. Acad. Sci. USA*, **99**: 10237-10239.

Robinson, M.B., Blakely, R.D., Couto, R. and Coyle, J.T., 1987, Hydrolysis of the brain dipeptide N-acetyl-L-aspartyl-L-glutamate. Identification and characterization of a novel N-acetylated alpha-linked acidic dipeptidase activity from rat brain. *J. Biol. Chem.* **262**: 14498-14506

Saadoun, S., Papadopoulos, M.C., Davies, D.C., Krishna, S. and Bell, B.A., 2002, Aquaporin-4 expression is increased in oedematous human brain tumours. *J Neurol. Neurosurg. Psychiatry* **72**: 262-265

Schoepp, D.D., Jane, D.E. and Monn, J.A., 1999, Pharmacological agents acting at subtypes of metabotropic glutamate receptors. *Neuropharmacol.* **38**: 1431-1476

Smith, A.J., Blumenfeld, H., Behar, K.L., Rothman, D.L., Shulman, R.G. and Hyder, F., 2002, Cerebral energetics and spiking frequency: the neurophysiological basis of fMRI. *Proc. Natl. Acad. Sci.* **99**: 10765-10770

Thomas, D.L., Lythgoe, Pell, G.S., Calamante, F. and Ordidge, R.J., 2000, The measurement of diffusion and perfusion in biological systems using magnetic resonance imaging. *Phys. Med. Biol.* **45**: R97-R138

Tsacopoulos, M., 2002, Metabolic signaling between neurons and glial cells: a short review. *J. Physiol.* **96**: 283-288

Tyson, R.L. and Sutherland, G. R., 1998, Labeling of N-acetylaspartate and N-acetylaspartylglutamate in ratneocortex, hippocampus and cerebellum from [1-^{13} C] glucose. *Neurosci. Lett.* **251**: 181-184

Vanzetta, I., and Grinvald, A., 1999, Increased cortical oxidative metabolism due to sensory stimulation: implications for functional brain imaging. *Science* **286**: 1555-1558

Zarahn, E., 2001, Spatial localization and resolution of BOLD fMRI. Curr. Opin. In Neurobiol. 11: 209-212

Zonta, M., Angulo, M.C., Gobbo, S., Rosengarten, B., Hossmann, K-A., Pozzan, T. and Carmignoto, G., 2003, Neuron-to-astrocyte signaling is central to the dynamic control of brain microcirculation. *Nature Neurosci.* **6** (1): 43-50

CONTROL OF BRAIN VOLUME DURING HYPOOSMOLALITY AND HYPEROSMOLALITY

Joseph G. Verbalis[*]

1. INTRODUCTION

Hypoosmolality and hyperosmolality are relatively common clinical problems (1). Many different factors contribute to the morbidity and mortality known to occur during states of altered osmotic homeostasis. The most serious complications are associated with pathological changes in brain volume: brain edema during hypoosmolar states and brain dehydration during hyperosmolar states. This chapter will summarize what is known about the changes that occur in brain fluid and solute composition during hypoosmolar and hyperosmolar states, which are responsible for the compensatory process of brain volume regulation. Most experimental and clinical studies have used serum sodium concentration as an indicator of osmolality, and throughout this chapter the terms hyponatremia and hypoosmolality are used interchangeably, as are hypernatremia and hyperosmolality.

2. HYPOOSMOLALITY

Hypoosmolality is well known to cause a variety of neurological symptoms, including disorientation, confusion, obtundation, seizures and death from tentorial herniation, called hyponatremic encephalopathy (4,5), but the incidence and severity of such symptoms in hyponatremic patients is quite variable (6). It is not unusual to find patients with low serum sodium concentrations ($[Na^+]$) who are relatively asymptomatic, while others exhibit severe neurological dysfunction at equivalently low levels of serum $[Na^+]$. Such clinical observations indicate that the brain can successfully adapt to even severe degrees of hypoosmolality in many cases. Knowledge of how the brain regulates

---

[*] Georgetown University Medical Center, 4000 Reservior Rd., Washington DC 20007. USA Phone: 202-687-2818, Email: verbalis@georgetown.edu.

its volume in response to hypoosmolality has been crucial to understanding this sometimes perplexing spectrum of clinical presentations of hypoosmolar patients.

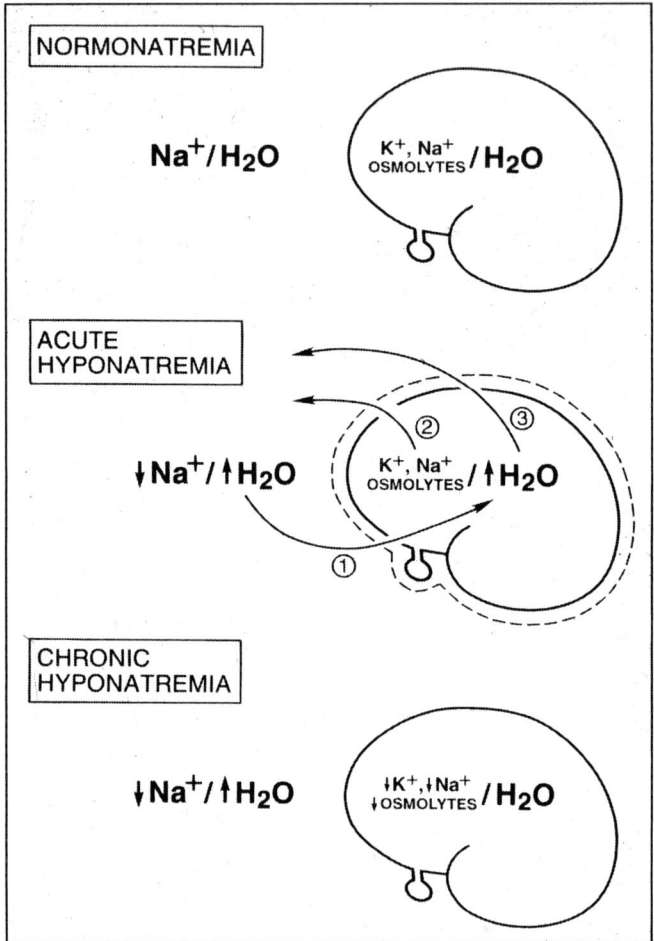

Figure 1. Brain Adaptation to Hypoosmolality: Schematic diagram of brain volume adaptation to hyponatremia. Under normal conditions brain osmolality and extracellular fluid (ECF) osmolality are in equilibrium (top panel; for simplicity the predominant intracellular solutes are depicted as K^+ and organic osmolytes, and the extracellular solute as Na^+). Following the induction of ECF hypoosmolality, water moves into the brain in response to osmotic gradients producing brain edema (middle panel, #1, dotted lines). However, in response to the induced swelling the brain rapidly loses both extracellular and intracellular solutes (middle panel, #2). As water losses accompany the losses of brain solute, the expanded brain volume then decreases back toward normal (middle panel, #3). If hypoosmolality is sustained, brain volume eventually normalizes completely and the brain becomes fully adapted to the ECF hyponatremia (bottom panel). Reproduced with permission from (1).

Experimental studies in animals over the last century have elucidated many of the physiological mechanisms underlying brain adaptation to hypoosmolality (7-9).

Following decreases in plasma osmolality, water moves into the brain along osmotic gradients, causing cerebral edema. In response, the brain loses solute from the extracellular (10) and the intracellular (7,8) fluid spaces, thereby decreasing brain water content back toward normal levels (Fig. 1). The marked variability in the presenting neurological symptoms of hyponatremic patients can be understood in the context provided by this process of brain volume regulation. Most of the neurological symptoms associated with hyponatremia are thought to reflect brain edema as a consequence of osmotic water movement into the brain (5). However, once the brain has volume-adapted through solute losses, thereby reducing brain edema, neurological symptoms will not be as prominent, and in some cases may even be totally absent (Fig. 1).

It has also long been appreciated that the rate of fall of serum [Na^+] is generally more strongly correlated with morbidity and mortality than the actual magnitude of the decrease. This is due to the fact that brain volume regulation occurs over a finite period; the more rapid the fall in serum [Na^+], the more water will be accumulated before the brain is able to lose solute, and with it the increased water. This temporal association explains the much higher incidence of neurological symptoms in patients with acute hyponatremia compared to those with chronic hyponatremia. It is therefore important to understand the mechanisms underlying brain volume regulation during both acute and chronic hypoosmolality.

2.1. Adaptation to Acute Hypoosmolality.

The clinical distinction between acute and chronic hypoosmolality is somewhat arbitrary, but generally hypoosmolality is considered to be acute when it develops over 24 to 48 hours. Such patients are indeed at high risk for neurological complications, with mortality rates as high as 50% in some studies. Induction of rapid hyponatremia has similarly been shown to cause severe neurological dysfunction in rabbits, and virtually all animals so treated die with marked brain edema. In these animals, brain water content increased by an amount equivalent to the fall in their serum [Na^+], and brain electrolyte contents did not decrease significantly, indicating an absence of brain volume regulation (9).

Thus, when hypoosmolality develops at a rate that exceeds the brain's ability to regulate its volume by electrolyte losses, severe brain edema results, potentially leading to neurological dysfunction and sometimes death. It is therefore important to define the time course over which brain volume regulation can occur. This has been studied in rats by measuring brain water and electrolyte contents at various times after induction of an acute dilutional hyponatremia (10). Na^+ and Cl^- losses began very rapidly, generally within 30 minutes, whereas brain K^+ losses were somewhat more delayed. Nonetheless, all electrolyte losses were found to be maximal by 3 hours, and they completely accounted for the degree of brain volume regulation that was achieved over this period. Although brain edema still occurred, with measured increases in brain water from 6 to 9%, the ability of the brain to lose electrolytes rapidly within several hours limited the severity of brain swelling. These results are consistent with many experimental studies in animals that have reported variable neurological symptoms and survival rates following induction of acute hyponatremia, since over short periods of time (i.e., several hours) relatively small differences in the rates of loss of electrolytes can have profound effects on the resulting brain edema and neurological dysfunction.

2.2. Adaptation to Chronic Hypoosmolality

In contrast to acute hyponatremia, many experimental studies of chronic hyponatremia have been characterized by a relative absence of neurological symptoms and mortality. These findings suggest that more complete degrees of brain volume regulation occur after longer periods of sustained hyponatremia. Studies in rats in which hyponatremia was maintained for 21 days confirmed virtually complete normalization of brain water content (11). However, in these and other studies the measured electrolyte losses accounted for only 60 to 70% of the observed brain volume regulation, which suggested a potential contribution from losses of other brain solutes as well. Subsequent studies confirmed that brain content of most organic osmolytes also decreases markedly during induced hyponatremia in mice (12) and rats (13,14). The organic osmolytes involved in volume regulation are amino acids, methylamines, and polyols. The major organic osmolytes in the brain are glutamine, glutamate, taurine, and myoinositol. Organic osmolytes of lesser significance include several other amino acids, two methylamines glycerophosphorylcholine (GPC) and betaine], phosphocreatine/creatine, and the neurotransmitter γ-aminobutyric acid (GABA). N-acetylaspartate (NAA) is included among the solutes that are lost during volume regulation, but this amino acid represents a relatively minor component of the total brain solute losses (2). Figure 2 shows the relative brain losses of organic osmolytes compared to electrolytes after 14 days of sustained hyponatremia in rats (14). Total brain electrolyte losses are larger, as expected, nonetheless the measured brain organic osmolyte losses accounted for roughly one third of the measured brain solute losses during sustained hypoosmolality. Such coordinate losses of both electrolytes and organic osmolytes from brain tissue enable very effective regulation of brain volume during chronic hyponatremia (Fig. 1). Consequently, it is now clear that cellular volume regulation *in vivo* occurs predominantly through depletion, rather than intracellular osmotic "inactivation," of a variety of intracellular solutes (2). Studies using NMR spectroscopy in hyponatremic patients have confirmed that similar mechanisms occur in humans with hyponatremia (15).

In addition to physiological implications for brain volume regulation, the large decrease in brain organic osmolyte contents over relatively short periods (i.e. <48 hours) has potentially important functional implications. Such relatively rapid decreases suggest the possibility that some of these losses are occurring via effluxes of intracellular osmolytes from brain cells during the process of volume regulation. This could result in transiently increased local brain extracellular fluid concentrations of organic osmolytes, which in the case of amino acids could produce significant effects on neuronal membrane potential. In particular, given the known actions of glutamate as an excitatory neurotransmitter, locally increased brain glutamate concentrations occurring at the time of the active phase of volume regulation during hyponatremia could potentially account for some of the neurological abnormalities known to occur during this period, especially the increased incidence of seizure activity (16,17). This hypothesis could also explain, at least in part, the observation that when hypoosmolality is maintained for longer periods of time, both animals and patients become less symptomatic, since increased brain neurotransmitter concentrations would likely occur only transiently during the initial development of hyponatremia and then return to more normal levels after the completion of brain volume regulation.

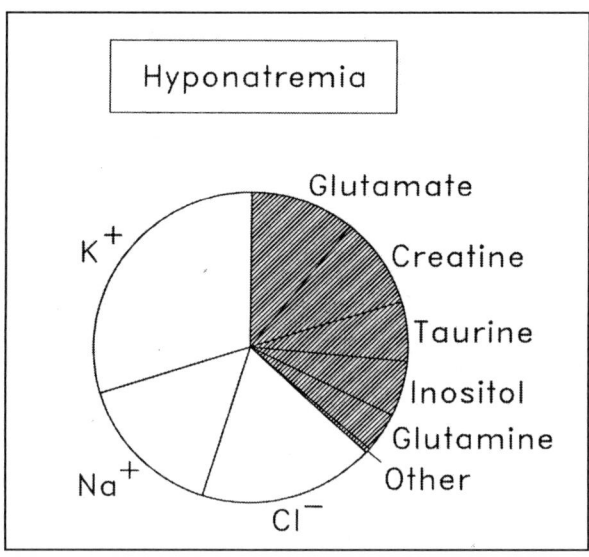

Figure 2. Relative decreases in individual brain electrolytes and organic osmolytes during adaptation to chronic (14 days) hyponatremia in rats. The category "other" represents GPC, urea, and several other amino acids. Reproduced with permission from (2).

2.3. Cellular Mechanisms Underlying Brain Adaptation to Hypoosmolality

Although brain volume regulation in response to perturbations of extracellular osmolality represents the most dramatic demonstration of volume regulation in response to changes in extracellular osmolality, the ability to regulate intracellular volume is an evolutionarily conserved mechanism inherent to variable degrees in most cells. Abundant *in vitro* experimentation has yielded important insights into the cellular mechanisms that underlie this important adaptive process (18).

With acute decreases in external osmolality, cells initially behave as osmometers and swell in proportion to the reduction in extracellular osmolality as a result of movement of water into the cells along osmotic gradients. Very soon thereafter, a process known as volume regulatory decrease (VRD) in cell volume begins, in which intracellular solutes are extruded together with osmotically obligated water (19). The time necessary to activate RVD and restore normal, or near-normal, cell volume is variable across different cell types. RVD occurs very rapidly *in vitro,* with a 70-80% recovery of normal cell volume reached within a few minutes in most brain and epithelial cells (19,20).

RVD has been studied in detail in astrocytes and neurons from primary cultures (20,21), in neuroblastoma (22) and glioma (23) cells lines. The osmolytes responsible for RVD are essentially the same in most cell types and can be grouped into two broad categories: electrolytes (predominantly K^+ and Cl^-) and organic osmolytes (amino acids, polyalcohols, sugars, and methylamines). In most cells examined to date, electrolyte fluxes appear to occur by diffusive pathways, i.e., K^+ and Cl^- efflux through separate volume-sensitive channels, and organic osmolytes through "leak pathways" with no significant contribution from energy-dependent carriers (18). In brain cells, swelling activates at least two different types of K^+ channels, both a large and a small conductance channel (24). The volume-sensitive Cl^- channel (VSCC) has high selectivity of anions over cations, but exhibits broad anion selectivity, displaying permeability to the majority of monovalent anions (25,26). Although the molecular species of VSCC are as yet unidentified, recent evidence has supported the ClC3 channel gene as encoding the channel protein responsible for the volume-sensitive Cl^- current (27), but different types of VSCC and other anion-permeating molecules coincide in the same cell allowing for participation of more than one VSCC in RVD (28).

Although many different organic osmolytes are also released by cells during RVD, their efflux pathways have been characterized for only a few, particularly taurine and myoinositol. In general, these are bidirectional leak pathways with net solute movement depending on concentration gradient direction (18,29). Organic osmolyte pathways commonly exhibit a pharmacologic profile similar to that of the VSCC, suggestive of a common pathway with Cl^-, or of a close connection between the two pathways (29,30). Other amino acids also responsive to swelling are glycine, GABA, glutamate, and aspartate, which contribute to correction of osmotic disturbance. Recent evidence of hyposmolality-induced glutamate release that is insensitive to Cl^- channel blockers is different from the pattern found with most other organic osmolytes (17). This suggests either different pathways, or different stimuli and mechanisms for release, of this amino acid.

Exactly how cells sense volume changes is a critical step in the reactions activated to achieve volume correction. Among possible mechanisms considered to play this role are membrane receptors such as integrins or receptors with intrinsic tyrosine kinase activity, cytoskeleton rearrangements, dilution of cytosolic macromolecules, decrease in intracellular ionic strength, stretch-induced activation of adhesion molecules, activation of phospholipases, or changes in the concentration of signaling molecules such as calcium or magnesium (31). Calcium and protein kinases are among the most likely candidates to act as osmotransductory elements. One of the most constant features of hyposmolar swelling is an increase in systolic Ca^{++}. Despite this, the main corrective osmolyte efflux pathways and consequently RVD are Ca^{++}-independent in a large variety of cell types. This is the case for brain cells, in which VSCC, VSKC, and organic efflux pathways are largely Ca^{++}-independent (24). This is an area of active ongoing research, and the reader is referred to excellent recent reviews of this topic for more details (18,31,32).

2.4. Recovery from Hypoosmolality (Deadaptation)

Compensatory adaptations that enable organisms to survive chronic perturbations of body homeostasis must be reversed after recovering from the underlying abnormality. In some cases reversal of the adaptive process, or "de-adaptation," may be more problematical than the initial adaptation itself. This appears to be true for correction of chronic hyponatremia (33). Multiple studies have shown that rapid correction of chronic hyponatremia causes dehydration of brain tissue (34-36), and in some cases demyelination of white matter in various parts of the brain (37-39). Because this dehydration occurs to a greater degree in hyponatremic rats than in normonatremic rats following similarly large increases in osmolality, it has been suggested that this phenomenon reflects a loss of osmotic buffering capacity by brain tissue as a consequence of the initial brain solute losses that allowed survival despite hypoosmolar conditions (33).

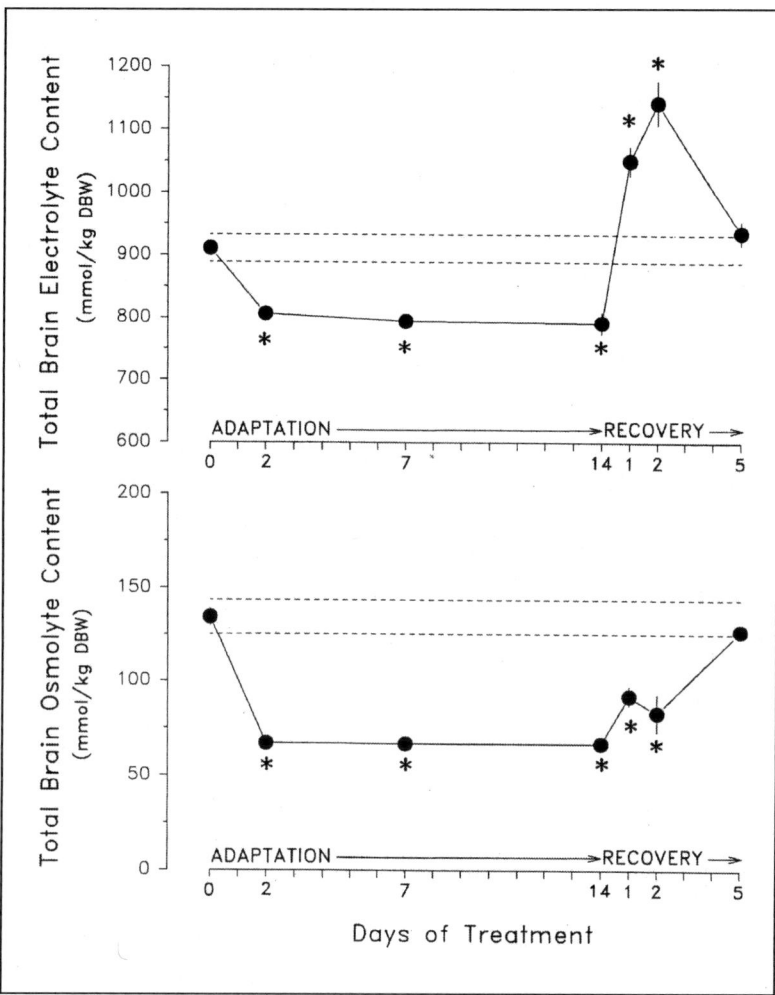

Figure 3. Time course of changes in brain electrolytes (top panel) and organic osmolytes (bottom panel) during adaptation to chronic hyponatremia and following correction of hyponatremia in rats. Dotted lines depict normal brain levels; *p<0.05 compared to day 0. Reproduced with permission from (3).

Several studies have now demonstrated markedly different rates of reaccumulation of brain solutes after normalization of hypoosmolality in hyponatremic mice and rats (13,40). In response to hyponatremia, brain tissue rapidly loses all classes of osmotically active solutes, including both electrolytes and organic osmolytes, thereby allowing the brain to efficiently regulate its volume. On the other hand, after recovery from hyponatremia organic osmolytes, with the exception of glutamate, return to normal brain contents very slowly over a period of many days, while electrolytes reach normal or supranormal contents in the brain within 24 hours after correction of hyponatremia (Fig. 3). This slow reaccumulation of organic solutes is very analogous to the similarly slow increases in osmolytes that occur during chronic hypernatremia (see below), and suggests that in general the brain is much better able to lose organic solutes than to reaccumulate them. Furthermore, the rapid electrolyte reaccumulation after correction of hypoosmolality consists mainly of the extracellular electrolytes Na^+ and Cl^-, and these significantly overshoot brain contents necessary to achieve normal volume regulation (Fig. 3). This again is quite analogous to the rapid increases in brain Na^+ and Cl^- that occur in response to acute hyperosmolality (41,42), and it suggests that in this situation these electrolytes are similarly gaining access to the brain rapidly via the CSF and are acting to stabilize intracellular volume (43). Consequently, the mechanisms that enable the brain to adapt to hypoosmolar conditions and those that accomplish de-adaptation after subsequent normalization of plasma osmolality are not simply mirror images of each other.

Knowledge of this greater inefficiency of brain solute reaccumulation and volume regulation following correction of chronic hyponatremia is very relevant to understanding the pathological sequelae known to be associated with rapid correction of chronic hypoosmolality, namely the occurrence of osmotic demyelination. Every adaptation made by the body in response to a perturbation of homeostasis bears within it the potential to create new problems. Although the mechanism(s) by which rapid correction of hyponatremia leads to brain demyelination remain unproven, this pathological disorder likely results from the brain dehydration that has been demonstrated to occur following correction of plasma $[Na^+]$ toward normal ranges in animal models of chronic hyponatremia. Because the degree of osmotic brain shrinkage is greater in animals that are chronically hyponatremic than in normonatremic animals undergoing similar increases in plasma osmolality, by analogy the brains of human patients adapted to hyponatremia are likely to be particularly susceptible to dehydration following subsequent increases in osmolality. This, in turn, can lead to pathological demyelination. Further support for dehydration-induced demyelination has come from recent reports that acute hyperosmolality can also cause demyelination in experimental animals (44), though larger increases in plasma osmolality are required than in hyponatremic rats. Although the exact mechanisms responsible for production of brain demyelination following correction of hyponatremia remain uncertain, one possibility is that acute brain dehydration produced by rapid correction could potentially disrupt the tight junctions of the blood-brain barrier. Recent magnetic resonance studies in animals have shown that chronic hypoosmolality predisposes rats to opening of the blood-brain barrier following rapid correction of hyponatremia, and that the disruption of the blood-brain barrier is highly correlated with subsequent demyelination (45). A potential mechanism by which blood-brain barrier disruption might lead to subsequent myelinolysis is via an influx of complement, which is toxic to the oligodendrocytes that manufacture and maintain myelin sheaths of neurons, into the brain (46).

3. HYPEROSMOLALITY

Hyperosmolality and hypernatremia usually occurs as a result of hypotonic fluid losses that are not compensated by sufficient water intake to maintain body fluid homeostasis. Less commonly, excess NaCl ingestion or administration can cause hyperosmolality as a result of solute excess. Although hyperosmolality can develop in association with a broad spectrum of disease processes in people of all ages, infants and elderly individuals are particularly susceptible (4). The neurological symptoms of hyperosmolar states are a result of the cellular dehydration produced by osmotic shifts of water from the intracellular fluid space to the more hypertonic extracellular fluid space. The clinical symptomatology is related both to the severity of the hyperosmolality and also to the rate at which it develops (47). The symptoms of hyperosmolar states are a consequence of neurological dysfunction resulting from cellular dehydration; these include irritability, restlessness, stupor, muscular twitching, hyperreflexia, spasticity, and in severe cases, seizures, coma, and ultimately death (4).

Because cell membranes are relatively more permeable to water than to electrolytes, a rapid increase in plasma osmolality causes the brain to shrink. The brain subsequently undergoes an adaptation process involving the accumulation of solutes to restore the brain volume to its normal level. This adaptation process involves rapid accumulation of inorganic ions and slower accumulation of organic osmolytes, traditionally termed "idiogenic osmoles" (48). Marked differences in symptoms, and hence in recovery from hyperosmolality, exist because of the time-dependent nature of this complex brain adaptation. Optimal treatment of hyperosmolar patients is facilitated by knowledge of the basic mechanisms underlying the process of adaptation to the hyperosmolar state. Just as with hypoosmolality, neurological symptoms and mortality are generally higher in patients with acute rather than chronic hypernatremia. Consequently, it is useful to consider brain adaptation to these different pathological states separately.

3.1. Adaptation to Acute Hyperosmolality

Acute hypernatremia, generally defined as the development of serum $[Na^+]$ >145 mmol/L in 24 to 48 hours, is relatively uncommon. It can, however, be seen in infants as a result of accidental salt poisoning or severe gastroenteritis. It occurs less commonly in adults, although patients with untreated diabetes insipidus who are unable to drink can develop severe hypernatremia very rapidly. Despite the relative rarity of acute hyperosmolality, it is important to understand the pathophysiology underlying the neurological symptoms associated with this disorder because of its marked morbidity and mortality, which can be as high as 75% in adults and 45% in children (4,47).

Acute hypernatremia is typically induced in animals by intraperitoneal injections of hypertonic NaCl, which causes a prompt reduction in brain water content. However, the rapid loss of brain water is less than would be expected if the brain behaved as a perfect osmometer, because the brain is capable of rapidly accumulating solute to stabilize its volume. In a study using rats, three hours of hypernatremia (serum $[Na^+]$ >200 mmol/liter) decreased brain water by 14% and promoted increases in contents of brain Na^+ and Cl^- of 34% and 60%, respectively; K^+ content was unaltered (8). Other studies of acute (15-120 min) hypernatremia in rats showed that the reduction in brain volume was proportional to the increase in plasma osmolality and generally stabilized by 15 to 30 minutes after the NaCl injection. However, by longer periods, i.e., 30 and 120 minutes, the brain water loss was only 35% of that predicted, which indicates that significant volume regulation had already occurred (49,50). This acute but partial volume regulation was due to rapid increases in tissue electrolytes. The accumulation of Na^+ and Cl^- was attributed to influx from the CSF, whereas the slower rise in tissue K^+ content was related to an influx from plasma across the blood-brain barrier.

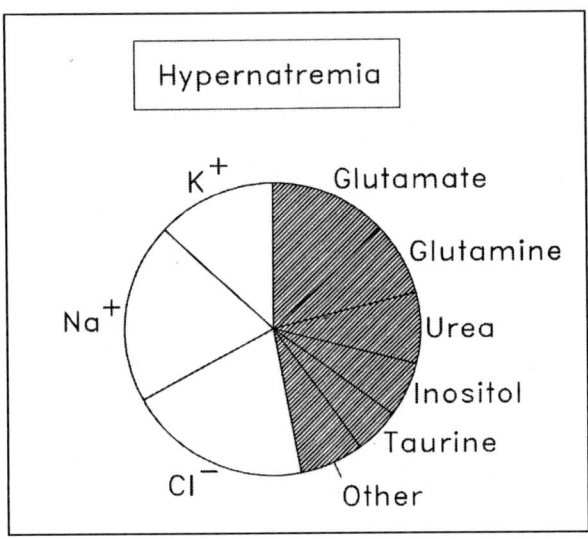

Figure 4. Relative increases in individual brain electrolytes and organic osmolytes during adaptation to chronic hypernatremia in rats. The category "other" represents GPC, urea, and several other amino acids. Reproduced with permission from (2).

A central problem in studies of brain volume regulation has been an inability to distinguish the changes that occur in the intracellular versus the extracellular spaces. A seminal study by Cserr and coworkers used ion-selective electrodes to resolve this issue (43). Rats given a single intraperitoneal injection of NaCl experienced a 7% loss of brain water after 30-90 minutes, but this loss was related entirely to a decrease in and extracellular water content; intracellular water content remained at basal levels. Estimates of intracellular and extracellular ion contents indicated that extracellular Na^+, Cl^-, K^+ decreased by 32, 21, and 42%, respectively, whereas intracellular contents of these ions increased by 100, 169, and 5%, respectively. In contrast, studies of organic osmolytes have indicated that acute hypernatremia is not associated with increases in organic solutes that are sufficient to appreciably contribute to brain volume regulation (41,48). Thus, acute hypernatremia is characterized by a rapid loss of total brain water, but protection of intracellular volume by an almost equivalently rapid accumulation of electrolytes from the extracellular fluid, CSF, and plasma.

3.2. Adaptation to Chronic Hyperosmolality

In most hypernatremic patients, hyperosmolality develops gradually over a period of several days regardless of the etiology (47). Although morbidity and mortality rates are reported to be high in both adults and children, interpretation of these findings is difficult because death is often a result of the underlying disease that caused the fluid imbalance (4). Nonetheless, chronic hypernatremia is generally better tolerated with less neurological symptomatology than occurs during acute hypernatremia of comparable magnitude, which indicates that the brain is able to adapt to hyperosmolar conditions over longer periods of time. This has been attributed to a slower accumulation of organic solutes by the brain, and recent studies have provided greater insights into this adaptation process.

In animals, when hypernatremia persists beyond several days, total brain tissue water content slowly returns to normal levels (8,41). This restoration of total brain water does not result from continued accumulation of electrolytes but rather from accumulation of specific organic osmolytes (41,42,48,51). The organic osmolyte accumulation generally accounts for 30-50% of the solute accumulation in hypernatremic animals. The organic osmolytes involved with volume regulation to hyperosmolar conditions are essentially the same substances found to be involved with adaptation to chronic hypoosmolality, as described earlier. No single study has quantified the major electrolyte and organic osmolyte changes in the brains of chronically hypernatremic animals. However, using data from several sources, one can estimate the relative contributions of major osmolytes during adaptation to chronic hypernatremia (2) (Fig. 4). As for adaptation to hypoosmolality, NAA is included among the solutes that accumulate during volume regulation in response to hyperosmolality, but this amino acid represents a relatively minor component of the total brain solute losses; studies of in rats indicated that NAA accounted for only 16% of the total increase in brain amino acids after chronic salt loading (42).

Organic osmolytes accumulate relatively slowly in brain following induction of hypernatremia. Indirect measurements of the brain contents of undetermined solutes (i.e., total osmolality minus the sum of tissue electrolytes) indicated that organic osmolytes begin to accumulate after 9 to 24 hours but do not reach a steady-state level until 2 to 7 days (52). *In vitro* studies of cultured brain cells have corroborated this delayed and slow rate of organic osmolytes accumulation.

3.3. Cellular Mechanisms Underlying Brain Adaptation to Hyperosmolality

With acute increases in external osmolality, cells initially behave as osmometers and shrink in proportion to the reduction in extracellular osmolality as a result of movement of water out of cells along osmotic gradients. Soon thereafter, a process known as *volume regulatory increase (VRI)* in cell volume begins, in which cells accumulate intracellular solutes together with osmotically obligated water (19). Similar to VRD, the time necessary to activate RVI and restore normal, or near-normal, cell volume is variable across different cell types. However, in general RVI occurs more slowly than VRD in most cell types where this has been carefully studied.

The mechanisms responsible for brain and cell organic osmolytes accumulation during hyperosmolar conditions remain inadequately understood, but likely represent a combination of cellular uptake (e.g., myoinositol and taurine (53)) and synthetic mechanisms (e.g., glutamate), as has been well described in renal medulla where hypertonic conditions predominate (54). Circulating blood levels of many amino acids are increased during hypernatremia, and these may serve as a precursor pool for brain osmolytes (48). One important issue that remains to be resolved is the intracellular/extracellular distribution of organic and inorganic osmolytes in the brain during chronic hyperosmolality. Studies of other systems would suggest that organic osmolytes are preferentially accumulated intracellularly and thus replace the inorganic solutes, which are primarily responsible for the acute phase of brain cell volume regulation.

3.4. Recovery from Hyperosmolality (Deadaptation)

Accumulation of solutes enables the brain to adapt to hyperosmolar states, and thus is life saving, but during correction of the hyperosmolality this increase in total brain solute content can lead to neurological dysfunction due to osmotic shifts of water into the now more hypertonic intracellular fluid space. As with the adaptation process, both the duration of the hyperosmolality and the rate of the correction determine the degree of brain of edema that occurs. When acute hyperosmolar animals are given access to water, they recover relatively rapidly. The recovery phenomenon involves a transient but small increase in tissue water and a relatively rapid loss of the electrolytes that accumulated during the hyperosmolar episode. In contrast, during recovery from chronic hypernatremia, restoration of brain organic solute contents to normal levels occurs slowly over 24 to 48 hours. Studies in rats found that betaine fell to normal levels within 24 hours whereas glutamine, glutamate, taurine, phosphocreatine, and GPC took two days to achieve normal levels, and myoinositol remained significantly elevated even after two days (41). The mechanisms responsible for organic and inorganic solute loss during recovery from hypernatremia are not known, but are likely similar to those responsible for VRD during adaptation to hypoosmolality.

This slow dissipation of accumulated organic solutes is the basis for clinical recommendations that chronic hypernatremia be corrected relatively slowly over 48 hours (4,47). In support of a slow correction, studies of children have reported a high incidence of seizures following rapid correction of severe hypernatremia, presumably caused by brain edema (55,56). Although well-controlled studies of optimal correction rates in adults do not exist, based on what is known about the rates of organic solute losses from brain tissue in animals it seems prudent to continue to recommend the more cautious approach of prompt but gradual correction of chronic hypernatremia and hyperosmolality (57).

4. SUMMARY

Multiple studies over several decades have provided evidence that both electrolytes and organic osmolytes play crucial roles in regulating brain volume, both during increases as well as during decreases in extracellular fluid osmolality. In both situations, however, changes in brain electrolyte contents appear to occur more rapidly, and represent the first line of defense of brain volume during acute perturbations of body fluid tonicity, while organic osmolytes allow adaptation to more chronic perturbations. For both hyperosmolality and hypoosmolality, the rate of development of the disorder is an important determinant of neurological morbidity and mortality, since sufficiently rapid changes in tonicity can exceed the brain's capacity to regulate its volume leading to more severe degrees of brain edema or dehydration.

Recovery from both hyper- and hypoosmolality requires reversal of the adaptive processes that enabled regulation of brain volume in response to the initial insult. However, adaptation and recovery are not symmetrical processes. Marked differences occur in the speed with which the brain is able to lose or to reaccumulate different types of solutes after recovery from chronic disturbances of body fluid tonicity. In general, accumulation, or reaccumulation, of organic solutes by brain tissue is a much slower process than volume regulatory losses of such solutes. As with the adaptation process, the rate of recovery is an important determinant of subsequent morbidity and mortality, since rapid corrections of osmolality can also exceed the capacity of the brain to readjust its solute content, and consequently its volume, back to normal levels.

Whether or not transient excesses or deficiencies of either electrolytes or specific organic osmolytes in brain intracellular or extracellular fluid contribute to functional disturbances independently of changes in brain volume is an intriguing question that has not been sufficiently evaluated. Also remaining to be answered are questions regarding other physiological, pathophysiological, and pharmacological factors that either impair or enhance volume regulatory processes, and thereby modify the neurological manifestations accompanying disorders of body fluid osmolality in humans. Finally, a complete understanding of the cellular mechanisms underlying adaptation to and deadaptation from acute and chronic perturbations of body osmolality will be essential to design the most enlightened, and therefore appropriate, treatments for these disorders.

5. QUESTION AND ANSWER SESSION

SESSION CO-CHAIR NAMBOODIRI: Thank you, Dr. Verbalis. Now we will have a few questions.

DR. PHILIPPART: Philippart, Los Angeles. I wonder if you also look at plasma bicarbonate.

DR. VERBALIS: Plasma bicarbonate is basically unchanged, as it is in humans with hyponatremia. It doesn't appear to be a major factor in this disorder.

SESSION CO-CHAIR KOLODNY: Brian Ross?

DR. ROSS: First is just a comment. I presume since we can measure all of these osmolytes with MR spectroscopy in the human brain, that you now routinely use spectroscopy in your patients.

DR. VERBALIS: You presume wrong, but it has been done. The same kind of decreases in osmolytes shown in the animal models has also been documented in human brains at UCLA using magnetic resonance spectroscopy. But it's not being used routinely for therapeutic purposes at this time.

DR. ROSS: I am challenging you to do so since you see these patients. But, my next point is a comment which I discussed briefly with you earlier and then hearing Dr. Baslow's presentation, you note that creatine is a major osmolyte in the brain.

If creatine is truly a major osmolyte and fluctuates with this huge 15-20 percent decrease with hypoosmolality, how do we explain the maintenance of energy metabolism because creatine and phosphocreatine are one and the same thing in your hands? And in Dr. Baslow's theory, he needs to demonstrate changes in energy metabolism.

DR. VERBALIS: I actually went one step further than that. I showed you my rats playing basketball, despite a 30 percent decrease in brain glutamate in those rats.

DR. ROSS: I understand.

DR. VERBALIS: So, how can a rat have normal motor transmission and normal motor function with a 30 percent decrease in a major excitatory amino acid in addition to a 20% decrease in creatine?

DR. ROSS: Right.

DR. VERBALIS: And I think that I gave you my explanation to both of those points, which is I believe that one of the factors that has not been taken into account in understanding brain energy metabolism, or even neurotransmission, are the separate pools of these various osmolytes that are involved with distinct cellular processes.

So, I believe that creatine, glutamate, aspartate, and NAA all have an osmoregulatory role which is separate from their other cellular roles, both metabolic and neurotransmission, and that it is possible to decrease one of these pools without necessarily impacting on the functions of the other pools.

That's the only way, for example, that I could explain normal motor function of these animals with those kinds of decreases in amino acids that we think are crucial for normal neural transmission, certainly in motor neurons but probably in all types of neurons for that matter. It's the only way one could understand it.

DR. ROSS: I had one more point to make. In all seriousness, I think you're absolutely right, of course, but we still do have the problems of these equilibria.

The proposal that was made - not in recent years, and this may have, therefore, not been brought to any of our attention, was actually made, in part, by R. L. Veech from 1979 or '80, in which he combines the Nernst equation, the Gibbs equation, and the osmotic equilibrium, the Donnan equilibrium, into one huge equation, saying that, actually, all of these things are linked.

So I think that there is a way in which Dr. Baslow's hypothesis and yours could really be saying the same thing.

DR. VERBALIS: Certainly, the water movement in association with the movement of both ions and organic particles across cell membranes is potentially very large. And so I am not by any means saying that I don't believe that inhibition of one part of that system might not result in what would be predicted, which is retention of water by the cell.

What I'm saying, however, is that based on our studies in this field, the brain has many other systems that are able to compensate for deficiencies, or excesses, of individual osmolytes.

So I'm not saying that that excess NAA might not produce exactly the effects that Dr. Baslow, and others, are postulating. What I am questioning is, why wouldn't the same kinds of mechanisms that allow brains to adapt to an even more severe osmotic stresses allow them to regulate their volume in this situation as well? Why wouldn't the same mechanisms come into play in a situation in which one molecular water pump was paralyzed, nonexistent, or somehow impaired? Why wouldn't these other osmolytes be able to regulate the volume of these cells with increased water retention?

That's where I have trouble fitting the molecular water pump hypothesis together as a viable single hypothesis to explain the observations in Canavan disease.

PARTICIPANT: Sir, there is also hypernatremia. So can you give an explanation for that in this way?

DR. VERBALIS: Yes, you're right that severe hypernatremia has also produced pontine and extrapontine myelinolysis. And the answer is that hypernatremia is also an osmotic stress. If you take an animal, or a human, from a normal plasma osmolality up to 360 or 370 mOsm/kg H_2O, then the blood brain barrier will be disrupted; complement and other immune proteins can then gain access into the brain and potentially can produce the same kind of demyelinating syndrome.

The difference between correction of hyponatremia and induction of hypernatremia is that in the chronic hyponatremic state, the brain has lost its pool of excess osmolytes, its osmotic "buffering capacity", so it's more susceptible to shrinkage and breaking the blood brain barrier with a lesser increase in plasma osmolality.

But if you take a normal human or animal up high enough in plasma osmolality, yes, you would break the blood brain barrier as well. If the increased osmolality is prolonged and sufficient, then the same pathophysiology as is seen with rapid correction of hyponatremia can and does occur.

The fact that this does occur is further proof for the blood-brain barrier disruption mechanism underlying the pathogenesis of pontine and extrapontine myelinolysis.

DR. COYLE: I just want to say that I concur with your suggestion that there are probably at least two pools of glutamate. The lesion studies indicate that probably less than ten percent of the total glutamate is in the neurotransmitter pool and the other ninety percent is in another pool. I'm not surprised that there's still a slam-dunk even when you have thirty percent of the brain glutamate decreased.

DR. VERBALIS: But would you expect, at least over prolonged periods of time, that this might, at least in some ways, impact upon the transmitter pool? Again, we don't see any evidence for that, not that we have measured those pools separately, nor would I know how to at this point. But it is impressive that, even for long periods of time, with severe depletion of these excitatory amino acids from the brain, there still is no apparent functional neurological effect of that dramatic phenomenon.

But, if I'm right and you're right, then just separate pools could explain that and allow one component to be markedly decreased without really impacting too severely on the others. If this is true, then it would imply preferential shunting into the neurotransmitter pool at the expense of the free cytosolic pool, which comprises the osmotically active component of cells.

6. REFERENCES

1. Verbalis JG, The syndrome of inappropriate antidiuretic hormone secretion and other hypoosmolar disorders. In: Schrier RW (ed). *Diseases of the Kidney and Urinary* (Tract.Lippincott Williams & Wilkins, Philadelphia) 2511-2548, (2001).
2. Gullans SR, Verbalis JG, Control of brain volume during hyperosmolar and hypoosmolar conditions. Annual Review of *Medicine* **44**, 289-301, (1993).
3. Verbalis JG, Adaptation to acute and chronic hyponatremia: implications for symptomatology, diagnosis, and therapy. *Semin. Nephrol.* **18**, 3-19, (1998).
4. Arieff AI, Central nervous system manifestations of disordered sodium metabolism. *Clinics Endocrin. Metab.* **13**, 269-294, (1984).
5. Fraser CL, Arieff AI, Epidemiology, pathophysiology, and management of hyponatremic encephalopathy. *Am. J. Med.* **102**, 67-77, (1997).
6. Sterns RH, Severe symptomatic hyponatremia: treatment and outcome. A study of 64 cases. *Ann. Int. Med.* **107**:656-664, (1987).
7. Yannet H Changes in the brain resulting from depletion of extracellular electrolytes. *Am. J. Physiol.* **128**, 683-689, (1940).

8. Holliday MA, Kalayci MN, Harrah J, Factors that limit brain volume changes in response to acute and sustained hyper- and hyponatremia. *J.Clin. Invest.* **47**, 1916-1928, (1968).

9. Arieff AI, Llach F, Massry SG Neurological manifestations and morbidity of hyponatremia: correlation with brain water and electrolytes. *Medicine* **55**, 121-129, (1976)

10. Melton JE, Patlak CS, Pettigrew KD, Cserr HF, Volume regulatory loss of Na, Cl, and K from rat brain during acute hyponatremia. Am. J. Physiol. **252**, F661-F669, (1987).

11. Verbalis JG, Drutarosky MD, Adaptation to chronic hypoosmolality in rats. Kidney Int. **34**, 351-360, (1988).

12. Thurston JH, Hauhart RE, Nelson JS, Adaptive decreases in amino acids (taurine in particular), creatine, and electrolytes prevent cerebral edema in chronically hyponatremic mice: rapid correction (experimental model of central pontine myelinolysis) causes dehydration and shrinkage of brain. *Metab. Brain Dis.* **2**, 223-241 (1987).

13. Lien YH, Shapiro JI, Chan L, Study of brain electrolytes and organic osmolytes during correction of chronic hyponatremia. Implications for the pathogenesis of central pontine myelinolysis. *J. Clin. Invest.* **88**, 303-309, (1991).

14. Verbalis JG, Gullans SR Hyponatremia causes large sustained reductions in brain content of multiple organic osmolytes in rats. *Brain Res.* **567**, 274-282, (1991).

15. Videen JS, Michaelis T, Pinto P, Ross BD, Human cerebral osmolytes during chronic hyponatremia. J. Clin. Invest. **95**, 788-793, (1995).

16. Verbalis JG, Adler S, Hoffman GE, Martinez AJ, Brain adaptation to hyponatremia: physiological mechanisms and clinical implications. In: Saito T, Kurokawa K, Yoshida S (eds). *Neurohypophysis: Recent Progress of Vasopressin and Oxytocin Research.* (Elsevier Science, Amsterdam) 615-626, (1995).

17. Pasantes-Morales H, Franco R, Ochoa L, Ordaz BOsmosensitive release of neurotransmitter amino acids: relevance and mechanisms. *Neurochem. Res.* **27**,59-65, (2002).

18. Pasantes-Morales H, Franco R, Ordaz B, Ochoa LD, Mechanisms counteracting swelling in brain cells during hyponatremia. *Arch. Med. Res.* **33**, 237-244, (2002).

19. Grantham JJ, Pathophysiology of hyposmolar conditions: a cellular perspective. In: Andreoli TE, Grantham JJ, Rector FC, (eds). *Disturbances in Body Fluid Osmolality.* (American Physiological Society, Bethesda), 217-225, (1977).

20. Pasantes-Morales H, Maar TE, Moran J, Cell volume regulation in cultured cerebellar granule neurons. *J. Neurosci. Res.* 34:219-224, (1993).

21. Sanchez-Olea R, Pena C, Moran J, Pasantes-Morales H, Inhibition of volume regulation and efflux of osmoregulatory amino acids by blockers of Cl- transport in cultured astrocytes. *Neurosci. Lett.* **156**, 141-144, (1993).

22. Basavappa S, Huang CC, Mangel AW, Lebedev DV, Knauf PA, Ellory JC, Swelling-activated amino acid efflux in the human neuroblastoma cell line CHP-100. *J. Neurophysiol.* **76**, 764-769, (1996).

23. Strange K, Morrison R, Volume regulation during recovery from chronic hypertonicity in brain glial cells. *Am. J.Physiol.* **263**, C412-C419, (1992).

24. Pasantes-Morales H, Morales MS, Influence of calcium on regulatory volume decrease: role of potassium channels. *Nephron* **86**, 414-421, (2000).

25. Nilius B, Eggermont J, Voets T, Buyse G, Manolopoulos V, Droogmans G 1997 Properties of volume-regulated anion channels in mammalian cells. Prog Biophys Mol Biol 68:69-119

26. Okada Y 1997 Volume expansion-sensing outward-rectifier Cl- channel: fresh start to the molecular identity and volume sensor. Am J Physiol 273:C755-C789

27. Hermoso M, Satterwhite CM, Andrade YN, Hidalgo J, Wilson SM, Horowitz B, Hume JR 2002 ClC-3 is a fundamental molecular component of volume-sensitive outwardly rectifying Cl- channels and volume regulation in HeLa cells and Xenopus laevis oocytes. J Biol Chem 277:40066-40074

28. Sardini A, Amey JS, Weylandt KH, Nobles M, Valverde MA, Higgins CF 2003 Cell volume regulation and swelling-activated chloride channels. Biochim Biophys Acta 1618:153-162

29. Kirk K 1997 Swelling-activated organic osmolyte channels. J Membr Biol 158:1-16

30. Pasantes-Morales H 1996 Volume regulation in brain cells: cellular and molecular mechanisms. Metab Brain Dis 11:187-204

31. Hoffmann EK 2000 Intracellular signalling involved in volume regulatory decrease. Cell Physiol Biochem 10:273-288

32. Pasantes-Morales H, Cardin V, Tuz K 2000 Signaling events during swelling and regulatory volume decrease. Neurochem Res 25:1301-1314

33. Verbalis JG 1998 Adaptation to acute and chronic hyponatremia: implications for symptomatology, diagnosis, and therapy. Semin Nephrol 18:3-19

34. Berl T 1990 Treating hyponatremia: damned if we do and damned if we don't. Kidney International 37:1006-1018

35. Adler S, Verbalis JG, Williams D 1994 Brain buffering is restored in hyponatremic rats by correcting their plasma sodium concentration. Journal of the American Society of Nephrology 5:85-92
36. Sterns RH, Thomas DJ, Herndon RM 1989 Brain dehydration and neurologic deterioration after rapid correction of hyponatremia. Kidney International 35:69-75
37. Sterns RH, Riggs JE, Schochet SS, Jr. 1986 Osmotic demyelination syndrome following correction of hyponatremia. New England Journal of Medicine 314:1535-1542
38. Kleinschmidt-DeMasters BK, Norenberg MD 1981 Rapid correction of hyponatremia causes demyelination: relation to central pontine myelinolysis. Science 211:1068-1070
39. Laureno R 1980 Experimental pontine and extrapontine myelinolysis. Transactions of the American Neurological Association 105:354-358
40. Verbalis JG, Gullans SR 1993 Rapid correction of hyponatremia produces differential effects on brain osmolyte and electrolyte reaccumulation in rats. Brain Research 606:19-27
41. Lien YH, Shapiro JI, Chan L 1990 Effects of hypernatremia on organic brain osmoles. J Clin Invest 85:1427-1435
42. Heilig CW, Stromski ME, Blumenfeld JD 1989 Characterization of the major brain osmolytes that accumulate in salt-loaded rats. American Journal of Physiology 257:F1108-F1116
43. Cserr HF, DePasquale M, Nicholson C, Patlak CS, Pettigrew KD, Rice ME 1991 Extracellular volume decreases while cell volume is maintained by ion uptake in rat brain during acute hypernatremia. J Physiol 442:277-295
44. Soupart A, Penninckx R, Namias B, Stenuit A, Perier O, Decaux G 1996 Brain myelinolysis following hypernatremia in rats. J Neuropathol Exp Neurol 55:106-113
45. Adler S, Martinez J, Williams DS, Verbalis JG 2000 Positive association between blood brain barrier disruption and osmotically-induced demyelination. Mult Scler 6:24-31
46. Baker EA, Tian Y, Adler S, Verbalis JG 2000 Blood-brain barrier disruption and complement activation in the brain following rapid correction of chronic hyponatremia. Exp Neurol 165:221-230
47. Palevsky PM 1998 Hypernatremia. Semin Nephrol 18:20-30
48. Chan PH, Fishman RA 1979 Elevation of rat brain amino acids, ammonia and idiogenic osmoles induced by hyperosmolality. Brain Research 161:293-301
49. Cserr HF, DePasquale M, Patlak CS 1987 Volume regulatory influx of electrolytes from plasma to brain during acute hyperosmolality. American Journal of Physiology 253:F530-F537
50. Cserr HF 1988 Role of secretion and bulk flow of brain interstitial fluid in brain volume regulation. Annals of the New York Academy of Sciences 529:9-20
51. Trachtman H, Futterweit S, Hammer E, Siegel TW, Oates P 1991 The role of polyols in cerebral cell volume regulation in hypernatremic and hyponatremic states. Life Sciences 49:677-688
52. Arieff AI, Guisado R 1976 Effects on the central nervous system of hypernatremic and hyponatremic states. Kidney International 10:104-116
53. Strange K, Morrison R, Heilig CW, DiPietro S, Gullans SR 1991 Upregulation of inositol transport mediates inositol accumulation in hyperosmolar brain cells. American Journal of Physiology 260:C784-C790
54. Garcia-Perez A, Burg MB 1991 Renal medullary organic osmolytes. Phys Rev 71:1081-1115
55. Kahn A, Blum D, Casimir G, Brachet E 1981 Controlled fall in natremia in hypertonic dehydration: possible avoidance of rehydration seizures. Eur J Pediatr 135:293-296
56. Hogan GR, Dodge PR, Gill SR, Pickering LK, Master S 1984 The incidence of seizures after rehydration of hypernatremic rabbits with intravenous or ad libitum oral fluids. Pediatr Res 18:340-345
57. Blum D, Brasseur D, Kahn A, Brachet E 1986 Safe oral rehydration of hypertonic dehydration. J Pediatr Gastroenterol Nutr 5:232-235

PHYSIOLOGICAL ROLE OF
N-ACETYLASPARTATE

Contribution to Myelinogenesis

Robert W. Ledeen, Jianfeng Wang, Gusheng Wu, Zi-Hua Lu,
Goutam Chakraborty, Markus Meyenhofer, Stephen K. Tyring, and
Reuben Matalon[*]

1. INTRODUCTION

It was close to half a century ago that *N*-acetylaspartate (NAA) was first discovered as a major amino acid derivative in mammalian brain (Tallan et al., 1956). It was subsequently shown to occur in all regions of the CNS with highest concentrations in cerebral gray matter (Tallan, 1957), estimated at close to 10 mM (Marcucci et al., 1966; Burri et al., 1991). During development, NAA is found in both neurons and oligodendrocytes (OLs) but at maturity it becomes localized in the former where its intraneuronal concentration reaches an estimated 10-14 mM (Simmons et al., 1991; Urenjack et al., 1992; 1993; Moffett and Namboodiri, 1995). Owing to this feature, together with its distinctive chemical shift in magnetic resonance spectroscopy, it has become widely used as a non-invasive *in vivo* indicator of neuronal viability (or lack thereof) in a variety of neurological disorders (Tsai and Coyle, 1995). This in turn has stimulated interest in the biological function of NAA which has remained elusive. Aspartoacylase (ASPA), the enzyme that converts NAA to aspartate and acetate, showed highest activity in OLs among cultured rat macroglia (Baslow et al., 1999). The biomedical significance of this enzyme became evident with the discovery

[*] Robert W. Ledeen, Jianfeng Wang, Gusheng Wu, Zi-Hua Lu, Goutam Chakraborty, Markus Meyenhofer, Dept. Neurology & Neurosciences, New Jersey Medical School, UMDNJ; 185 So. Orange Ave., Newark, NJ 07103; Stephen K. Tyring and Reuben Matalon, Dept. Pediatrics, Univ. Texas Medical Branch, Children's Hospital, 301 University Boulevard, Galveston, TX 77555-0359. Corresponding author, RWL: ledeenro@umdnj.edu

that Canavan disease (hyperacetylaspartia), resulting from autosomal recessive defect in ASPA, gives rise to NAA accumulation in brain accompanied by edema and progressive loss of OLs and myelin (Matalon et al., 1988).

Among the earlier studies that focused on biological function were those of D'Adamo and coworkers who showed that the acetyl group of NAA is incorporated into rat brain lipids (D'Adamo and Yatsu, 1966; D'Adamo et al., 1968). Extending that finding, Burri et al. (1991) showed that NAA is a major source of acetyl groups for lipid synthesis during brain development. Both groups found NAA to be incorporated more efficiently than free acetate. In the same vein, Mehta and Namboodiri (1995) proposed that NAA serves as a storage form of acetate for acetyl-CoA formation. Whereas these various studies employed extracellular application of NAA in the form of intracerebral injection or tissue incubation, a subsequent report (Chakraborty et al., 2001) demonstrated NAA-acetyl incorporation into myelin lipids from NAA originating in the neuron. This suggested axon-to-myelin transfer of NAA and hydrolysis within the membrane by myelin-localized ASPA. That study provided biochemical evidence for occurrence of ASPA in myelin, consistent with the latter model. To further highlight the role of this enzyme in myelin lipid synthesis we have utilized an ASPA-null knockout (KO) mouse created by Matalon et al. (2000) as a model of Canavan disease to show that inability to liberate acetyl groups from NAA-acetyl results in reduced myelin content in brain and specific lipid deficit within myelin.

As an additional test of the hypothesis, we have compared the NAA-synthase activity in neurons that extend myelinated axons with those whose axons remain unmyelinated, on the assumption that the latter would be programmed to produce less NAA than the former. One system we employed for that purpose was retina, which has only one cell type, retinal ganglion neurons, that extend myelinated axons and several, including photoreceptor cells, that do not. Another was comparison of cell cultures of cerebellar granule neurons (CGN) that do not extend myelinated axons with mixture of cortical neurons (CN), some of which do. Neurons lacking myelinated axons might still have need for some (presumably less) NAA as precursor to N-acetylaspartylglutamate (NAAG), an abundant dipeptide neurotransmitter (Gehl et al., 2004). This study required a reliable assay for NAA-synthase (L-aspartate N-acetyltransferase = ANAT), and because of our unexpected finding that this enzyme is bimodally expressed in both microsomes and mitochondria of brain (Lu et al., 2004), we briefly review the assay method and subcellular results obtained with rat brain.

2. METHODS

2.1. Lipid Analyses

For this study we employed knockout (KO) mice ~9 weeks of age and normal mice ~6 weeks old. Two brains of each were separately pooled and homogenized in 0.32 M sucrose with 20 mM Tris buffer (pH 7.2); aliquots were subjected to protein assay by the method of Lees and Paxman (1972), designed for highly insoluble proteins. Separate portions were used for whole brain lipid analysis (see below) and myelin isolation, the latter fraction also being assayed for protein and lipid phosphorus (Mrsny et al., 1986). Both portions were extracted with chloroform-methanol and the solubilized lipids passed through DEAE-Sephadex to yield the mixture of neutral and zwitterionic lipids (Ledeen and Yu, 1982). These were analyzed by thin-layer chromatography (TLC) on silica gel 60 plates (Merck), the amounts of lipid corresponding to equal amounts of protein (whole brain, ~120 μg; myelin ~30 μg). The plate was developed first with chloroform/methanol/acetic acid/formic acid/water (35/15/5/1/1, by vol.) ascending 6 cm, followed by drying and redevelopment in the same direction to the top with hexane/ethyl acetate (7/3, by vol.). Lipid bands were visualized by spraying with 3% cupric acetate in 8% phosphoric acid and heating at 150°C for 10 min; densitometric scanning was carried out with a FluoChem™ digital imaging system (Alphainnotech.com). Cerebroside bands were individually quantified by comparing to standard curve obtained with bovine brain cerebrosides (SIGMA) on the same TLC.

2.2. Method for Assay of NAA-Synthase (ANAT)

This reaction was carried out as described (Lu et al., 2004) on 3 subcellular fractions: microsomes, mitochondria, and an intermediate fraction obtained by subjecting the supernatant from the crude mitochondrial preparation (obtained at 10,000g x 10 min) to centrifugation at 35,000g x 20 min; the latter proved to be a mixture of microsomes and mitochondria. Each of these subfractions was dispersed in reaction buffer consisting of (in mM) 40 Na-PO_4 (pH 7), 600 sorbitol, 60 NaCl, 1 DTT; the protein concentration was ~5 mg/ml. Variable amounts of this dispersion representing 250-350 μg protein were added to 1.5 ml Eppendorf tubes containing the above substances at the concentrations indicated together with (in mM) 2 acetyl-CoA, 2.7 [^{14}C]L-aspartate (50 x 10^3 DPM), 2 CHAPS, in total volume of 60 μl. The tubes were placed in a 37°C water bath and shaken 1 hr. Reaction was stopped by placing in ice and 4 μl of 30 mM NAA was added. After centrifugation at 16,000g x 5 min, half (30 μl) of the supernatant was applied to an aluminum-backed silica gel 60 TLC plate (10 x 20 cm); standards of [^{14}C]NAA and [^{14}C]aspartate were also applied The plate was first developed with ethyl acetate/methanol/water (70/20/2, by vol.), dried and subjected to a second run in ethyl acetate/methanol/formic acid (60/30/7, by vol.). After drying, the plate was exposed to a storage phosphor screen for 2 days and radioactivity quantified by PhosphorImager (Amersham Biosciences, Typhoon™). The readings for background zones were subtracted from the NAA bands. Standard [^{14}C]NAA of varying radioactivity gave a linear response that was used to convert PhosphorImager readings to DPM or nmoles.

2.3. Assay of NAA-Synthase (ANAT) in Pig Retina

Eyes were obtained from freshly euthanized adult pigs as a gift from Dr. Stephen Vatner (animals were employed by his group in an unrelated experiment). Rear eyecups containing retina were cut into pieces approximately 4 x 4 mm in area and 200 μm thick, and quickly frozen at -80°C. Those pieces were flattened and the retinal layers (~200 μm thick) were frozen-sectioned into 9 sections, ~20 μm thick. The 3 top-most sections were retinal ganglion cell-enriched and the 3 bottom- most were photoreceptor cell-enriched. These were separately pooled and homogenized in sodium phosphate buffer (40 mM, pH 7.0) containing sorbitol (600 mM) and dithiothreitol (1 mM). Microsomal and mitochondrial fractions were isolated as described above. These were dispersed in the above buffer to a concentration of ~5 mg protein/ml, then supplemented with NaCl (60 mM) and assayed for ANAT as above. ANAT activities expressed as nmol/mg protein/hr are presented as average ± SD from 2-3 independent experiments.

2.4. Comparison of ANAT in Cerebellar Granule Neurons and Cortical Neurons

Cell cultures of cortical neurons (CN) were prepared from 2-day old rat pups as described (Xie et al., 2002) and cerebellar granule neurons (CGN) from 7-day old rat pups also as described by us (Wu et al., 1995). Both cell types were cultured over 6-7 days; glial cell growth was suppressed with FUdR. The cells were collected and homogenized in the above buffer. ANAT assay was carried out as above, applied to whole cells rather than subcellular fractions.

3. RESULTS

3.1. Lipid Analyses

Biochemical analysis (Table) revealed myelin yield to be decreased ~30% for the KO mouse based on protein. Myelin phospholipid, measured as total myelin phosphorus relative to myelin protein, was reduced ~20% in both trials. Most striking was reduction of cerebroside, one of the characteristic lipids of myelin (Fig. 1A). This was significantly more pronounced for the upper band of the doublet (cerebroside 1), which possesses unsubstituted long chain fatty acids within ceramide in contrast to the lower band which has 2-hydroxyfatty acids in the ceramide unit (Norton and Cammer, 1984). Quantfication of these 2 cerebroside classes was carried out by densitometry, using standard brain cerebrosides of variable amounts applied to the same TLC; this gave a linear response curve (Fig. 1B). Cerebroside 1 was depleted ~65% in myelin relative to myelin protein, and somewhat more (87-90%) in whole brain relative to brain protein (Table). The greater decrease in brain was evidently the result of both reduced myelination plus deficient synthesis of that lipid within myelin. Cerebroside 2 showed moderate (~22%) depletion in myelin and somewhat more (35-44%) in brain (Table). The higher weight of the KO brains, despite equivalent protein, was attributed to edema characteristic of the Canavan condition.

Table: Reduction of myelin level and myelin lipids in ASPA-null mice

		Trial 1		Trial 2	
		Normal	KO	Normal	KO
Brain weight (mg)/brain		355	475	373	473
Total protein (mg)/brain		42.6	43.4	48.1	44.6
Myelin protein (mg)/brain		1.12	0.75	1.09	0.74
Myelin protein/brain protein (%)		2.62	1.72	2.27	1.66
Myelin phospholipid P					
(μg/mg myelin protein)		53.0	41.8	55.0	43.8
Cerebroside (μg/mg protein)					
Brain:	Cerebroside 1	17.4	1.7	17.5	2.2
	Cerebroside 2	32.8	21.3	35.3	19.9
Myelin:	Cerebroside 1	202	73.0	175	55.0
	Cerebroside 2	317	249	322	254

Summary of lipid and protein quantifications of KO and normal brain and myelin. KO mice gave decreased myelin yield based on myelin protein per brain and brain protein. Myelin phospholipid phosphorus relative to myelin protein was decreased in KO myelin. Cerebrosides were decreased in KO brain relative to brain protein and in KO myelin relative to myelin protein.

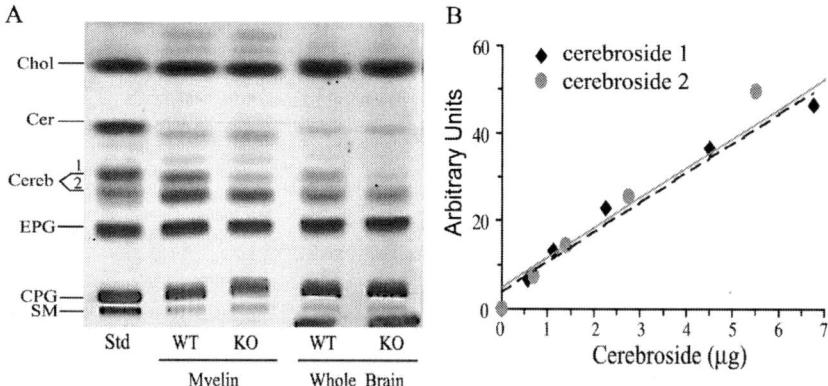

Figure 1. Thin-layer chromatography of lipids from brain and myelin of normal and ASPA KO mice. **A:** Neutral and zwitterionic lipids were subjected to TLC on silica gel 60 plates as described. The main observed difference was significant depletion of the upper cerebroside band (Cereb 1) and moderate reduction of the lower band (Cereb 2) in the KO samples (see Table). Chol, cholesterol; Cer, ceramide; EPG, ethanolamine phosphoglycerides; CPG, choline phosphoglycerides; SM, sphingomyelin. **B:** Standard curves obtained with known amounts of brain cerebrosides; Cereb 1 and Cereb 2 were separately quantified.

In addition to the major alterations of cerebroside content, less pronounced changes were suggested in other lipids, e.g. modest decrease in ethanolamine phosphoglycerides and possible increase in ceramide of KO myelin. The latter might represent unutilized precursor for cerebroside synthesis. However, those and other changes require more detailed study.

3.2. Method for Assay of ANAT

Subcellular fractions were isolated from whole brains of 30-60 day old rats as described (Lu et al., 2004). In addition to microsomes and mitochondria, this included an intermediate fraction obtained by centrifuging the supernatant from crude mitochondria (obtained by

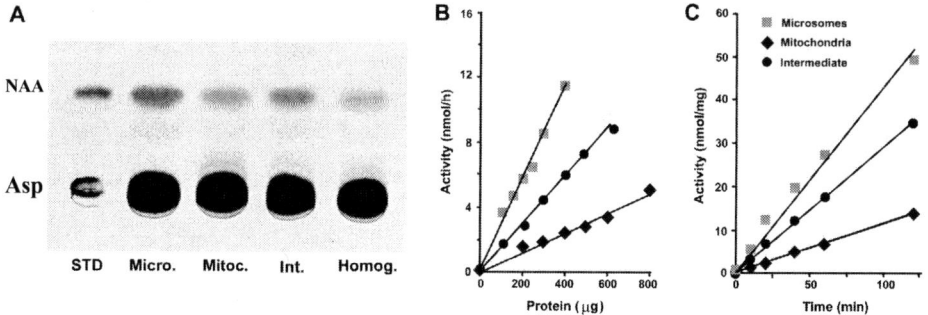

Figure 2. Application of ANAT assay to brain homogenate and subfractions. **A:** Representative image of TLC plate on storage phosphor screen. **B, C:** Variation of ANAT activity with protein and time, respectively. Asp = aspartate; STD = standard. Reproduced from Lu et al. (2004) with permission of Elsevier.

centrifugation at 10,000g x 10 min) at 35,000g x 20 min. As previously described (Lu et al., 2004), the microsomal fraction had the highest specific activity, ~4-5x that of mitochondria, while the intermediate fraction showed intermediate activity indicative of a mixture of these 2 fractions (Fig. 2). The latter result indicated ANAT was not localized in a subfraction such as small mitochondria that occur in axons and nerve endings. As shown, ANAT activity varied linearly with protein content and time for the 3 fractions.

3.3. Assay of ANAT in Pig Retina and Brain Cells

Applying the above assay to the various sections of pig retina revealed ANAT activity in the ganglion cell-enriched layer to be several times greater than that enriched in photoreceptor cells (Fig. 3A). This was especially true for microsomes, whose specific activity in the ganglion cell-enriched layer was ~8x that of microsomes in the photoreceptor cell-enriched layer. As with whole brain, ANAT activities in mitochondria were significantly lower in all cases. As indicated, the tissue sections analyzed in this manner represented enriched- rather than pure cell populations.

Analysis of isolated brain cells (without subcellular fractionation, in this case) also revealed significant differences in ANAT activity between cells that do not extend myelinated axons (CGN) and cell cultures of mixed types (CN) that include some projection neurons that do ultimately extend myelinated axons (Fig. 3B). That the difference in this case was not as striking as with retinal neurons was likely due to the presence of non-myelinating interneurons in the CN as well as the early stage of development. Even though myelinated axons have not yet formed in the analyzed cultures, upregulation of ANAT has likely begun at that stage.

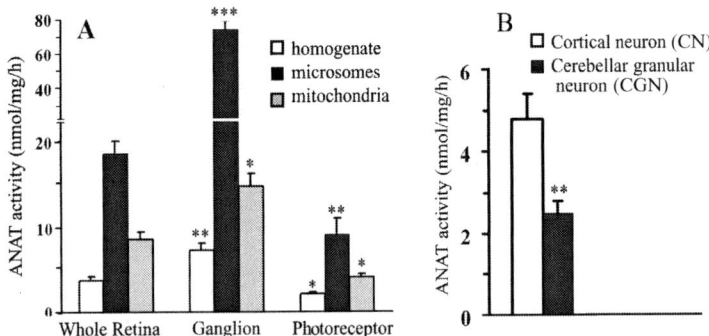

Figure 3. Comparison of ANAT activity in various neuron types. **A:** Sections of pig retina. **B:** Cultures of cortical neurons (CN) and cerebellar granule neurons (CGN). Two-tailed student t test: * $p < 0.05$; ** $p < 0.01$; *** $p < 0.001$; compared to value of corresponding fraction in whole retina (panel **A**) or to value of CN (panel **B**).

4. DISCUSSION

The importance of NAA and its metabolism in CNS function is highlighted by the fatal outcome of Canavan disease in which ASPA activity is disrupted. Several physiological roles have been proposed for this CNS-unique substance (Tsai and Coyle, 1995), including neuronal molecular water pump (Baslow, 2000; 2003), neuronal osmoregulation (Taylor et al., 1995), intracellular anion pool (McIntosh and Cooper, 1965), a storage form of aspartate (Birken and Oldendorf, 1989), both precursor and metabolite of NAAG (Cangro et al., 1987), a storage form of acetate for acetyl-CoA formation (Mehta and Namboodiri, 1995), and source of acetyl groups for myelin lipid synthesis (Chakraborty et al., 2001). Biochemical studies showing ASPA occurrence predominantly in white matter of brain (McIntosh and Cooper, 1965; D'Adamo et al., 1973; Goldstein, 1976; Kaul et al., 1991) provided early suggestions of a role in myelin formation and/or turnover, consonant with developmental data in the rat showing maximal enzyme activity (Goldstein, 1976) and *in situ* hybridization (Kirmani et al., 2003) at the peak of myelination. Initially described as a soluble enzyme (D'Adamo et al., 1973; 1977), ASPA was also claimed to be membrane-bound as well (Goldstein, 1976), a result supported by our findings (Chakraborty et al., 2001). Kaul et al. (1991) pointed to the requirement of detergent for solubilization as evidence of membrane association. This may also explain our finding of an hour or so delay in biochemical detection of ASPA activity in myelin (Fig. 4); this was not observed with brain homogenate which included cytosolic- as well as membrane-bound activity (Chakraborty et al., 2001). A recent study in which biochemical analysis failed to detect ASPA in myelin employed only 1 hr of incubation (Madhavarao et al., 2004). Our quantitative biochemical analyses indicated similar specific activities of ASPA in myelin and brain cytosol. These likely refer to subfractions of OLs, the cell type in which ASPA is localized (Baslow et al., 1999).

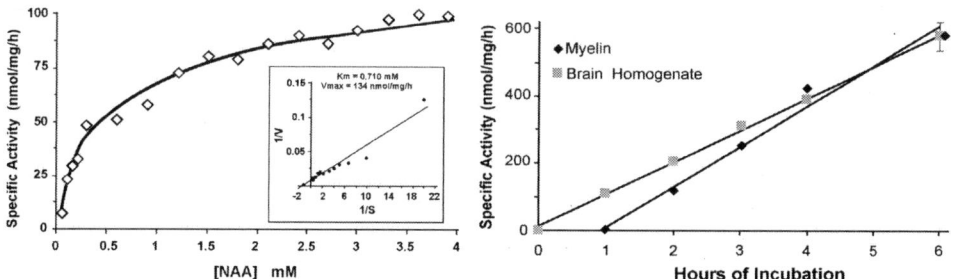

Figure 4. Biochemical detection of ASPA in myelin. Left panel: variation of myelin ASPA activity with NAA concentration; inset = Lineweaver-Burk plot. Right panel: variation of ASPA activity with time. Myelin required one hr of contact with detergent before activity commenced; it achieved parity with brain homogenate at 4 hr. Reproduced from Chakraborty et al. (2001) with permission from Blackwell Publishing.

Evidence supporting a role for myelin-associated ASPA in providing acetyl groups for myelin lipid synthesis was provided in the current study of myelin lipids from brains of

ASPA-null (KO) mice, previously created by Matalon et al. (2000). Absence of this enzyme in brain would preclude liberation and hence utilization of acetyl groups from NAA. The yield of isolated myelin from the brains of such animals was ~30% less than from normal brain, consistent with the myelin aberration found in Canavan disease brain (Matalon et al., 1995). In addition to that deficit we found ~20% reduction of phospholipid relative to protein in the KO myelin; preliminary studies suggest this may apply to specific phospholipids rather than phospholipids in general. Of particular interest was the significant reduction of cerebroside, pronounced deficit occurring in the subgroup of cerebrosides containing unsubstituted fatty acids. This group (Cereb 1, Fig. 1A) was depleted ~65% in myelin and somewhat more in whole brain, attributed to reduced myelination plus deficient synthesis of that lipid relative to myelin protein. The final step in cerebroside synthesis, catalyzed by galactosyltransferase, was previously shown to occur bimodally in myelin and microsomes (Neskovic et al., 1973; Costantino-Ceccarini and Suzuki, 1975; Koul and Jungalwala, 1986); our finding suggests the same may apply to synthesis of the ceramide unit of cerebrosides, the component that would utilize NAA-acetyl.

The above mentioned early studies showing incorporation of the acetyl group of NAA into brain and myelin lipids employed extracellular application of NAA in the form of intracerebral injection or tissue slice incubation, leaving open the question of intercellular transfer. A more recent study utilizing the rat optic system demonstrated that [^{14}C-acetyl]-NAA originating in retinal ganglion neurons provided acetyl for lipid synthesis within myelin of the optic system (Chakraborty et al., 2001). This suggested transaxonal movement of NAA from neuron to myelin, following by hydrolysis via myelin-localized ASPA. The presence of ASPA in myelin provides an interesting parallel to the numerous other lipid-synthesizing and metabolizing enzymes that have been detected as integral components of purified myelin (for review: Norton and Cammer, 1984; Ledeen, 1992). Especially significant in the present context are the fatty acid synthesizing enzymes, acetyl-CoA carboxylase and fatty acid synthase (Chakraborty and Ledeen, 2003); the latter enzyme complex in purified myelin showed a level of activity approximately half that of its cytosolic counterpart, and differed from the latter in requiring detergent. These are the enzymes that would utilize ASPA-liberated acetyl groups to form fatty acids, which then become incorporated into myelin phospholipids. Several myelin intrinsic enzymes that catalyze the latter reactions have been identified, including those that mediate the Kennedy pathway for synthesis of choline- and ethanolamine-phosphoglycerides (Ledeen, 1992). These findings indicate NAA can be added to the list of precursors shown to undergo axon-to-myelin transfer with subsequent incorporation into myelin lipids. These include choline (Droz et al., 1978; 1981), phosphate (Ledeen and Haley, 1983), acyl chains (Toews and Morell, 1981; Alberghina et al., 1982), and serine (Haley and Ledeen, 1979). Yet to be elucidated is the mechanism by which such axon-to-myelin transfer proceeds. Figure 5 depicts the proposed pathways for NAA metabolism, transfer, and utilization within myelin.

This study has provided additional correlative evidence for a role of NAA in myelinogenesis in comparing rates of NAA synthesis (ANAT) in various neuronal types. This was seen most clearly in the robust rate of NAA synthesis in adult pig retinal sections containing retinal ganglion neurons, the one retinal neuron type whose axon is myelinated; ANAT rate was ~8-fold greater in those sections than in retinal sections enriched in photoreceptor cells whose axons are not myelinated (Fig. 3A). A lesser though still significant difference was seen in ANAT activity in CN vs CGN (Fig. 3B), the former containing some neurons destined to extend myelinated axons in contrast to CGN that do not. That the difference was not greater was likely due to the mixed neuronal types in CN and the early developmental stage. Complete absence of ANAT in neurons with non-myelinated axons would not be expected since such cells might well require ANAT for synthesis of NAAG, an abundant neurotransmitter (Gehl et al., 2004). This study further highlights the bimodal distribution of ANAT in microsomes and mitochondria, reported in our recent study (Lu et al., 2004). These findings suggest the possibility of compartmentalization, microsomal ANAT producing NAA for myelin and the mitochondrial product perhaps being utilized for NAAG synthesis. However, there is no supporting evidence as yet for this speculative idea.

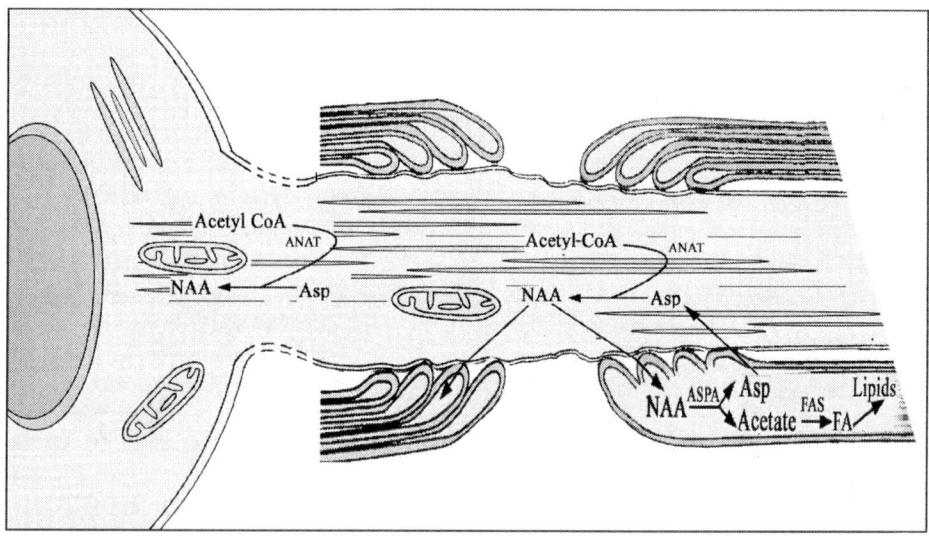

Figure 5. Cartoon depiction of our working hypothesis showing neuronal synthesis of NAA (by ANAT) in both microsomes and mitochondria, followed by transaxonal transfer to myelin and release of acetyl groups within myelin by aspartoacylase (ASPA). Myelin-associated enzymes, such as fatty acid synthase (FAS), utilize the liberated acetyl groups for fatty acid synthesis followed by incorporation into myelin lipids by other myelin-localized enzymes. Liberated aspartate (Asp) is recycled back to the neuron. Absence of ASPA is postulated to block this pathway of myelin lipid synthesis.

In proposing this role for NAA in relation to myelinogenesis it is appropriate to emphasize that this is not conceived as the sole or even major pathway for myelin lipid synthesis. Myelination proceeds to a significant degree in the absence of ASPA-generated, NAA-derived acetyl groups, obviously through parallel pathways residing within the OL. Nevertheless inability to utilize NAA-acetyl leads to significant lipid deficits within myelin that may contribute to the observed pathology and nervous system dysfunction in Canavan disease. Analysis of other myelin lipids from ASPA-null animals may further elucidate this deficit as it affects myelin not only of Canavan patients but also victims of other neurological disorders characterized by subnormal levels of NAA in brain.

5. QUESTION AND ANSWER SESSION

DR. BASLOW: Now we're open to questions. I have one question, if I may. Would you say that this acetate is necessary for the buildup of myelin?

DR. LEDEEN: Oh, no. I should make that point clear. This is a pathway that is not the primary source of myelin. We see it more as a later addition to the repertoire of myelin synthesis, perhaps maintenance and turnover of myelin.

After the CNS myelin has been synthesized, it would be a tremendous job for one little oligodendrocyte to maintain these 50 internodes or so that it has put out. And we think in that case, there is the need for trophic support from the axon.

So we feel it's more of a later phenomenon, although we don't preclude the possibility that it might also have a role earlier on but not the sole or even primary role.

DR. BASLOW: Thank you.

PARTICIPANT: As a clinician, I've never been able to satisfy myself of the answer to what that rate of myelin turnover is. Do you have a sense of the rate of myelination?

DR. LEDEEN: Oh, my. There's a long history of studies on myelin turnover rates. And there is no clear picture except we now know that the early idea that myelin does not turn over, that it is an inert membrane, is not true, that both the proteins and the lipids do turn over at variable rates. Some of the lipids turn over more rapidly than others. And one that does not turn over quite so rapidly is the one I mentioned, the ethanolamine phosphoglyceride, which is made up largely of plasmalogen form, that lipid. And that has a longer half-life. And that is the one that seems to be labeled by NAA more than the others that are turning over more rapidly.

So I suspect that's going to be part of the story that will have to be clarified as time goes on, what is the relation between this pathway and turnover, as opposed to other pathways that lead to myelin synthesis.

DR. NAMBOODIRI: I have one question regarding that high molecular weight form of your enzyme. The ASPA you find associated with myelin, you find that the molecular weight is around the 53-54?

DR. LEDEEN: Or higher, yes.

DR. NAMBOODIRI: What type of antibody did you use? Was it a peptide-based antibody or was it an antibody against the protein?

DR. LEDEEN: Well, I got the antibody from Reuben Matalon. So he can probably answer that.

SESSION CO-CHAIR MATALON: This is an antibody that we prepared from purified protein. Actually, the antibody was made with the 313 amino acids, about 36 kilodalton protein, just what we used to generate the antibody, this class of antibody. We had in the past a different antibody. And we had similar finding to what Dr. Ledeen had. In our first publication on aspartoacylase, we said that the molecular weight was about 56 kilodaltons.

When we isolated the gene and we found that it codes for 313 amino acids, we went back and tried to find out what was the cause for that big protein. And we found out that when we changed the isolation technique, we got a smaller protein. We decreased the salt concentration in the isolation prep. And we got about 36 kilodalton.

DR. KOLODNY: Is it possible, then, that there are really two different proteins, one in the cytosol and the 56, the larger protein and because aspartoacylase is an esterase and has structure that is similar to other esterases? So could the antibody, then, be reacting to more than one esterase?

SESSION CO-CHAIR MATALON: This is a polyclonal antibody.   So it can have various epitopes.

So I can't tell you. It detected the 56 kilodalton in your case and also the 36 kilodalton.

DR. MADHAVARAO: It is possible that the 56 kilodalton is only in myelin, and is not the cytosoic isoform.   So unless the cytosolic isoform is also purified, which has no contamination from the myelin form, it is not possible; you should see some residual activity with the cytosolic form.

SESSION CO-CHAIR MATALON: Okay. Our isolation of aspartoacylase, we used two types of preparations. One is human skin fibroblasts, a large quantity of them, and the other one is bovine brain.

I cannot really remember which is which, and which gave what results. That was about ten years ago or more.

DR. ROSS: Just a question. Is there any way of translating? You gave us the relative rates of incorporation from NAA compared with acetate as the gold standard.   Can we translate that into rates in micromoles per gram or something which you might --

DR. LEDEEN: Oh, it would be difficult to make that calculation because of the methodology. I can talk to you about that later if you like. It's difficult to do that.

DR. TRAPP: It appeared to me that the 57-kilodalton band is a little higher than that. Have you considered the possibility that incorporation of this enzyme into myelin in the membrane requires dimerization of the molecule? And are you potentially looking at a dimer, rather than a separate molecule?

DR. LEDEEN: That we're looking at a dimer? Well, then we should see some of the monomer, too, as well as the dimer if that's the case, shouldn't we? You might not expect it to. Well, maybe it could. Well, that is an interesting possibility, yes. So sometimes it's difficult to get a really accurate measure of molecular weights, but it's within the range anyway.

SESSION CO-CHAIR MATALON: I think this is a good question because this same question came up when we presented the data at another meeting; the Society for Inborn Metabolic Diseases. And the question came up, could this be a dimer? I have no answer for that, really. It could be.

DR. MADHAVARAO: In SDS-PAGE, unless they are covalently linked, you cannot expect to have a dimer, right?

PARTICIPANT: Or just to comment that it could be glycosylated, that the protein that went into myelination or in that component, could it be glycosylated?

DR. LEDEEN: In fact, we're looking at that. I should mention that my colleague who did a lot of this work is here, Dr. Wu. He is in the process of doing it. Unfortunately, we don't have a clear answer, but it would take too many carbohydrates to make that much difference in molecular weight.

But we know that some glycoproteins migrate aberrantly. They don't conform exactly to their true molecular weight after they acquire carbohydrate. So it's possible.

6. REFERENCES

Alberghina, M.M., Viola, M., and Giuffrida, A.M., 1982, Transfer of axonally transported phospholipids into myelin isolated from rabbit optic pathway, *Neurochem. Res.* **7**:139-149.

Baslow, M.H., Suckow, R., Saperstein, V., and Hungund, B.L., 1999, Expression of aspartoacylase activity in cultured rat macroglial cells is limited to oligodendrocytes, *J. Molec. Neurosci.* **13**:47-53.

Baslow, M.H., 2000, Functions of N-acetylaspartate and N-acetyl-L-aspartylglutamate in the vertebrate brain: role in glial cell-specific signaling, *J. Neurochem.* **75**:453-459.

Baslow, M.H., 2003, N-Acetylaspartate in the vertebrate brain: metabolism and function, *Neurochem. Res.* **28**:941-953.

Birken, D.L. and Oldendorf, W.H., 1989, N-acetyl-L-aspartic acid: a literature review of a compound prominent in the ^{1}H-NMR spectroscopic studies of brain, *Neurosci. Biobehav. Rev.***13**:23-31.

Burri, R., Steffen, C., and Herschkowitz, N., 1991, N-Acetyl-L-aspartate is a major source of acetyl groups for lipid synthesis during rat brain development, *J. Molec. Neurosci.* **13**:403-411.

Cangro, C.B., Namboodiri, M.A.A., Sklar, L.A., Corigliano-Murphy, A., and Neale, J.H., 1987, Biosynthesis and immunohistochemistry of N-acetylaspartylglutamate in spinal sensory ganglia, *J. Neurochem.* **49**:1579-1588.

Chakraborty, G., Mekala, P., Yahya, D., Wu, G., and Ledeen, R.W., 2001, Intraneuronal N-acetylaspartate supplies acetyl groups for myelin lipid synthesis: evidence for myelin-associated aspartoacylase, *J. Neurochem.* **78**:736-745.

Chakraborty, G. and Ledeen, R.W., 2003, Fatty acid synthesizing enymes intrinsic to myelin, *Molec. Brain Res.* **112**:46-52.

Costantino-Ceccarini, E. and Suzuki, K., 1975, Evidence for the presence of UDP-galactose:ceramide galactosyltransferase in rat myelin, *Brain Res.* **93**:358-362.

D'Adamo, A.F. and Yatsu F.M., 1966, Acetate metabolism in the nervous system. N-acetyl-L-aspartic acid and the biosynthesis of brain lipids, *J. Neurochem.* **13**:961-963.

D'Adamo, A.F., Gidez, L.I., and Yatsu, F.M., 1968, Acetyl transport mechanisms. Involvement of N-acetyl aspartic acid in *de novo* fatty acid biosynthesis in the developing rat brain, *Exp. Brain Res.* **5**:267-273.

D'Adamo, A.F., Smith, J.C., and Woiler, C., 1973, The occurrence of N-acetylaspartate amidohydrolase (aminoacylase II) in the developing rat, *J. Neurochem.* **20**:1275-1278.

D'Adamo, A.F., Peisach, J., Manner, G., and Weiler, C.T., 1977, N-Acetyl-aspartate amidohydrolase : purification and properties, *J. Neurochem.* **28**:739-744.

Droz, B., Di Giamberardino, L., Koenig, H.L., Boyenval, J., and Hassig, R., 1978, Axon-myelin transfer of phospholipid components in the course of their axonal transport as visualized by radioautography, *Brain Res.* **155**:347-353.

Droz, B., Di Giamberardino, L., and Koenig, H.L., 1981, Contribution of axonal transport to the renewal of myelin phospholipids in peripheral nerves. I. Quantitative radioautographic study, *Brain Res.* **219**:57-71.

Gehl, L.M., Saab, O.H., Bzdega, T., Wroblewska, B., and Neale, J.H., 2004, Biosynthesis of NAAG by an enzyme-mediated process in rat central nervous system neurons and glia, *J. Neurochem.* **90**:989-997.

Goldstein, F.B., 1976, Amidohydrolases of brain: enzymatic hydrolysis of N-acetyl-L-aspartate and other N-acyl-L-amino acids, *J. Neurochem.* **26**:45-49.

Haley, J.E. and Ledeen, R.W., 1979, Incorporation of axonally transported substances into myelin lipids, *J. Neurochem.* **32**:735-742.

Kaul, R., Casanova, J., Johnson, A.B., Tang, P., and Matalon, R., 1991, Purification, characterization, and localization of aspartoacylase from bovine brain, *J. Neurochem.* **56**:129-135.

Kirmani, B.F., Jacobowitz, D.M., and Namboodiri, M.A.A., 2003, Developmental increase of aspartoacylase in oligodendrocytes parallels CNS myelination, *Dev. Brain Res.* **140**:105-115.

Koul, O. and Jungalwala, F.B., 1986, UDP-galactose:ceramide galactosyltransferase of rat central nervous system myelin during development, *Neurochem. Res.* **11**:231-239.

Ledeen, R.W., 1992, Enzymes and receptors of myelin, in: *Myelin: Biology and Chemistry*, R.E. Martenson, ed., CRC Press, Boca Raton, pp. 531-570.

Ledeen, R.W. and Yu, R.K., 1982, Gangliosides: structure, isolation and analysis, *Methods Enzymol.* **83**:139-191.

Ledeen, R.W. and Haley, J.E., 1983, Axon-myelin transfer of glycerol-labeled lipids and inorganic phosphate during axonal transport, *Brain Res.* **269**:267-275.

Lees, M. and Paxman, S., 1972, Modification of the Lowry procedure for analysis of proteolipid protein, *Analy. Biochem.* **47**:184-192.

Lu, Z.-H., Chakraborty, G., Ledeen, R.W., Yahya, D. and Wu, G., 2004, N-Acetylaspartate synthesis is bimodally expressed in microsomes and mitochondria of brain, *Molec. Brain Res.* **122**:71-78.

Madhavarao, C.N., Moffett, J.R., Moore, R.A., Viola, R.E., Namboodiri, M.A.A., and Jacobowitz, D.M., 2004, Immunohistochemical localization of aspartoacylase in the rat central nervous system, *J. Compar. Neurol.* **472**:318-329.

Marcucci, F., Mussini, E., Valzelli, L., and Garattini, S., 1966, Distribution of N-acetyl-L-aspartic acid in rat brain, *J. Neurochem.* **13**:1069-1070.

Matalon, R., Michals, K., Sebasta, D., Deanching, M., Gashkoff, P., and Casanova, J., 1988, Aspartoacylase deficiency and N-acetylaspartic aciduria in patients with Canavan disease, *Am.. J.Med. Genet.* **29**:463-471.

Matalon, R., Michals, K., and Kaul, R., 1995, Canavan disease: From spongy degeneration to molecular analysis, *J. Pediat.* **127**:511-517.

Matalon, R., Rady, P.L., Platt, K.A., et al., 2000, Knock-out mouse for Canavan disease: a model for gene transfer to the central nervous system, *J. Gene Med.* **2**: 165-175.

McIntosh, J.C. and Cooper, J.R., 1965, Studies on the function of N-acetylaspartic acid in the rat brain, *J. Neurochem.* **12**:825-835.

Mehta, V. and Namboodiri, M.A.A., 1995, N-Acetylaspartate as an acetyl source in the nervous system, *Mol. Brain Res.* **31**:151-157.

Moffett, J.R. and Namboodiri, M.A.A., 1995, Differential distribution of N-acetylaspartylglutamate and N-acetylaspartate immunoreactivities in rat brain, *J. Neurocytol.* **24**:409-433.

Mrsny, R.J., Volwerk, J.J., and Griffith, O.H., 1986, A simplified procedure for lipid phosphorus analysis shows that digestion rates vary with phospholipid structure, *Chem. Ph;ys. Lipids* **39**:185-191.

Neskovic, N.M., Sarlieve, L.L., and Mandel, P., 1973, Subcellular and sub-microsomal distribution of glycolipid synthesizing transferases in young rat brain, *J. Neurochem.* **20**:1419-1430.

Norton, W.T. and Cammer, W., 1984, Isolation and characterization of myelin, in: *Myelin*, P. Morell, ed., Plenum Press, New York, pp. 147-195.

Simmons, M.L., Frondoza, C.G., and Coyle, J.T., 1991, Immunocytochemical localization of N-acetyl-aspartate with monoclonal antibodies, *Neurosci.* **45**:37-45.

Tallan, H. H., Moore, S., and Stein, W. H., 1956, N-Acetyl-L-aspartic acid in brain, *J. Biol. Chem.* 219:257-264.

Tallan, H.H., 1957, Studies on the distribution of N-acetyl-L-aspartic acid in brain, *J. Biol. Chem.* **224**:41-45.

Taylor, D.L., Davies, S.E.C., Obrenovitch, T.P., Doheny, M.H., Patsalos, P.N., Clark, J.B., and Symon, L., 1995, Investigation into the role of N-acetylaspartate in cerebral osmoregulation, *J. Neurochem.* **65**:275-281.

Toews, A.D. and Morell, P., 1981, Turnover of axonally transported phospholipids in nerve endings of retinal ganglion cells, *J. Neurochem.* **37**:1316-1323.

Tsai, G. and Coyle, J.T., 1995, N-Acetylaspartate in neuropsychiactric disorders, *Prog. Neurobiol.* **46**:531-540.

Urenjak, J., Williams, S.R., Gadian, D.G., and Noble, M., 1992, Specific expression of N-acetyl-aspartate in neurons, oligodendrocyte-type-2 astrocyte progenitors, and immature oligodendrocytes *in vitro*, *J. Neurochem.* **59**:55-61.

Urenjack, J., Williams, S.R., Gadian, D.G., and Noble, M., 1993, Proton magnetic resonance spectroscopy unambiguously identifies different neural cell types, *J. Neurosci.* **13**:981-989.

Wu, G., Lu, Z.-H., and Ledeen, R.W., 1995, Induced and spontaneous neuritogenesis are associated with enhanced expression of ganglioside GM1 in the nuclear membrane, *J. Neurosci.* **15**:3739-3746.

Xie, X., Wu, G., Lu, Z.-H., and Ledeen, R.W., 2002, Potentiation of a sodium-calcium exchanger in the nuclear envelope by nuclear GM1 ganglioside, *J. Neurochem.* **81**:1185-1195.

DEFECTIVE MYELIN LIPID SYNTHESIS AS A PATHOGENIC MECHANISM OF CANAVAN DISEASE

Aryan M. A. Namboodiri, John R. Moffett, Peethambaran Arun, Raji Mathew, Sreela Namboodiri, Asha Potti, Jeremy Hershfield, Batool Kirmani, David M. Jacobowitz[†] and Chikkathur N. Madhavarao[*]

1. INTRODUCTION

Canavan disease was first reported by Myrtelle Canavan in 1931 and was recognized as a distinct disease by Van Bogaert and Bertrand in 1949.[1] The clinical symptoms of CD include poor head control, macrocephaly, marked developmental delay, optic atrophy, seizures, hypotonia and death in early childhood.[2-4] Three clinical variants of CD are recognized: 1) the neonatal form in which the disease is more severe and is recognizable in the first few weeks of life, 2) the infantile form, the most common form in which the disease is apparent by 6 months of age and 3) the juvenile form in which the disease manifests only by age 4 or 5.[4-6]

The pathologies associated with Canavan disease include cortical and subcortical spongy degeneration, a lack of myelination, accumulation of water in the brain, and hypertrophy and hyperplasia of astrocytes.[1,7-9] Ultrastructural studies have demonstrated intramyelinic vacuolation, intense astrocyte swelling and unusually elongated mitochondria within those astrocytes, and most of these changes are detectable in the recent mouse model.[4,10-13]

Earlier, diagnosis of CD was established by brain biopsy demonstrating spongy degeneration of the white matter with vacuoles within myelin sheaths, astrocyte swelling and deformed mitochondria. Biochemical analyses have shown that hypomyelination is a characteristic feature of Canavan disease.[14] Patients with CD are found to excrete 10-100 fold higher amounts of NAA in their urine, and deficiency of the NAA degradative enzyme ASPA was demonstrated in their cultured skin fibroblasts.[15] NAA levels are also elevated in the blood and cerebrospinal fluid of CD patients. Proton nuclear magnetic resonance

[*] Uniformed Services University of the Health Sciences, 4301 Jones Bridge Rd, Bethesda MD, 20814 USA,
[†] Laboratory of Clinical Science, NIMH, NIH, Bethesda MD 20892, USA.

spectroscopy (MRS) of CD patients has revealed increased NAA levels in the brain.[16,17] However, increased urinary NAA remains the most reliable marker for CD, especially to distinguish it from other leukodystrophies.

Cloning of the human ASPA gene has enabled molecular genetic studies of CD.[18] Two mutations were found to be prevalent among Ashkenazi Jewish patients with CD.[9,19,20] A missense mutation in codon 285 causing substitution of glutamic acid to alanine accounts for 83.6 % of the mutations identified in 104 alleles from 52 unrelated Ashkenazi Jewish patients. A nonsense mutation on codon 231, which converts tyrosine to a stop codon, was found in 13.4% of the alleles from the Jewish patients. Among non-Jewish patients, the mutations are different and more diverse (see Zeng *et al.* this volume).[6] The most common is in codon 305, a missense mutation substituting alanine for glutamic acid. This mutation was observed in 35.7% of the 70 alleles from 35 unrelated non-Jewish patients. Fifteen other mutations were detected in 24 other CD patients. More recently, additional mutations, some with the children dying immediately after birth, have been reported[6]. The diverse mutations limit the use of DNA analysis for prenatal diagnosis to couples who are both carriers and their mutations are known.

The precise pathogenesis of CD is not known, primarily due to our lack of understanding the role of NAA in the nervous system. NAA is synthesized in neuronal mitochondria by acetylation of aspartate by acetyl CoA. Aspartate *N*-acetyl transferase (Asp-NAT), the enzyme that catalyzes this reaction, is highly specific to aspartate, and is found only in the nervous system.[21] In contrast, ASPA, which degrades NAA to aspartate and acetate with high specificity, has a ubiquitous distribution in the body[22]. In the brain, ASPA is distributed predominantly in oligodendrocytes, the myelin synthesizing cells of the CNS. However, immunohistochemical localization studies have shown that NAA is distributed primarily in neurons.[23-25]

2. ACETATE DEFICIENCY HYPOTHESIS OF CD.

Despite the established connection between mutations in the gene for ASPA in CD, and the lost capacity to deacetylate NAA, the specific connection between ASPA deficiency and the failure of proper CNS development, and axonal myelination remains unclear.[9,26] In addition, the precise roles that NAA plays in the development of the CNS, and its proper functioning, remain a matter of study. A number of different hypotheses have been proposed for the various roles played by NAA in the nervous system. The high concentration of NAA in the brain, its lack of known actions on neurons or glia, and its high concentration gradient from neurons to the extracellular space have led to the proposal that NAA is an organic osmolyte which removes excess water from neurons by acting as a molecular water pump.[27] In this regard it has been proposed that excess NAA leads to osmotic dysregulation, or has other cytotoxic effects which are responsible for the pathology observed in CD patients.[28,29]

Another theory about the role of NAA for which there is mounting evidence is that NAA is essential for lipid synthesis and myelination in the CNS, especially during the period of peak, postnatal myelination that occurs in the CNS. In 1987, Hagenfeldt and colleagues proposed that the dismyelination observed in the brains of Canavan disease patients was due to the failure of NAA to act as an acetate carrier from mitochondria to the cytosol, thus impairing lipogenesis.[15] The evidence in favor of this theory include the facts that the levels of NAA, ASPA and Asp-NAT rise with a temporal course which is highly similar to the time-course of myelin protein synthesis.[30,31] Further, it has been shown that NAA contributes acetyl groups for the synthesis of lipids, which in turn are incorporated into myelin.[32,33] Finally, radiolabeled NAA is transported down optic nerve axons and is incorporated into their ensheathing myelin lipids.[34] Based on these and other observations we hypothesize that mutations in the gene encoding ASPA result in a deficiency in the supply of NAA-derived acetate, which in turn results in decreased synthesis of myelin-related fatty acids and lipids. We propose that it is this lipogenic deficiency that compromises CNS myelination, impairs CNS development, and ultimately results in the white matter degeneration observed in CD.[35,36] We have recently tested the acetate deficiency hypothesis in the mouse model of CD by studying myelin lipid synthesis using the tritiated water incorporation method. The results showed significant decreases in 6 classes of newly synthesized myelin-associated lipids in the brains of these mice at the time of peak postnatal CNS myelination. Furthermore, brain acetate levels decreased by about 80% in CD mice, whereas acetate levels in the liver and kidney remained unchanged. These results demonstrate that myelin lipid synthesis is significantly compromised in CD, and provide the first direct evidence that defective myelin synthesis, resulting from a deficiency of NAA-supplied acetate, is involved in the pathogenesis of CD. These issues are discussed in detail in the following sections.

3. MULTIPLE LINES OF EVIDENCE INDICATING A STRONG RELATIONSHIP BETWEEN NAA AND MYELIN LIPID SYNTHESIS

3.1. Incorporation Studies

The first report showing that NAA provides acetate groups for lipid synthesis during myelination was published by D'Adamo and Yatsu in 1966. In a subsequent study in 1968, D'Adamo et al. showed that injection of acetyl labeled NAA into rats of various ages resulted in maximum incorporation into fatty acids just before and during myelination.[37] Since that time, these observations have been confirmed and extended by at least three other research groups. In 1991, Burri et al. performed a detailed study on incorporation of acetyl labeled NAA into various lipids in the rat brain, and showed that NAA is a major source of acetyl groups for lipid synthesis during rat brain development.[33] In 1995, we performed similar incorporation studies using rat brain slices and showed that free acetate and acetyl CoA are formed from the radiolabeled NAA under this condition, indicating acetyl CoA route for this NAA-lipid pathway.[35] Finally, in 2001, Chakraborty et al. have shown by intra ocular injection studies that neuronally derived NAA supplies acetyl groups for myelin lipid synthesis.[34] Taken together, these results provided strong evidence for the involvement of NAA-derived acetyl groups for lipid synthesis during myelination. However, the quantitative significance of this acetate source for myelin synthesis remained unclear until recently.

3.2. Cellular Localization Studies of ASPA

Earlier enzyme activity studies had indicated that ASPA is enriched in the white matter in the CNS.[38] Two recent studies have reported cellular localization of ASPA using polyclonal antibodies against recombinant ASPA.[39,40] Both studies showed predominant localization of ASPA in white matter. The study by Madhavarao et al.[40]

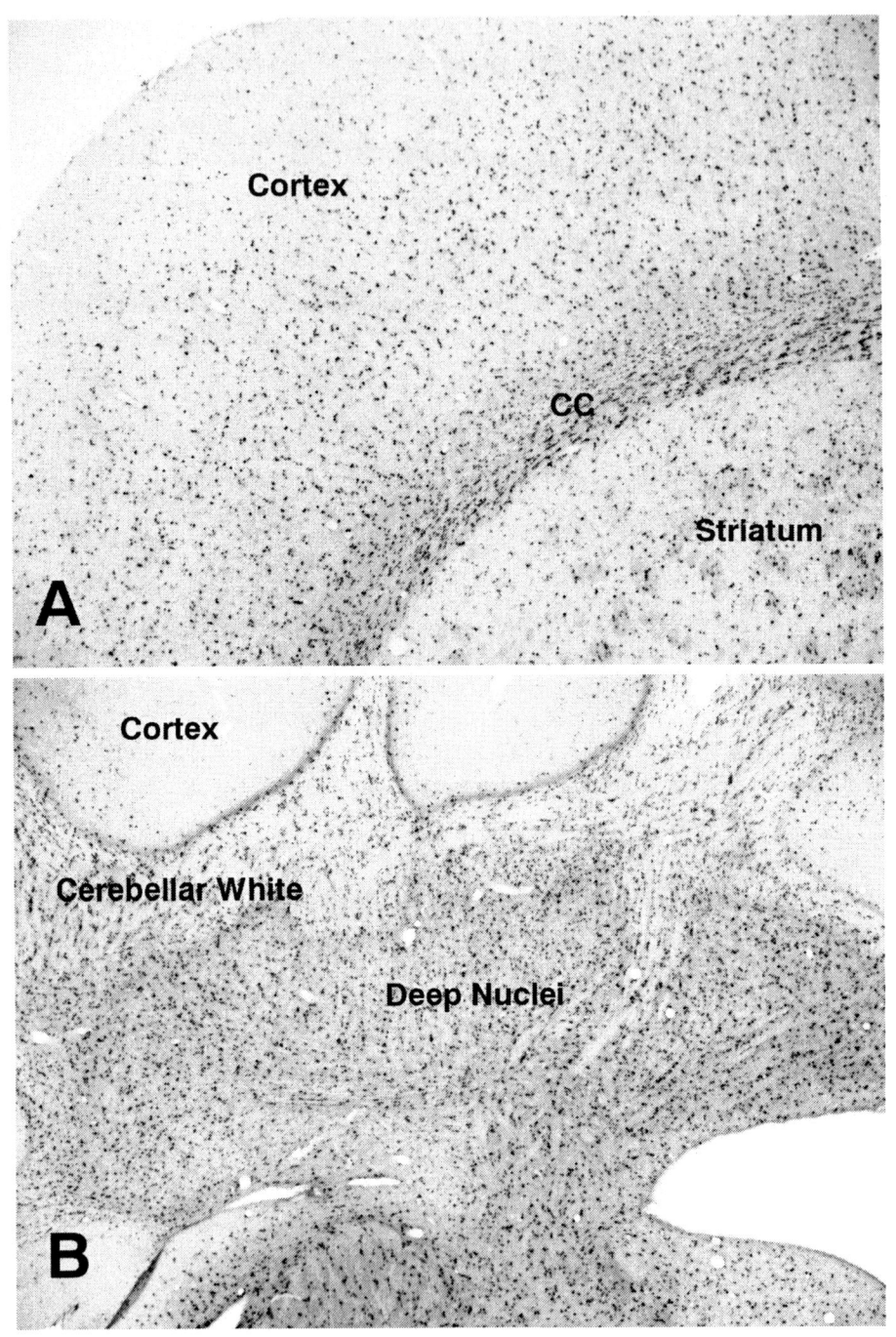

Figure 1. ASPA immunoreactivity in the rat forebrain (A) and hindbrain (B). In forebrain, ASPA was present in oligodendrocytes throughout the corpus callosum (CC), and in fiber bundles in the striatum. Fewer ASPA-positive oligodendrocytes were observed in the superficial layers of cortex. In the hindbrain, ASPA-immunoreactive oligodendrocytes were present in great numbers throughout the cerebellar deep nuclei, white matter, and medulla, but were very sparse in the cerebellar cortex.

is detailed here. The polyclonal antibody preparation against ASPA showed a single band at 37kD by Western blot analysis. Immunohistochemical studies with the antibodies showed that ASPA is localized in the CNS predominantly in oligodendrocytes (Figure 1). ASPA was found to be co-localized throughout the brain with CC1, a marker for oligodendrocytes. Many ASPA-positive cells were observed in white matter, including cells in the corpus callosum and cerebellar white matter. Relatively, fewer cells were labeled in gray matter, particularly in the superficial layers of cerebral cortex. In all the stained cells, ASPA staining was restricted to the cell body and proximal processes. Interestingly, the ASPA antibodies labeled not only the cell bodies and proximal processes of oligodendrocytes, but also labeled their cell nuclei, indicating that ASPA is not restricted to the cytoplasm of these cells (Figure 2). No astrocytes were labeled for ASPA, and neurons were unstained in the forebrain, although a small number of reticular and motor neurons were faintly to moderately stained in the brain stem and spinal cord. Finally, it should be mentioned that microglial cells were faintly stained throughout the brain.

The extensive localization of ASPA in oligodendrocytes strengthens the case for the NAA/ASPA system being required for myelination during development. However, it also raised additional questions because of the prominent localization of the substrate NAA in neurons.

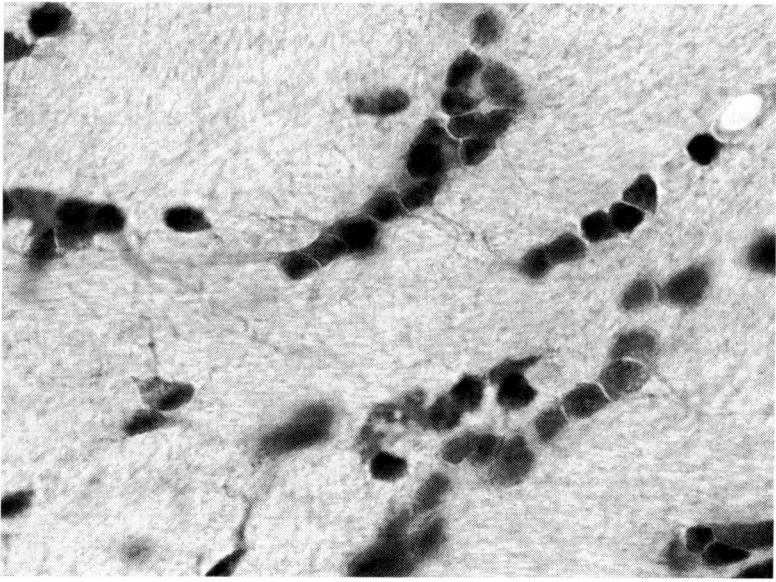

Figure 2. ASPA immunoreactivity in the rat corpus callosum. Strongly stained oligodendrocytes are arranged in rows between myelinated axons, with immunoreactive processes visible on many of the stained cells. An unusual finding was that ASPA was present not only in the cytoplasm of oligodendrocytes, but was also strongly expressed in their nuclei.

3.3. Developmental Increases in ASPA

Another line of evidence indicating a connection between ASPA and myelination is the parallelism between the developmental increases in ASPA activity and myelination, which was first studied by D'Adamo's group in 1973.[22] They showed that the increase of ASPA activity in the rat brain with development almost paralleled the pattern of myelination. In 2001, Bhakoo *et al.* performed a more detailed developmental study covering different rat brain regions, and have confirmed the similarity between the developmental increases in ASPA enzyme activity and myelination.[38] More recently, we used *in situ* hybridization to study ASPA gene expression in the rat brain during development using a riboprobe based on a 400-bp cDNA fragment of murine ASPA.[30] These studies also confirmed a temporal correlation between the developmental increase in ASPA and myelination.

3.4. Myelin Lipid Synthesis Study in CD Mice

Despite the evidence that the acetyl moiety of NAA is used for acetyl CoA/lipid synthesis during myelination, and other evidence that ASPA is associated with myelination, the quantitative significance of this pathway for myelination remained unclear. In most cell types, such as hepatocytes, the enzyme citrate lyase provides the acetyl groups for fatty acid synthesis.[41] It remained to be demonstrated that the NAA-ASPA system contributes significantly to lipid synthesis, most likely myelin lipid synthesis, in the CNS. This is a central issue in the context of the pathogenesis of CD, so we addressed this question by examining the rate of myelin lipid synthesis during the period of maximal postnatal myelination in ASPA knockout (*ASPA -/-*) mice.[42] These mice exhibit pathology similar to that of human CD patients (See chapter by Matalon *et al.*, this volume).

In the ASPA knockout mouse lipid synthesis study, the control group consisted of homozygous wild type (*ASPA+/+*) mice and the experimental group consisted of homozygous knockout (*ASPA-/-*) mice.[42] Comparison of a number of general parameters between the control and knockout mice showed that the two groups differed in only two measures; brain weight and ASPA activity. Other parameters, including animal weight, liver weight and kidney weight were not significantly different between wild type and ASPA knockout mice. The brain weight of the mutant mice was significantly greater than the normal group (461mg ± 17.8 vs. 416mg ± 22.3 respectively; p<0.05). In addition, ASPA activity was undetectable in *ASPA-/-* mice, as compared with ASPA activity of 18.3 ± 5.6 nmol/h/mg protein in control mice (p<0.001).

The level of lipid synthesis was determined in wild type and *ASPA-/-* mice using tritiated water incorporation, which has previously been demonstrated to be a reliable method for determining the rate of lipid synthesis,[43] cholesterol synthesis[44] and myelin synthesis.[45] Table 1 shows the incorporation of tritium into lipids from liver, kidney and myelin in control and *ASPA-/-* mice 5 hours after intraperitoneal administration of 20 mCi $[^3H]_2O$ per animal. The specific activity of $[^3H]_2O$ in the serum samples from the control and *ASPA-/-* mice did not differ significantly (p<0.05).[46] There was no difference between the two groups with respect to lipids synthesized in liver, however total myelin lipids were decreased in the brains of *ASPA-/-* mice by approximately 30% (p<0.005). Total lipid incorporation in kidney was increased by approximately 18% in the *ASPA-/-* group.

Table 1. Specific activity of ^{3}H-H$_2$O in serum, and radioactivity in lipids from various tissues 5 hours after administration of 20mCi ^{3}H-H$_2$O per mouse.

Lipid Category	Control	ASPA -/-
Specific activity of [^{3}H]$_2$O (dpm/nmol water)	90298 ± 6816	97290 ± 8930 NS
Total lipids-liver (cpm/mg protein)	60056 ± 6850	60499 ± 4402 NS
Total lipids-kidney (cpm/mg protein)	95611 ± 13007	113233 ± 8438 **
Total myelin lipids (cpm/mg myelin protein)	26946 ± 3864	18991 ± 2099 ***

The differences between the mean values (± standard deviations) were assessed for significance by two-tailed paired and unpaired t-test. (** $p<0.05$; *** $p<0.001$; NS = not significant).

Myelin lipids were further analyzed by two-dimensional thin layer chromatography (2D-TLC). Table 2 shows the changes in the various lipid classes from the myelin samples of control and *ASPA-/-* mice. Among nonpolar lipids separated on the first dimension, four spots showed significant decreases in the mutant mice, corresponding to glycerol 1-fatty acids (decreased by ~35%), cholesterol, (decreased by ~22%), cholesteryl fatty acids, (decreased by ~35%), and glycerol tri fatty acids (trimyristin, tripalmitin, trilaurin, tristearin: decreased by ~21%). Glycerol 1,2 fatty acids (dimyristin, dipalmitin, dilaurin and distearin) did not show statistically significant change.

In the second TLC dimension for the separation of polar lipids, two out of three lipid spots showed significant reductions ($p<0.05$) in the *ASPA-/-* mice. The Rf of 0.75 corresponded to phospholipids and sulfatides (phosphatidylinositol, phosphatidyl choline, phosphatidyl glycerol, phosphatidic acid and cerebroside sulfate), which were decreased by approximately 38% in *ASPA-/-* mice. The Rf of 0.88 corresponded to phosphatidyl ethanolamine, galactocerebroside and hydroxy fatty acid ceramide, which were decreased approximately 35% in the experimental group.

These data show that CNS myelin lipid synthesis is decreased in the murine model of CD, whereas lipid synthesis in other organs such as kidney and liver was either increased, or not affected. This indicates that lipid synthesis in the brain, in part, requires an intact ASPA enzyme, providing the first direct evidence for deficiency of NAA-derived acetate as an etiological mechanism of CD. As mentioned above, earlier studies have demonstrated that the acetate moiety from neuronally-derived NAA is incorporated into myelin in the CNS[34], and the present study establishes that the acetate contribution to myelin synthesis from NAA is quantitatively sufficient to decrease myelin synthesis during the period of elevated postnatal myelination in the murine model of CD.

Table 2. Comparison of the myelin lipids in the control and mutant mice[1].

(a) Nonpolar lipids[2]

Rf	Control	ASPA -/-	Comigrating lipid standards
0.16 ±0.01	342 ± 48 **	224 ± 62	Glycerol 1-fatty acids[3]
0.45 ± 0.02	3837 ± 564 **	2989 ± 351	Cholesterol
0.53 ± 0.03	548 ± 208 NS	556 ± 98	Glycerol 1,2-fatty acids
0.61 ± 0.02	1699 ± 112 *	1346 ± 330	Glycerol tri fatty acids
0.66 ± 0.02	500 ± 62 **	379 ± 85	Cholesteryl fatty acids (myristate, palmitate)

(b) Polar lipids

0.60 ± 0.06	821 ± 400 NS	788 ± 807	Unknown
0.75 ± 0.02	9995 ± 2525 **	6153 ± 1320	PC, PI, PG, Sulfatides, PA[4]
0.88 ± 0.02	7513 ± 1114 **	4904 ± 1157	PE, GC, Ceramide

[1] All lipid spots are given as cpm/mg myelin protein.
[2] Myelin lipids were separated by TLC.
[3] Fatty acids include laurate, myristate, palmitate and stearate.
[4] PC, phosphatidyl choline; PI, phosphatidyl inositol; PG, phosphatidyl glycerol; Sulfatides, 3-sulfate ester of galactosyl cerebroside; PA, phosphatidic acid; PE, phosphatidyl ethanolamine; GC, galactocerebroside; Ceramide, hydroxy fatty acid ceramides.

4. CELLULAR LOCALIZATION STUDIES OF ASPA VS. NAA

The significance of the expression of NAA in oligodendrocytes assumes added importance in view of the increasing evidence that the NAA/ASPA system plays a major role in the myelin lipid synthesis in the brain. Earlier cell culture studies have shown both the presence and absence of NAA in oligodendrocytes based on the culture conditions.[47] Optic nerve transaction experiments have suggested that between 5% and 20% of NAA in developing white matter is derived from proliferating oligodendrocyte progenitor cells.[48] We have addressed this issue in the rat brain by immunohistochemistry using polyclonal antibodies against NAA as previously described.[23-25]

Figure 3 shows the immunohistochemical localization of ASPA and NAA in the rat corpus callosum. ASPA staining was moderate to strong in oligodendrocytes throughout corpus callosum (Fig 3A), whereas NAA immunoreactivity was low in oligodendrocytes (Fig 3B). Strong staining for NAA in neurons, and low levels of staining in mature oligodendrocytes is likely to be related in part to the absence of ASPA activity in neurons, and strong ASPA expression and activity in oligodendrocytes. Based on these results, it is not clear if oligodendrocytes synthesize sufficient NAA for myelin synthesis[47], or if they acquire a significant proportion of it by neuronal-glial transfer.[34] The key question is whether the acetate derived from NAA is essential for postnatal myelin synthesis, or acts as a secondary lipogenic pathway which contributes acetate, but

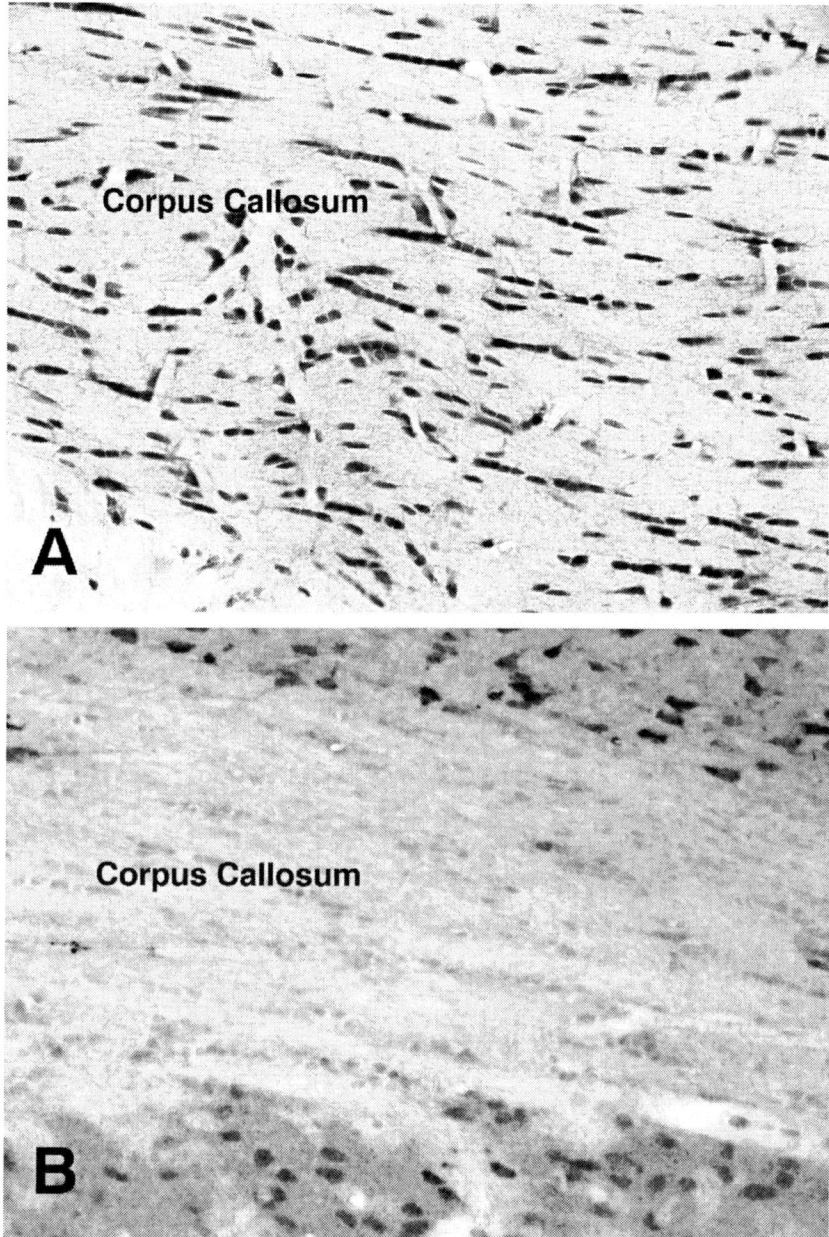

Figure 3. ASPA immunoreactivity and NAA immunoreactivity in adult rat corpus callosum (CC). ASPA is strongly expressed in oligodendrocytes in the CC (A), whereas NAA is expressed strongly in neurons (above and below the corpus callosum), but only at low levels in oligodendrocytes in the CC (B).

is not critical for myelin synthesis during CNS maturation. To help answer this question, we looked at brain acetate levels in the ASPA -/- mice used for the lipid synthesis studies described above.

5. ACETATE DEFICIENCY IN CD MICE

Very few metabolic pathways other then ASPA can generate free acetate in mammals, and most serum acetate is thought to be derived from bacterial metabolism in the gut.[49,50] We had proposed previously that NAA is a major source of acetate in the brain,[35] and in order to confirm this we determined free acetate levels in the brains of *ASPA -/-* mice using an enzymatic assay. The results showed that there was nearly an 80% reduction in free acetate levels in the brains of *ASPA-/-* mice, but that the levels in kidney and liver were not reduced relative to controls (Table 3).[42] Taken together, the above findings demonstrate that in the brain, NAA is a major source of free acetate, and that myelin lipid synthesis in the brain derives a substantial portion of the requisite acetate (acetyl CoA) from NAA via ASPA-mediated catalysis. One of the key issues concerning the etiological role of reduced NAA-derived acetate in CD is the concentration of acetate in oligodendrocytes during the period of intense postnatal axonal myelination. The current data only pertain to whole brain homogenate levels in the developing CD mice, but considering that ASPA is predominantly localized in oligodendrocytes, it is likely that their acetate levels would be substantially reduced in the ASPA deficient animals.

Table 3. Acetate concentrations are decreased in brain, but not in liver or kidney in the mutant mouse.[1]

Tissue	ASPA -/-	Control
Brain (n=5)	0.31 ± 0.11	1.51 ± 0.34 ***
Liver (n=5)	0.68 ± 0.03	0.81 ± 0.13 NS
Kidney (n=4)	1.41 ± 0.11	1.30 ± 0.08 NS

[1] Acetate concentrations (μmol/g wet tissue) were measured by a coupled enzymatic assay. The differences between the mean values ($\pm$ standard deviations) were assessed for significance by two-tailed paired and unpaired t-test (*** $p<0.001$).

6. A COMPREHENSIVE FUNCTIONAL MODEL FOR NAA

The discrepant expression of NAA and ASPA in neurons and oligodendrocytes respectively, has been proposed as a mechanism for channeling NAA-associated acetate from neurons to oligodendrocytes.[34] Based on this, and many findings from laboratories around the world, as well as the results of ongoing studies in our laboratory, we propose an expanded model of NAA metabolism wherein NAA has multiple roles in the nervous system (Figure 4).[42,51] In neurons, the NAA biosynthetic enzyme Asp-NAT acts to remove excess aspartate from mitochondria via acetylation, which would favor α-ketoglutarate formation from glutamate, and energy production via the citric acid cycle. By this mechanism, the extra demand for ATP in neurons is met in part by oxidation of glutamate via the aspartate aminotransferase pathway.[52] The high concentrations of NAA

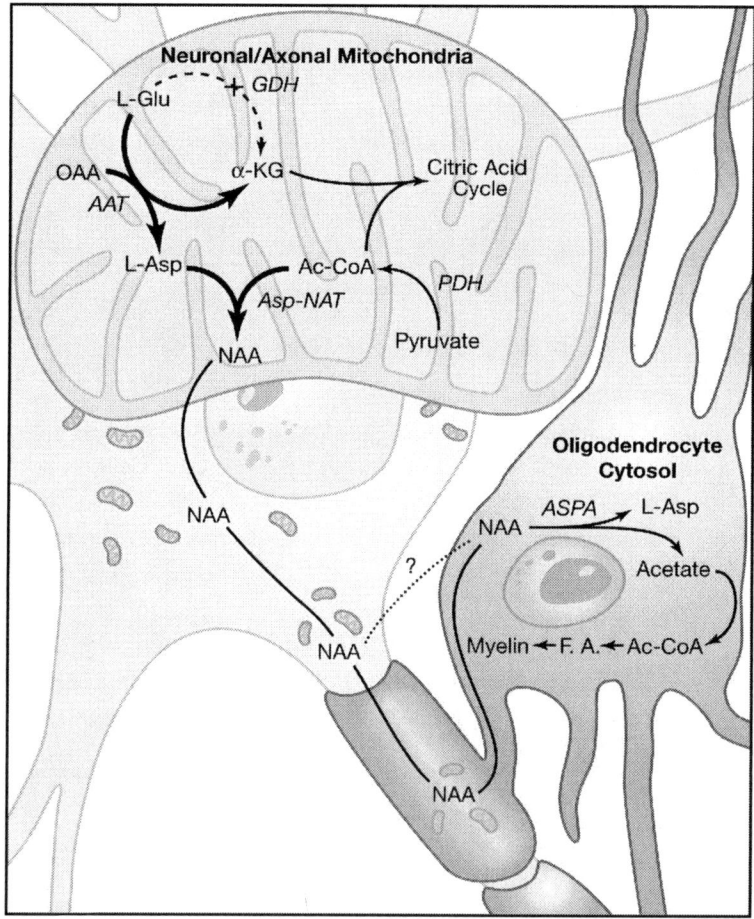

Figure 4. Diagrammatic representation of NAA synthesis in neurons and degradation in oligodendrocytes. See text for details. OAA = oxaloacetic acid, AAT = aspartate aminotransferase, Asp-NAT = aspartate N-acetyltransferase, PDH = pyruvate dehydrogenase, α-KG = alphaketoglutarate, GDH = glutamate dehydrogenase, F.A. = fatty acids.

in neurons, and the steady-state levels maintained there, would reflect the metabolic state of neuronal mitochondria because of the direct coupling of NAA production to α-ketoglutarate formation from glutamate. By preferentially using the aspartate aminotransferase reaction instead of the glutamate dehydrogenase reaction to generate α-ketoglutarate, neuronal mitochondria would prevent ammonia production associated with the glutamate dehydrogenase reaction, and this might avoid additional metabolic stress on neurons. In this model, NAA synthesis is intimately associated with the proper functioning of neuronal energy metabolism via the aspartate aminotransferase reaction in neuronal mitochondria. Further, NAA synthesized in neuronal mitochondria is transferred to oligodendrocytes by an as yet unknown mechanism, where ASPA liberates the acetate moiety to be used for myelin lipid synthesis. This hypothesis emphasizes the metabolic coupling of myelinated axons to oligodendrocytes, where axons provide major biochemical precursors for the demanding task of myelination during early postnatal CNS development. Whether the liberated aspartate is predominantly utilized in oligodendrocytes for metabolism and protein synthesis, or is instead recycled in great part back to neurons for a new cycle of NAA synthesis, is presently unknown.

7. PRECLINICAL EFFORTS TOWARD ACETATE SUPPLEMENTATION THERAPY

In view of the evidence presented above that brain acetate levels and myelin lipid synthesis are reduced in CD mice, correcting the acetate deficit by acetate supplementation would appear to be an obvious therapeutic approach for CD. In our preclinical efforts toward such a therapy for CD, we are currently examining glyceryl triacetate (Triacetin) and calcium acetate as potential exogenous acetate sources for delivering acetate to the brain. Glyceryl triacetate is a non-toxic glyceryl tri-ester of acetic acid that is widely used as a solvent and plasticizer in perfumery, tanning, dyes, as a food additive, a gelatinizing agent in cosmetics and is also used in external medicine. Biochemical studies on glyceryl triacetate have shown that glyceryl triacetate is hydrolyzed *in vivo* by all tissues of mammals including the gastrointestinal tract.[53] Calcium acetate is currently approved as a drug for the treatment of kidney disease to control high blood phosphate levels.[54] In preliminary studies we have found that calcium acetate is not as effective as glyceryl triacetate as an acetate source to the brain (Fig. 5). Therefore, efforts are underway for a systematic preclinical study with glyceryl triacetate.

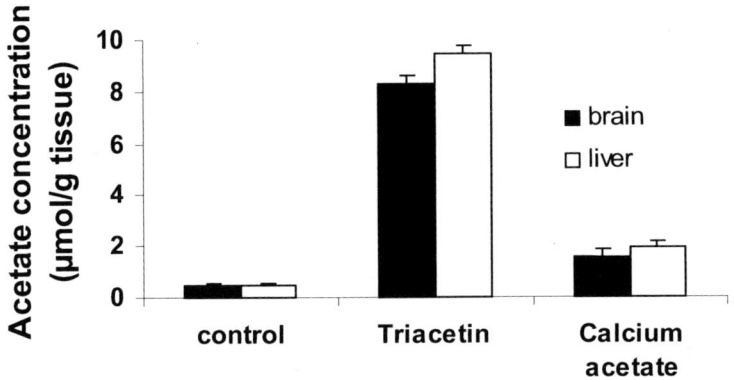

Figure 5. Effect of equimolar GTA and calcium acetate (CA) on acetate levels in the liver and brain. Mice (20-21day old) were fed (26.5 mmol/kg) of either GTA or CA and acetate levels were determined in tissues 1hr following administration. Values are means ± SD of samples from 3-5 animals per group. GTA administration increased brain and liver acetate levels significantly more than calcium acetate administration ($p<0.01$). Statistical analysis was carried out using Student's t test.

8. FINAL COMMENTS

It should be stressed that the question of which mechanisms are primarily involved in the pathogenesis of CD, and which are less critical, remains unclear at the present time. It seems clear now that an acetate deficiency caused by dysfunction in a specific enzyme in oligodendrocytes has an etiological role in CD. It is also possible that osmotic dysregulation in the CNS is mediated by excessive extracellular NAA concentrations, and that high NAA concentrations lead to seizures, which further contribute to the pathogenesis. However, we should not neglect the fact that ASPA is extensively expressed in the kidney, and therefore it is possible that renal pathologies could also contribute to disease progression by as yet undetermined mechanisms. It is also not clear whether or not the primary substrate for ASPA in the kidney is NAA, because NAA levels are very low there. If the function of ASPA in the kidney and other peripheral tissues is unrelated to NAA, which seems likely, then other approaches may need to be developed to deal with the pathological consequences of the loss of ASPA activity in those tissues.

One approach to testing if an acetate deficiency is a primary etiological mechanism of CD would be to determine whether myelin synthesis can be promoted in newborn CD infants by increasing brain acetate levels through dietary supplementation. CD pathogenesis develops predominantly after birth, and therefore this approach should be feasible. An orally administered form of acetate in a supplemented infant formula could directly provide the required substrate for the rapid myelination that takes place during early, postnatal neural development. Early diagnosis of CD using urinalysis to detect high NAA levels could be followed by immediate acetate supplementation of the diet, which may provide an extremely safe, simple and low cost treatment for CD.

In conclusion, the findings presented above provide direct support for the proposed etiology of CD as a deficiency of NAA/ASPA-derived acetate, resulting in reduced lipid synthesis, and a failure of proper myelination during CNS development. Based upon these and other findings, early postnatal acetate supplementation trials appear warranted in confirmed cases of CD.

9. ACKNOWLEDGEMENTS

This work supported by grants from the NIH (ROI, NS39387), and The Samueli Institute for Information Biology, CA, USA (G170ON).

10. QUESTION AND ANSWER SESSION

SESSION CO-CHAIR MATALON: Any questions for Dr. Namboodiri? Michel?

DR. PHILIPPART: I have one comment on the triglycerides. Many years ago, I did an experiment by exposing cultured fibroblasts to various lipids, including triglycerides. You get very good uptake, but EM on these cells shows it's mostly in the form of droplets. So we get it in, and then probably it remains quite inert. So that may not be the whole story.

Of course, we certainly were not using -- I forgot what kind of triglyceride we were using. Definitely it was not Triacetin, as you used, which may have some advantage. But we used probably some medium chain triglyceride. We were pretty sure that's what it was.

And then we got something which sat within the fibroblasts as droplets. And, actually, we could load the whole fibroblast with an enormous amount of droplets, which were fairly inert.

DR. NAMBOODIRI: We measured acetate levels in the brain after Triacetin supplementation, and we found that the levels were increased significantly. So we know it is getting into brain. Our plan is to start early on, soon after birth. In our studies, we find that free acetate is going up in the liver and brain after Triacetin (glyceryl triacetate) administration, so we know that the glyceryl triacetate is not inert.

And the way we are giving it is intra-gastric. So we can do that without too much of a problem and start early on and see what happens. And if we succeed, then we can try to find out if there is a time window where we have to start the administration, beyond which it will not be effective.

All of this can be done in the CD mice the way we are now planning. This acetate source is inexpensive and readily available, and it's, as I said, quite nontoxic. So we are very interested and ready to go ahead with this.

SESSION CO-CHAIR MATALON: Dr. Ledeen?

DR. LEDEEN: Were you surprised that the brain took up as much as the liver of your Triacetin? That seems a little bit surprising.

DR. NAMBOODIRI: We were not really surprised because it is hydrophobic. In half an hour, it should reach every tissue in the body.

DR. LEDEEN: Okay.

DR. NAMBOODIRI: If you give glyceryl triacetate by a gastric route, most of it goes through the liver. So we thought that we would have to saturate the liver before it would get into the brain, but that is not what we are seeing. We see a similar pattern, both in the liver and the brain.

DR. LEDEEN: My next question is, did you check to see whether label from triacetin got into myelin lipids?

DR. NAMBOODIRI: We will look at that.

DR. LEDEEN: That is really a key question, I think.

DR. NAMBOODIRI: The Triacetin experiments were just started, and we haven't checked incorporation into myelin yet. And also so far we have been doing it for very short durations, you know, one hour, two hours. We will need to do a time course.

SESSION CO-CHAIR MATALON: Any question or comment?

PARTICIPANT: The purpose of using Triacetin was to get the acetate across the blood-brain barrier? When you give it orally, does it cross the gastrointestinal tract intact? I would think that most of it would be hydrolyzed in the intestinal lumen. So in order to see if it can enter the brain, you need to see that it can enter the blood intact. Do you have any evidence that the Triacetin gets across the gastrointestinal lumen intact?

DR. NAMBOODIRI: We don't know – when we looked at acetate at two-hours after Triacetin treatment, we do see dramatically increased acetate in the brain.

DR. GANAPATHY: But you are not comparing acetate with the tri-acetate yet. Do you not know the tri-acetate is better than acetate itself?

DR. NAMBOODIRI: We have tried calcium acetate, but you know, we were not getting good uptake in the brain. That is the reason we moved to Triacetin, as opposed to calcium acetate. With calcium acetate, we detected much lower increases in brain acetate levels at very high concentrations, when compared with glyceryl triacetate. So to that extent, Triacetin is more effective.

SESSION CO-CHAIR MATALON: Thank you. Dr. Leone?

DR. LEONE: Yes. I would like to make three points. And perhaps you can comment on it. I think what you are doing is very relevant. And following Dr. Kolodny's suggestion on administering calcium acetate or sodium acetate, depending upon the patients and their baseline physiological problems, we have been following something like, let's see, about 14 patients, looking at brain imaging, looking at clinical development during the years.

You could argue that, of course, you don't know if calcium acetate can deliver as much acetate as required to make myelin, but while we definitely reported just an improvement in ability, especially the young patients, younger than 12 months old, we never saw an increase in myelin looking at the MRIs and T-1 quantitative measures. This is with calcium acetate.

The second point that I have is in the tremor rat where the absolute total deletion of the aspartoacylase we don't even have any nonfunctional enzyme. We do have a little bit of NAA, but in the rat model, we do have myelin; much more than the patients. The patients have, even the four-month-olds have, just a severe total hypomyelination, perhaps one or two percent normal myelination. We can say that in the best patient, we have about two percent of the myelin at the same age the child would have.

And the third point about your hypothesis is that there was a report of a patient with zero NAA. And that patient based on this part, if this was so clear one way straight, then we would have zero myelin. The patient had a mild retardation. And the patient also had hypomyelination, not as severe, not even close to any of the Canavan patients.

So I think your hypothesis and Dr. Ledeen's hypothesis and your group may be very possible, but I think within the context of Canavan disease, we need to be able to look at other potential hypotheses. One of them is that NAA has the active role in toxicity, specifically during development. Dr. Burlina is going to present the case with a low NAA or zero NAA.

DR. NAMBOODIRI: Is the lack of myelin correlated with the severity of the disease?

DR. LEONE: The lack of myelin in our patients? The Canavan patients I follow are a very homogeneous group. They never develop to the point where they can walk, or talk, for example.

You see near zero myelin in the brain of some patients, although you have to know that some of the patients have up to approximately five percent of normal myelin levels in the brain. So perhaps a 5% increase in acetate in the brain should be able to increase myelin in the brain.

DR. BURLINA: We were following a patient, he was supplemented with acetate, and he was not improving at all after one year, we had an opportunity to perform some tests and we found that NAA was increased. So I think we have to be very careful with the acetate supplementation.

SESSION CO-CHAIR MATALON: I think the mouse would be a good idea to check safety of the compound.

Yes. Michel?

DR. PHILIPPART: You mentioned some discrepancy between the size of the enzymes, and I was wondering if you had considered the ratio of the difference between some of this data. Is it possible that one form is an apoenzyme which is then made into the active enzyme in oligodendrocytes

DR. NAMBOODIRI: With our antibody, we don't see a different size. We find the same size, so we don't know why there is the difference between the two findings, whether there is something else going on.

DR. PHILIPPART: Maybe the antibody is already transformed into something else before –

DR. NAMBOODIRI: No. When we do the same experiment with our antibody, with the cytosol, we find 37, but with the myelin fraction, we don't detect a band at all, and we also don't see any activity. So there are some differences between our results and Dr. Ledeen's results.

The problem we have is if it is high molecular weight, small covalent modifications cannot explain that. You may expect a change of 5,000, but not 20,000. So there is something else going on.

SESSION CO-CHAIR MATALON: Brian Ross had a comment.

DR. ROSS: I just have one comment to the question of acetate uptake in the brain, sodium acetate. We can see with radiolabeled acetate that acetate is taken up by the brain, and is metabolized quickly, we believe by the glia. Of course the down side is we can see incorporation into many other molecules in the brain is greater, so that the question is not really will the acetate will get into the bran, but will there be enough around for myelin synthesis after all the other processes it is involved in?

DR. NAMBOODIRI: Yes, that's always a major question because the oxidative path to CO_2 is far more significant, and that could cause a problem in terms of acidity and all of that. Those are the questions that we need to look at in terms of toxicity.

SESSION CO-CHAIR MATALON: Ed Kolodny has a comment.

DR. KOLODNY: Perhaps I'll answer the question with regard to acetate in a patient population. When we began to use it years ago, it wasn't clear whether it would do any good, but parents encouraged us because we got very positive feedback with respect to their behavior; their alertness; the children's ability to move.

However, what has I think been a stumbling block in trying to carry out any kind of careful study is that you need clinical markers. In Canavan's disease, it brings them evoked potentials, MR spectroscopy for NAA, head circumference measurements. Perhaps these are some of the markers that could be used, but no study has really been done.

In our experience, the head circumference seems to level off. And the increasing growth of the head seems to be arrested with the use of calcium acetate.

But I would be the first to say that I'm not sure whether it's doing any good. We only have the parents' word. And most parents are looking to any solution as a positive one.

SESSION CO-CHAIR MATALON: I am not sure that the acetate will reduce the NAA level because it doesn't break it down. So that may not be really a good marker. And head size, that's a good one, but I have seen patients with Canavan. Their head does not grow forever.

DR. NAMBOODIRI: Our goal is not to reduce the NAA concentration, but to replace the acetate that was trapped in the NAA, and that was not being liberated.

SESSION CO-CHAIR MATALON: Thank you Dr. Namboodiri. And now we move to Dr. Kolodny's talk.

11. REFERENCES

1.   L. van Bogaert, I. Bertrand, Sur une idiote familiale avec degenerescence spongieuse de neuraxe (note preliminaire). *Acta Neurol. Belg.*, **12**, 572-587 (1949).
2.   G. R. Hogan, E. P. Richardson, Jr., Spongy degeneration of the nervous system (Canavan's disease). *Pediatrics*, **35**, 284-294 (1965).
3.   H. Pratt, Canavan's disease and spongiform encephalopathy. *Br. Med. J.*, **4** (837), 427 (1972).
4.   M. Adachi, L. Schneck, J. Cara and B. W. Volk, Spongy degeneration of the central nervous system (van Bogaert and Bertrand type; Canavan's disease). A review. *Hum. Pathol.*, **4** (3), 331-347 (1973).

5. E. C. Traeger, I. Rapin, The clinical course of Canavan disease. *Pediatr. Neurol.*, **18** (3), 207-212 (1998).
6. B. J. Zeng, Z. H. Wang, L. A. Ribeiro, P. Leone, R. De Gasperi, S. J. Kim, S. Raghavan, E. Ong, G. M. Pastores and E. H. Kolodny, Identification and characterization of novel mutations of the aspartoacylase gene in non-Jewish patients with Canavan disease. *J. Inherit. Metab Dis.*, **25** (7), 557-570 (2002).
7. S. Kamoshita, I. Rapin, K. Suzuki and K. Suzuki, Spongy degeneration of the brain. A chemical study of two cases including isolation and characterization of myelin. *Neurology*, **18** (10), 975-985 (1968).
8. R. Matalon, R. Kaul and K. Michals, Canavan disease: biochemical and molecular studies. *J. Inherit. Metab. Dis.*, **16**, 744-752 (1993).
9. R. Matalon, K. Michals and R. Kaul, Canavan disease: from spongy degeneration to molecular analysis. *J Pediatr.*, **127** (4), 511-517 (1995).
10. R. Matalon, P. L. Rady, K. A. Platt, H. B. Skinner, M. J. Quast, G. A. Campbell, K. Matalon, J. D. Ceci, S. K. Tyring, M. Nehls, S. Surendran, J. Wei, E. L. Ezell and S. Szucs, Knock-out mouse for Canavan disease: a model for gene transfer to the central nervous system. *J. Gene Med.*, **2** (3), 165-175 (2000).
11. S. Surendran, K. M. Matalon, S. Szucs, S. K. Tyring and R. Matalon, Metabolic changes in the knockout mouse for Canavan's disease: implications for patients with Canavan's disease. *J. Child Neurol.*, **18** (9), 611-615 (2003).
12. S. Surendran, E. L. Ezell, M. J. Quast, J. Wei, S. K. Tyring, K. Michals-Matalon and R. Matalon, Mental retardation and hypotonia seen in the knock out mouse for Canavan disease is not due to succinate semialdehyde dehydrogenase deficiency. *Neurosci. Lett.*, **358** (1), 29-32 (2004).
13. S. Surendran, G. A. Campbell, S. K. Tyring and R. Matalon, Aspartoacylase gene knockout results in severe vacuolation in the white matter and gray matter of the spinal cord in the mouse. *Neurobiol. Dis.*, **18** (2), 385-389 (2005).
14. S. Kamoshita, I. Rapin, K. Suzuki and K. Suzuki, Spongy degeneration of the brain. A chemical study of two cases including isolation and characterization of myelin. *Neurology*, **18** (10), 975-985 (1968).
15. L. Hagenfeldt, I. Bollgren and N. Venizelos, N-acetylaspartic aciduria due to aspartoacylase deficiency- a new aetiology of childhood leukodystrophy. *J. Inherit. Metab. Dis.*, **10**, 135-141 (1987).
16. H. J. Wittsack, H. Kugel, B. Roth and W. Heindel, Quantitative measurements with localized 1H MR spectroscopy in children with Canavan's disease. *J. Magn Reson. Imaging*, **6** (6), 889-893 (1996).
17. N. Aydinli, M. Caliskan, M. Calay and M. Ozmen, Use of localized proton nuclear magnetic resonance spectroscopy in Canavan's disease. *Turk. J. Pediatr.*, **40** (4), 549-557 (1998).
18. R. Kaul, G. P. Gao, K. Balamurugan and R. Matalon, Cloning of the human aspartoacylase cDNA and a common missense mutation in Canavan disease. *Nat. Genet.*, **5** (2), 118-123 (1993).
19. R. Matalon, K. Michals-Matalon, Molecular basis of Canavan disease. *Eur. J. Paediatr. Neurol.*, **2** (2), 69-76 (1998).
20. A. Feigenbaum, R. Moore, J. Clarke, S. Hewson, D. Chitayat, P. N. Ray and T. L. Stockley, Canavan disease: carrier-frequency determination in the Ashkenazi Jewish population and development of a novel molecular diagnostic assay. *Am. J. Med. Genet. A*, **124** (2), 142-147 (2004).
21. M. E. Truckenmiller, M. A. Namboodiri, M. J. Brownstein and J. H. Neale, N-Acetylation of L-aspartate in the nervous system: differential distribution of a specific enzyme. *J. Neurochem.*, **45**, 1658-1662 (1985).
22. A. F. D'Adamo, Jr., J. C. Smith and C. Woiler, The occurrence of N-acetylaspartate amidohydrolase (aminoacylase II) in the developing rat. *J. Neurochem.*, **20** (4), 1275-1278 (1973).
23. J. R. Moffett, M. A. Namboodiri, C. B. Cangro and J. H. Neale, Immunohistochemical localization of N-acetylaspartate in rat brain. *Neuroreport*, **2**, 131-134 (1991).
24. J. R. Moffett, M. A. Namboodiri and J. H. Neale, Enhanced carbodiimide fixation for immunohistochemistry: Application to the comparative distributions of N-acetylaspartylglutamate and N-acetylaspartate immunoreactivities in rat brain. *J. Histochem. Cytochem.*, **41**, 559-570 (1993).
25. J. R. Moffett, M. A. Namboodiri, Differential distribution of N-acetylaspartylglutamate and N-acetylaspartate immunoreactivities in rat forebrain. *J. Neurocytol.*, **24**, 409-433 (1995).
26. R. Matalon, K. Michals, D. Sebesta, M. Deanching, P. Gashkoff and J. Casanova, Aspartoacylase deficiency and N-acetylaspartic aciduria in patients with canavan disease. *Am. J. Med. Genet.*, **29**, 463-471 (1988).
27. M. H. Baslow, Evidence supporting a role for N-acetyl-L-aspartate as a molecular water pump in myelinated neurons in the central nervous system. An analytical review. *Neurochem. Int.*, **40** (4), 295-300 (2002).
28. M. H. Baslow, T. R. Resnik, Canavan disease. Analysis of the nature of the metabolic lesions responsible for development of the observed clinical symptoms. *J. Mol. Neurosci.*, **9** (2), 109-125 (1997).
29. P. Leone, C. G. Janson, L. Bilaniuk, Z. Wang, F. Sorgi, L. Huang, R. Matalon, R. Kaul, Z. Zeng, A. Freese, S. W. McPhee, E. Mee, M. J. During and L. Bilianuk, Aspartoacylase gene transfer to the

mammalian central nervous system with therapeutic implications for Canavan disease. *Ann. Neurol.*, **48** (1), 27-38 (2000).

30. B. F. Kirmani, D. M. Jacobowitz and M. A. Namboodiri, Developmental increase of aspartoacylase in oligodendrocytes parallels CNS myelination. *Brain Res. Dev. Brain Res.*, **140** (1), 105-115 (2003).

31. A. F. D'Adamo, L. I. Gidez and F. M. Yatsu, Acetyl transport mechanisms. Involvement of N-acetyl aspartic acid in de novo fatty acid biosynthesis in the developing rat brain. *Exp. Brain Res.*, **5**, 267-273 (1968).

32. A. F. D'Adamo, Jr., F. M. Yatsu, Acetate metabolism in the nervous system. N-acetyl-L-aspartic acid and the biosynthesis of brain lipids. *J. Neurochem.*, **13** (10), 961-965 (1966).

33. R. Burri, C. Steffen and N. Herschkowitz, N-acetyl-L-aspartate is a major source of acetyl groups for lipid synthesis during rat brain development. *Dev. Neurosci.*, **13**, 403-412 (1991).

34. G. Chakraborty, P. Mekala, D. Yahya, G. Wu and R. W. Ledeen, Intraneuronal N-acetylaspartate supplies acetyl groups for myelin lipid synthesis: evidence for myelin-associated aspartoacylase. *J. Neurochem.*, **78** (4), 736-745 (2001).

35. V. Mehta, M. A. Namboodiri, N-acetylaspartate as an acetyl source in the nervous system. *Brain Res. Mol. Brain Res.*, **31** (1-2), 151-157 (1995).

36. B. F. Kirmani, D. M. Jacobowitz, A. T. Kallarakal and M. A. Namboodiri, Aspartoacylase is restricted primarily to myelin synthesizing cells in the CNS: therapeutic implications for Canavan disease. *Brain Res. Mol. Brain Res.*, **107** (2), 176-182 (2002).

37. A. F. D'Adamo, Jr., L. I. Gidez and F. M. Yatsu, Acetyl transport mechanisms. Involvement of N-acetyl aspartic acid in de novo fatty acid biosynthesis in the developing rat brain. *Exp. Brain Res.*, **5** (4), 267-273 (1968).

38. K. K. Bhakoo, T. J. Craig and P. Styles, Developmental and regional distribution of aspartoacylase in rat brain tissue. *J. Neurochem.*, **79** (1), 211-220 (2001).

39. M. Klugmann, C. W. Symes, B. K. Klaussner, C. B. Leichtlein, T. Serikawa, D. Young and M. J. During, Identification and distribution of aspartoacylase in the postnatal rat brain. *Neuroreport*, **14** (14), 1837-1840 (2003).

40. C. N. Madhavarao, J. R. Moffett, R. A. Moore, R. E. Viola, M. A. Namboodiri and D. M. Jacobowitz, Immunohistochemical localization of aspartoacylase in the rat central nervous system. *J. Comp Neurol.*, **472** (3), 318-329 (2004).

41. J. M. Lowenstein, Effect of (-)-hydroxycitrate on fatty acid synthesis by rat liver in vivo. *J. Biol. Chem.*, **246** (3), 629-632 (1971).

42. C. N. Madhavarao, P. Arun, J. R. Moffett, S. Szucs, S. Surendran, R. Matalon, J. Garbern, D. Hristova, A. Johnson, W. Jiang and M. A. Namboodiri, Defective N-acetylaspartate catabolism reduces brain acetate levels and myelin lipid synthesis in Canavan's disease. *Proc. Natl. Acad. Sci. U. S. A*, **102** (14), 5221-5226 (2005).

43. R. L. Jungas, Fatty acid synthesis in adipose tissue incubated in tritiated water. *Biochemistry*, **7** (10), 3708-3717 (1968).

44. J. M. Dietschy, D. K. Spady, Measurement of rates of cholesterol synthesis using tritiated water. *J. Lipid Res.*, **25** (13), 1469-1476 (1984).

45. E. D. Muse, H. Jurevics, A. D. Toews, G. K. Matsushima and P. Morell, Parameters related to lipid metabolism as markers of myelination in mouse brain. *J. Neurochem.*, **76** (1), 77-86 (2001).

46. H. A. Jurevics, P. Morell, Sources of cholesterol for kidney and nerve during development. *J. Lipid Res.*, **35** (1), 112-120 (1994).

47. J. Urenjak, S. R. Williams, D. G. Gadian and M. Noble, Specific expression of N-acetylaspartate in neurons, oligodendrocyte-type-2 astrocyte progenitors, and immature oligodendrocytes in vitro. *J. Neurochem.*, **59**, 55-61 (1992).

48. C. Bjartmar, J. Battistuta, N. Terada, E. Dupree and B. D. Trapp, N-acetylaspartate is an axon-specific marker of mature white matter in vivo: a biochemical and immunohistochemical study on the rat optic nerve. *Ann. Neurol.*, **51** (1), 51-58 (2002).

49. S. E. Knowles, I. G. Jarrett, O. H. Filsell and F. J. Ballard, Production and utilization of acetate in mammals. *Biochem. J.*, **142** (2), 401-411 (1974).

50. D. W. Pethick, D. B. Lindsay, P. J. Barker and A. J. Northrop, Acetate supply and utilization by the tissues of sheep in vivo. *Br. J. Nutr.*, **46** (1), 97-110 (1981).

51. C. N. Madhavarao, C. Chinopoulos, K. Chandrasekaran and M. A. Namboodiri, Characterization of the N-acetylaspartate biosynthetic enzyme from rat brain. *J. Neurochem.*, **86** (4), 824-835 (2003).

52. M. Yudkoff, D. Nelson, Y. Daikhin and M. Erecinska, Tricarboxylic acid cycle in rat brain synaptosomes. Fluxes and interactions with aspartate aminotransferase and malate/aspartate shuttle. *J. Biol. Chem.*, **269** (44), 27414-27420 (1994).

53. A. Bach, P. Metais, [Fats with short and medium chains. Physiological, biochemical, nutritional, and therapeutic aspects]. *Ann. Nutr. Aliment.*, **24** (5), 75-144 (1970).
54. L. R. Schiller, C. A. Santa Ana, M. S. Sheikh, M. Emmett and J. S. Fordtran, Effect of the time of administration of calcium acetate on phosphorus binding. *N. Engl. J. Med.*, **320** (17), 1110-1113 (1989).

MUTATION ANALYSIS OF THE ASPARTOACYLASE GENE IN NON-JEWISH PATIENTS WITH CANAVAN DISEASE

Bai-Jin Zeng, Gregory M. Pastores, Paola Leone, Srinivasa Raghavan, Zhao-Hui Wang, Lucilene A. Ribeiro, Paola Torres, Elton Ong, and Edwin H. Kolodny[*]

1. INTRODUCTION

Canavan disease (OMIM # 271900) is an inherited leukodystrophy caused by deficiency of aspartoacylase (ASPA) (Matalon, et al., 1988) a cytosolic enzyme found in oligodendrocytes (Madhavarao, et al., 2004). It hydrolyzes N-acetylaspartic acid (NAA) to acetate and aspartic acid. In Canavan disease (CD), NAA concentrations in brain increase causing vacuolization in the lower layers of the cerebral cortex and subcortical white matter with intramyelinic swelling and myelin loss. Symptoms of this rare antosomal recessive disease appear within two to four months of birth, consisting of poor head control, truncal hypotonia, developmental delay, and lack of visual tracking. Subsequently, progressive macrocephaly, limb spasticity and seizures develop. With gastrostomy feedings and close nursing care, some of these children will remain interactive for a number of years although they are unable to sit alone, speak or walk.

The human ASPA gene maps to 17pter-p13 and covers a span of ~30 kb of genomic DNA. The cDNA is 1435 bp in length and contains 6 exons encoding a protein of 313 amino acids (Kaul et al., 1993). CD is especially prevalent among individuals of Ashkenazi Jewish ancestry in which the carrier frequency approximates 1 in 37-57 (Feigenbaum et al., 2004). Nearly all (98%) of the mutant alleles in this population are due to two founder mutations, E285A and Y231X. A third mutation, A305E, accounts

[*] Department of Neurology, New York University School of Medicine   550 First Avenue, New York, NY 10016 USA

for approximately 40 percent of mutant alleles in non-Jewish patients. Most other mutations appear to be private, confined to single families or to small geographic areas.

A total of 53 mutations of the ASPA gene are listed in the Human Gene Mutation Database (http://archive.uwcm.ac.uk/uwcm/mg/search/231014.html). Our laboratory first reported 14 of these mutations (Zeng et al., 2002), and has recently discovered another 10 novel mutations, all in non-Jewish families. These mutations were found in the course of our study of 38 non-Jewish families with CD. This report reviews the ASPA mutations we have observed in these families and illustrates the importance of molecular genotyping for the prenatal diagnosis of CD.

2. METHODS

Blood, and in some cases cultured skin fibroblasts were obtained from patients following informed written consent. In one case, only fetal skin and cartilage were available. In this particular case, and in many others, blood was also obtained from the parents. The diagnosis of CD was based on typical clinical, neuroradiologic and biochemical criteria. The clinical manifestations of the patients are summarized in Table 1. Their ethnic origins are included in tables 2 and 3.

Table 1. Clinical manifestations 38 non-Jewish CD patients

Manifestations		n
Age of onset	<2m	14
	3-4m	20
	>6m	4
Poor head control		38
Macrocephaly		38
Visual failure		38
White-matter disease		38
Seizures		13
Alive	no	2
	yes	31 (one is >17y)
	unknown	5

Procedures for cell culture, isolation of peripheral blood lymphocytes and extraction of total RNA and genomic DNA are described in Zeng et al. (2002). Genomic DNA was amplified using primers previously described (Kaul et al. 1993, 1994, 1996; Zeng et al., 2002). The procedure of Zeng et al. (2002) was also followed for the synthesis and amplification of cDNA. DNA and cDNA PCR products were either directly sequenced or cloned in pGEM-T Easy-Vector according to the manufacturer's suggested protocols and then sequenced. Automated sequencing was used.

In vitro mutagenesis, ASPA cDNA expression and assay of ASPA activity were performed as described by Zeng et al. (2002).

3. RESULTS AND DISCUSSION

Our general strategy involved, first, a search for the three common mutations (E285A, Y231X, A305E) using site-specific restriction endonuclease digestion of PCR products from genomic DNA. Then PCR products from both cDNA and genomic DNA were sequenced. To confirm the alterations found, specific restriction endonuclease digestion analysis of genomic DNA was performed. When possible, the carrier status of the parents was also confirmed. *In vitro* mutagenesis and expression of mutant cDNA's was performed for many of the mutations reported in our earlier series (Zeng et al., 2002) but have not yet been done for the mutations identified more recently.

3.1. Mutations in 39 Non-Jewish Patients with Canavan Disease

The mutations found are shown in Tables 2 and 3. Novel mutations included 12 missense mutations, 2 nonsense mutations, 5 deletions, 1 insertion mutation, 1 case of two variations in a single allele, 1 elimination of a stop codon and 2 splice accepter site mutations. Table 4 list all mutations found by class.

The A305E mutation is known to be present primarily in patients of European origin and that was the case in our cohort; 16/17 alleles with this mutation were of European ancestry. The smaller percentage of our cases with the A305E mutation (21.7% 17/78) than in other case series (39.5-60%; Elpeleg & Shaag, 1999, Sistermans et al., 2000) may reflect the large number of non-European patients in our series.

3.2. Characteristics of 24 Novel Mutations

From 39 patients, we identified 24 novel mutations (tables 2-4). Within human ASPA are esterase-like sequences including amino acid motifs GGTHGNE, DCTV and VNEAAYY. Two of the mutations identified, G18R and E24G, produce substitutions of invariable amino acid residues within the first esterase catalytic domain consensus sequence GGTHGNE in the first and last residues. No ASPA activity was detected in COS-7 cells transfected with mutant cDNA containing the E24G mutation whereas the activity was markedly increased after transfection with normal ASPA cDNA. This suggested that the E24G mutation caused malfunction of ASPA.

Of the two splice accepter site mutations found, one (IVS 1-2A→T) caused retention of 40 nucleotides of intron 1 on the upstream side of exon 2 while the other (IVS 4-1G→C) resulted in skipping of exons 5 and 6. In the case of the IVS 1-2A→T, intron retention occurred due to the presence of a cryptic splice acceptor sequence within the intron introducing new amino acids (Zeng et al., 2002).

Table 2. Mutations in 22 non-Jewish CD Patients*

Patient	Mutation 1	Mutation 2	Ethnic origin
1.	E24G	P181T	German
2	D68A	D249V	British
3	923delT	245insA	Jamaican
4	D249V	unidentified	British
5	Q184X	A305E	German/British
6	C152W	C152W	Yemenite
7	E285A	33del13	German/Italian
8	G27R	H244R	Italian
9	A305E	E214X	German/British/Italian
10	A287T	10T→GG	German
11	698insC	698insC	Yemenite
12	A305E	244delA	African American
13	IVS1-2A→T	not identified	German
14	X314W	X314W	Guam
15	A305E	not identified	European/Greek/Irish
16	G27R	not identified	Italian
17	P181T	not identified	German/French
18	A305E	A305E	British
19	A305E	876del4	German/American Indian
20	A305E	I16T	Italian
21	D114E	D114E	Turkish
22	A305E	G123E	European

*Zeng et al, 2002; Novel mutations are underlined.

Within exon 2, a run of seven adenines occurs at positions 238-244 that is prone to slipped mispairing at the replication fork. The seventh adenine at nucleotide 244 was deleted in one of our patients and in another, 244-245$_{AT}$ was deleted. In two other patients, we encountered insertion of an adenine at position 245, previously described by others. Hence, this site tends to be associated with both insertion and deletion, leading in both cases to a truncated non-functional gene product.

We identified a normal mutation in the cystine residue at position 152 (C152W) which resulted in complete deficiency of ASPA activity on *in vitro* mutagenesis and expression in COS-7 cells. As noted (Zeng et al., 2002) the C152 residue participates in the creation of a strong β-sheet structure and may be required for maintaining ASPA in a conformationally active state (Kaul et al., 1995). The existence of two other mutations in this same residue, i.e. C152R (Kaul, et al., 1995) and C152Y (Kaul, et al., 1996), highlights the importance of this site for the structural integrity of ASPA.

Several missense mutations resulted in a change in electron charge of the substituted amino acid or conversion of a polar amino acid to a hydrophilic one. These included D68A (aspartate →alanine), N121I (asparagine→isoleucine) T166I (threonine→isoleucine), and D249V (aspartate→valine). In the case of D68A expression of the mutant cDNA did not produce any ASPA activity in COS-7 cells. All 12 novel missense mutations are in conserved regions of the ASPA gene.

Homozygosity for these rare novel mutations was found in the cases of G18R C152W, H244R, 244 del AT, X314W and 698 ins C. Consanguinity was present only for the 698 ins C mutation (Yemenite). While an ethnic predilection for each of these mutations is a possibility, only for two of the mutations, H244R and D249V, were they present in more than one patient of the same ethnic origin.

Table 3. Mutations in non-Jewish CD Patients*

Patient	Mutation 1	Mutation 2	Ethnic origin
1	I143F	IVS4-1G→C	Guatemala
2	I143F	IVS4-1G→C	Guatemala
3	A305E	N121I	German/Polish
4	A305E	A305E	German
5	A305E	P181L	Dutch/German
6	H244R	H244R	Italian
7	G18R	G18R	Persian/Iran
8	A305E	L272P	Italian
9(Fetus)	E285A	IVS4+1G→T	Jewish/Turkish
10	A305E	A305E	German
11	R168C	382delC	Italian
12	244delAT	244delAT	Venezuelan
13	245insA	T166I	Mexican
14	P181T	V14G	German/Polish-Dutch
15	A305E	unidentified	German/British
16	A287T	unidentified	German/European
17	H244R	unidentified	Italian

*Mutations identified subsequent to Zeng et al, 2002; Novel mutations are underlined.

3.3. Genotype: Phenotype Correlations

While all patients presented with severe psychomotor delay, poor head control and truncal hypotonia and developed progressive spasticity and macrocephaly, the clinical onset and time appearance of seizures varied. Clinical manifestations were noted at birth in two youngsters of British ancestry, both of who possessed one allele containing the D249V substitution. Both died early. Another infant with an onset at birth carried the E214X mutation which produces a stop codon.

Several other novel mutations were associated with an onset before two or three months and early seizures. These include V14G, C152W, P181L, 244 del AT, X314W, 698 inc C and 923 del T. In several of these examples, either a stop codon is introduced or eliminated or a frameshift occurs. In the case of the C152W mutation, a disulphide bond critical for the molecular conformation of ASPA is disrupted. Therefore, clinical variations in Canavan disease may result from differing effects exerted on the expression of ASPA activity by specific mutations.

Table 4. Results of Mutation Analysis in 39 Non-Jewish CD Patients*

Summary	n	Mutations
Missense mutations	21	V14G, I16T, G18R, E24G, G27R, D68A, D114E, N121I, G123E, I143F, C152W, T166I, R168C, P181T, P181L, H244R, D249V, L272P, E285A, A287T, A305E.
Nonsense mutations	2	Q184X, E214X
Deletions	6	33del13, 876del4, 244delA, 244delAT, 382delC, 923delT.
Insertion mutations	2	245insA, 698insC.
Two variations in one allele	1	10TindelGG (10T→G and 11insG)
Elimination of the stop code	1	X314W (941A→G, TAG→TGG)
Splice acceptor site mutation	2	IVS1-2A→T, IVS4-1G →C.

*Novel mutations are underlined.

3.4. Prenatal Diagnosis of Canavan disease

Aspartoacylase activity is low to undetectable in chorionic villus and amniocytes. Therefore, a prenatal diagnosis of Caravan disease cannot be made reliably by enzyme activity. As an alternative, the concentration of NAA in amniotic fluid can be determined using a stable-isotype dilution technique (Kelley, 1993). This method was used to detect Canavan disease in a pregnancy originally thought to be at low risk.

The mother was one-quarter Jewish and a carrier for the E285A mutation. Her husband, who was non-Jewish and of Turkish, Persian and Indian ancestry, tested negative for the three common Canavan disease mutations. Nevertheless, the couple elected to have the amniotic fluid analyzed at 17 weeks of pregnancy. The amniotic fluid NAA was reported as elevated and was again elevated when repeated at 21 weeks of gestation (Dr. Richard Kelley, Kennedy-Krieger Institute, Baltimore, Maryland). The fetus was aborted. The father's genomic DNA was sequenced and a mutation at IVS4+1G→T (Rady et al., 2002) was found. Both this mutation and the mother's E285A mutation were confirmed in the fetal tissue.

Mutation analysis of fetal cells is the "gold standard" for prenatal diagnosis of Canavan disease but may not always be possible. In another family with an affected child, we were able to positively identify a fetus as unaffected using polymorphisms within the ASPA gene. The mother carried the A305E mutation but it was not possible to identify the father's mutated allele since neither the father nor the affected child produced mRNA from this allele. The 693C→T polymorphism could not be used because the father appeared to be homozygous whereas the affected sibling was heterozygous for this polymorphism. Sequencing of the father's genomic DNA revealed that he also carried a new variation at IVS2-284A→T on one of his ASPA alleles. His wife also carried this variation on one of her alleles and the affected sibling was homozygous for the new polymorphism. However, the fetus was heterozygous for the novel variation. With cDNA, the affected child had a one allele pattern, whereas the fetal cDNA contained two alleles confirming the diagnosis of an unaffected fetus.

Consequently, when one mutation remains unidentified in either parent, the search for polymorphisms may be a useful alternative to determine the genotype of an at risk fetus.

4. CONCLUSIONS

Whereas two mutations in the ASPA gene account for more than 98% of all mutant alleles causing Canavan disease in the Ashkenazi Jewish population, many different mutations can be found in non-Jewish individuals with Canavan disease. In our investigation of 40 non-Jewish patients with Canavan disease, we have found 24 novel mutations and one new polymorphism in the ASPA gene.

On the basis of this experience, it is concluded that the diagnosis of non-Jewish persons with Canavan disease requires sequencing of all exons and their splice sites as well as a search for insertions and deletions. When mRNA cannot be found so that cDNA is unavailable for sequencing, it may be possible to use polymorphisms in place of the actual mutations for prenatal diagnosis.

5. ACKNOWLEDGEMENTS

These studies were supported by grants from the Morris and Alma Shapiro Foundation, the Margaret Enoch Fund, the Freddy Teft Memorial Fund, and the Genzyme Corporation.

6. QUESTION AND ANSWER SESSION

SESSION CO-CHAIR MATALON: Any comments? We fixed the microphone. You can talk now. Let's give it a try. Good.

PARTICIPANT: I'm fairly new to the Canavan's field. I'm a spectroscopist. Do the carriers have any manifestations? And have the carriers been evaluated with spectroscopy or imaging?

DR. KOLODNY: Perhaps others in the audience can answer this question. We haven't imaged any of our carriers. It is said that if you do psychological testing that carriers might have visual, spatial, or other maybe executive function abnormalities. I haven't really seen that in our patient cohort.

DR. ROSS: In response to the question that occasionally parents will hop into the magnet when their children aren't around. We haven't ever seen any hint, but, of course, we haven't usually known which parent was the carrier; double-double blind study.

DR. KOLODNY: Looks better.

PARTICIPANT: The reason why I ask, we have discovered -- it's SLC6AA -- a creatine transporter deficiency. And the parents, the female carriers, the grandmother, mother, and aunt all have psychological deficits; and I did spectroscopy. And they did have relatively normal creatine levels, but what was interesting was we imaged a nine-day-old carrier who had a deficiency of creatine.

Now, with her mother and her aunt and her grandmother, as I just said, it normalizes with the functioning allele. So that's why I was wondering if anyone had looked to see if there were any kind of myelin disruptions or maybe a reduced -- or, yes, an NAA that was somewhere between a normal and the Canavan's might be interesting.

DR. KOLODNY: Carriers don't normally have elevations in NAA.

PARTICIPANT: Hello. Yes. In CSF analysis, can that be more informative in an early diagnosis or if possible to avoid genotyping for that matter in a Canavan's disease patient suspected of hydrocephalus?

DR. KOLODNY: You're asking whether CSF NAA can be used for the early diagnosis of Canavan disease?

PARTICIPANT: We had a child with hydrocephalus, and we were suspecting CD. So CSF analysis, can that be more informative? Or directly going on to the genotyping that according to the early detection can --

DR. KOLODNY: Using CSF rather than urine analysis of organic acids? I can't answer that because we haven't done CSF exams. Has anyone? Reuben? CSF, Canavan children?

DR. LEONE: Yes, we did it.

DR. BURLINA: Of course, we are also concerned because, you know, you have to make a lumbar puncture in a small child. But the response is very high. And in the ten cases we had until now, eight also had a very high level of NAAG. It was very high because it went from 100 micromoles to 250 micromoles per liter in the CSF.

DR. KOLODNY: It is interesting that for some of our patients, we have had normal organic acids. Then after making the diagnosis of Canavan, we have asked the lab to repeat it, and it has been elevated. I think you have to be very careful when you review laboratory results for urine NAA levels.

DR. BURLINA: There is a question for that one because usually NAA was always considered an organic acid, not an acetylated amino acid. So you went through the analysis with the organic acid with the usual analysis for organic acid, but you risked overestimating with the organic acid. And you lose the higher peak.

So you risk that you don't have a real number. So this is the reason that sometimes you have to reject it and use a different system. So we use capillary electrophoresis for that reason.

DR. KOLODNY: So that's an important caution.

Dr. Tsipis, you had a question?

DR. TSIPIS: Just a quick question. With genotype/phenotype, is there anything in there that is interesting?

DR. KOLODNY: There are three children that we reported in our paper in 2002 where seizures and irritability and failure to thrive really began immediately after birth, very soon, before two months.

But that isn't the story in most patients. And in others who have the same genotype, there was a more typical clinical picture. So it's very hard to make a story, even when you have termination stop codons early in the gene so that protein isn't made. Still we have children surviving and doing quite well.

DR. NAMBOODIRI: Can you comment on any kidney complications in your patients?

DR. KOLODNY: Well, with regard to that, I can only answer that we really stress the kidney by creating a strong acidosis with Diamox and calcium acetate driving the bicarb level way down, as low as 12 milliequivalents. And families know enough that when the child starts hyperventilating with a respiratory acidosis, that they will back off on the drug.

So I think that both the lungs and the kidneys seem to be working very well, in spite of the metabolic stress that we put on them with these two drugs.

SESSION CO-CHAIR MATALON: Maybe just before the break, I'll throw something. You can think about it when we come back. Among all of the patients I have seen, which is quite a few, and that I have heard of, there was one patient who had nephritic syndrome. So I don't really know if this has any relation.

Thank you.

7. REFERENCES

Elpeleg, O.N., Shaag, A., 1999, The spectrum of mutations of the aspartoacylase gene in Canavan disease in non-Jewish patients. *J. Inherit. Metab. Dis.* 22:531.

Feigenbaum, A., Moore, R., Clarke, J., et al., 2004, Canavan Disease: carrier-frequency determination in the Ashkenazi Jewish population and development of a novel molecular diagnostic assay. *Am. J. Med. Genet.* 124A:142.

Kaul, R., Gao, G.P., Aloya, M., et al., 1994, Canavan disease: mutations among Jewish and non-Jewish patients. *Am. J. Hum. Genet.* 55:34.

Kaul, R., Gao, G.P., Balamurugan, K., Matalon, R., 1993, Cloning of the human aspartoacylase cDNA and a common missense mutation in Canavan disease. *Nature Genet.* 5:118.

Kaul, R., Gao, R., Matalon, R., et al., 1996, Identification and expression of eight novel mutations among non-Jewish patients with Canavan disease. *Am. J. Hum. Genet.* 59:95

Kaul, R., Gao, G.P., Michals, K. et al., 1995, Novel (cys 125 arg) missense mutation in an Arab patient with Canavan disease. *Hum. Mutat.* 5:269.

Kelley, R.I., 1993, Prenatal detection of Canavan disease by measurement of N-acetyl-L-aspartate in amniotic fluid. *J. Inherit. Metab. Dis.* 16:918.

Matalon, R., Michals, K., Sebasta, D., et al., 1988, Aspartoacylase deficiency and N-acetylaspartic aciduria in patients with Canavan disease. *Am J Med Genet.* 29:463.

Madhavarao, C.N., Moffett, J.R., Moore, R.A., 2004, et al. Immunohistochemical localization of aspartoacylase in the rat central nervous system. *J. Comp. Neurol.* 472:318.

Rady, P.L., Penzieu, J.M., Vargas, T., et al., 2000, Novel splice site mutation of aspartoacylase gene in a Turkish patient with Canavan disease. *Eur. J. Pediatr. Neurol.* 4:27.

Sistermans, E. A., de Coo, R. F., van Beerendonk, H. M., Poll-The, B. T., Kleijer, W. J., and van Oost, B. A., 2000, Mutation detection in the aspartoacylase gene in 17 patients with Canavan disease: four new mutations in the non-Jewish population, *Eur. J. Hum. Genet.* 8:557.

Zeng, B. J., Wang, Z. H., Ribeiro, L, A,, Leone, P., De Gasperi, R., Kim, S. J., Raghavan, S., Ong, E., Pastores, G. M., and Kolodny, E. H., 2002, Identification and characterization of novel mutations of the aspartoacylase gene in non-Jewish patients with Canavan disease, *J. Inherit. Metab. Dis.* 25:557.

DOES ASPA GENE MUTATION IN CANAVAN DISEASE ALTER OLIGODENDROCYTE DEVELOPMENT?

A Tissue Culture Study of ASPA KO Mice Brain

Shalini Kumar, Rasika Sowmyalakshmi, Sarah L. Daniels, Ruth Chang, Sankar Surendran, Reuben Matalon, and Jean de Vellis*

1. INTRODUCTION

Canavan Disease (CD) is a devastating disease, which is caused by a recessive mutation in the enzyme aspartoacylase (ASPA). ASPA hydrolyzes N-acetylaspartate (NAA) to generate L-aspartate and acetate. In mammalian tissues, NAA is one of the most abundant metabolites and its level is detected highest in the brain as compared to other organs (1, 2). It is found mostly concentrated in the cerebral cortex area, while much lower concentration can be detected in the medulla. For more than a decade, NAA has served as an important tool in clinical application due to its high resonance property, which is captured by NMR to examine the status of neurons and underlying pathology in CNS. In the CNS, the substrate NAA and its converting enzyme, ASPA, have recently gained much attention due to their implication in CD. The lack of functional ASPA causes an accumulation of NAA and leukodystrophy that can be visualized in the white matter regions of the CD brain (3, 4). The finding of ASPA deficiency as the basic biochemical defect in CD, had, for the first time, suggested that hydrolysis of NAA by ASPA is crucial for maintenance of functionally intact white matter in brain. It was also reported that the ASPA activity occurs predominantly in the white matter (5, 6). Further work in this area has validated both localization of ASPA protein and enzyme activity (7-9) and NAA (10, 11) in cells of oligodendrocyte lineage. A recent study showed that ASPA activity appears at a high level in purified myelin and that neuronal NAA contributes acetyl groups for myelin lipid biosynthesis, perhaps via transaxonal movement in the optic system (12). A subsequent study reported the presence of fatty acid synthesizing enzymes necessary for the production of acyl chain within myelin and

* Mental Retardation Research Center, Department of Neurobiology and Psychiatry, Brain Research Institute, David Geffen School of Medicine, University of California, Los Angeles. Sankar Surendran and Reuben Matalon; Division of Genetics, Department of Pediatrics, University of Texas Medical Branch, Galveston.

thus supported the transaxonal movement on NAA in neurons (13). It is not clearly understood, however, if increase in the subtrate NAA or the lack of individual product, aspartate and actetate, or combination of all, are responsible for observed neuropathology of CD. And most importantly, the casual target gene functions and temporal onset of events during development are not understood at this time.

In this preliminary study we report preparation of primary cultures of mixed glia from the newborn rat and ASPA knockout (KO) mice cerebral cortex for characterization of glial cell types on the basis of cell and lineage specific marker gene expressions. The generation of ASPA KO mice strain and its characterization has been well described by Matalon et al., (14). The ASPA KO mouse serves as a model for CD to address metabolic and developmental targets in the brain that can also throw light into ASPA mutation-mediated events in CD. The ASPA KO mouse characterization had revealed a lack of ASPA enzyme activity and spongy degeneration in cerebral regions with neuropathology similar to CD (14). While the underlying relationship of oligodendrocyte development to ASPA deficiency has not been reported at this time, a profound decrease in expression of GABA-A receptor has been seen in ASPA KO mice (15). In addition, a reduced level of GABA and glutamate, an over-expression of Spi1 transcription factor has been observed in adult ASPA KO brain (15). For the purpose of establishing primary cell culture as a tool to study developmental events in ASPA mutation, a characterization of ASPA gene expression, protein and enzyme levels were examined along with glial lineage-specific markers in mixed glial cultures from normal newborn rat cortex. Since the subventricular zone gives rise to early glial progenitors, we cultured cortical tissue for analysis. The primary cell cultures of ASPA wild type (wt) and KO newborn mice were similarly prepared and characterized for early development of oligodendrocytes.

Our laboratory has pioneered the preparation of cerebral cortex mixed glial cell cultures, pure cultures of oligodendrocytes and astrocytes from newborn rodent brains (16, 17). Utilizing these cultures, a wide range of developmental studies have been possible based on cellular, biochemical and molecular approach. The oligodendroglial development procedes through several stages from progenitor cell to the mature, myelin-synthesizing oligodendrocyte (reviewed in 18, 19). The onset of oligodendrogenesis, its survival and maturation is a multi-step process, which is dependent upon availability of several essential factors.

The postnatal genesis of oligodendrocyte progenitors in newborn cultures can be identified by immunocytochemical staining for the early oligodendrocyte markers such as platelet-derived growth factor receptor alpha (PDGFRα, 20, 21), the proteoglycan NG2, GD3 and A2B5 (Reviewed: 22, 23). Working with cyclic nucleotide phosphohydrolase (CNP)-enhanced green fluorescent protein (EGFP) transgenic mice, Aguirre and Gallo (24) demonstrated that postnatal subventricular zone NG2+ progenitors can migrate distances and develop into interneurons and oligodendrocytes in a region-specific manner. The study also shows that the cortical, olfactory bulb and cerebellar NG2+ cells have limited migration ability and they give rise to glia. Morphologically, oligodendrocyte progenitors appear as bipolar or cells with bright soma, developing more processes and becoming less motile with maturity. These early oligodendrocytes also are responsive to mitogens, PDGF-AA, NT3 and bFGF. Expression of functional TrkC receptors have been characterized in these early progenitors and are responsive to NT-3 via PTK pathways to promote cell proliferation, i.e., transition of cell-cycle from G1 to S phase, and in cell survival by preventing PARP fragmentation (25, 26). The NT3 KO mice show a severe reduction in oligodendrocyte development in their spinal cord at birth and are still born (27). The next step in oligodendrocyte progenitor differentiation occurs when they adopt a phenotype referred to as "pre-oligodendrocytes" (preOL). These cells are multipolar and still express PDGFαR. At this stage, these cells stop the expression of the A2B5 marker and acquired immunoreactivity for sulfatides recognized by the mab O4. PreOLs become immature, non-mitotic oligodendocytes that express galacto-cerebroside (GalC), CNP and some isoforms of myelin basic proteins (MBPs) but still do not synthesize myelin membranes (28, 29). The final stage of oligodendrocyte differentiation is characterized by the appearance of the myelinating oligodendrocyte, which expresses 4 MBP isoforms, MAG, PLP/DM20 proteins, and MOG (30, 31). We have begun characterizing these stages in ASPA wt and KO mice cultures, to delineate if observed myelin deficiency is the result of a disruption at any one of these stages.

2. RESULTS

For the characterization of mixed glial cells from newborn rat cerebral cortex, we examined ASPA gene expression by in situ hybridization with digoxigenin (DIG)-labeled ASPA sense and antisense probes (Fig. 1A, B). A full-length mouse ASPA cDNA was cloned in Bluescript (SKII) riboprobe vector and was labeled in the opposite direction using a kit and hybridization was carried out as per manufacturer's (Roche Molecular Biochemicals, Germany) protocol. Sense riboprobe was used as a negative control. The expression of ASPA is visible with anti-sense probe within round cell bodies (Fig. 1B arrows) on the bed layer of astrocytes not stained for ASPA. The ASPA activity was determined in cell lysates prepared from rat cortical cultures of mixed glial cells, neurons, and pure cultures of oligodendrocytes and astrocytes as described (14). The highest level of ASPA activity, 1.3mU/mg protein, was observed in oligodendrocyte cell lysates, approximately 6-times higher than that reported in a whole brain tissue lysate for wt mouse (14). The astrocytes and neuronal cultures showed less than 0.01mU/mg protein level of ASPA activity (Fig.1C). The cultures of mixed glial cells exhibited nearly 29% ASPA activity of pure cultures of oligodendrocytes.

The mixed glial cultures from normal rat brain cortex were immunostained, examined for expression and localization of ASPA protein. A rabbit polyclonal antibody generated against mouse ASPA in the laboratory of Dr. Matalon was IgG purified with a kit (Pierce, Rockford, IL) and used at 1:1000 dilution. The oligodendrocyte cells showed co-localization of ASPA protein with oligodendrocyte specific cell markers, A2B5 (Fig. 1D), and cyclic nucleotide phosphohydrolase (CNP) (Fig. 1F) in a mixed glial cell culture. The double immunostaining of ASPA with glial fibrillary acidic protein (GFAP), a marker for astrocyte, showed a discrete staining for ASPA and GFAP in mixed glial cultures (Fig. 1G).

Our preliminary observations show an early arrival of ASPA KO (-/-) oligodendrocyte cells on the bedlayer of astrocytes by 1-2 days, as compared to the wt (+/+) oligodendrocyte cells. In some instances upto 50% more oligodendrocyte-like cells can be seen in the KO cultures within 3-4 days of plating. Once arrived, ASPA KO oligodendrocyte cells appear to be less adherent to the bedlayer of astrocytes and become unattached.

We have analyzed a few early (NG2, GD3, A_2B_5, GPDH), mid (O4) and late (MBP, PLP) oligodendrocyte cell markers and GFAP in mixed glial cultures (Figure 2). The preliminary results show a robust staining for proteoglycan, NG2 and GD3, characterized as progenitor oligodendrocyte cell marker (Fig. 2A, B). A strong signal for A_2B_5 and cytosolic glycerol-3- phosphate dehydrogenase (GPDH) expressed early in development

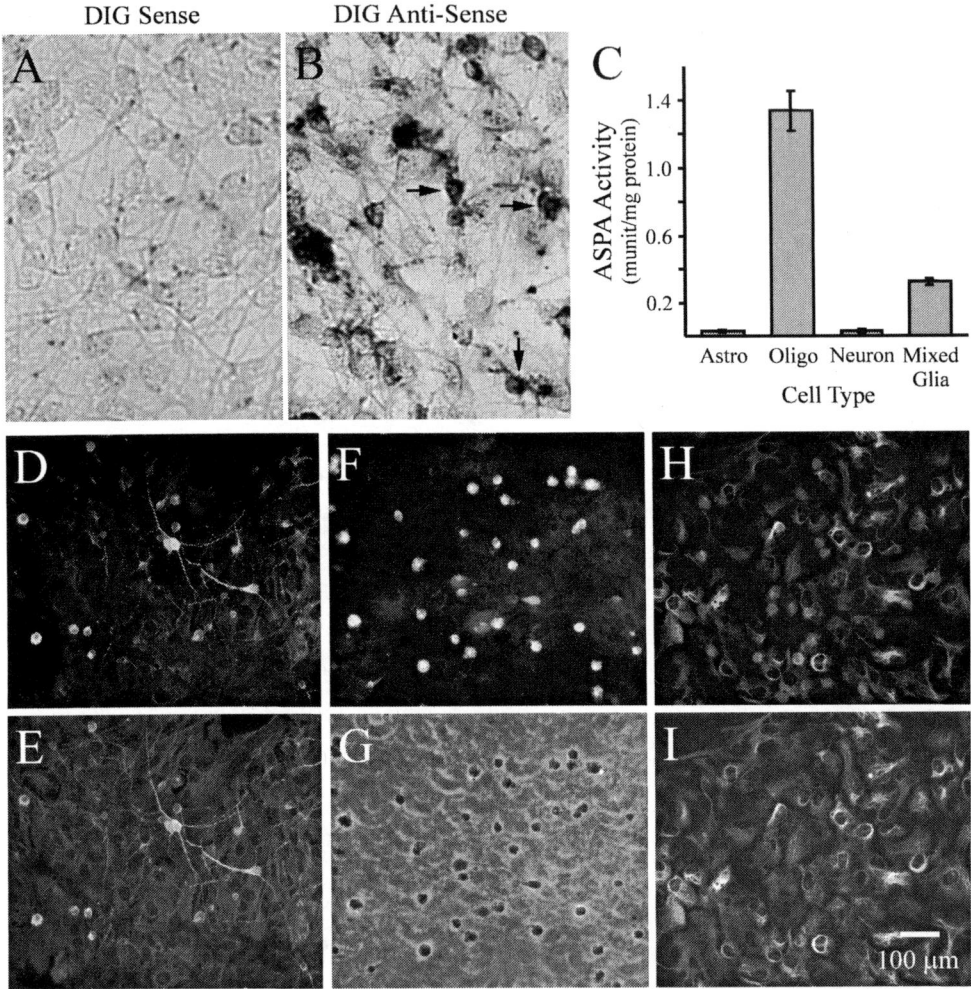

Figure 1. Characterization of primary cultures of neural cells from newborn rat cortex for expression of ASPA mRNA, enzyme activity and co-immunostaining of ASPA protein with neural cells markers. The in situ detection of ASPA mRNA in oligodendrocytes cells in cultures mixed glia probed with sense (A) and anti-sense (B) ASPA digoxigenin labeled probes. The oligodendrocyte cells show dark staining (arrows) on the bed layer of unstained astrocytes. Optimal ASPA enzyme activity (mU/mg protein) can be seen in culture of oligodendrocytes (C) and about 29% ASPA activity can also be seen in mixed glial cultures. The pure cultures of astrocytes and cortical neurons exhibit insignificant level of ASPA activity. Immunostaining of mixed glial cultures from normal rat brain shows a co-localization of ASPA (red) with oligodendrocyte cell specific markers A_2B_5 (D) and CNP (F), both green, or exhibiting a yellow staining for co-expression. The immunostaining of ASPA with GFAP (green), an astrocyte cell specific marker shows a discrete localization of the two markers (H). (Epifluorescence images D, F, H; Phase image E, G, I; see color insert that appears between pages 364 and 365)

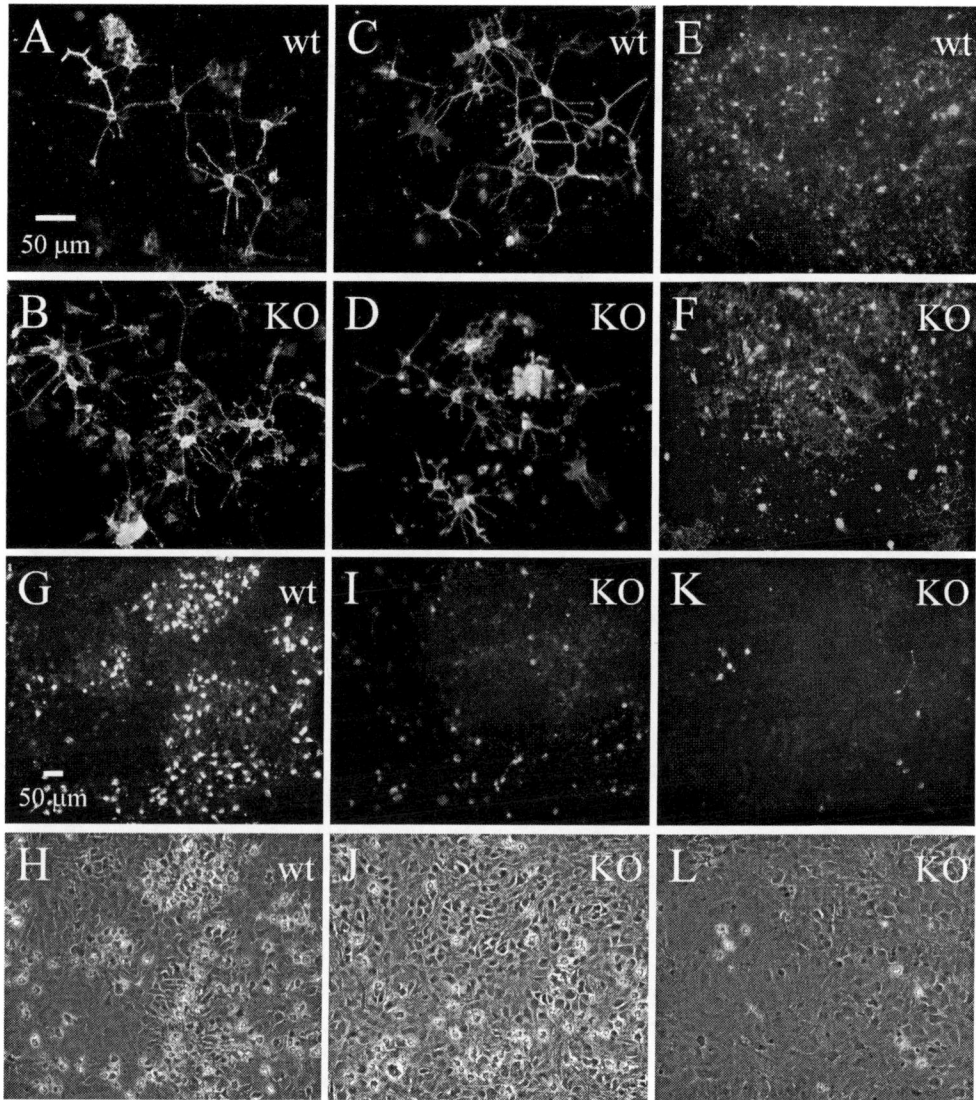

Figure 2. Immunocytochemistry of mixed glia cultures from ASPA wt and KO mice cerebral cortex. The immunostaining of markers for progenitor and mature oligodendrocytes were carried out in 6 and 15d-old cultures. The immunostaining for marker NG2 (green) and GD3 (Red; A and B), A_2B_5 (red) and GPDH (green; C and D), and O4 (red; E and F) showed the status of oligodendrocytes, while GFAP (green; E and F), the marker for astrocyte shows no co-localization with oligodendrocyte marker O4. A delay in MBP (green) and PLP (red) expressing oligodendrocyte cells was detected in 15day old ASPA KO mice cortical cultures (G, I and K). The double immunostaining of MBP astrocyte, showed a discrete staining for ASPA (red) and GFAP (green) in mixed glial cultures (Fig. 1G and panel H as phase image; see color insert that appears between pages 364 and 365)

(32) and marker for immature oligodendrocyte, O4 was also detected in both wt and KO cultures (Fig. 2 C-F). The immunostaining of these early markers was significantly higher in cultures of KO mouse as compared to wt, suggesting accumulation of oligodendrocyte progenitors in KO cultures, while progenitors in the wt cultures may have progressed to different level of development. In all these instances KO oligodendrocyte cell morphology appeared to be distinctly different. While wt oligodendrocyte cells showed fine processes with a round cell body, the KO oligodendrocytes appear to have more globular structure with rough processes. The cause and effect of these alterations are unknown at this time. In general, a fewer oligodendrocytes stayed attached on the bedlayer of astrocytes by day 15 in the KO cultures. Of those, only a few cells showed expression of late markers, MBP and PLP as compared to the wt (Fig. 2 G-K).

Our findings suggest an overall delay in maturation of ASPA KO oligodendrocytes. It is, however, evident that the surviving oligodendrocytes in the KO cultures can potentially express MBP and PLP. More comprehensive study is needed such as cell proliferation analysis and cell survival vs. cell death that may help delineate the outcome of oligodendrocyte cell developmental progression. Most importantly, in combination to the postnatal culture study, an age matched in vivo study will help elucidate the causal and temporal development of oligodendrocytes in ASPA KO mouse. These studies are currently in progress.

3. QUESTION AND ANSWER SESSION

DR. KOLODNY: Thank you, Reuben. A very simple question in cell biology which has perplexed me. If you take a mature somatic cell and grow it in culture, after 50 generations, it dies and the culture is lost because the telomeres are shortening with each cell cycle. Are the telomeres preserved in the primitive stem cells?

DR. De VELLIS: In neural stem cells, yes, they can. There is a little bit of controversy about whether they have the finite life or not. People like Terry Vendercoy think that maybe there is still a finite life. But some of it could be questioned. When you do it *in vitro,* it could be just a question of also technique. But we potentially have an indefinite life or so indefinite that you can work actually with the same cells. I mean, you can expand them and then freeze them and then just work with essentially the same passage.

So in a way, we have come full circle. We started 30 years ago or something with cell lines in everybody, tumor cell lines, Gordon Sato type stuff. And now we come back working with stem cells. And they are being treated almost like cell lines.

DR. COYLE: I was struck by the rampant programmed neuronal cell death in the knockout. But what struck me even more were the changes in expression of the markers for neurogenesis in your neurospheres in the knockout. Did I get that right?

DR. De VELLIS: Yes. The neurospheres were all from the knockout. Yes.

DR. COYLE: Right. So these neurospheres are grown in tissue culture, where you have an infinite dilution of the N-acetylaspartate. So this is different than the brain and the human body, which suggests to me that the lack of the ASPA and the inability to deacetylate the NAA, NAA must be having some specific cytotoxic and cell communicational properties aside from just osmolarity or not being a source of acetate for myelin synthesis. Would you agree?

DR. De VELLIS: Yes. Actually, Shalini has done that experiment. And also she knows not just the experiment. She knows a lot of details. And I could have said a few things wrong anyway. But she has done that. Why don't you give her the microphone? Oh, I thought you had the microphone. I'm sorry.

DR. KUMAR: So let me get the question first just to see if the -- please repeat one more time so I could have it clear in my mind what you are asking, just the second part of it.

DR. COYLE: My basic question is when you grow these neurospheres, they are grown in tissue culture, where there is --

DR. De VELLIS: Right, yes.

DR. COYLE: And, yet, you see these changes that are consistent with cytotoxicity, which means that it must be the very local environment itself where the cells are reacting to the NAA or the inability to deacetylate and not simply secretion of very high levels of NAA resulting in these abnormalities in osmolarity or --

DR. De VELLIS: That's one. Yes, that's one of the very logical questions. There are others also, I think, but --

DR. KUMAR: There are two aspects to it. First of all, you are removing these cells from the brain and from the entire environment that it has been. And you are putting it in a medium which is pretty conducive of their survival and also being flushed at least.

Now, when we take them and we put them back into the brain, we noticed one thing that they didn't quite differentiate. If we had taken some stem cells within eight days, we will have seen NBP expression, NBP protein expression. So we didn't in these cells. Okay?

And they showed olpolyg-2. They showed PGF receptor expression, both of which and quite a few others that we have seen are mostly seen very early in development or in progenitor cells. I wouldn't expect them to still be progenitors 8 days or 11 days later in brain. So some level of myelination has just begun.

So if this cell was in the milieu or in the matrix for the cells, whether myelination has begun, it should have caught up and started to express at least NBP protein within the CMP, within CPLP. We only could see CMP, which we have seen before, in the progenitor cells, in addition to PGF receptor and others.

But yes, your point is well-taken that at least they are surviving and they are there with at least a few of the markers expressed. So they're a combination of things, yes.

4. REFERENCES

1. Tallan, H., 1957, Studies on the distribution of N-acetyl-L-aspartic acid in brain. *J. Biol. Chem.* 224:41.
2. Miyake, M., Kakimoto, Y., and Sorimachi, M., 1980, A gas chromatographic method for the determination of N-acetyl-L-aspartic acid, N-acetyl-aspartylglutamic acid and beta-citryl-L-glutamic acid and their distributions in the brain and other organs of various species of animals. *J. Neurochem.* 36:804.
3. Matalon, R., Michals, K., Sebesta, D., Deanching, M., Gashkoff, P., and Casanova, J., 1988, Aspartoacylase deficiency and N-acetylaspartic aciduria in patients with Canavan disease. *Am. J. Med. Genet.* 29:463.
4. Matalon, R., Kaul, R., Casanova, J., Michals, K., Johnson, A., Rapin, I., Gashkoff, P., and Deanching, M., 1989, Aspartoacylase deficiency: The enzyme defect in Canavan disease. *Inher. Metab. Dis.* 12:463.
5. D'Adamo, A.J., Smith, J., and Woiler, C., 1973, The occurrence of N-acetylaspartate amidohydrolase (aminoacylase II) in the developing rat. *J. Neurochem.* 20:1275.
6. Kaul, R., Casanova, J., Johnson, A., Tang, P., and Matalon, R., 1991, Purification, characterization, and localization of aspartoacylase from bovine brain. *J. Neurochem.* 56:129.
7. Baslow, M., Suckow, R., Sapirstein, V., and Hungund, B., 1999, Expression of aspartoacylase activity

incultures rat macroglial cells is limited to oligodendrocytes. *J. Mol. Neurosci.***13**:47.

8. Bhakoo, K., and Pearce, D., 2000, In vitro expression of N-acetyl aspartate by oligodendrocytes: Implications for proton magnetic resonance spectroscopy signal in vivo. *J. Neurochem.* **74**:254.

9. Madhavarao, C.N., Hammer, J.A., Quarles, R.H., Namboodiri, M.A., 2002, A radiometric assay for aspartoacylase activity in cultured oligodendrocytes. *Anal. Biochem.* **308**:314.

10. Baslow, M., 2000, Functions of N-acetyl-L-aspartate and N-acetyl-L-aspartylglutamate in the vertebrate brain: Role in glial cell-specific signaling. *J. Neurochem.* **75**:453-459.

11. Bhakoo, K., Craig, T., and Styles, P., 2001, Developmental and regional distribution of aspartoacylase in rat brain tissue. *J. Neurochem.* **79**:211.

12. Chakraborty, G., Mekala, P., Yahya, D., Wu, G., and Ledeen, R., 2001, Intraneuronal N-acetylaspartate supplies acetyl groups for myelin lipid synthesis: Evidence for myelin-associated aspartoacylase. *J. Neurochem.* **78**:736.

13. Chakraborty, G., Ledeen R., 2003, Fatty acid synthesizing enzymes intrinsic to myelin. *Brain Res. Mol. Brain Res.* **112**:46.

14. Matalon, R., Rady, P., Platt, K., Skinner, H., Quast, M., Campbell, G., Matalon , K., Ceci, J., Tyring, S., Nehls, M., Surendran, S., Wei, J., Ezell, E., and Szucs, S., 2000, Knock-out mouse for Canavan disease: a model for gene transfer to the central nervous system. *J. Gene Med.* **2**:165.

15. Surendran, S., Michals-Matalon, K., Quast, M.J., Tyring, S.K., Wei, J., Ezell, E.L., Matalon, R., 2003, Canavan disease: a monogenic trait with complex genomic interaction. *Mol. Genet. Metab.* **80**:74.

16. McCarthy, K., and de Vellis, J., 1980, Preparation of separate astroglial and oligodendroglial cell cultures from rat cerebral tissue. *J. Cell Biol.* **85**:879.

17. Cole, R., and de Vellis, J., 1998, Preparation of astrocyte and oligodendrocyte cultures from primary rat glial cultures, in: A Dissection and Tissue Culture Manual of the Nervous System. Shahar, A., de Vellis, J.,Vernadakis, A., and Haber, B., ed., Alan R. Liss, Inc. pp 121-133. 18. Lee, J., Mayer-Proschel, M., and Rao, M., 2000, Gliogenesis in the central nervous system. *Glia* **30**:105.

19. Richardson, W.D., Smith, H.K., Sun, T., Pringle, N.P., Hall, A., and Woodruff, R., 2000, Oligodendrocyte lineage and the motor neuron connection. *Glia* **29**:136.

20. Ellison, J., and de Vellis, J., 1994 Platelet-Derived growth factor receptor is expressed by cells in the early oligodendrocyte lineage. *J. Neurosci. Res.* **37**, 116.

21. Ellison, J., Scully, S., and de Vellis, J., 1996, Evidence for Neuronal Regulation of Oligodendrocyte Development: Cellular Localization of Platelet-Derived Growth Factor ɔ Receptor and A-Chain mRNA During Cerebral Cortex Development in the Rat. *J. Neurosci. Res.* **44**:28.

22. Liu, Y., Rao, M.S., 2004, Glial progenitors in the CNS and possible lineage relationships among them. *Biol. Cell* **96**:279.

23. Rogister, B., Ben Hur, T., and Dubois-Dalcq, M., 1999, From neural stem cells to myelinating oligodendrocytes. *Mol. Cell. Neurosci.* **14**:287.

24. Aguirre, A., Gallo, V., 2004, Postnatal neurogenesis and gliogenesis in the olfactory bulb from NG2-expressing progenitors of the subventricular zone. *J. Neurosci.* **24**:10530.

25. Kumar, S., and de Vellis, J., 1996, Neurotrophin activates signal transduction in oligodendroglial cells: expression of functional TrkC receptor isoforms. *J. Neurosci. Res.* **44**:490.

26. Kumar, S., Kahn, M., Dinh, L., and de Vellis, J., 1998, NT-3-mediated TrkC receptor activation promotes proliferation and cell survival of rodent progenitor oligodendrocyte cells in vitro and in vivo. *J. Neurosci.Res.* **54**:754.

27. Kahn, M., Kumar, S., Liebl, D., Chang, R., Parada, L., and de Vellis J. 1999 Mice lacking NT-3 and its Receptor TrkC exhibit profound deficiencies in CNS glial cells. *Glia* **26**:153.

28. Hardy, R., and Reynolds, R., 1991, Proliferation and differentiation potential of rat forebrain oligodendroglial progenitors both in vitro and in vivo. *Development* **111**:1061.

29. Monge, M., Kadiiski, D., Jacque, C.M., and Zalc, B., 1986, Oligodendroglial expression and deposition of four major myelin constituents in the myelin sheath during development. An in vivo study. *Dev Neurosci.* **8**:222.

30. Pfeiffer, S., Warrington, A., and Bansal, R., 1993, The OL and its many cellular processes. *Trends Cell Biol.* **3**:191.

31. Gardinier, M., Amiguet, P., Linington, C., and Matthieu, J., 1992 Myelin/oligodendrocyte glydoprotein is a unique member of the immunoglobulin superfamily. *J. Neurosci. Res.* **33**:177.

32. Kumar, S., Sachar, K., Huber, J., Weingarten, D.P., de Vellis, J., 1985, Glucocorticoids regulate the transcription of glycerol phosphate dehydrogenase in cultured glial cells. *J. Biol. Chem.* **260**:14743.

QUANTITATION OF NAA IN THE BRAIN BY MAGNETIC RESONANCE SPECTROSCOPY

Peter B. Barker, David Bonekamp, Gerard Riedy and Mari Smith*

1. INTRODUCTION

Since the early development of *in vivo* magnetic resonance spectroscopy (MRS) in the 1980's, a recurring problem associated with this technique has been the issue of how to analyze spectra in order to calculate metabolite concentrations. Since the magnitude of the nuclear magnetization is directly proportional to the number of nuclei from which it originates, in principle there is a linear relationship between the voltage which is induced in the spectrometer receiver coil and the nuclear (i.e. molecular) concentration. However, a large number of additional factors, many of which are unknown, will also affect the amplitude of the detected signal, preventing the direct calculation of molecular concentrations from first principles.

Therefore, to date, all approaches to spectroscopic quantitation that have been published make use of the comparison of the amplitude of the signal to be detected with that of a known reference signal. Ideally, the reference signal originates from a well-characterized compound whose concentration is accurately known. Multiple different choices of reference signal have been proposed, either indigenous or exogenous compounds, with each method having its own particular advantages and disadvantages (1).

*Russell H Morgan Department of Radiology and Radiological Science, Johns Hopkins University School of Medicine, 600 N Wolfe Street, Baltimore, MD 21287 and F.M Kirby Research Center for Functional Brain Imaging, Kennedy Krieger Institute, Baltimore, MD 21205. Author for Correspondence, Peter B. Barker: Phone (410) 955-1740, FAX (410) 614-2535, email: barker@mri.jhu.edu.

This article reviews some of the more commonly used methods for quantifying proton spectra of the human brain, with a particular emphasis on determining the concentration of N-acetyl aspartate (NAA). The article does not cover methods for quantifying spectra from other organ systems, or for heteronuclear spectroscopy, although many analysis methods for proton brain spectra may well be applicable in these other instances also. The article also does not cover, except in passing, the many numerical methods which have been developed for estimating spectral peak areas (2-4), which is a pre-requisite for quantitative analysis. Finally, a method for quantitative analysis of spectroscopy data from multi-coil receiver arrays is discussed, and an example of NAA measurements in low- and high-grade human brain tumors is presented.

2. METHODS

The voltage detected by the spectrometer from a sample containing NAA can be expressed as:

$$S_{NAA} = \beta \times [NAA] \times v \times f(T_1^{NAA}, T_2^{NAA}, TR, TE, B_1) \qquad [1]$$

where v is the MRS voxel size (in cm^3), [NAA] is the millimolar concentration of NAA, f() is a "pulse sequence modulation" factor that will depend on the pulse sequence used, its repetition and echo times (TR and TE), the T_1 and T_2 relaxation times of NAA, and B_1 is the strength of the radiofrequency field within the MRS voxel. β is a scaling factor that can be expressed as:

$$\beta \propto NS \times G \times \omega_0 \times Q \times n\xi V_c/a$$

$$[2]$$

where NS is the number of scans (averages) performed, G is the receiver gain, ω_0 is the spectrometer operating frequency (e.g. 64 MHz for a 1.5 Tesla magnet) and Q, n, ξ, V_c, and a are all factors related to the geometry and quality of the radiofrequency receiver coil (5). The proportionality constant for Eq. [2] is unknown, prohibiting the calculation of [NAA] without additional calibration measurements. However, Eqs. [1] and [2] do indicate how the signal (and hence sensitivity) of NAA can be enhanced, by using large numbers of scans, larger voxel size, high field systems, sensitive RF coils (high B_1), and maximizing the modulation factor f (by choosing optimum TR and flip angles, and short TE). Sensitivity is usually largely independent of the receiver gain, G, since this usually increases both signal and noise equally.

In order to calculate NAA concentrations, a reference signal must be acquired, ideally under identical conditions to those used to record the NAA signal, so that all proportionality factors in Eqs. [1] and [2] are identical, allowing the ratio equation to be written:

$$[NAA] = [Reference] \times \frac{S_{NAA}}{S_{reference}} \qquad [3]$$

where we assume from hereon that S_{NAA} and $S_{reference}$ have already been corrected for possible differences in T_1 and T_2 relation times according to standard equations (6).

The choice of the reference compound is of key importance for the accuracy of the quantitation procedure. The most commonly used method is to select some other compound in the brain spectrum (e.g. most often creatine (Cr)) and report ratios of NAA/Cr. This is an example of an *internal* intensity reference, namely one that comes from the same voxel as the signal to be measured (i.e. NAA). Internal references have the advantage that many of the factors in Eqs. [1] and [2] are virtually identical for both signals (e.g. volume of tissue, B_1 (and B_0) field strength, flip angle and other pulse sequence related factors), and so are insensitive to systematic errors associated with these parameters. Using a reference signal from the same spectrum (as the compound to be measured) also has the advantage that no additional scan time is required. However, while initially it was hoped that Cr levels might be relatively constant throughout the brain and invariant with pathology, subsequent studies have shown both substantial regional (6,7) and pathology-related changes in Cr (8), so that in general it is somewhat unsafe to infer changes in NAA from the measurement of only the NAA/Cr ratio. The same comment also applies to other compounds detected in the brain spectrum (e.g. choline (Cho)) that might be considered as potential reference signals. An alternative, and widely used, internal intensity reference is the unsuppressed tissue water signal (9-11), which can be easily and quickly (at least for single voxel spectroscopy) recorded by turning off the sequence water suppression pulses. Brain water content is relatively well known, and pathology-associated changes are relatively small. Furthermore, it is possible to estimate voxel water content from appropriate MRI sequences (12). Finally, the unsuppressed water signal may also be helpful for phase- and eddy current-correction of the water-suppressed spectrum (13,14). For all these reasons, quantitation of single-voxel spectra using the internal tissue water signal has become a popular technique over the last few years. Studies have demonstrated that this is a reliable method, e.g. for multi-site trials of brain spectroscopy (15).

However, there may be situations where water referencing is not optimal, for instance where brain water content is variable or not well known (for instance in neonatal studies, or pathologies involving major changes in brain water content). Also, water referencing may not be convenient in certain MR spectroscopic imaging (MRSI) studies. For instance, it may be prohibitively time consuming to record both water suppressed and non-suppressed MRSI datasets. In these instances, other approaches to quantitation should be considered. These approaches can be considered to fall into two classes, either *external* references, or the *phantom replacement* technique. External references involve the recording of a reference signal from a region outside of the primary region of interest. Often, a vial of known concentration compound (or simply water) is placed next to the head, and the signal from this region measured, either before or after the brain spectrum is measured. While external standards have the advantage that the reference concentration is exactly known, their use is complicated by the fact that some of the factors in Eq. [1] may no longer be constant. The biggest source of error is likely differences in B_1 field strength (and probably also RF pulse flip angles) due to inhomogeneities in the RF coils used for reception and/or transmission. The external standard may also induce magnetic susceptibility effects that degrade the B_0 field homogeneity, again leading to systematic errors because of suboptimal spectral quality. Also care has to be taken to ensure that the spatial localization sequence interrogates the same volume of tissue and reference sample. Sometimes a spectrum from a different (e.g.

contralateral) brain region can be used as a reference, e.g. in patients with focal brain disease, if the contralateral hemisphere is known to be normal. This approach avoids the susceptibility problems associated with the placement of an external vial, but does share the other potential problems of external references. Also, in many patients, it may be unsafe to assume that metabolite concentrations in apparently uninvolved brain regions (i.e. with normal brain MRI) are in fact the same as in the normal populations (16).

The phantom replacement technique can be regarded as a hybrid of external and internal intensity references. The basic idea is to record a spectrum from a patient, remove the patient from the magnet, insert a standard sample, and then record its spectrum using as closely matched experimental conditions as possible. The method has the advantage that the reference concentration is exactly known, and most of the factors in Eqs. [1] and [2] are also known and can be controlled for. One major difference between the two acquisitions, however, is that the RF coil "quality" factor (Q) will almost certainly be different, because the electrical properties (impedance) of the phantom will be different from the human head. Fortunately, this loading factor (F) can be readily determined from the RF power (voltage) calibration required to obtain a 90° pulse, and applied as a correction factor:

$$[NAA] = [Reference] \times F \times \frac{S_{NAA}}{S_{reference}} \qquad [4]$$

It may also sometimes be necessary to include correction factors if the brain and reference scans are collected with different numbers of scans and/or different receiver gains. The phantom replacement technique is convenient to use for MRSI scans, since it does not require any additional *patient* scan time, and, in fact, on stable clinical scanners, the reference scan may change little from one day to the next, so that it does not need to be recorded for every patient. Potential errors can occur due to B_1 inhomogeneity if the brain and reference scans are from different locations. The method therefore works best with highly homogeneous transmit and/or receive coils (such as quadrature birdcage head coils found on most 1.5 Tesla scanners), otherwise care has to be taken to attempt to match the reference scan location as close as possible to those in the brain. Alternatively, an approach (described below) for correcting MRSI scans for B_1 inhomogeneity prior to quantification (e.g. for use with phased-array receiver coils) can be used.

Table 1 contains a summary of the different quantitation methods available for MR spectroscopy and spectroscopic imaging, and lists their relative advantages and disadvantages.

Table 1. Summary of relative advantages and disadvantages of commonly used quantitation techniques for proton MR spectroscopy of the human brain.

INTERNAL	ADVANTAGES	DISADVANTAGES
Creatine	Simple, no extra scan time, no B_0 or B_1 errors	Pathological and regional variations common
Water	Minimal extra scan time, simple, no B_0 or B_1 errors	May change up to $\approx 20\%$ in pathological conditions
EXTERNAL		
Contralateral hemisphere	Simple, no coil loading errors	Extra scan time required, sensitive to B_0 or B_1 errors, can't be used in global/diffuse diseases, or in midline structures
External Standard	Reference concentration exactly known, no coil loading errors	Extra scan time required, sensitive to B_0 or B_1 errors, may cause susceptibility effects, can't be used with all pulse sequences
Phantom Replacement	Reference concentration exactly known, no extra patient scan time required	Coil loading correction required, requires stable system

2.1. Units and Tissue Compartmentalization

There are several different units which can be used to express *in vivo* tissue metabolite concentrations. From Eq. [3], it can be seen that the *in vivo* concentration will be expressed in the same units as the reference concentration, since the ratio of signal intensities is unitless. Therefore, for instance, in the external reference approach, if the external reference concentration is measured in millimolar (i.e. millimoles solute per liter solution) then the *in vivo* concentration will be returned in the same units (equivalent to millimoles per liter brain volume). Traditionally, metabolite concentrations in tissue determined by conventional biochemical techniques are more often expressed in units such as millimoles per kg tissue wet or dry weight. To convert millimolar to millimoles per kg wet weight, it is necessary to divide by the tissue density (1.05 kg/liter for normal brain). To convert to millimoles per dry weight, it is necessary to know the wet/dry weight ratio. Note that these approaches assume that the entire volume of sample is occupied by solid brain tissue; in practice, this may well not be the case, since the large voxel sizes used for in vivo MRS often contain appreciable cerebrospinal fluid (CSF) contamination.

CSF normally contains much lower levels of metabolites than brain; therefore CSF contamination (without appropriate correction methods) will lead to underestimation of brain metabolite concentrations. Fortunately, several methods now exist for estimating voxel CSF content and applying appropriate correction factors. One method makes use of measuring the voxel water signal as a function of multiple different echo times (17). Since CSF has a much longer T_2 than brain water, bi-exponential fitting of the echo signal versus time can estimate the relative fractions of brain and CSF water. This information can then be used to estimate true tissue metabolite concentrations. Alternatively, MR imaging based segmentation methods can also be used to estimate brain and CSF volumes within the MRS voxel. One particularly simple approach is to use long echo time (for instance, TE = 500 msec) fast-spin echo (FSE) MRI, which essentially only contains signal from the long T_2 CSF, the brain signal having decayed to the noise level at this TE (18). More sophisticated approaches use multiple FSE scans with different contrast in combination with appropriate multi-spectral post-processing methods, in order to estimate not only CSF content, but also fractional gray and white matter content as well (19). An example of brain and CSF segmentation using rapid FSE imaging is shown in Figure 1.

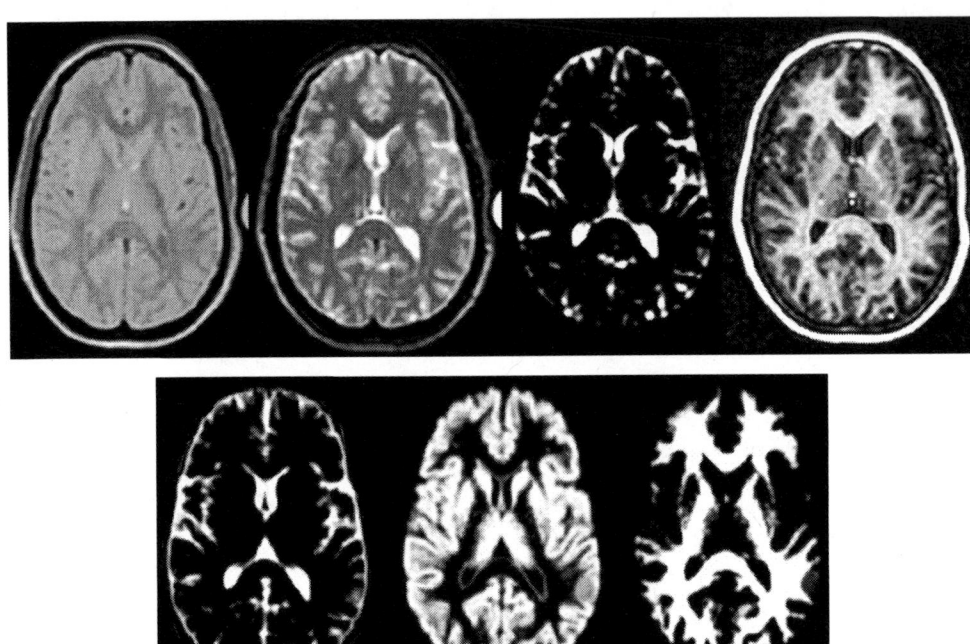

Figure 1. An example of multi-spectral brain and CSF segmentation for use with MRSI. Top row: rapid fast-spin echo sequences are acquired with different degrees of T2-weighting (proton density (TE 20 msec), T2-weighting (TE 100 msec), CSF only (TE 500 msec), and a T1-weighted (TI 500, TE 20 msec) . Through the use of region of interest measurements of individual tissue types and subsequent "Eigenimage" filtering (50), these images can be processed to generate maps (bottom row) corresponding to pure CSF, gray matter and white matter. This information can be used to correct metabolite concentrations determined by MRS(I) for partial volume effects.

2.2. Determination of Peak Areas

Equation [1] indicates that the signal induced in the receiver coil is directly proportional to the NAA concentration. From the Fourier transformation, it can be shown that the area under the curve in the spectrum (frequency domain) is equal to the amplitude of the first point of the time domain signal ("free induction decay", or FID). Therefore, quantitative analysis either requires the determination of peak areas in the frequency domain, or direct time-domain analysis to estimate the amplitude of the first data point of the different frequency components that comprise the FID.

A variety of methods have been developed for *frequency domain* analysis (3,6,20). The simplest approach is simply to use numerical integration, although this method will work poorly when spectral overlap occurs. More sophisticated analysis methods include parametric curve-fitting routines, using various model functions (e.g. Lorentzian, Gaussian, Voigt, others (21,22)) and fitting algorithms (simplex, non-linear least squares, etc..). The most sophisticated method, and one which is becoming widely used, is the so-called linear combination model ("LCModel") that fits the spectrum as a linear combination of the pure compound spectra known to exist in the spectrum (4). The LCModel is particularly attractive for several reasons; (a) it makes full use of all the resonances in the molecule, (b) it is fully automated and user independent, including both baseline and phase correction, (c) with appropriate calibration data, it can give absolute metabolite concentrations, and an estimate of the uncertainty. An example of a LCModel analysis of a short echo time spectrum recorded at 3 Tesla in the normal human brain is shown in Figure 2.

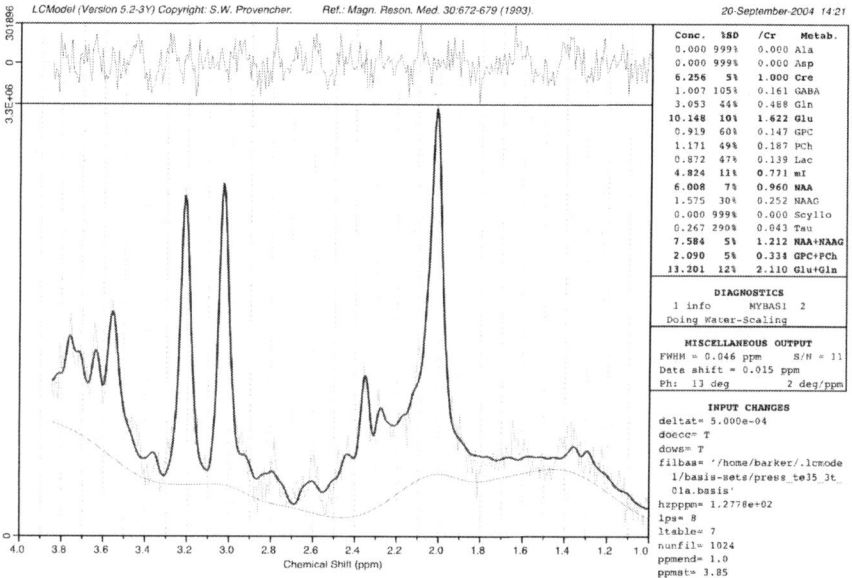

Figure 2. An example of the "LCModel" analysis of a short echo proton spectrum from the anterior cingulated gyrus of the normal human brain recorded at 3 Tesla. The original data (black line) is overlaid with the results of the linear combination analysis (red line), with the difference (noise) plotted at the top of the display. Metabolite concentrations (determined using brain water as an internal intensity reference), their % standard deviations, and other imformation is tabulated on the right hand side.

A variety of *time-domain* fitting methods have been proposed, usually using parametric models based on the exponentially decaying oscillations (10,23). One of the main perceived advantages of time-domain fitting is that it can avoid artifacts that may be induced by Fourier transformation of incomplete of partially corrupted time-domain data (e.g. missing or incorrect data points at the beginning of the FID). While this is true, since the properties of the Fourier transform are well known, it is usually possible to account for these problems using appropriate functions in frequency domain fitting as well, so that the choice of method may depend as much on numerical convenience, rather than any fundamental difference in approach.

2.3. Literature Review of Human Brain NAA Concentrations

There have been numerous papers over the last 12 years on the determination of NAA concentrations in different regions of the human brain (4,6,10,12,15,24-34). Generally, NAA concentrations have been found to be in the range of 7-13 mM in the adult human brain, with relatively small regional variations. It should be remembered that the signal at 2.02 ppm has components from both NAA and N-acetyl aspartyl glutamate (NAAG), which is typically present at about 1 mM concentration (35). Since these compounds are difficult to resolve, in most papers the quoted "NAA" concentration usually is in fact the sum of NAA and NAAG.

NAA concentrations are low at birth, and increase rapidly over the first one to two years of life (36). NAA appears to relatively stable in adults, with some investigators finding no change with age, while others find slight decreases in NAA in elderly subjects (6,37,38).

3. RESULTS

3.1. Metabolite Concentrations in Human Brain Tumors

Quantitation measurements of metabolite concentrations are becoming increasingly common in proton spectroscopy studies of human brain pathology (39,40). Presented here is one example of a pilot study of quantitative MRSI in patients with low- and high-grade, primary, untreated brain tumors.

All patients were studied at 1.5T using a multi-slice MRSI technique with TR/TE = 2300/280 msec and a nominal 0.8 cm^3 voxel size. Metabolite concentrations were estimated using the phantom replacement technique as described previously (6). The patient population consisted of two cases of oligodendroglioma (grade II and grade III), one astrocytoma grade II, and four cases of glioblastoma multiforme (GBM). Metabolic images were reconstructed and abnormal spectra from the lesion voxels with the highest visible Cho signal were selected for analysis. A mirror-image region of interest in the contralateral hemisphere was also analyzed as an internal control. Two groups were defined, low-grade tumors (oligodendrogliomas and grade II astrocyotoma) and high-grade tumors (GBM). Examples of spectroscopic images and selected spectra are shown in Figures 3 and 4 for low- and high-grade tumors respectively. Figure 5 summarizes the results for Cho, Cr and NAA. It can be seen that NAA is lower in both high and low grade brain tumors compared to the contralateral hemisphere. However, the Cho metabolite is only significantly elevated for the high-grade tumors. No consistent changes were observed in the Cr resonance.

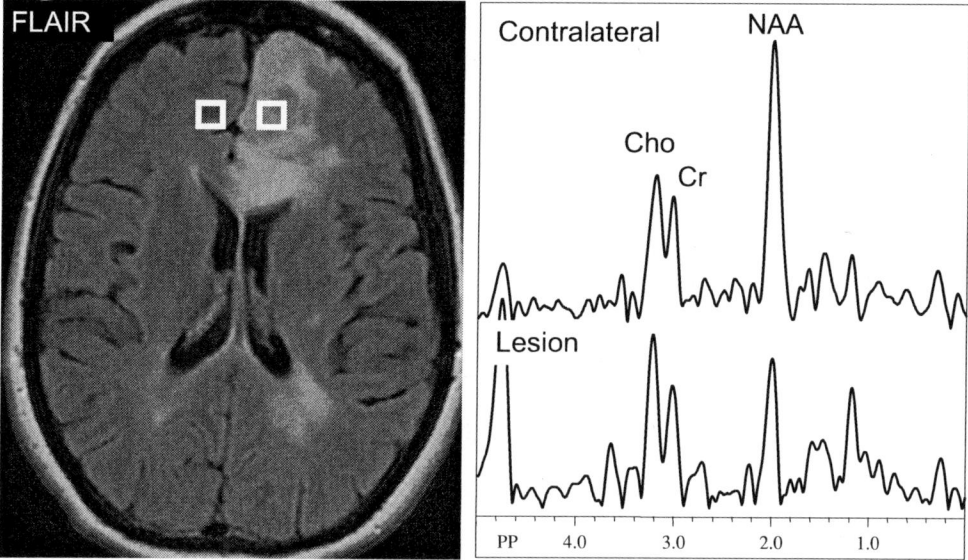

Figure 3. FLAIR MRI and proton MRSI of a low grade brain tumor (oligodendroglioma) involving the mesial left frontal lobe. The lesion exhibits a decreased signal from N-acetyl aspartate, but only minor changes in choline and creatine.

Although this is only a small pilot study, it does suggest that there are spectral differences between low- and high-grade tumors, and that brain tumors virtually always show reduced NAA levels. This is consistent with the concept and experimental data that suggest that NAA is primarily located in neuronal tissue. These results are consistent with some, but not all, previous reports that high-grade brain tumors show higher levels of Cho than low-grade tumors (41,42). However, some of the prior studies differed in terms of patient population (e.g. including patients with prior treatments, or tumors of different type) and/or technique (e.g. single voxel spectroscopy). One feature of the current study was that MRSI scans were inspected for the greatest metabolic abnormality (highest Cho signal), which in some cases was located on the rim of the lesion. This may partially account for differences with prior SV-MRS studies that placed the MRS voxel over the center of the lesion, which in high grade primary or metastatic tumors may be necrotic (and therefore have low levels of all metabolites (except lipids)) (40). Further studies in larger numbers of patients will be required to substantiate these preliminary results.

3.2. Quantitation Using Inhomogeneous Receiver Coil Arrays

In recent years, the use of phased-array receiver coils for brain MRI has become quite common (43). Phased-array coils are attractive in that they provide substantial SNR gains compared to conventional volume resonators (typically about 100% improvement in superficial brain locations closest to the coil elements), and can also be used to reduce scan time through parallel imaging methodologies such as "SENSE" (44). These same imaging advantages also apply to MRS and MRSI applications.

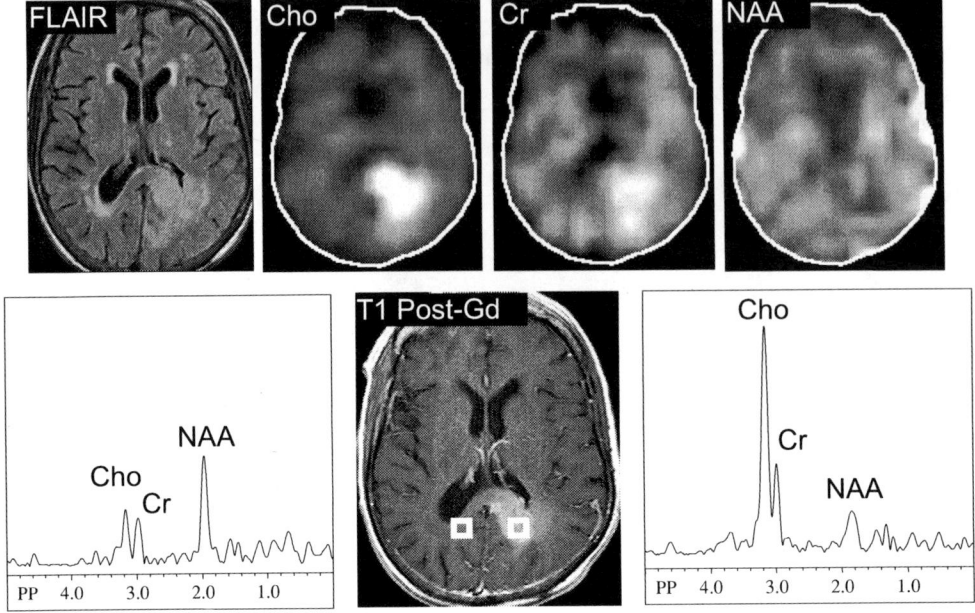

Figure 4. Proton MRSI of a high grade brain tumor (glioblastoma multiforme) involving the left parietal white matter and extending into the splenium of the corpus callosum. The lesion exhibits a dramatically elevated choline signal, as well as mildly increased creatine, and decreased signal from N-acetyl aspartate. The lesion shows enhancement post administration of gadolinium-DTPA contrast agent.

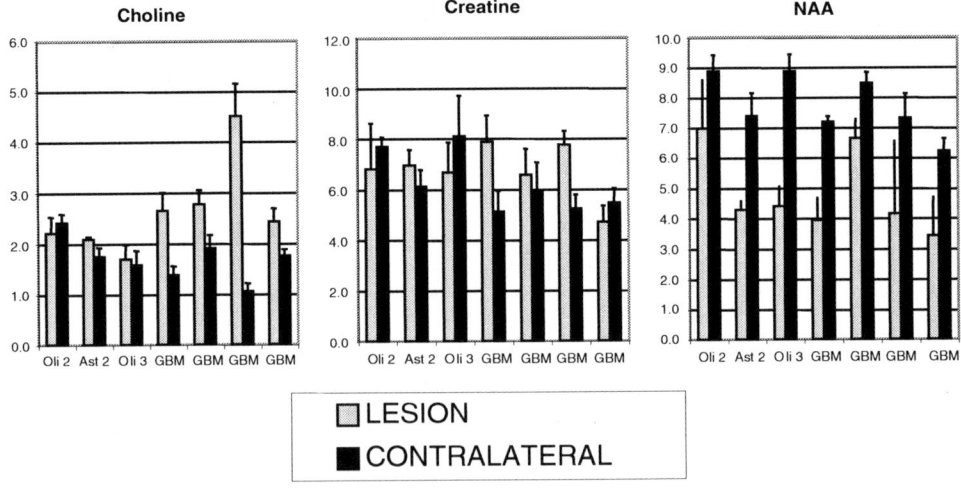

Figure 5. Concentrations (mM) of choline, creatine and NAA in 7 primary, untreated human brain tumors, and in the contralateral hemisphere of each patient. All tumors were subsequently biopsied and evaluated histologically. There are 3 low-grade tumors (Oli 2 - oligodendroglioma grade 2, Oli 3 - oligodendroglioma grade 3, Ast 2 - astrocytoma grade 2) and 4 high grade (GBM - glioblastoma multiforme, grade 4). All tumors have reduced NAA compared to the contralateral hemisphere, while only the high grade tumors have elevated Cho. Cr showed no consistent patterns - variations in Cr in the contralateral hemisphere reflect differences in tumor locations.

However, spectroscopy performed with phased-array coils does increase complexity of the study, especially in terms of spectral analysis and quantitation. Using phased-array coils, each receiver channel generates its own spectrum that has a different sensitivity (depending on coil geometry and voxel location) and phase. In order to produce a single MRS or MRSI dataset with optimum sensitivity, phase-correction and optimal combination of channels must be performed. This can readily be performed using information either in the spectra themselves or through the use of MRI scans to measure the coil sensitivity profiles (45,46).

Local, phased-array coils are intrinsically associated with inhomogenous B_1 (RF) fields, and may also exhibit variable sample loading, further complicating quantitation methods, such as the phantom replacement or external reference methods, based on equations (1) and (2). Fortunately, techniques for combining spectra from different channels with uniform sensitivity are available which can correct for variable loading and inhomogenous B_1 fields. One example is the SENSE-MRSI method which uses reduced phase-encoding to significantly decrease scan time (45). Multi-channel time-domain MRSI data ($\mathbf{b(t,x,y)}$) can be combined and un-folded using the relationship:

$$\mathbf{s}(t, x, y) = \mathbf{A}(x, y)^{-1} \mathbf{b}(t, x, y) \qquad [5]$$

where $\mathbf{s(t,x,y)}$ are combined, uniform spatial sensitivity MRSI data, and $\mathbf{A(x,y)}$ is the complex coil sensitivity matrix for each element of the phased-array. $\mathbf{A(x,y)}$ can be readily calculated from rapid gradient echo images collected alternatively using the body coil and the phased-array coil.

To calculate metabolite concentrations, a SENSE-MRSI scan must be performed and reconstructed under identical conditions on a calibration phantom of known concentration (in our case a 4 liter sample containing 65 mM NAA). *In vivo* metabolite concentrations [M] (where M = Choline (Cho), Creatine (Cr) or NAA) can then be calculated using the following expression:

$$[M] = [NAA] \times (S_M/S_{NAA}) \times (LF_i/LF_{NAA}) \times f(T_1, T_2) \qquad [6]$$

where S is the peak area and LF_i and LF_{NAA} are the body coil transmitter load factors for the *in vivo* and phantom scans, respectively. $f(T_1,T_2)$ applies a correction factor to account for differences in relaxation times between the phantom and *in vivo* metabolites. *In vivo* relaxation times for normal human brain at 3T can either be measured or taken from the literature.

An example dataset using this approach is described below; all scans were performed on a Philips Intera 3.0 Tesla system using a 6-channel phased-array receiver coil. RF pulses were transmitted using the body coil. A 3-slice, spin-echo circularly-encoded 2D-MRSI pulse sequence with water/lipid suppression (47) and OVS, covering from the basal ganglia to the vertex, was collected (TR/TE 2000/144 msec, FOV 230x115 mm, matrix size 32x16, SENSE factor 2, scan time 12 minutes). Prior to MRSI, field homogeneity was optimized using high order shimming. After MRSI, additional rapid gradient echo MRI scans were recorded to calculate the coil sensitivity matrix. The protocol was tested in 5 normal adult subjects (age 34 ± 11 years, 4 male). Bilateral metabolite peak areas were measured in 6 representative white and gray matter regions.

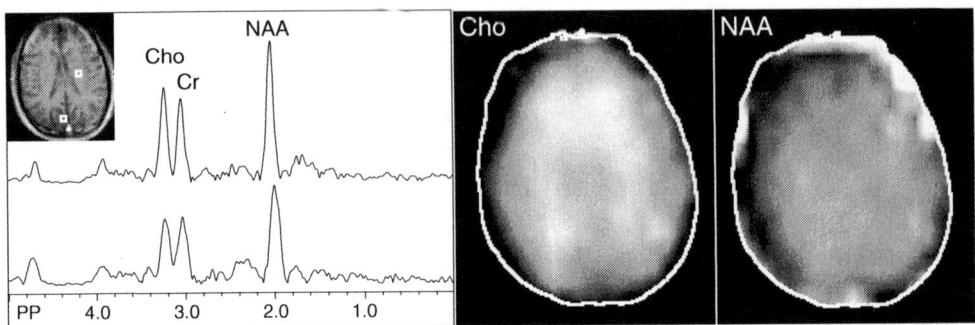

Figure 6. An example 3 Tesla SENSE-MRSI data set from a normal human brain, showing representative spectra from white and gray matter locations, as well as metabolic images of choline and N-acetyl aspartate.

Figure 6 shows a representative SENSE-MRSI dataset with 2 of the 6 voxel locations selected for quantitative analysis. Metabolite concentrations for all 6 brain regions analyzed (millimolar, mean ± st.dev.) are given in Table 2. Metabolite concentrations are in reasonable agreement with those from prior MRSI and single-voxel studies that have used conventional head coils (6,29). MRSI with phased-array coils offers improved SNR compared to conventional volume head coils, and allows for shorter scan times using SENSE encoding. However, because of the spatially varying sensitivity (and loading) of the individual receiver coil elements, quantitation is more complex. Fortunately, collection of B_1 sensitivity maps and use of SENSE processing allows the reconstruction of uniform sensitivity metabolic images. Since the B_1 sensitivity maps are derived relative to the body coil, a global loading correction based on the body coil transmitter gain required for a 90° pulse can be applied. The phantom replacement technique has previously been demonstrated to be a reliable method for quantifying multi-slice MRSI data recorded with homogeneous, quadrature transmit-receive coils (6). The procedure described here allows this method to be extended for use with any type of inhomogeneous coil array, and can be used with either conventional or SENSE-encoding provided that appropriate reconstruction techniques are applied.

4. DISCUSSION

In summary, the routine non-invasive determination of regional brain NAA concentrations is possible in the human brain using localized proton magnetic resonance spectroscopy. Since direct biochemical validation is not an option for human studies, verification of MRS measurements of NAA need to be performed in animal models where rapidly frozen biopsy samples can be collected and analyzed by conventional techniques. Generally, studies of this type have found reasonable agreement between MRS and biochemical determinations of NAA concentrations (48).

Table 2. Millimolar metabolite concentrations (mean ± standard deviation) using SENSE-MRSI at 3.0 Tesla in 5 normal adult human subjects.

Region	[Cho]	[Cr]	[NAA]
Thalamus	2.48 ± 0.65	8.44 ± 1.44	11.71 ± 4.83
Corona radiata	2.85 ± 0.73	9.72 ± 1.94	9.72 ± 1.94
Parietal periventricular white matter	2.83 ± 0.74	10.70 ± 1.68	13.69 ± 1.97
Mesial occipital gray matter	2.39 ± 0.28	10.53 ± 2.01	12.06 ± 3.15
Centrum semiovale	2.77 ± 0.41	11.03 ± 1.25	12.31 ± 1.46
Posterior cortical gray matter	2.54 ± 0.48	10.24 ± 1.19	8.05 ± 2.91

The choice of quantitation technique depends somewhat on the localization method used; for single voxel spectroscopy, quantitation of metabolites using the brain internal water signal is a popular method which generally yields reproducible results, particularly when used in combination with the LCModel spectral analysis method. However, it should always be kept in mind with this method that pathological variations in brain water content can occur, and partial volume with CSF should be estimated using segmentation methods. For MRSI acquisitions, water referencing can also be used, although for many protocols there may not be sufficient time available to collect an additional unsuppressed MRSI (in addition to the water suppressed scan). In this case, other quantitation techniques are required, for instance, the phantom replacement method, which does not require any additional patient scan time. Finally, quantitative measurements of brain NAA can be made with this method using either conventional transmit-receive head coils, as well as the newer generation of multi-channel, receive-only phased-array head coils employed in parallel imaging and spectroscopic techniques (49).

5. ACKNOWLEDGEMENTS

Supported in part by P41RR15241, R21CA/RR91798, R21-EB00991 and Philips Medical Systems.

6. REFERENCES

1. Tofts PS, Wray S. A critical assessment of methods of measuring absolute metabolite concentrations by nmr spectroscopy. *NMR Biomed* 1988;**1**(1):1-10.
2. Vanhamme L, Van Huffel S, Van Hecke P, van Ormondt D. Time-domain quantification of series of biomedical magnetic resonance spectroscopy signals. *J Magn Reson* 1999;**140**(1):120-130.
3. Soher BJ, Young K, Govindaraju V, Maudsley AA. Automated spectral analysis III: application to in vivo proton MR spectroscopy and spectroscopic imaging. *Magn Reson Med* 1998;**40**(6):822-831.
4. Provencher SW. Estimation of metabolite concentrations from localized in vivo proton NMR spectra. *Magn Reson Med* 1993;30(6):672-679.
5. Freeman R. *A Handbook of Nuclear Magnetic Resonance.* Longman Scientific and Technical; Harlow, England: 1987. 312 p.
6. Soher BJ, van Zijl PC, Duyn JH, Barker PB. Quantitative proton MR spectroscopic imaging of the human brain. *Magn Reson Med* 1996;**35**(3):356-363.

7. Hetherington HP, Mason GF, Pan JW, Ponder SL, Vaughan JT, Twieg DB, Pohost GM. Evaluation of cerebral gray and white matter metabolite differences by spectroscopic imaging at 4.1T. *Magn Reson Med* 1994;**32**(5):565-571.
8. Stockler S, Holzbach U, Hanefeld F, Marquardt I, Helms G, Requart M, Hanicke W, Frahm J. Creatine deficiency in the brain: a new, treatable inborn error of metabolism. *Pediatr Res* 1994;**36**(3):409-413.
9. Thulborn KR, Ackerman JJH. Absolute molar concentrations by NMR in inhomogeneous B_1. A scheme for analysis of *in vivo* metabolites. *J Magn Reson* 1983;**55**:357-371.
10. Barker PB, Soher BJ, Blackband SJ, Chatham JC, Mathews VP, Bryan RN. Quantitation of proton NMR spectra of the human brain using tissue water as an internal concentration reference. *NMR Biomed* 1993;**6**(1):89-94.
11. Christiansen P, Henriksen O, Stubgaard M, Gideon P, Larsson HB. In vivo quantification of brain metabolites by ^{1}H-MRS using water as an internal standard. *Magn Reson Imaging* 1993;**11**(1):107-118.
12. Alger JR, Symko SC, Bizzi A, Posse S, DesPres DJ, Armstrong MR. Absolute quantitation of short TE brain ^{1}H-MR spectra and spectroscopic imaging data. *J Comput Assist Tomogr* 1993;**17**(2):191-199.
13. Ordidge RJ, Cresshull ID. The correction of transient B0 field shifts following the application of pulsed gradients by phase correction in the time domain. *J Magn Reson* 1986;**69**:151-155.
14. Webb PG, Sailasuta N, Kohler SJ, Raidy T, Moats RA, Hurd RE. Automated single-voxel proton MRS: Technical development and multisite verification. *Magn Reson Med* 1994;**31**:365-373.
15. Soher BJ, Hurd RE, Sailasuta N, Barker PB. Quantitation of automated single-voxel proton MRS using cerebral water as an internal reference. *Magn Reson Med* 1996;**36**(3):335-339.
16. Mathews VP, Barker PB, Blackband SJ, Chatham JC, Bryan RN. Cerebral Metabolites in patients with acute and subacute strokes: concentrations determined by quantitative proton MR spectroscopy. *Am J Roentgenol* 1995;**165**:633-638.
17. Ernst T, Kreis R, Ross B. Absolute quantitation of water and metabolites in the human brain. i. compartments and water. *J Magn Reson B* 1993;**102**:1-8.
18. Horská A, Calhoun VD, Bradshaw DH, Barker PB. Rapid method for correction of csf partial volume in quantitative proton mr spectroscopic imaging. *Magn Reson Med* 2002;**48**(3):555-558.
19. Horska A, Jacobs MA, Calhoun V, Arslanoglu A, Barker PB. A fast method for image segmentation: application to quantitative proton MRSI at 3 Tesla. 2003; Toronto, Canada.
20. Mierisova S, Ala-Korpela M. MR spectroscopy quantitation: a review of frequency domain methods. *NMR Biomed* 2001;**14**(4):247-259.
21. Marshall I, Bruce SD, Higinbotham J, MacLullich A, Wardlaw JM, Ferguson KJ, Seckl J. Choice of spectroscopic lineshape model affects metabolite peak areas and area ratios. *Magn Reson Med* 2000;**44**(4):646-649.
22. Soher BJ, Maudsley AA. Evaluation of variable line-shape models and prior information in automated ^{1}H spectroscopic imaging analysis. *Magn Reson Med* 2004;**52**(6):1246-1254.
23. de Beer R, van den Boogaart A, van Ormondt D, Pijnappel WW, den Hollander JA, Marien AJ, Luyten PR. Application of time-domain fitting in the quantification of in vivo ^{1}H spectroscopic imaging data sets. *NMR Biomed* 1992;**5**(4):171-178.
24. Barker PB, Szopinski K, Horska A. Metabolic heterogeneity at the level of the anterior and posterior commissures. *Magn Reson Med* 2000;**43**(3):348-354.
25. Henriksen O. In vivo quantitation of metabolite concentrations in the brain by means of proton MRS. *NMR Biomed* 1995;**8**(4):139-148.
26. Hennig J, Pfister H, Ernst T, Ott D. Direct absolute quantification of metabolites in the human brain with in vivo localized proton spectroscopy. *NMR Biomed* 1992;**5**(4):193-199.
27. Michaelis T, Merboldt KD, Bruhn H, Hanicke W, Frahm J. Absolute concentrations of metabolites in the adult human brain in vivo: quantification of localized proton MR spectra. *Radiology* 1993;**187**(1):219-227.
28. Kreis R, Ernst T, Ross BD. Absolute quantitation of water and metabolites in the human brain. II. Metabolite concentrations. *J Magn Reson* 1993;B **102**:9-19.
29. Pouwels PJ, Frahm J. Regional metabolite concentrations in human brain as determined by quantitative localized proton MRS. *Magn Reson Med* 1998;**39**(1):53-60.
30. Jacobs MA, Horska A, van Zijl PC, Barker PB. Quantitative proton MR spectroscopic imaging of normal human cerebellum and brain stem. *Magn Reson Med* 2001;**46**(4):699-705.
31. Arslanoglu A, Bonekamp D, Barker PB, Horska A. Quantitative proton MR spectroscopic imaging of the mesial temporal lobe. *J Magn Reson Imaging* 2004;**20**(5):772-778.
32. Frahm J, Bruhn H, Gyngell ML, Merboldt KD, Hanicke W, Sauter R. Localized proton NMR spectroscopy in different regions of the human brain in vivo. Relaxation times and concentrations of cerebral metabolites. *Magn Reson Med* 1989;**11**(1):47-63.

33. Danielsen ER, Henriksen O. Absolute quantitative proton NMR spectroscopy based on the amplitude of the local water suppression pulse. Quantification of brain water and metabolites. *NMR Biomed* 1994;**7**(7):311-318.
34. Christiansen P, Toft P, Larsson HB, Stubgaard M, Henriksen O. The concentration of N-acetyl aspartate, creatine + phosphocreatine, and choline in different parts of the brain in adulthood and senium. *Magn Reson Imaging* 1993;**11**(6):799-806.
35. Pouwels PJ, Frahm J. Differential distribution of NAA and NAAG in human brain as determined by quantitative localized proton MRS. *NMR Biomed* 1997;**10**(2):73-78.
36. Kreis R, Ernst T, Ross BD. Development of the human brain: in vivo quantification of metabolite and water content with proton magnetic resonance spectroscopy. *Magn Reson Med* 1993;**30**(4):424-437.
37. Chang L, Ernst T, Poland RE, Jenden DJ. In Vivo Proton magnetic resonance spectroscopy of the normal aging human brain. *Life Sci* 1996;**58**:2049-.
38. Lundbom N, Barnett A, Bonavita S, Patronas N, Rajapakse J, Tedeschi, Di Chiro G. MR image segmentation and tissue metabolite contrast in ^{1}H spectroscopic imaging of normal and aging brain. *Magn Reson Med* 1999;**41**(4):841-845.
39. Helms G. Volume correction for edema in single-volume proton MR spectroscopy of contrast-enhancing multiple sclerosis lesions. *Magn Reson Med* 2001;**46**(2):256-263.
40. Howe FA, Barton SJ, Cudlip SA, Stubbs M, Saunders DE, Murphy M, Wilkins P, Opstad KS, Doyle VL, McLean MA, Bell BA, Griffiths JR. Metabolic profiles of human brain tumors using quantitative in vivo 1H magnetic resonance spectroscopy. *Magn Reson Med* 2003;**49**(2):223-232.
41. Gill SS, Thomas DG, Van BN, Gadian DG, Peden CJ, Bell JD, Cox IJ, Menon DK, Iles RA, Bryant DJ. Proton MR spectroscopy of intracranial tumours: in vivo and in vitro studies. *J Comput Assist Tomogr* 1990;**14**(4):497-504.
42. Preul MC, Caramanos Z, Collins DL, Villemure JG, Leblanc R, Olivier A, Pokrupa R, Arnold DL. Accurate, noninvasive diagnosis of human brain tumors by using proton magnetic resonance spectroscopy. *Nat Med* 1996;**2**(3):323-325.
43. Hayes CE, Tsuruda JS, Mathis CM. Temporal lobes: surface MR coil phased-array imaging. *Radiology* 1993;**189**(3):918-920.
44. Pruessmann KP, Weiger M, Scheidegger MB, Boesiger P. SENSE: sensitivity encoding for fast MRI. *Magn Reson Med* 1999;**42**(5):952-962.
45. Dydak U, Weiger M, Pruessmann KP, Meier D, Boesiger P. Sensitivity-encoded spectroscopic imaging. *Magn Reson Med* 2001;**46**(4):713-722.
46. Brown MA. Time-domain combination of MR spectroscopy data acquired using phased-array coils. *Magn Reson Med* 2004;**52**(5):1207-1213.
47. Smith M, Gillen J, McMahon MT, Barker PB, Golay X. Simultaneous Water And Lipid Suppression For In Vivo Brain Spectroscopy In Humans. Magn Reson Med 2005;(accepted for publication).
48. Barker PB, Breiter SN, Soher BJ, Chatham JC, Forder JR, Samphilipo MA, Magee CA, Anderson JH. Quantitative proton spectroscopy of canine brain: in vivo and in vitro correlations. *Magn Reson Med* 1994;**32**(2):157-163.
49. Natt O, Bezkorovaynyy V, Michaelis T, Frahm J. Use of phased array coils for a determination of absolute metabolite concentrations. *Magn Reson Med* 2005;**53**(1):3-8.
50. Windham JP, Abd-Allah MA, Reimann DA, Froelich JW, Haggar AM. Eigenimage filtering in MR imaging. *J Comput Assist Tomogr* 1988;**12**(1):1-9.

N-ACETYL-L-ASPARTATE IN MULTIPLE SCLEROSIS

Gerson A. Criste and Bruce D. Trapp[*]

1. INTRODUCTION

Multiple sclerosis (MS) is a chronic inflammatory demyelinating disease of the central nervous system affecting nearly 350,000 people in the United States and an estimated two million people worldwide. It is the most common cause of non-traumatic neurologic disability in young and middle aged adults[1]. The etiology is unknown, although the disease commonly is regarded as an autoimmune process triggered in susceptible individuals by an early environmental exposure.

Clinically MS is characterized by attacks of neurologic dysfunction, which may be few and far between with little or no impact on a person's ability to function. At times though, they may cause a rapid deterioration leading to complete disability. Most people with MS fall between these extremes. Most MS patients live for decades after their diagnosis. MS reduces life expectancy after onset by about 6-7 years[2], and about half of the patients survive 30 years or more from onset[3].

In the past few years, progress in MS research has accelerated. Novel approaches have elucidated many aspects of its pathophysiology. For example, genetics of susceptibility, identification of myelin antigens, inflammation, the roles of T-cells macrophages and astrocytes, mechanisms of demyelination and limited remyelination have all been under focus. Disease modifying drugs such as interferon beta and glatiramer acetate exert documented, albeit modest, effects during attacks in relapsing-remitting MS and are now widely used.

In addition, renewed interest in axonal damage in MS has opened new directions to approach the understanding and treatment of this disorder. Although MS is primarily an

[*] Department of Neurosciences, Lerner Research Institute, Cleveland Clinic Foundation, Cleveland, Ohio 44195; Email, trappb@ccf.org.

inflammatory demyelinating disease, it has become evident that axonal degeneration plays an important role in the pathogenesis of disability for MS patients[4-8]. More importantly axonal transection has been shown to begin at disease onset[9], but remains clinically silent probably due to the brain's remarkable plasticity in the face of injury. Considering the role of axonal injury in the pathogenesis of MS, non-invasive axonal monitoring could be used for prognostication, for the study of disease progression, as well as for evaluation of ongoing therapy. In vivo magnetic resonance spectroscopic determination of N-acetyl-L-aspartate (NAA), a specific neuronal/axonal marker, may provide such a measure[10, 11]. This chapter reviews current data on axonal pathology and the role of NAA in MS.

2. AXONAL INJURY IS THE MAJOR DETERMINANT OF PERMANENT NEUROLOGIC DISABILITY IN MULTIPLE SCLEROSIS

Even the earliest literature on MS mentioned axonal degeneration. Studies using contemporary technology such as magnetic resonance imaging and confocal microscopy more definitively demonstrated that axonal transection begins at disease onset and that cumulative axonal loss provides the pathologic substrate for the progressive disability that most long-term MS patients experience. Moreover, postmortem studies have shown that several histopathologic abnormalities including axonal loss, can be detected in the normal appearing white matter (NAWM)[4] and cortical gray matter[12] of patients with MS, suggesting a more diffuse pathology than previously thought.

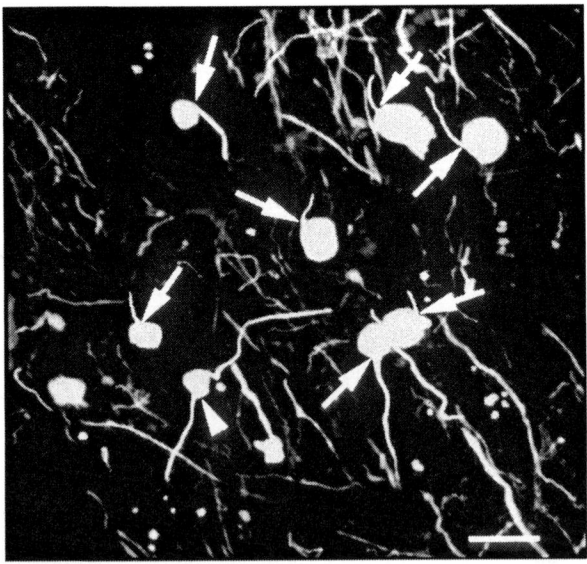

Figure 1. Axons end in large terminal ovoids (arrow) indicating axonal transection during demyelination (From Trapp et al,[4] with permission).

2.1 Axonal Injury Begins at the Early Stage of the Disease

Axonal amyloid precursor protein (APP) was demonstrated on acute MS lesions[13]. Accumulation of this protein is considered a marker for axonal dysfunction or injury since it is detected immunohistochemically only in axons with impaired axonal transport[14]. Many APP-immunoreactive structures resembled axonal ovoids, characteristic of newly transected axons. Hence, these results suggested axonal dysfunction within inflammatory MS lesions and indicated that many of these axons were transected. These observations were confirmed and extended by morphological investigation of lesions from MS brains with various degrees of inflammation and disease duration[4]. Axonal ovoids were identified through confocal microscopy as terminal ends of transected axons immunostained for non-phosphorylated neurofilaments (Figure 1). Over 11,000 transected axons were found per mm^3 in active lesions and over 3,000 per mm^3 at the edge of chronic active lesions. The core of chronic active lesions contained on average 875 transected axons per mm^3. In contrast, less than one axonal ovoid per mm^3 was detected in control white matter. Kornek and colleagues reported a similar correlation between activity of MS lesions and density of APP-positive axons[15]. The occurrence of axonal ovoids in active lesions at an early stage of the disease supports axonal transection from the onset of MS.

The mechanism of axonal damage remains to be elucidated, and we can only speculate on the possibilities. Some of the possible mechanisms are summarized here. First, since the extent of axonal damage in active MS lesions is proportional to the degree of inflammatory activity within the lesion, axonal injury could be a direct result of inflammation per se (Figure 2). Substances such as free radicals, proteolytic enzymes, oxidative products and cytokines produced by activated immune and glial cells are potential mediators of such damage[16]. Oxidative damage to mitochondrial DNA and impaired activity of mitochondrial enzyme complexes in MS lesions indicate that inflammation can affect energy metabolism, ATP synthesis, and viability of affected cells[17]. Recently, data indicating that cytotoxic $CD8^+$ T cells can mediate axonal transection in active MS lesions were provided in MS tissue[18], in EAE mice[19] and in vitro[20]. Another observation is that

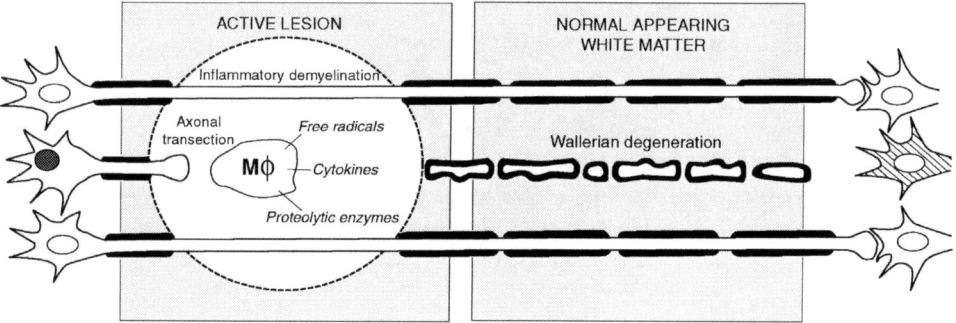

Figure 2. Axonal injury cause by inflammatory demyelination in an active MS lesion. Inflammatory substances secreted by activated immune and glial cells may mediate tissue damage including axonal transection. The distal axonal segment undergoes Wallerian degeneration but CNS myelin can persist for a long time. Thus, white matter distal to the active lesion may appear normal despite considerable axonal dropout (From Bjartmar et al.,[32] with permission).

treatment with the AMPA/kainate glutamate receptor antagonist NBQX resulted in increased oligodendrocyte survival and reduced axonal damage in experimental autoimmune encephalomyelitis (EAE), an animal model of MS. This suggests that excitotoxicity mediated by glutamate is involved in tissue damage in acute lesions[21]. Inflammatory edema is a possible culprit as well. This may cause increased extracellular pressure that results in axonal damage, particularly in anatomical locations of the CNS where space for tissue expansion is limited such as the spinal cord[22]. In support of this hypothesis, the spinal cord cross-sectional area of relapsing-remitting EAE mice was shown to increase by 9% at first attack, but returned to normal at end-stage disease[23]. Finally, genes involved in axonal responses to inflammation and demyelination could determine the extent of axonal injury in individual patients.

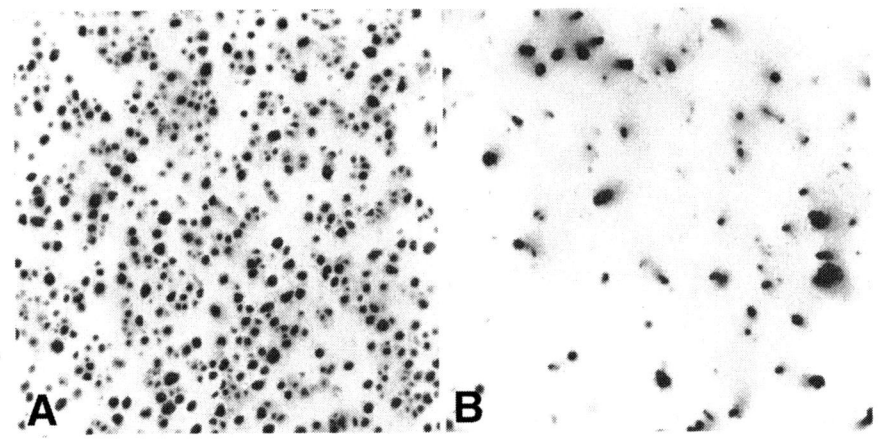

Figure 3. Axonal loss in the spinal cord of a paralyzed patient with MS for 22 years. Neurofilament staining demonstrates axonal density in control (A) and in a demyelinated area (B), in the gracile fasciculus of MS cervical spinal cord. This chronic inactive lesion exhibits obvious axonal loss. (From Bjartmar et al,[24] with permission).

2.2 Cumulative Axonal Loss is Seen in Chronic MS

Despite evidence of early axonal injury, most patients undergo a relapsing-remitting course. For years or decades after the first attack, they show no evidence of residual neurologic dysfunction in between relapses until a certain point when permanent disability ensues. From that point on, they all take a predictable downhill course. This interesting phenomenon is peculiar to MS and raises questions regarding the magnitude of cumulative axonal loss during chronic MS. To quantify total axonal loss in MS lesions, an axonal sampling protocol that accounts for both tissue atrophy and reduced axonal density was developed using spinal cord cross sections[24]. Total axonal loss was quantified in 10 chronic inactive lesions from 5 MS patients with significant functional impairment (EDSS $\geq$ 7.5) and long disease duration. In these lesions, axons were reduced by an average of 68% (45-84%) compared to controls, while average axonal density (number of axons per unit area) was decreased by 58%. This demonstrates that axonal loss constitutes a significant part of the pathology that many chronic MS lesions develop (Figure 3).

Given that the patients are permanently disabled, the data also support axonal degeneration as the main cause of irreversible neurological disability in non-ambulatory MS patients. A similar reduction in axonal density, 61%, was reported in spinal cord lesions from patients with SP-MS[25].

In SP-MS, extensive axonal loss and progression of disability occur in the absence of overt inflammatory activity. This suggests that mechanisms other than inflammatory demyelination contribute to axonal degeneration. Recently, it was proposed that abnormal expression of sodium channel subtypes in response to demyelination may render axons vulnerable to degeneration, raising the possibility that MS may involve an acquired channelopathy[26]. More importantly, a number of genes coding for myelin related proteins such as MAG, PLP, PMP22, P_0 and connexin 32, are being studied in relation to axonal pathology[27]. In one report, late onset axonal pathology such as atrophy or swelling, cytoskeleton alterations, organelle accumulation and degeneration was observed in mice lacking MAG[27] and PLP[28]. In PLP-null mice, the axonal pathology was accompanied with progressive clinical disability including impaired gait, tremor and spasticity. Hence, it is postulated that the lack of trophic support from myelin or myelin forming cells may cause degeneration of chronically demyelinated axons[7, 29].

2.3 Axonal Pathology is Present in Normal Appearing White Matter

It is well established that axons once severed will undergo relatively rapid Wallerian degeneration distal to the site of transection. Unlike axons, CNS myelin can persist for a long time after proximal fiber transection. Histologically, such remaining myelin sheaths may appear as empty tubes or as degenerating ovoids. Despite this microscopic pathology, however, the white matter may appear normal grossly and on conventional neuroimaging studies.

Immunohistochemical evidence suggesting Wallerian degeneration, such as discontinuous staining of axonal neurofilaments and presence of terminal axonal ovoids, has been demonstrated in normal appearing white matter from MS brains[4]. The extent of axonal loss in this region has been addressed quantitatively. Ganter et al.[30] working in areas without plaque, reported reductions in axonal density by 19-42% at the lateral corticospinal tract of MS patients with lower limb weakness. Lovas and colleagues compared axonal density in lesions and in normal appearing white matter (NAWM) from the cervical spinal cords of SP-MS patients. The average reduction in axonal density in lesions from lateral and posterior columns was 61%[25]. In normal appearing white matter, however, the average decrease in axonal density was as much as 57%. They also noted that axons with diameter smaller than approximately 3 μm were more affected than larger axons. In a study that accounted for both decreased axonal density and changes in tissue volume, total axonal loss in the corpus callosum of MS patients with disease durations between 5 and 34 years and various degree of functional impairment was determined[31]. An average total axonal loss of 53% in normal appearing corpus callosum was reported. Note however that in the same material, the reduction in axonal density was only 34%, emphasizing the need to consider both tissue volume and axonal density to properly assess the degree of axonal loss. These studies suggest that white matter may appear normal upon immunohistochemistry for myelin, or on MRI scans, but may still exhibit a considerable axonal dropout, especially in chronic patients with long disease duration.

Wallerian degeneration in normal appearing white matter has been observed by immunohistochemistry in an MS patient with short disease duration[32]. The patient succumbed to a fatal brain stem lesion just after a 9-month-history of relapsing-remitting MS (RR-MS) with few permanent neurologic signs. Demyelinated lesions were not found in the spinal cord postmortem. However, the ventral spinal cord column, containing tracts projecting from the brainstem lesion, exhibited a 20% axonal loss. Microscopy revealed myelin ovoids (Figure 4) and signs of myelin degradation by activated microglia, characteristic of Wallerian degeneration. Since much of the myelin remains, these are 'invisible lesions' as far as MRI and immunostaining for myelin are concerned.

3. NAA IS A SURROGATE MARKER OF NEURONAL HEALTH

NAA is an abundant free amino acid present in the vertebrate brain and is enriched only in neurons and its processes[33]. This neuronal specificity makes it an ideal marker for monitoring neuronal and axonal health. It has been shown to be consistently reduced in

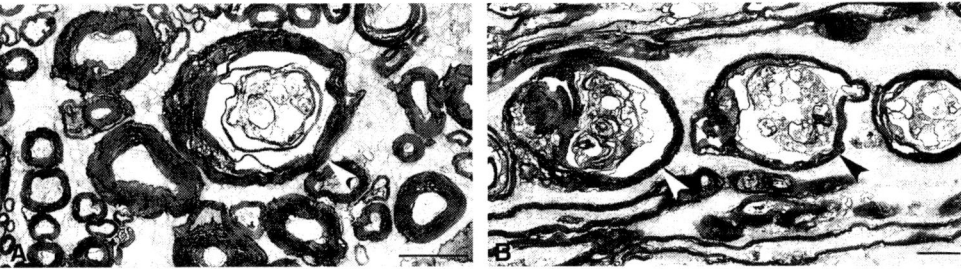

Figure 4. Wallerian degeneration in normal appearing white matter from a patient diagnosed with relapsing-remitting MS for 9 months. In both cross section (A), and longitudinal section (B), myelin ovoids lacking axons (arrows) were detected. In longitudinal section (B), these myelin ovoids often lay in rows. (From Bjartmar et al,[32] with permission).

neurodegenerative diseases, including MS. In vivo it can be reliably measured non-invasively by MRS. In tissues it can easily be detected and quantified by high performance liquid chromatography (HPLC).

3.1 NAA is Specific for Mature White Matter Axons

NAA as measured by MRS, is currently the best and most specific noninvasive marker of axonal pathology in multiple sclerosis. An issue has been raised, however, that NAA is expressed also by oligodendroglial lineage cells[34-36]. In order to investigate NAA specificity for white matter axons, transected and contralateral non-transected mature and developing rat optic nerves were analyzed by HPLC and immunohistochemistry[37]. In adult transected optic nerves, axons degenerated rapidly reflected by decreasing NAA levels (Figure 5A) while myelin profiles, oligodendrocytes and NG2+ oligodendrocyte progenitor cells (OPCs) remained abundant. Because NAA became undetectable in these axon-free nerves, the data suggest that neither differentiated oligodendrocytes nor adult OPCs contribute to detectable NAA levels in mature CNS white matter tracts in vivo[37].

Interestingly, NAA levels increased significantly over time in contralateral non-transected nerves in operated animals. The mechanism(s) behind this is unclear, but the observation raises several possibilities regarding NAA and neuronal compensation.

Urenjak et al.[34] reported that O-2A cells isolated from developing rodent CNS express NAA in vitro. As numerous progenitor cells proliferate and differentiate in transected developing rat optic nerves[38], NAA levels were examined following transection of P4 optic nerves. At P14 and P20, average NAA per axon-free nerve was reduced by 80% and 94%, respectively, compared to age- matched control nerves. These data indicate that a minor proportion (<20%) of total NAA in developing nerves is located outside

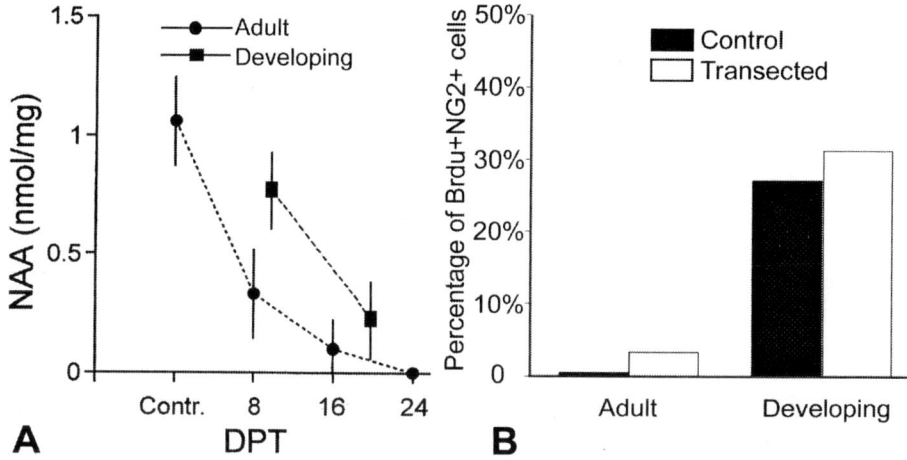

Figure 5. (A) Axonal degeneration correlates with NAA reduction in transected adult optic nerve and is undetected 24 days post transaction (DPT). In contrast 10-20% of NAA values in developing optic nerve appear to be non-axonal. (B) *The percentage of BrdU labeled (proliferating) NG2+oligodendroglial lineage cells in developing nerves are markedly higher than the adult nerves and may account for the non-axonal NAA.* (From Bjartmar et al,[37] with permission).

axons. In developing nerves as much as 25 to 33%of NG2+OPCs were proliferating compared to 0.6 to 3.3% in adult nerves. The number of proliferating OPCs is therefore markedly higher in transected developing optic nerves compared to that of adult rats (Figure 5B). Considering the relatively high NAA expression by O-2A OPCs in vitro[34], the NAA pool detected in vivo in these developing nerves without axons is likely to be associated with proliferating progenitor cells. Whether OPCs present in some MS lesions can contribute to NAA as measured by MRS remains to be determined[39-41]. In chronic MS lesions, however, most OPCs appear non-proliferating[39], and the density of these cells is generally similar or reduced compared to non-lesion white matter[40,41] suggesting that OPCs do not contribute significantly to total NAA in chronic lesions.

The silent nature of early axonal loss and the lack of reliable measures of disease progression make biologic markers that reflect early axonal damage and treatment efficiency in patients with MS essential. In this respect, the present data support NAA, as measured by MRS, as a specific in vivo marker for adult myelinated axons.

3.2 Magnetic Resonance Spectroscopy Detects NAA in Vivo

Clinical parameters such as relapses and progression underestimate the actual damage to tissue that occurs in MS. Hence, conventional magnetic resonance imaging (cMRI) has been an important tool for following disease activity and evolution in MS. However despite the correlation between axon loss and hypointensity in T1-weighted images of MS brain[42,43] cMRI generally lacks pathological specificity as many factors can influence image contrast[10,11]. Though sensitive in detecting lesions, it cannot distinguish inflammation, gliosis, demyelination, and more importantly axon loss—the pathologic substrate of irreversible disability[44]. To adequately understand disease progression and monitor efficacy of newer therapeutic strategies in clinical trials, it would be useful to be able to assess the accumulated load of irreversible damage. MRS determination of NAA may provide such an index.

MRS is one of the modern quantitative MR techniques that have the potential to overcome some of the limitations of cMRI. Other modern MR techniques like magnetization transfer and diffusion weighted MRI enable one to quantify the extent of structural changes occurring within and outside the MS lesion with increase specificity[45]. MRS can add information on the biochemical nature of such changes, with the potential to improve significantly our ability to monitor inflammatory demyelination and axonal injury. It facilitates noninvasive assessment of neurons, axons, and membrane integrity through the quantification of N-acetylaspartate (NAA), total choline (Cho), and creatine (Cr) levels. Decreased NAA, coupled with Cho and Cr variations, have been reported in MS lesions, normal-appearing white matter (NAWM), and cortical gray matter[10,44].

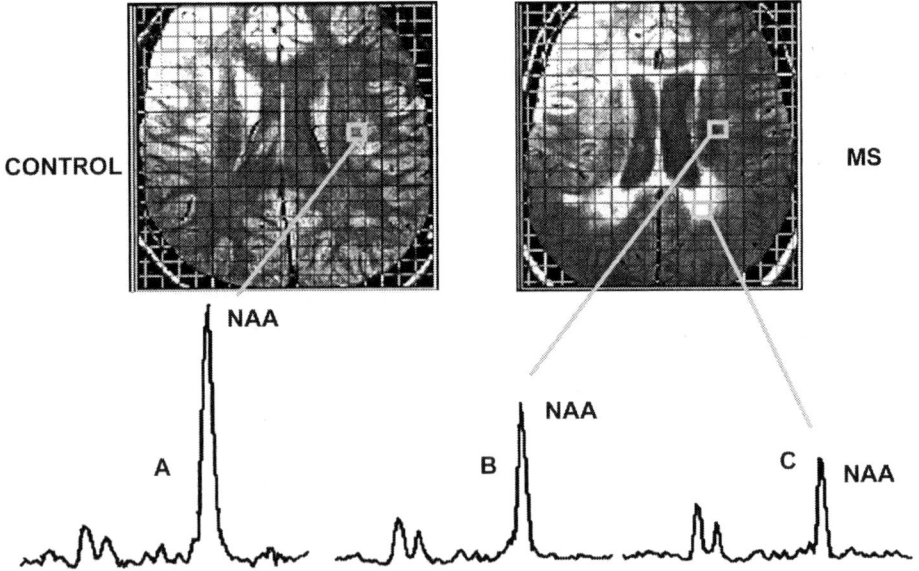

Figure 6. Proton MR spectra from a normal brain (A), NAWM of a patient with MS (B), and chronic periventricular plaque of the same patient with MS. (From Arnold, 1999,[8] with permission).

That MRS detection of NAA concentration is an accurate measure of axonal density has been confirmed by histology of biopsied samples[46]. To determine NAA levels in MS spinal cords, HPLC analysis of whole cord cross sections was performed postmortem. At cervical and lumbar levels, average NAA levels were significantly decreased by 53% and 55% respectively.   Since these patients were severely disabled (Expanded Disability Status Scale ≥ 7.5) the data indicates that reduced NAA levels in chronic MS as detected by MRS, can reflect irreversible functional impairment[24].

Falini et al.[47] tested the utility of MRS in defining the extent of metabolic changes in benign versus secondary progressive MS and found significant differences in NAA pattern according to the phase (acute vs. chronic) and the clinical form (benign vs. progressive) of the disease. MRS also detected reduced NAA levels in NAWM, substan- tiating the histopathologic findings of axonal loss or Wallerian degeneration outside MS lesions[32]. The extent of NAWM changes varies between different patient groups[48], but the changes are invariably present in all major MS phenotypes[49] and correlate with the level of physical disability[48] and cognitive impairment[50].

Gray matter involvement at cortical and subcortical levels in MS is gaining increased recognition[12, 51]. MRS was able to document substantial NAA reduction in this area as well[52, 53] corroborating reports of diffuse axonal pathology starting at the early stage of the disease[54].

The concept that MS has a diffuse pathology poses questions for the current localized MRS detection of NAA[55]. It assumes that relevant metabolic changes occur only at sites of imaging abnormality. This problem of partial coverage was addressed by recent studies utilizing Whole Brain NAA (WBNAA). One study showed that the inter-individual WBNAA variation in a cohort of normal, middle-aged women (range 42+/-5 years of age) is small, 3% (p<0.01), and the intra-individual temporal variation is smaller still[56]. Comparison of the WBNAA of 12 rapidly remitting MS patients and age matched controls showed that the MS group had markedly reduced WBNAA. This difference was noted to be greater among older than younger patients. The linear prediction of equations of WBNAA with age indicate a faster decline in the patients, ~0.8% per year of age (p=0.22)[55]. In a study of WBNAA dynamics in RR-MS patients, measurement defined 3 subgroups: stable, exhibiting an insignificant change in NAA level per year of clinically definite disease duration (p=0.54); moderate decline, -2.8% per year (<0.01); and rapid decline, -27.9 % per year (p<0.01). Thus, WBNAA offers a quick, highly reproducible measure of disease progression and may help stratify patients for active therapeutic intervention in MS[57]. This is particularly compelling now, in the light of the observation that axonal injury starts early in the course of the disease and that partially effective treatment for MS is available for certain group of patients.

3.3 Dynamics of NAA in Multiple Sclerosis

It was initially thought that NAA reduction was always secondary to axonal loss. However, as demonstrated by in vivo MRS, reduced NAA in acute lesions is at least partly reversible, indicating that early axonal damage due to inflammatory demyelination can be reversible[10]. This view is compatible with clinical recovery during remissions in MS, which is attributed to axonal recovery due to mechanisms such as resolution of inflammation, remyelination or redistribution of axolemmal ion-channels. Given the limited levels of neuronal regeneration within the CNS, reversible levels of NAA in MS white matter also suggest that mechanisms other than axonal loss contribute to overall NAA reductions in MS.

To address whether axonal loss is the only contributor to decreased levels of NAA in chronic MS spinal cords, NAA per axonal volume was calculated (Figure 7. A-G)[24]. Compared to myelinated axons in control white matter, average NAA per axonal volume was reduced in myelinated axons in MS non-lesion white matter by 30% ($p = 0.05$), and in demyelinated axons in MS lesions by 42% ($p = 0.01$). These data raise several interesting interpretations of MRS data. One possibility is that myelin related molecules and/or oligodendrocytes influence axonal levels of NAA. When NAA was expressed per axonal volume, demyelinated axons contained 42% less NAA than myelinated axons in control spinal cords. It is well established that myelination and demyelination dynamically influence axonal neurofilament phosphorylation by regulating kinase and phosphatase activity locally[58]. NAA metabolism appears to be related to neuronal activity in a tract as well. Since these MS patients were paralyzed, the 30% reduction of NAA in normal appearing

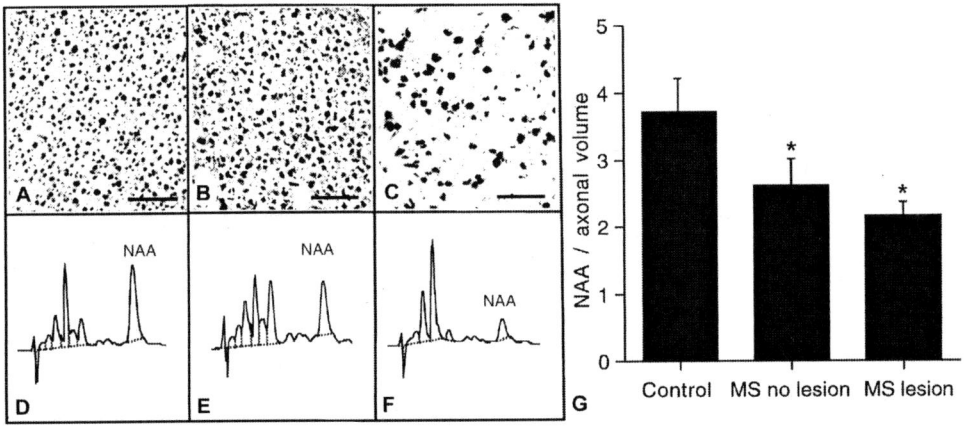

Figure 7. Correlation between axonal loss and levels of N-acetyl-aspartate (NAA) in MS spinal cord white matter. Axonal density was determined in neurofilament-stained cryostat sections, and NAA levels were determined by HPLC in adjacent sections. Top panel: representative micrographs showing axons in control white matter (*A*), MS white matter without lesion (*B*), and in an MS lesion (*C*). Scale bars = 40-μm. Lower panels: HPLC chromatograms (*D, E, and F*) from tissue sections adjacent to those immunostained in (*A*), (*B*) and (*C*) respectively. Levels of NAA per axonal volume in myelinated and demyelinated axons in control and MS samples (*G*). Average NAA per axonal volume was significantly reduced in MS non-lesion white matter samples (30%, $p = 0.05$) and in samples from chronic MS lesions (42%, $p = 0.01$). (From Bjartmar et al.,[24] with permission).

myelinated spinal cord axons provides direct support for reduced NAA in the absence of demyelination in functionally impaired axons. It is also possible that denervation of a neuron due to damage of afferent axons, for example in one or several remote lesions, causes altered NAA levels. Indeed, acute deafferentiation in the CNS causes transsynaptic decreases of NAA levels without ultrastructural abnormalities, indicating that denervation and/or impaired function reduces neuronal NAA[59]. It is also possible that myelination and demyelination may modulate activity of enzymes involved in NAA metabolism, thus influencing axonal NAA levels locally. Since NAA is synthesized in mito-chondria[60, 61], changes in NAA could reflect mitochondrial dysfunction[62]. In patients with acute mitochondrial encephalopathy with lactic acidosis and stroke-like episodes (MELAS), a primary mitochondrial disorder, NAA levels in the lesions recover partly after the acute illness[63]. These lesions, and acute demyelinating lesions, exhibit an inverse relationship between reduced NAA and lactate production, suggesting that reduced NAA production due to mitochondrial dysfunction might be associated with increased anaero-bic glycolysis[62, 63]. Local inflammation could indirectly influence mitochondrial function. In acute EAE, an animal model of MS characterized by inflammation and demyelina-tion, mitochondrial dysfunction and reduced NAA has been reported in the absence of neuronal loss[62, 64, 65]. Finally, in theory, edema (dilution effects) or inflammation-induced metabolic effects may also cause reduction in NAA levels. For example, decrease in NAA levels of up to 50% have been observed in NAWM[5] and up to 80% in lesions[66, 67]. However the observed reductions in NAA levels in these regions are greater in proportion than the volume changes, indicating that the reversible NAA change is unlikely to result merely from edematous dilution effects[68].

In summary, reduced NAA levels in MS could reflect several mechanisms such as reversible neuronal/axonal damage due to inflammatory demyelination, altered neu-ronal/axonal metabolism related to activity, axonal atrophy, or axonal loss[10, 63].

3.4. Correlation of NAA with other Measures of Neurodegeneration

Brain atrophy reflects the net result of irreversible and destructive pathological proc-esses in MS. Axonal damage and loss, chronic demyelination, and gliosis all contribute to a reduction in brain parenchymal tissue volume and a corresponding expansion of cere-brospinal fluid (CSF) spaces. Hence, measures of brain atrophy (e.g. brain parenchymal volume, index of brain atrophy, whole brain ratio) are a reliable, and meaningful global surrogate for the destructive pathologic process in MS patients [69, 70]. In studies correlating NAA and brain atrophy, consistent direct correlations have been observed. The observed correlations between atrophy and decreased NAA/creatine levels suggest that axonal loss is a substantial contributor to atrophy and that MRS can assess that component, even in the very early stages of the disease[71].

Efforts to correlate NAA levels to disability status like Expanded disability Status Scale (EDSS) however, have been contradictory. While most studies found a correlation[5, 24] other studies did not[57, 72]. While disappointing, it is not surprising that despite the speci-ficity of NAA as an index of neuronal health, correlation between NAA levels and EDSS has been inconsistent. There are several reasons for this discrepancy. First, lesion location was not considered in these correlations. Understandably, disability or lack thereof result-ing from any CNS lesion, do vary with lesion location. Second, it reflects the inherent weakness of EDSS as a clinical instrument[72, 73]. While it is a useful clinical measure of neurologic impairment, EDSS score consistently fails to reflect the full burden of disease because it is weighted toward cerebellar and spinal cord deficits[73, 74]. Hence, spinal cord NAA level is well correlated with EDSS score[24] while WBNAA is inconsistent. Third, this disparity reflects the brains ability to compensate for accumulating injury and to con-ceal its extent. Cortical reorganization is known to occur in MS[75, 76] and underlies the difficulty in using clinical criteria such as EDSS to predict disease course. In contrast, NAA dynamics yield a direct measure of the brain's pathologic structure load without the distorting overlay of function, thereby more objectively predicting the course of organic pathology which may be more appropriate for monitoring disease progression in clinical trials.

4. CONCLUSION

It is well established that axonal pathology is the basis of permanent disability in MS from its earliest stage. The promise of NAA measurement is that, it provides a specific and readily quantifiable index of neuronal and axonal dysfunction or loss. Therefore, monitoring NAA dynamics provides an index of change associated with irreversible stages in the evolution of diffuse MS lesions central to determining disability. Use in the proper context and together with other modern MR techniques, understanding NAA concentration dynamics may enable early forecast of disease course, reflect disease load and so influence treatment decision; improve clinical trial efficiency, and enhance further understanding of this complex disease.

5. ACKNOWLEDGMENTS

This work was supported by NIH grants NS35058, NS38667 (B.D.T.). The authors thank Dr. Grahame Kidd for the figures and Susan De Stefano for editorial assistance.

6. QUESTION AND ANSWER SESSION

DR. LIEBERMAN: I would like to suggest that you be careful about the specificity of any -- One of the things that is clear from a number of studies is that, as soon as you destroy the axon and glial cells, that metabolically they may be quite different, and where they may have expressed NAA or any other number of substances, they will not do that following transection.

One of the things that would be very interesting for you to test to see that is look at just oxygen consumption of the optic nerve just after transection, and then look at it after a period of time to see if there is a significant change; because the change would be in the metabolic capacity of the glial cells.

DR. TRAPP: I see. So the question was that it is possible that the transection itself is going to change the expression of NAA. Certainly, I think that must happen.

So we know NAA is transferred to the oligodendrocytes to be degraded. Certainly, if you transect the axon, that probably can't happen. But, you know, whether these cells are making NAA -- the oligodendrocytes -- certainly, I don't think it is at significant levels.

I didn't show it here, but we do immunostaining for NAA and really cannot detect any NAA or very little NAA within the oligodendrocytes. But certainly, I don't think it is something that one could prove conclusively, because we know NAA gets transferred to the myelin-oligodendrocyte unit. So there's got to be some there.

I think, for the purposes of this, though, in looking at spectroscopy, I still think that the glia cells are probably contributing in the adult brain very little NAA to the total amount that is there. So I don't think they are going to significantly change the values that the spectroscopists are obtaining within the tissue.

DR. ROSS: Thank you. It is very challenging data, and it really asks the question: Can we measure NAA in the spinal cord? Peter Barker with his phased ray coils or you in Cleveland actually do that.

DR. TRAPP: Yes. I am not the one to answer that question. So Peter or Doug probably -- we certainly would like to.

DR. BARKER: There are two or three papers now on spectroscopy in the spinal cord. It is definitely possible, at least in the cervical spine.

7. REFERENCES

1.   R. B. Johnston, *Multiple Sclerosis: Current Status and Strategies for the Future*, (National Academy Press, Washington, D.C., 2001).
2.   A. D. Sadovnick, G. C. Ebers, R. W. Wilson, and D. W. Paty, Life expectancy in patients attending multiple sclerosis clinics. *Neurology* **42**, 991-994 (1992).
3.   N. Koch-Henriksen, H. Bronnum-Hansen, and E. Stenager, Underlying cause of death in Danish patients with multiple sclerosis: results from the Danish Multiple Sclerosis Registry. *J. Neurol. Neurosurg. Psychiatry* **65**, 56-59 (1998).
4.   B. D. Trapp, J. Peterson, R. M. Ransohoff, R. Rudick, S. Mork, and L. Bo, Axonal transection in the lesions of multiple sclerosis. *N. Engl. J. Med.* **338**, 278-285 (1998).
5.   N. De Stefano, P. M. Matthews, L. Fu, S. Narayanan, J. Stanley, G. S. Francis, J. P. Antel, and D. L. Arnold, Axonal damage correlates with disability in patients with relapsing- remitting multiple sclerosis. Results of a longitudinal magnetic resonance spectroscopy study. *Brain* **121**, 1469-1477 (1998).
6.   C. Bjartmar, X. Yin, and B. D. Trapp, Axonal pathology in myelin disorders. *J Neurocytol.* **28**, 383-395 (1999).
7.   B. D. Trapp, R. M. Ransohoff, E. Fisher, and R. A. Rudick, Neurodegeneration in multiple sclerosis: Relationship to neurological disability. *The Neuroscientist* **5**, 48-57 (1999).
8.   D. L. Arnold, Magnetic resonance spectroscopy: imaging axonal damage in MS. *J Neuroimmunol* 98[1], 2-6. 1999.
9.   C. Bjartmar and B. D. Trapp, Axonal degeneration and progressive neurologic disability in multiple sclerosis. *Neurotox. Res.* **5**, 157-164 (2003).
10. P. M. Matthews, N. De Stefano, S. Narayanan, G. S. Francis, J. S. Wolinsky, J. P. Antel, and D. L. Arnold, Putting magnetic resonance spectroscopy studies in context: axonal damage and disability in multiple sclerosis. *Semin. Neurol.* **18**, 327-336 (1998).
11. W. Bruck, A. Bitsch, H. Kolenda, Y. Bruck, M. Stiefel, and H. Lassmann, Inflammatory central nervous system demyelination: correlation of magnetic resonance imaging findings with lesion pathology. *Ann. Neurol.* **42**, 783-793 (1997).
12. J. W. Peterson, L. Bo, S. Mork, A. Chang, and B. D. Trapp, Transected neurites, apoptotic neurons and reduced inflammation in cortical MS lesions. *Ann. Neurol.* **50**, 389-400 (2001).
13. B. Ferguson, M. K. Matyszak, M. M. Esiri, and V. H. Perry, Axonal damage in acute multiple sclerosis lesions. *Brain* **120**, 393-399 (1997).
14. E. H. Koo, S. S. Sisodia, D. R. Archer, L. J. Martin, A. Weidemann, K. Beyreuther, P. Fischer, C. L. Masters, and D. L. Price, Precursor of amyloid protein in Alzheimer disease undergoes fast anterograde axonal transport. *Proc. Natl. Acad. Sci. USA* **87**, 1561-1565 (1990).
15. B. Kornek, M. K. Storch, R. Weissert, E. Wallstroem, A. Stefferl, T. Olsson, C. Linington, M. Schmidbauer, and H. Lassmann, Multiple sclerosis and chronic autoimmune encephalomyelitis: a comparative quantitative study of axonal injury in active, inactive, and remyelinated lesions. *Am J Pathol.* **157**, 267-276 (2000).
16. R. Hohlfeld, Biotechnological agents for the immunotherapy of multiple sclerosis. Principles, problems and perspectives. *Brain* **120**, 865-916 (1997).
17. K. Strigard, P. Larsson, R. Holmdahl, L. Klareskog, and T. Olsson, In vivo monoclonal antibody treatment with Ox19 (anti-rat CD5) causes disease relapse and terminates P_2-induced immunospecific tolerance in experimental allergic neuritis. *J. Neuroimmunol.* **23**, 11-18 (1989).
18. H. Babbe, A. Roers, A. Waisman, H. Lassmann, N. Goebels, R. Hohlfeld, M. Friese, R. Schroder, M. Deckert, S. Schmidt, R. Ravid, and K. Rajewsky, Clonal expansions of CD8(+) T cells dominate the T cell infiltrate in active multiple sclerosis lesions as shown by micromanipulation and single cell polymerase chain reaction. *J. Exp. Med.* **192**, 393-404 (2000).

19. E. S. Huseby, D. Liggitt, T. Brabb, B. Schnabel, C. Ohlen, and J. Goverman, A pathogenic role for mye-lin-specific CD8(+) T cells in a model for multiple sclerosis. *J. Exp. Med.* **194**, 669-676 (2001).

20. I. Medana, M. A. Martinic, H. Wekerle, and H. Neumann, Transection of major histocompatibility com-plex class I-induced neurites by cytotoxic T lymphocytes. *Am. J. Pathol.* **159**, 809-815 (2001).

21. D. Pitt, P. Werner, and C. S. Raine, Glutamate excitotoxicity in a model of multiple sclerosis. *Nat. Med.* **6**, 67-70 (2000).

22. R. Shi and A. R. Blight, Compression injury of mammalian spinal cord in vitro and the dynamics of action potential conduction failure. *J Neurophysiol.* **76**, 1572-1580 (1996).

23. J. R. Wujek, C. Bjartmar, E. Richer, R. M. Ransohoff, M. Yu, V. K. Tuohy, and B. D. Trapp, Axon loss in the spinal cord determines permanent neurological disability in an animal model of multiple sclerosis. *J. Neuropathol. Exp. Neurol.* **61**, 23-32 (2002).

24. C. Bjartmar, G. Kidd, S. Mork, R. Rudick, and B. D. Trapp, Neurological disability correlates with spinal cord axonal loss and reduce *N*-acetyl aspartate in chronic multiple sclerosis patients. *Ann Neurol* **48**, 893-901 (2000).

25. G. Lovas, N. Szilagyi, K. Majtenyi, M. Palkovits, and S. Komoly, Axonal changes in chronic demyeli-nated cervical spinal cord plaques. *Brain* **123**, 308-317 (2000).

26. S. G. Waxman, Acquired channelopathies in nerve injury and MS. *Neurology* **56**, 1621-1627 (2001).

27. X. Yin, T. O. Crawford, J. W. Griffin, P.-H. Tu, V. M. Y. Lee, C. Li, J. Roder, and B. D. Trapp, Myelin-associated glycoprotein is a myelin signal that modulates the caliber of myelinated axons. *J. Neurosci.* **18**, 1953-1962 (1998).

28. I. Griffiths, M. Klugmann, T. Anderson, D. Yool, C. Thomson, M. H. Schwab, A. Schneider, F. Zimmermann, M. McCulloch, N. Nadon, and K.-A. Nave, Axonal swellings and degeneration in mice lacking the major proteolipid of myelin. *Science* **280**, 1610-1613 (1998).

29. S. Scherer, Axonal pathology in demyelinating diseases. *Ann. Neurol.* **45**, 6-7 (1999).

30. P. Ganter, C. Prince, and M. M. Esiri, Spinal cord axonal loss in multiple sclerosis: a post-mortem study. *Neuropathol. Appl. Neurobiol.* **25**, 459-467 (1999).

31. N. Evangelou, M. M. Esiri, S. Smith, J. Palace, and P. M. Matthews, Quantitative pathological evidence for axonal loss in normal appearing white matter in multiple sclerosis. *Ann. Neurol.* **47**, 391-395 (2000).

32. C. Bjartmar, R. P. Kinkel, G. Kidd, R. A. Rudick, and B. D. Trapp, Axonal loss in normal-appearing white matter in a patient with acute MS. *Neurology* **57**, 1248-1252 (2001).

33. M. L. Simmons, C. G. Frondoza, and J. T. Coyle, Immunocytochemical localization of *N*-acetylaspartate with monoclonal antibodies. *Neuroscience* **45**, 37-45 (1991).

34. J. Urenjak, S. R. Williams, D. G. Gadian, and M. Noble, Specific expression of N-acetylaspartate in neu-rons, oligodendrocyte-type-2 astrocyte progenitors, and immature oligodendrocytes *in vitro. J. Neuro-chem.* **59**, 55-61 (1992).

35. J. Urenjak, S. R. Williams, D. G. Gadian, and M. Noble, Proton nuclear magnetic resonance spectroscopy unambiguously identifies different neural cell types. *J. Neurosci.* **13**, 981-989 (1993).

36. K. K. Bhakoo and D. Pearce, In vitro expression of N-acetylaspartate by oligodendrocytes: implications for proton magnetic resonance spectroscopy signal in vivo. *J Neurochem* **74**, 254-262 (2000).

37. C. Bjartmar, J. Battistuta, N. Terada, E. Dupree, and B. D. Trapp, N-acetylaspartate is an axon-specific marker of mature white matter in vivo: a biochemical and immunohistochemical study on the rat optic nerve. *Ann. Neurol.* **51**, 51-58 (2002).

38. H. Ueda, J. M. Levine, R. H. Miller, and B. D. Trapp, Rat optic nerve oligodendrocytes develop in the absence of viable retinal ganglion cell axons. *J. Cell Biol.* **146**, 1365-1374 (1999).

39. G. Wolswijk, Chronic stage multiple sclerosis lesions contain a relatively quiescent population of oli-godendrocyte precursor cells. *J. Neurosci.* **18**, 601-609 (1998).

40. N. Scolding, R. Franklin, S. Stevens, C.-H. Heldin, A. Compston, and J. Newcombe, Oligodendrocyte progenitors are present in the normal adult human CNS and in the lesions of multiple sclerosis. *Brain* **121**, 2221-2228 (1998).

41. A. Chang, A. Nishiyama, J. Peterson, J. Prineas, and B. D. Trapp, NG2-positive oligodendrocyte progeni-tor cells in adult human brain and multiple sclerosis lesions. *J. Neurosci.* **20**, 6404-6412 (2000).

42. M. A. van Walderveen, W. Kamphorst, P. Scheltens, J. H. van Waesberghe, R. Ravid, J. Valk, C. H. Pol-man, and F. Barkhof, Histopathologic correlate of hypointense lesions on T1-weighted spin- echo MRI in multiple sclerosis. *Neurology* **50**, 1282-1288 (1998).

43. J. H. van Waesberghe, W. Kamphorst, C. J. De Groot, M. A. van Walderveen, J. A. Castelijns, R. Ravid, G. J. Nijeholt, d. van, V, C. H. Polman, A. J. Thompson, and F. Barkhof, Axonal loss in multiple scle-rosis lesions: magnetic resonance imaging insights into substrates of disability. *Ann. Neurol.* **46**, 747-754 (1999).

44. D. L. Arnold, P. M. Matthews, G. Francis, and J. Antel, Proton magnetic resonance spectroscopy of human brain in vivo in the evaluation of multiple sclerosis: assessment of the load of disease. *Magn Reson. Med* **14**, 154-159 (1990).
45. M. Filippi, C. Tortorella, and M. Rovaris, Magnetic resonance imaging of multiple sclerosis. *J Neuroimaging* **12**, 289-301 (2002).
46. A. Bitsch, H. Bruhn, V. Vougioukas, A. Stringaris, H. Lassmann, J. Frahm, and W. Bruck, Inflammatory CNS demyelination: histopathologic correlation with in vivo quantitative proton MR spectroscopy. *AJNR Am. J Neuroradiol.* **20**, 1619-1627 (1999).
47. A. Falini, G. Calabrese, M. Filippi, D. Origgi, S. Lipari, B. Colombo, G. Comi, and G. Scotti, Benign versus secondary-progressive multiple sclerosis: the potential role of proton MR spectroscopy in defining the nature of disability. *AJNR Am. J Neuroradiol.* **19**, 223-229 (1998).
48. M. Filippi, G. Iannucci, C. Tortorella, L. Minicucci, M. A. Horsfield, B. Colombo, M. P. Sormani, and G. Comi, Comparison of MS clinical phenotypes using conventional and magnetization transfer MRI. *neurology* **52**, 588-594 (1999).
49. D. J. Werring, C. A. Clark, G. J. Barker, A. J. Thompson, and D. H. Miller, Diffusion tensor imaging of lesions and normal-appearing white matter in multiple sclerosis. *neurology* **52**, 1626-1632 (1999).
50. M. Filippi, C. Tortorella, M. Rovaris, M. Bozzali, F. Possa, M. P. Sormani, G. Iannucci, and G. Comi, Changes in the normal appearing brain tissue and cognitive impairment in multiple sclerosis. *j neurol neurosurg psychiatry* **68**, 157-161 (2000).
51. A. Cifelli, M. Arridge, P. Jezzard, M. M. Esiri, J. Palace, and P. M. Matthews, Thalamic neurodegeneration in multiple sclerosis. *Ann. Neurol.* **52**, 650-653 (2002).
52. P. Kapeller, M. A. McLean, C. M. Griffin, D. Chard, G. J. Parker, G. J. Barker, A. J. Thompson, and D. H. Miller, Preliminary evidence for neuronal damage in cortical grey matter and normal appearing white matter in short duration relapsing-remitting multiple sclerosis: a quantitative MR spectroscopic imaging study. *J Neurol.* **248**, 131-138 (2001).
53. D. T. Chard, C. M. Griffin, M. A. McLean, P. Kapeller, R. Kapoor, A. J. Thompson, and D. H. Miller, Brain metabolite changes in cortical grey and normal-appearing white matter in clinically early relapsing-remitting multiple sclerosis. *Brain* **125**, 2342-2352 (2002).
54. N. De Stefano, S. Narayanan, S. J. Francis, S. Smith, M. Mortilla, M. C. Tartaglia, M. L. Bartolozzi, L. Guidi, A. Federico, and D. L. Arnold, Diffuse axonal and tissue injury in patients with multiple sclerosis with low cerebral lesion load and no disability. *Arch. Neurol.* **59**, 1565-1571 (2002).
55. O. Gonen, I. Catalaa, J. S. Babb, Y. Ge, L. J. Mannon, D. L. Kolson, and R. I. Grossman, Total brain N-acetylaspartate: a new measure of disease load in MS. *Neurology* **54**, 15-19 (2000).
56. O. Gonen, A. K. Viswanathan, I. Catalaa, J. Babb, J. Udupa, and R. I. Grossman, Total brain N-acetylaspartate concentration in normal, age-grouped females: quantitation with non-echo proton NMR spectroscopy. *Magn Reson. Med.* **40**, 684-689 (1998).
57. O. Gonen, D. M. Moriarty, B. S. Li, J. S. Babb, J. He, J. Listerud, D. Jacobs, C. E. Markowitz, and R. I. Grossman, Relapsing-remitting multiple sclerosis and whole-brain N-acetylaspartate measurement: evidence for different clinical cohorts initial observations. *Radiology* **225**, 261-268 (2002).
58. S. M. de Waegh, V. M. Lee, and S. T. Brady, Local modulation of neurofilament phosphorylation, axonal caliber, and slow axonal transport by myelinating Schwann cells. *Cell* **68**, 451-463 (1992).
59. M. Rango, D. Spagnoli, G. Tomei, F. Bamonti, G. Scarlato, and L. Zetta, Central nervous system trans-synaptic effects of acute axonal injury: A ^{1}H magnetic resonance spectroscopy study. *MRM* **33**, 595-600 (1995).
60. T. B. Patel and J. B. Clark, Synthesis of N-acetyl-L-aspartate by rat brain mitochondria and its involvement in mitochondrial/cytosolic carbon transport. *Biochem. J.* **184**, 539-546 (1979).
61. M. E. Truckenmiller, M. A. Namboodiri, M. J. Brownstein, and J. H. Neale, N-Acetylation of L-aspartate in the nervous system: differential distribution of a specific enzyme. *J. Neurochem.* **45**, 1658-1662 (1985).
62. J. B. Clark, N-acetyl aspartate: a marker for neuronal loss or mitochondrial dysfunction. *Dev. Neurosci.* **20**, 271-276 (1998).
63. N. De Stefano, P. M. Matthews, and D. L. Arnold, Reversible decreases in N-acetylaspartate after acute brain injury. *Magn Reson. Med.* **34**, 721-727 (1995).
64. M. Saragea, M. Clopotaru, M. Sica, A. Vladutiu, T. Negru, and N. Rotaru, Biochemical changes occurring in animals with experimental allergic encephalomyelitis. *Med. Pharmacol. Exp. Int. J Exp. Med.* **13**, 74-80 (1965).
65. R. E. Brenner, P. M. G. Munro, S. C. R. Williams, J. D. Bell, G. J. Barker, C. P. Hawkins, D. N. Landon, and W. I. McDonald, The proton NMR spectrum in acute EAE: The significance of the change in the Cho:Cr ratio. *MRM* **29**, 737-745 (1993).

66. N. De Stefano, P. M. Matthews, and D. L. Arnold, Reversible decreases in N-acetylaspartate after acute brain injury. *Magn Reson. Med.* **34**, 721-727 (1995).

67. C. A. Husted, D. S. Goodin, J. W. Hugg, A. A. Maudsley, J. S. Tsuruda, S. H. de Bie, G. Fein, G. B. Matson, and M. W. Weiner, Biochemical alterations in multiple sclerosis lesions and normal-appearing white matter detected by *in vivo* ^{31}P and ^{1}H spectroscopic imaging. *Ann. Neurol.* **36**, 157-165 (1994).

68. G. Helms, Volume correction for edema in single-volume proton MR spectroscopy of contrast-enhancing multiple sclerosis lesions. *Magn Reson. Med.* **46**, 256-263 (2001).

69. R. A. Rudick, E. Fisher, J. C. Lee, J. Simon, and L. Jacobs, Use of the brain parenchymal fraction to measure whole brain atrophy in relapsing-remitting MS. Multiple Sclerosis Collaborative Research Group. *Neurology* **53**, 1698-1704 (1999).

70. J. H. Simon, L. D. Jacobs, M. K. Campion, R. A. Rudick, D. L. Cookfair, R. M. Herndon, J. R. Richert, A. M. Salazar, J. S. Fischer, D. E. Goodkin, N. Simonian, M. Lajaunie, D. E. Miller, K. Wende, A. Martens-Davidson, R. P. Kinkel, F. E. Munschauer, III, and C. M. Brownscheidle, A longitudinal study of brain atrophy in relapsing multiple sclerosis. The Multiple Sclerosis Collaborative Research Group (MSCRG). *Neurology* **53**, 139-148 (1999).

71. N. De Stefano, S. Narayanan, G. S. Francis, R. Arnaoutelis, M. C. Tartaglia, J. P. Antel, P. M. Matthews, and D. L. Arnold, Evidence of axonal damage in the early stages of multiple sclerosis and its relevance to disability. *Arch. Neurol.* **58**, 65-70 (2001).

72. P. A. Narayana, J. S. Wolinsky, S. B. Rao, R. He, and M. Mehta, Multicentre proton magnetic resonance spectroscopy imaging of primary progressive multiple sclerosis. *Mult. Scler.* **10 Suppl 1**, S73-S78 (2004).

73. E. W. Willoughby and D. W. Paty, Scales for rating impairment in multiple sclerosis: a critique. *Neurology* **38**, 1793-1798 (1988).

74. M. Filippi, M. A. Horsfield, P. S. Tofts, F. Barkhof, A. J. Thompson, and D. H. Miller, Quantitative assessment of MRI lesion load in monitoring the evolution of multiple sclerosis. *Brain* **118**, 1601-1612 (1995).

75. M. A. Rocca, D. M. Mezzapesa, A. Falini, A. Ghezzi, V. Martinelli, G. Scotti, G. Comi, and M. Filippi, Evidence for axonal pathology and adaptive cortical reorganization in patients at presentation with clinically isolated syndromes suggestive of multiple sclerosis. *Neuroimage.* **18**, 847-855 (2003).

76. P. Pantano, G. D. Iannetti, F. Caramia, C. Mainero, S. Di Legge, L. Bozzao, C. Pozzilli, and G. L. Lenzi, Cortical motor reorganization after a single clinical attack of multiple sclerosis. *Brain* **125**, 1607-1615 (2002).

NAA AND HIGHER COGNITIVE
FUNCTION IN HUMANS

Ronald A. Yeo, William M. Brooks and Rex E. Jung*

1. INTRODUCTION

Over the past decade or so much interest has emerged in proton Magnetic Resonance Spectroscopy (^{1}H-MRS) investigations of the human brain. This largely reflects exciting findings from two major lines of research regarding N-acetyl-aspartate (NAA), the most prominent neurometabolite resolved *in vivo*. First, NAA concentrations appear to be reduced in diverse brain diseases, including both neurodevelopmental disorders and frank trauma [1]. Second, and the focus of this chapter, NAA concentrations have frequently been found to predict individual variation in higher cognitive function. Indeed, many studies have now identified rather impressive correlations between NAA and standard measures of cognitive function [2]. These relationships may eventually prove clinically useful, as for example in predicting long-term outcome from traumatic brain injury or the rate of progression to dementia in degenerative diseases. But they also raise compelling and important questions about the nature of the neurometabolic underpinnings of individual variation in cognition, and the specific ways in which NAA might be important for human brain function.

The central limitation of human studies linking NAA to cognition is readily apparent: we are permitted only investigations that are fundamentally correlational in nature. Progress in understanding the role of NAA in human cognition will clearly require studies of nonhuman animals, particularly nonhuman primates. Nonetheless, analyses of human studies may help generate research hypotheses, provide specificity and boundary conditions regarding NAA-cognition relationships, and suggest important

---

* Department of Neurology University of New Mexico, Albuquerque, NM 87131 ryeo@unm.edu.; William M. Brooks, Hogland Brain Imaging Center, The University of Kansas Medical Center wbrooks@kumc.edu.; Rex E. Jung, Department of Neurology, University of New Mexico School of Medicine, Albuquerque, NM 87131, rjung@salud.unm.edu; author for correspondence: Ronald A. Yeo.

clinical applications. In this chapter we will review two major domains of ^{1}H-MRS research suggesting a NAA-cognition link: studies of traumatic brain injury (TBI) and studies of normal controls. We will focus on spectroscopic investigations of cortical gray matter and underlying white matter tissue. First, though, we briefly review what ^{1}H-MRS may reveal about neurochemistry and salient issues in neurometabolite quantitation.

2. PROTON MAGNETIC RESONANCE SPECTROSCOPY

^{1}H-MRS detects neurometabolite variation of potential importance in understanding neuronal functioning and status. NAA plays a role in myelin synthesis in the developing brain [3], has been implicated in lipid repair [4], has recently been implicated as an anti-inflammatory agent [5], and appears to function as a "molecular water pump" in active neurons [6]. In the context of brain disorders, reduced NAA has often been construed as reflecting neuronal death [1]; however, a recent report of a child entirely lacking the NAA resonance [7, 8] has necessitated an adaptation of this conceptualization to include metabolic dysfunction without neuronal loss *per se*. The Cho peak reflects MRS-visible choline moieties (phosphocholine, glycerolphospho-choline, choline), likely reflective of tissue synthesis/breakdown or inflammatory processes. The Cre peak includes creatine and phosphocreatine resonances, reflective of tissue energetics. An important dimension of difference among MRS studies is whether concentrations of metabolites such as NAA or Cho are expressed relative to water levels (i.e., "absolute quantitation") or as a ratio with Cre as the denominator. As creatine and phosphocreatine levels are in dynamic equilibrium, their sum may be relatively stable across metabolic states, though this assumption has been questioned [9].

In MRS research, "reproducibility is the key to the successful application of spectroscopy to clinical medicine, just as it is the key to scientific research in general" [10]. For NAA to be accepted as an important marker of brain integrity it must be established as both reliable and valid. These psychometric properties are potentially mediated by acquisition techniques, quantitation algorithms, and group effects (i.e., normal *versus* patient populations). Two major acquisition techniques currently exist by which single voxel spectra are acquired: STimulated Echo Acquisition Method (STEAM) and Point REsolved Spin echo Sequence (PRESS). Although the STEAM technique has been used more often in single voxel studies, the PRESS technique has seen increased application primarily due to the roughly two times increase in signal-to-noise ratio [11]. Quantitation of spectroscopic measures is undertaken with two major fitting algorithms: Magnetic Resonance User Interface - MRUI [12] and LCModel [13]. Both fitting algorithms allow researchers to perform time-domain analysis of *in vivo* MR data. LCModel provides the additional benefit of full automation and use of known reference samples obtained to match individual echo time and machine characteristics. Using single voxel STEAM or PRESS spectroscopy (respectively), and MRUI analysis, our group has demonstrated reliability of NAA peaks measures in both normal populations (Coefficient of Variation = 3.3%) and in patients diagnosed with schizophrenia (Coefficient of Variation = 4%) in single voxels obtained within white matter volumes [14, 15].

3. TRAUMATIC BRAIN INJURY

TBI is the leading cause of death and morbidity in young, otherwise healthy young adults and children. Due to variability in the form and extent of the initial injury, as well as individual variation in response to injuries, prediction of outcome following TBI is challenging and generally inadequate [16]. Common clinical markers, such as the Glasgow Coma Scale (GCS) are linked statistically with outcome in large samples, but are less useful in predicting cognitive functioning in individual patients. CT is generally the preferred neuroimaging technique in the acute evaluation of TBI, while MRI studies are especially valuable in subacute or chronic forms due to its greater sensitivity to diffuse axonal injury. However, quantitative estimates of the extent of brain trauma with these methods have proven of only modest value in predicting concurrent cognitive status or cognitive outcome [17]. The most important limitation of imaging studies of TBI is a relative insensitivity to neuronal dysfunction as opposed to neuronal death.

We have noted in samples of both children and adults with TBI that NAA and Cho concentrations are highly predictive of cognitive deficits. In our initial study, nineteen adults with TBI (mean age = 32.1 years, SD = 13.2, 5% female, mean acute Glasgow Coma Scale = 8.4, SD = 3.4) were compared with 28 normal controls (mean age = 26.6 years, SD = 11.1, 39% female [18]). Most TBI patients were assessed at three time points (median = 38 days, 125 and 184 days post-injury). Figure 1 shows the location of the gray matter (GM) and white matter (WM) voxels studied, as well as representative spectra. Spectroscopic data were analyzed with MRUI (Leuven, Belgium) and normalized for CSF within the voxel, providing an estimate of absolute neurometabolite concentrations. A large battery of neurocognitive tests were administered, including measures of attention, memory, processing speed, and "executive" functioning. To reduce the number of variables examined and provide an overall measure of cognitive status, each test score was standardized and then averaged to create a "composite z-score". Additional methodological details are provided in [19, 20].

Greater GM NAA predicted better overall cognitive functioning at each time point (r's = .63, .70, and .57). Relevant to the clinical utility of ^{1}H-MRS, initial (i.e., sub-acute) GM NAA also predicted overall neuropsychological function at outcome, roughly six months post injury ($r = .70$): moreover, better cognitive performance was seen in patients whose GM NAA improved over time. For a smaller sample of patients we also assessed frontal WM NAA. This measure was highly correlated with overall cognitive function at the initial time point ($r = .89$), but did not predict subsequent neuropsychological performance or long-term outcome [unpublished data].

We believe these results suggest that MRS may be able to detect the neurobiological substrate of diffuse axonal injury producing neurocognitive deficit relatively early in the disease process. It is important to note that the voxels studied were free from overt abnormality, as determined by independent clinical neuroradiological examination. Consistent with this observation, Gasparovic and colleagues [21] concurrently demonstrated in the weight-drop animal model of TBI that NAA/Cre ratios were notably reduced in brain tissue remote from the impact site that prior histological studies have shown to be free from any cellular damage.

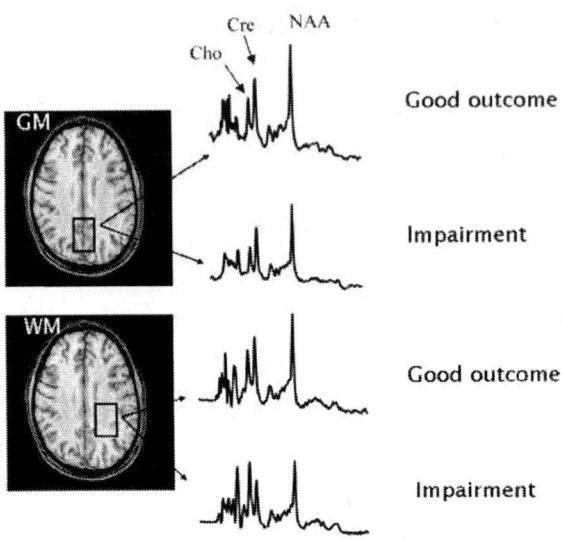

Figure 1. White matter (WM) and gray matter (GM) voxels analyzed in studies of adults with traumatic brain injuries.

Reduced NAA concentrations can reflect neuronal death, but this mechanism cannot account for the reversible decreases in NAA observed in multiple sclerosis [22] and stroke [23], or the possible increases over time, post-injury, observed in our TBI sample [18]. Given Gasparovic's results, and the fact that our NAA analyses were undertaken in normal appearing tissue, it may well be the case that individual variation in metabolism, rather than loss of neurons, underlies the well-established relationship between NAA and concurrent and long term cognitive function. Consistent with this possibility, Positron Emission Tomography analysis has shown that NAA concentrations correlate with overall metabolic rate [24], which is often reduced after TBI.

Our recent studies with children have utilized Spectroscopic Imaging (SI), rather than the single voxel methodology described above [25]. A spectroscopic imaging slice was selected above the lateral ventricles to extend from the frontal lobe to the occipital and parietal lobes sampling both white and gray matter. Water suppressed PRESS localization with outer voxel suppression bands was used to excite parenchyma and avoid lipid artifact from the skull. See Figure 2 (panel A) for an example MRI image with superimposed SI grid. Patient recruitment and data analyses are ongoing. Here we report on NAA/Cre ratios averaged across many voxels (predominantly white matter), with separate estimates for anterior and posterior halves of the data matrix. A large battery of cognitive tests was administered; individual scores were standardized and averaged into these composites: overall cognitive (as above), motor, language, visuo-motor, and working memory. The TBI children (N = 28, mean age = 13.6, SD = 3.6, 21% female; mean GCS = 7.8, SD = 4.7) were compared to healthy controls (N = 14, mean age = 15.2, SD = 1.7, 56% female). As expected, the TBI children had much lower NAA/Cre and greater Cho/Cre, overall and in each half of the brain. They also scored lower on each cognitive composite.

Table 1. Correlations of NAA/Cre from total supraventricular MRS slab, and anterior and posterior halves of the MRS slab, with age adjusted neurocognitive domain scores in pediatric TBI patients (N = 28).

	Total	Anterior	Posterior
Verbal	.38*	.45*	.26
Working Memory	.39*	.38*	.37*
Motor	.20	.26	.15
Visual-Motor	.42*	.51*	.23

*p < .05

Consistent with our single voxel studies of adults, reduced NAA/Cre and greater Cho/Cre predicted worse cognitive function in children with TBI (see Table 1). Two interesting trends are evident in these data. First, motor skills (manual dexterity and strength of grip) were not related to neurometabolite concentrations, while cognitive skills were. Second, frontal NAA/Cre generally predicted cognitive deficits better than posterior values. These ratio data are inherently more difficult to interpret than "absolute" concentrations due to the potential variability of the denominator, in this case Cre. But, as the NAA/Cre ratio was positively related to function, and the Cho/Cre ratio inversely related, we believe our results reflect the importance of the numerator rather than the denominator. Younger and older children with TBI showed the same pattern of results. Along with our results from adult studies, this suggests that NAA may serve an important clinical role across a very broad age range of individuals with TBI.

4. NORMAL ADULTS

Several studies from different laboratories have now reported that greater NAA is associated with better performance on some type of cognitive test, particularly broad measures of cognitive functioning such as those found in measures of intellectual functioning [2]. The details vary, as different brains regions have been studied with diverse tests, but the general finding seems fairly well established: *in vivo* measures of NAA are positively correlated with cognitive performance in normal individuals. In 1999 we reported that NAA concentrations from a left parieto-occipital white matter voxel predicted overall intellectual functioning in a young cohort of college aged subjects (Mean age = 22 +/-4.6) [26]. We have now conducted two independent replications, with different participants and slightly different imaging acquisition and processing methods, and found similar results [27, 28]. Across the three studies, NAA accounted for 27%, 36%, and 25% of variance in IQ scores. Among the diverse neurobiologic predictors of IQ (e.g., brain volume, evoked potentials), NAA seems to be one of the strongest biological correlate of intellectual functioning [29]. Importantly, this relationship is not observed when NAA is assessed in left frontal white matter. Across the same three data sets, frontal NAA accounted for 1%, 6%, and 6% of variance in IQ scores (each non-significant), and in the latter two data sets an inverse relationship was noted [unpublished data].

A core component skill underlying individual variation in intelligence is speed of cognitive processing [30]. As our NAA-intelligence findings emerged from voxels containing mostly myelinated axonal fibers, we hypothesized that NAA might be most related to neuropsychological measures tapping speeded cognitive performance [19]. Our battery of neuropsychological tests was split into those emphasizing rapid processing (i.e., timed performance measures) vs. those that did not (e.g., word finding ability). Left posterior white matter NAA accounted for 42% of the variance on a composite measure of speeded performance, vs. 8% (non-significant) on non-speeded tests. Interestingly, as also found in our TBI studies, correlations between NAA and motor tests were low (Grooved Pegboard Test, r = 0.18; Grip Strength, r = -0.05). Finally, one recent study reports that left frontal gray matter NAA predicts greater verbal intelligence, though only in women [31]. Thus, studies to date suggest that there may be both cognitive, anatomic, and sex specificity in the white matter NAA-cognition relationship. Obviously, we need to survey many more brain regions, especially gray matter voxels, in larger samples of individuals equally divided among males and females to confirm the hypothesis that regional NAA differences underlie cognitive ability in normal human brain.

The mechanism(s) underlying the WM NAA-cognition relationship are of great interest. It may be, however, that rather "global" cognitive measures such as described here (i.e., IQ, total z-score) are the ones that best map on to NAA, which has been linked to myriad metabolic functions [3-6]. As noted above, the correlational nature of human studies limits implications for understanding causal relationships. Human studies can, however, indicate possible covariance of NAA with other variables known to predict general intellectual functioning. The best known of these is overall brain volume, which correlates with IQ at about r = 0.3 or 0.4 [32]. In our sample (N = 28) we find this same relationship; total brain volume correlated with IQ at r = 0.49 (p < .01) [33]. When both NAA and brain volumetric values are regressed against FSIQ, only occipito-parietal NAA is retained. Thus, NAA levels appear to characterize important variance independent of sheer brain volume, and likely begin to capture subtle brain design variables underlying intellectual attainment: indeed, this point becomes manifest merely by noting that at least two brain designs (e.g., male, female) attain similar IQ's independent of brain size.

Recent fMRI [34] and voxel-based morphometry (VBM) studies [35] clearly show that some brain regions seem to be much more important for individual variation in intellectual functioning than others. In particular, frontal gray matter activations and volumes appear to be important to intellectual functioning [36] although both frontal and parietal regions tend to covary with IQ [29]. As we have recently demonstrated specific brain regions associated with intellectual functioning in a cohort of normal subjects [35] using Voxel Based Morphometry (VBM), we sought to determine the relationship between NAA levels within frontal and posterior white matter and gray matter morphometry in a cohort of normal subjects [37]. VBM is based on making regionally-specific (voxel-wise) inferences on the local relative concentrations of different tissue types after spatial normalization and segmentation of the underlying anatomical images [38]. The results of the procedure are presented in standard stereotactic (i.e., Talairach)

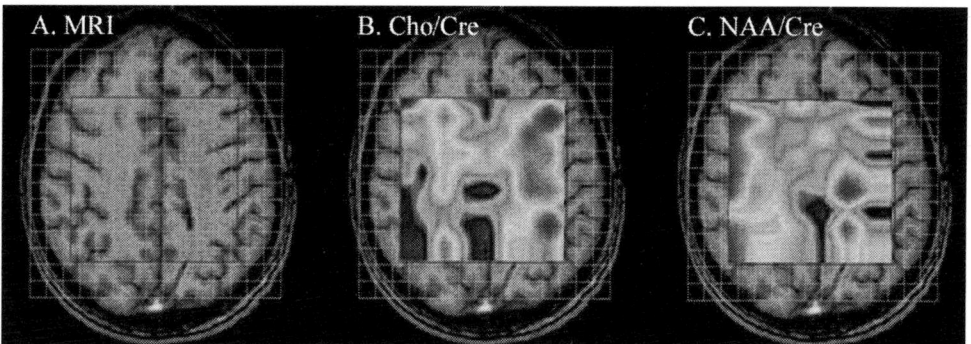

Figure 2. Spectroscopic Imaging (SI) of child with TBI maximally affecting the left cerebral hemisphere: (See color insert that appears between pages 364 and 365)

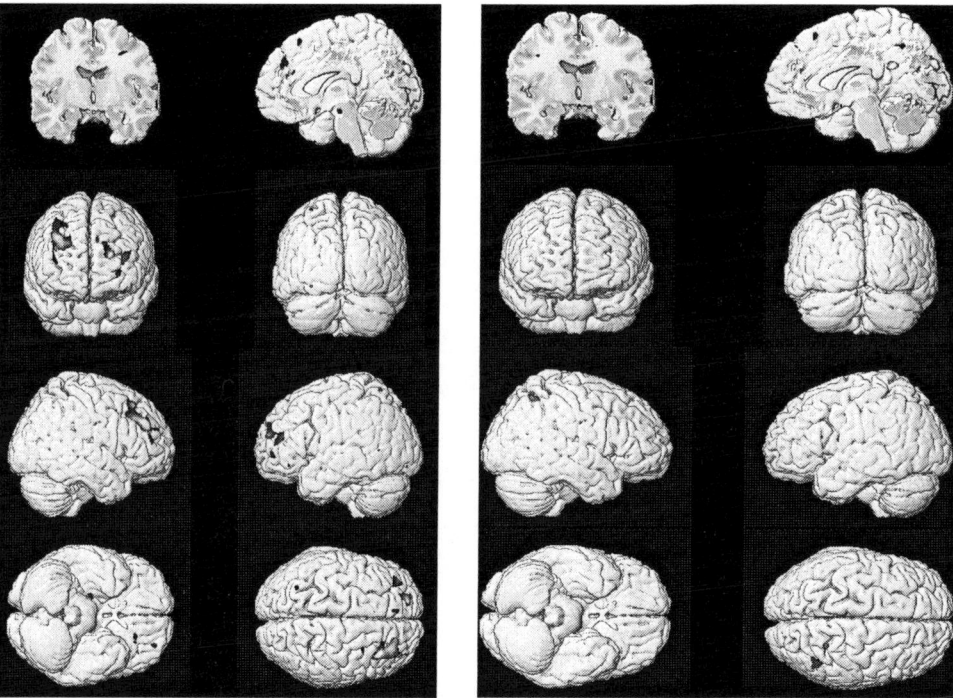

Figure 3 (left panel). Voxel Based Morphometry (VBM) analysis of GM cluster volumes covarying with posterior left hemisphere white matter NAA concentration: (See color insert that appears between pages 364 and 365)

Figure 4 (right panel). Voxel Based Morphometry (VBM) analysis of GM cluster volumes covarying with left hemisphere frontal white matter NAA concentration: (See color insert that appears between pages 364 and 365)

space that facilitates direct comparisons across studies. VBM has been extensively cross-validated with both ROI and functional analyses [e.g., 39].

Eighteen normal subjects (age range: 18 – 37), free of neurological or psychiatric disorders, were included in the analysis. MRIs were obtained with a clinical 1.5-T scanner, head coil, and software (Signa 5.4; General Electric Medical Systems, Waukesha, WI). Metabolite concentrations were determined using time-domain fitting (LC Model) to measure peak areas of NAA [13]. Significant positive correlations between occipito-parietal NAA and gray matter were found in predominantly frontal lobe regions including the bilateral middle and superior frontal gyrus (BAs 9, 10), left middle frontal gyrus (BA 46), and the anterior cingulate gyrus (BAs 24, 32). Positive correlations between frontal NAA and gray matter were limited to the posterior right cingulate gyrus (BA 31), and right parietal lobule (BA 7). This last result is particularly compelling, as the posterior cingulate cortex metabolic integrity has been linked to early detection of dementia in both AD and VCI [40], and suggests a critical interplay between frontal white matter metabolic integrity and posterior gray matter morphology. These results (Figure 3) demonstrate a compelling double dissociation between regional metabolic integrity and cortical volume, with occipito-parietal NAA related predominantly to frontal lobe gray matter volume, and frontal NAA related to predominantly posterior gray matter volumes. Moreover, the frontal regions in which posterior NAA predicts gray matter volume are well associated with cognitive activation elicited via functional magnetic resonance imaging and positron emission tomography [41]. Although preliminary in nature, this analysis highlights the importance of intact metabolic connectivity underlying frontal lobe morphology, the first time that this has been demonstrated in a cohort of normal subjects.

The studies described above have focused on relationships between white matter NAA and rather global measures of cognitive ability. ^{1}H-MRS may also prove quite useful in the investigation of functional differences in discreet gray matter nuclei. We have recently reported that NAA/Cre in the hippocampus predicts individual variation in hippocampal function [42]. The Morris Water Maze has been extensively employed in studying the relationship between hippocampal function and spatial learning in rodents, and more recently. Hippocampal damage reliably produces deficits on tasks such as this that require learning configural as opposed to elemental stimuli configurations. Use of the Virtual Water Maze Test (VWMT) in humans provides a solid basis for cross-species comparisons [43]. Both histological and volumetric studies of the hippocampus reveal age-related changes [44] that could potentially contribute to the well documented decline in memory. In this study we compared 16 young adults (mean age = 26.1 years) with a sample of 16 very healthy elderly individuals (mean age = 11.6 years), all of whom were homozygous APOE-3 genotypes, as the APOE-4 allele has been linked with risk of Alzheimers Disease. Magnetic resonance spectroscopic imaging data were obtained at the level of the hippocampus from a 15 mm thick axial slab, as shown in Figure 5. Concentraion was expressed as a ratio of NAA/Cre. As expected, age-related deficits were observed in VWMT measures reflecting configural memory, but not in VWMT measures reflecting visual-motor processing (and motivation). VWMT memory declines were accompanied by decreased hippocampal volumes and decreased NAA/Cre, though Cho/Cre showed no age effect. In a regression analysis of VWMT performance, NAA/Cre and age were retained as significant predictors, while hippocampal volume did not provide an independent contribution. These results suggest that normal aging is associated with hippocampal structural and biochemical changes, and that these changes

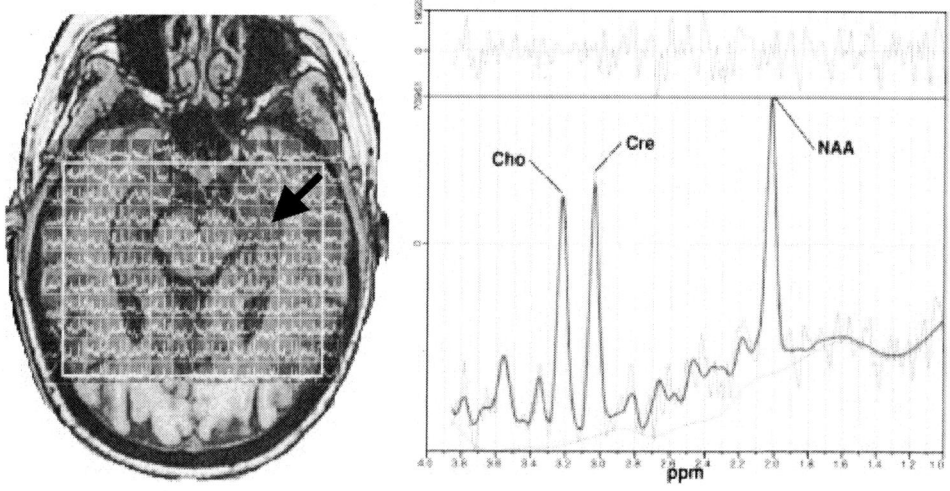

Figure 5. SI of adult brain showing location of hippocampal voxel (arrow) assessed in study of Virtual Water Maze performance.

may constitute an important component of age-related deficits in hippocampus-dependent learning and memory.

5. FUTURE DIRECTIONS

MRS investigations of cognitive function in both normal and clinical samples suggest that individual variation in cognitive function can be predicted by NAA concentrations and NAA/Cre ratios, but important caveats need to be considered. Much regional variation exists in both neurometabolite concentrations. And, of course, brain regions and tissue compartments differ markedly in their function. Most of the work described in this chapter reflects analyses of white matter NAA. Future investigations will clearly need assess multiple white and gray matter regions across the brain and evaluate relationships with diverse measures of cognitive, motor, and emotional functions. The variables of sex and age may be found to moderate NAA-cognition relationships.

As it true of every neuroimaging modality, the meaning of individual observations using MRS will become clearer when this technique in used in conjunction with other neuroimaging approaches. Let us offer two examples. We have suggested that NAA variation might be especially important for speed of cognitive processing. Magnetoencephalography (MEG) has excellent temporal resolution [45] and processing latencies in specific anatomic regions can potentially be linked with neurometabolite variation in those same locations. Diffusion Tensor Imaging (DTI) might prove especially valuable in evaluating white matter structural variation [46] covarying with NAA concentrations.

6. QUESTION AND ANSWER SESSION

DR. DUYN: Thank you. We have time for two questions.

DR. MADHAVARAO: In the data, are these -- In the majority of the data or the general data between NAA concentration and cognitive ability, are these subjects preconditioned? In other words, when were the activities subjected to this analysis? They were actually -- they have already taken a test or are preparing for a test?

DR. YEO: Yes, good question. In all of the studies that we have looked at here, the testing was done at more or less the same day as the imaging study was conducted, but the imaging is, of course, separate. They are just resting in the scanner. They are not performing any task while they are in the scanner. But for the normal controls as well as the traumatic brain injury folks, the testing is usually done the same day, but in some cases it is a day before or a day later.

DR. KOLODNY: In view of the studies of Monte de Leone and others showing enlargement of the temporal horn and atrophy of the hippocampus with aging and Alzheimer's Disease, how do you account for the fact that the NAA in the hippocampus apparently in your study does not show a decrease in NAA in aging, unless I have gotten your declaration wrong?

The other unrelated question is: In diffuse axonal injury, it seems to me that NAA measurements would be a very good way of determining the severity of the brain injury.

DR. YEO: In our aging study, I think it is important to note that the regression analysis I reported is across the age span. So we included the younger folks and the older folks, and thus there is a very broad age range there, from the twenties through the seventies.

The studies of aging per se, and especially of incipient dementia, I think, need focus much more -- solely on an older population. So I think our studies pertain more toward normal age changes over the whole adult span than that which might occur just in an elderly sample, which might be more critically significant.

In regard to the latter point, yes, I think indeed that spectroscopic measures of brain injury can add a lot to our clinical understanding. Certainly, Dr. Ross's data attest to that, too, and I can't help but argue that it should be more widespread.

DR. DUYN: Okay. Well, thank you very much, Dr. Yeo.

7. REFERENCES

1. Barker, P. B. N-acetyl aspartate--a neuronal marker? *Ann. Neurol.* **49**(4), 423-4 (2001).
2. Ross, A. J., and Sachdev, P. S. Magnetic resonance spectroscopy in cognitive research. *Brain Res. Rev.* **44**, 83-102 (2004).
3. Tallan, H.H., Studies on the distribution of N-acetyl-L-aspartic acid in brain. *J. Biol. Chem.* **224**, 41–45 (1957).
4. D'Adamo, A.F., Jr., and Yatsu, F.M. Acetate metabolism in the nervous system. N-acetyl-L-aspartic acid and the biosynthesis of brain lipids. *J. Neurochem.* **13**, 961–965 (1996).
5. Rael, L. T., Thomas, G. W., Bar-Or, R., Craun, M. L., and Bar-Or, D. An anti-inflammatory role for N-acetyl aspartate in stimulated human astroglial cells. *Biochem. Biophy. Res. Commun.* **319**, 847-853 (2004).
6. Baslow, M. H. N-acetylaspartate in the vertebrate brain: metabolism and function. *Neurochem. Res.* **28**(6), 941-53 (2003).
7. Martin, E., Capone, A., Schneider, J., Hennig, J., and Thiel, T. Absence of N-acetylaspartate in the human brain: impact on neurospectroscopy? *Ann. Neurol.* **49**(4), 518-21(2001).
8. Boltshauser, E., Schmitt, B., Wevers, R. A., Engelke, U., Burlina, A. B., and Burlina, A. P. Follow-up of a child with hypoacetylaspartia. *Neuroped.* **35**(4), 255-258 (2004).

9. Macmillan, C. S., Wild, J. M., Wardlaw, J. M., Marshall, I., and Easton, V. J. Traumatic brain injury and subarachnoid hemorrhage: in vivo occult pathology demonstrated by magnetic resonance spectroscopy may not be "ischaemic". A primary study and review of the literature. *Acta Neurochi.* **144**, 853-862 (2002).

10. Bottomley, P. A. The trouble with spectroscopy papers. *J. Magn. Reson. Imaging* **2**(1): 1-8 (1992).

11. Bottomley, P. A. Spatial localization in NMR spectroscopy in vivo. *Ann. NY Acad. Sci.* **508**, 333-48 (1987).

12. van den Boogaart A, Ala-Korpela M, Jikisaari J et al. Time and frequency-domain analysis of NMR data compared – an application to 1D H-1 spectra of lipoproteins. *Magn. Reson. Med.* **31**, 347-358 (1994).

13. Provencher, S. W. Estimation of metabolite concentrations from localized in vivo proton NMR spectra. *Magn. Reson. Med.* **30**(6), 672-9 (1993).

14. Brooks, W. M., Friedman, S. D., and Stidley, C. A. Reproducibility of 1H-MRS in vivo. *Magn. Reson. Med.* **41**(1), 193-7 (1999).

15. Mullins, P. G., Rowland, L., Bustillo, J., Bedrick, E. J., Lauriello, J., and Brooks, W. M. Reproducibility of 1H-MRS measurements in schizophrenic patients. *Magn. Reson. Med.* **50**(4), 704-707 (2003).

16. Levin, H. S. Prediction of recovery from traumatic brain injury. *J. Neurotrauma* **12**, 913-922 (1995).

17. Bigler, E. D. Quantitative magnetic resonance imaging in traumatic brain injury. *J. Head Trauma Rehabil.* **16**, 117-134 (2001).

18. Brooks, W. M., Friedman, S. D. and Gasparovic, C. (2001). Magnetic resonance spectroscopy in traumatic brain injury. *J. Head Trauma Rehabil.* **16**(2), 149-64.

19. Jung, R. E., Yeo, R. A., Chiulli, S. J., Sibbitt, W. L., Jr., Weers, D. C., Hart, B. L., and Brooks, W. M. (1999). "Biochemical markers of cognition: a proton MR spectroscopy study of normal human brain." Neuroreport **10**(16): 3327-31.

20. Jung, R. E., Yeo, R. A., Chiulli, S. J., Sibbitt, W. L., Jr. and Brooks, W. M. Myths of neuropsychology: intelligence, neurometabolism, and cognitive ability. *Clin. Neuropsychol.* **14**(4): 535-45 (2000).

21. Gasparovic, C., Arfai, N., Smid, N., and Feeney, D. M. Decrease and recovery of N-Acetylaspartate in rat brain remote from focal injury. *J. Neurotrauma* 18, 241-246 (2001).

22. Narayanan, S., De Stefano, N., Francis, G. S., Arnaoutelis, R., Caramanos, Z., Collins, D. L., Pelletier, D., Arnason, B. G. W., Antel, J. P. and Arnold, D. L. Axonal metabolic recovery in multiple sclerosis patients treated with interferon beta-1b. *J. Neurol.* **248**(11): 979-86 (2001).

23. De Stefano, N., Matthews, P. M. and Arnold, D. L. (1995). Reversible decreases in N-acetylaspartate after acute brain injury. *Magn. Reson. Med.* **34**(5): 721-7.

24. O'Neill, J., Eberling, I. L., Schuff, N., Jagust, W., Reed, B., Soto, G., Ezekiel, F., Klein, G., and Weiner, M. K. Method to correlate H01 MRSI and (18) FDG-PET. *Magn. Res. Med.* **43**(2), 244-250.

25. Weers, D. C., Yeo, R. A., Phillips, J. P., Campbell, R. C., Brooks, W. M., Petropoulos, H., Jung, R. E., Kernen, S., Brown, A. J., and Hart, B. Neurometabolic abnormalities in pediatric traumatic brain injury: A Magnetic Resonance Spectroscopy study. Under review.

26. Jung, R. E., Brooks, W. M., Yeo, R. A., Chiulli, S. J., Weers, D. C., and Sibbitt, W. L., Jr. Biochemical markers of intelligence: a proton MR spectroscopy study of normal human brain. *Proc. R. Soc. Lond. B Biol. Sci.* **266**(1426), 1375-1379 (1999).

27. Jung, R. E., Rowland, L. M., Yeo, R. A., Barrow, R. A., Petropoulos, H., Lauriello, J., Bustillo, J. and Brooks, W. M. Regional specificity of NAA - IQ relationships in schizophrenia and normal brain. *Proc. Int. Soc. Magn. Reson. Med.* **11**, 2006 (2003).

28. Jung, R.E., Rowland, L.M., Mullins, P.G., Lauriello, J., Bustillo, J.R., and Yeo, R.A. (2004). N-acetylaspartate within frontal and parietal white matter differentially predicts intellectual functioning in normal brain. *Proc. Int. Soc. Magn. Reson. Med.* **11**, 2431.

29. Gray, J. R. and Thompson, P. M. Neurobiology of intelligence: science and ethics. *Nat. Rev. Neurosci.* **5**(6), 471-82 (2004).

30. Jensen, A. R. Is speed of processing related to fluid or crystallized intelligence? *Intelligence* 7, 91-106 (1983).

31. Pfleiderer, B., Ohrmann, P., Suslow, T., Wolgast, M., Gerlach, A. L., Heindel, W., and Michael, N. N-acetylaspartate levels of left frontal cortex are associated with verbal intelligence in women but not in men: a proton magnetic resonance spectroscopy study. *Neurosci.* **123**(4), 1053-1058 (2004).

32. Anderson, B. (2003). Brain imaging and g. The scientific study of general intelligence. H. Nyborg. Oxford, Pergamon: 29-39.

33. Jung, R.E., Haier, R.J. Yeo, R.A., Rowland, L.M., Petropoulos H, Levine, A.S., Sibbitt, W.L., & Brooks, W.M. Sex Differences in N-acetylaspartate Correlates of General Intelligence: A 1H-MRS Study of Normal Human Brain. (In Press).

34. Gray, J. R., Chabris, C. F., and Braver, T. S. Neural mechanisms of general fluid intelligence. *Nat. Neurosci.* **6**(3), 316-22. (2003).

35. Haier, R., Jung, R. E., Yeo, R., Head, K., and Alkire, M. T. Structural brain variation and general intelligence. *Neuroimage* **23**(1), 425-433 (2004).

36. Duncan, J., Seitz, R. J., Kolodny, J., Bor, D., Herzog, H., Ahmed, A., Newell, F. N., and Emslie, H. A neural basis for general intelligence. *Science* **289**(5478), 457-60 (2000).

37. Jung, R.E., Haier R.J., Yeo, R.A., Rowland, L.M., Head, K. Alkire, M. and Gasparovic, C. Brain Volume Correlates of Frontal and Occipito-parietal N-acetylaspartate. Presented at the 1st International Symposium on N-acetylaspartate. Bethesda, Maryland (2004).

38. Ashburner, J., and Friston, K. J. Voxel-based morphometry--the methods. *Neuroimage* **11**(6 Pt 1), 805-21 (2000).

39. Good, C. D., Scahill, R. I., Fox, N. C., Ashburner, J., Friston, K. J., Chan, D., Crum, W. R., Rossor, M. N., and Frackowiak, R. S. Automatic differentiation of anatomical patterns in the human brain: validation with studies of degenerative dementias. *Neuroimage* **17**(1), 29-46 (2002).

40. Martinez-Bisbal, M. C., Arana, E., Marti-Bonmati, L., Molla, E., and Celda, B. Cognitive impairment: classification by 1H magnetic resonance spectroscopy. *Eur. J. Neurol.* **11**(3), 187-93 (2004).

41. Cabeza, R. and Nyberg, L. Imaging cognition II: An empirical review of 275 PET and fMRI studies. *J. Cogn. Neurosci.* **12**(1), 1-47 (2000).

42. Driscoll, I. Hamilton, D. A., Petropoulos, H., Yeo, R. A., Brooks, W. M., Baumgartner, R. N., and Sutherland, R. J. The aging hippocampus: Cognitive, biochemical and structural variations. *Cereb. Cortex* **13**, 1344-1351 (2003).

43. Hamilton, D. A., and Sutherland, R. J. Blocking in human place learning: Evidence from virtual navigation. *Psychobiol.* **27**, 453-461 (1999).

44. Raz, N., in: *Handbook of Aging and Cognition – II*, edited by F. I. M. Craik and T. A. Salthouse (Erlbaum, Mahwah, NJ, 1999), pp. 1-90.

45. Lewine, J. and Orrison, W.W. Magnetoencephalography and Magnetic Source Imaging. In W.W. Orrison, J. Lewine, J. Sanders & M. Hartshorne (Eds.), *Functional Brain Imaging*. St Louis: Mosby. Lewine, J. & Orrison, W.W. (1995).

46. Lim, K. O. and Helpern, J. A. Neuropsychiatric applications of DTI - a review. *NMR Biomed.* **15**(7-8), 587-93 (2002).

IN VIVO NMR MEASURES OF NAA
AND THE NEUROBIOLOGY
OF SCHIZOPHRENIA

Stefano Marenco, Alessandro Bertolino, and Daniel R. Weinberger[*]

1. INTRODUCTION

Measurements of *N*-Acetyl-Aspartate (NAA) with magnetic resonance proton spectroscopy (MRS) have contributed important clues to the arduous discovery of the pathophysiology of schizophrenia. In this chapter, we will describe the findings that have emerged from the Clinical Brain Disorders Branch of the NIMH over the last decade or so and try to give a critical context to the literature in this field. Given the space limitations and the vast literature available on MRS, the citation list will be far from exhaustive. Comprehensive reviews of NAA measures in schizophrenia have appeared elsewhere.[1-5] The findings will be articulated in three main threads: 1) evidence that measures of NAA index cortical changes in schizophrenia; 2) the changes shown in MRS studies converge with the information provided by other measurement modalities such as functional magnetic resonance imaging (fMRI) and positron emission tomography (PET), indicating that NAA concentrations predict the function of distributed cortical circuitry in schizophrenia; 3) variation in genes that are associated with increased risk for schizophrenia is also associated with changes in NAA concentrations, offering for the first time a window on the molecular mechanisms underlying reduction of NAA and the pathophysiology of schizophrenia.

[*] Genes, Cognition and Psychosis Program, Division of Intramural Research, National Institute of Mental Health, National Institutes of Health, Bethesda, Maryland USA. Correspondence to D.R Weinberger, 10 Center Drive, Rm. 4S-235, Bethesda, Maryland 20892, email: weinberd@mail.nih.gov.

2. METHODS

All studies of schizophrenia in our laboratory were conducted with an MRS imaging (MRSI) technique initially developed by Duyn *et al.*[6] and subsequently modified by van der Veen *et al.*[7] to allow the simultaneous acquisition of the water signal. It consisted of a spin echo technique (TR 2300 ms and TE 280 ms) from which 4 slices were acquired during approximately 30 min of examination. Outer volume suppression pulses were used to null the lipid signal from the skull. All of our human and animal studies were conducted on 1.5T GE systems (GE Medical Systems, Milwaukee, WI), but more recently we have been using a 3T system from the same manufacturer. At 1.5T, the nominal resolution of the voxels was 0.84 cc, with a slice thickness of 15 mm. Due to the increased susceptibility artifacts at 3T causing spectral line broadening, in the context of improved signal, the slice thickness at 3T was halved (voxel resolution of 0.42 cc), resulting in lines of the same width or narrower than at 1.5T for the same duration of exam. An example of spectra at 1.5T and 3T is shown in Figure 1. The analysis of the spectra consisted of integrating the values under the peaks corresponding to NAA (around 2 ppm), creatine (Cre: around 3 ppm) and choline (Cho: around 3.2 ppm) containing compounds. The values for each metabolite were then averaged over several voxels included in a region of interest (ROI). The ROIs were drawn on structural MRI images (T1 weighted) co-registered to the MRSI. Several ROIs were drawn on the MRSI scans, but our focus has been principally on the medial temporal lobe and the dorsolateral prefrontal cortex (DLPFC), areas that have been implicated in the pathophysiology of schizophrenia by multiple lines of converging evidence.[8,9] We have used ratios (NAA/Cre, NAA/Cho and Cho/Cre) as our main outcome measures. Although methods for absolute quantification of NAA are available, they are based on multiple assumptions and on the measurement of several quantities (e.g. the proportion of contamination with cerebrospinal fluid (CSF), the t2 of water in a certain tissue, etc) that allow for considerable uncertainty due to error propagation. The degree of error involved for these procedures has not been well characterized. All these technical issues are not a reason for concern when measuring ratios. However, the assumption underlying the use of ratios is that the denominator (in this case Cre) does not vary regionally or with pathological conditions. This seems to be generally true in the case of schizophrenia, although direct positive experimental evidence of this is lacking (on the other hand there is no available evidence to the contrary, to our knowledge).

For primate studies, a similar sequence to the one described above was used on the same 1.5T scanner equipment used for humans, but with a special surface coil for data acquisition. Only one slice was acquired through the frontal lobes.

Single voxel methodology was used for the study of rat models of a developmental lesion of the medial temporal lobe (described below). A 4.7T Varian Inova scanner was used with a 20 mm diameter circular surface coil. Spectra were acquired from a voxel of 36 mm^3 nominal volume with a stimulated echo acquisition sequence (STEAM: TR=3000 ms; TE=12 ms).

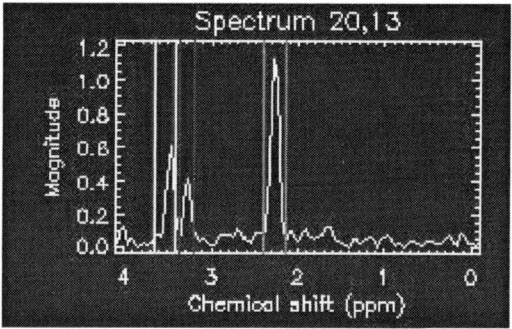

Representative spectrum for a 1.5T MRSI acquisition

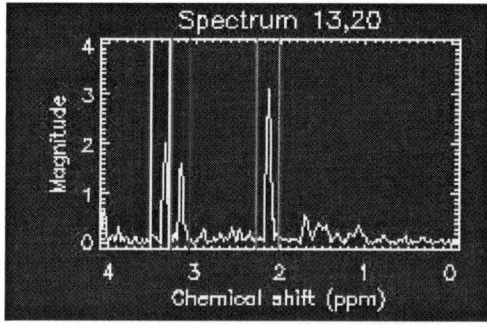

Representative spectrum for a 3T MRSI acquisition

Figure 1. Proton Spectra at 1.5T and 3T. Note that the 3T spectrum has larger magnitude and narrower line width.

3. RESULTS

Following our first line of enquiry in the attempt to show that NAA can index neuronal pathology in schizophrenia, Bertolino *et al.* found in two separate groups of patients that NAA/Cre ratios were reduced in the medial temporal lobe and in the DLPFC independent of the medication status of the patients.[10,11] The same phenomena was observed in patients with childhood onset schizophrenia.[12] These results have been confirmed by a number of other investigators around the world, but not by all.[1] It is important to note that NAA/Cre reductions in the DLPFC are not as consistently found as are changes in the hippocampal region, most likely due to variations in the degree of chronicity and age of the cohorts, as well as possible effects of medication.[13] Another variable that may be able to explain some of the variation in patient samples is clinical symptomatology: Callicott *et al.*[14] found that the lower the NAA/Cre in the DLPFC, the higher the negative symptoms in 36 patients with schizophrenia. This finding, which implicates NAA measures as having functional correlates, is shown in Figure 2. Further attesting to the sensitivity of NAA/Cre ratios to important functional aspects of the pathophysiology of schizophrenia, especially in areas that have been implicated by other functional and post-mortem approaches (e.g. Xe, PET, fMRI[9,15]) is the finding that neuroleptic treatment seems to increase NAA/Cre measurements selectively in the DLPFC.[16]

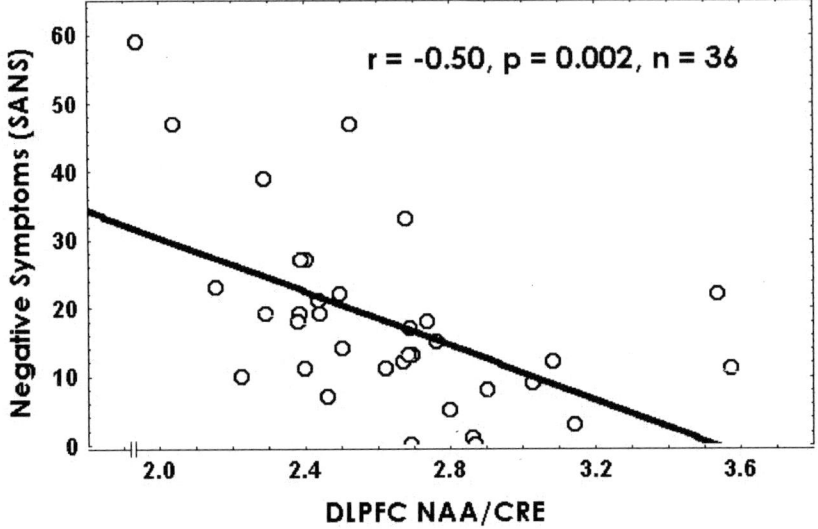

Figure 2. Correlation between negative symptoms and NAA/Cre in the DLPFC. (From Callicott *et al.*[14]).

Our second line of research has been directed at understanding the neuronal mechanisms of brain dysfunction associated with schizophrenia by studying the relationship of NAA alterations to other aspects of functional and neurochemical abnormalities. Prefrontal NAA/Cre has been shown to predict the functioning of distributed cortical circuits, as indexed by fMRI (blood oxygenation level dependent or BOLD) and PET (blood flow) activation. Bertolino *et al.*[17] examined two cohorts of patients with schizophrenia and matched normal controls performing two different tasks that critically engage the prefrontal cortex by activating a working memory network: the Wisconsin Card Sorting test (WCST) of executive cognition and the N-Back working memory test. These subjects received a spectroscopy study at rest and a PET cerebral blood flow study during the execution of either task. In patients with schizophrenia, NAA/Cre values specifically in the DLPFC predicted the degree of blood flow activation in an extended working memory network (Figure 3). Most patients were studied after a 2 week period off medications and the patients who did the WCST performed similarly to controls, ruling out non-specific performance-related explanations of the results. Callicott *et al.*[18] analyzed the relationship of NAA/Cre values in the DLPFC and changes in BOLD response during the execution of the N-Back with fMRI. They found an inverse correlation between NAA/Cre in the DLPFC and BOLD activation in patients with schizophrenia: the greater the apparent neuronal pathology (low NAA/Cre) the larger the change in BOLD signal during the N-Back task (Figure 4). The patients also had greater

BOLD responses than the controls, which was interpreted as lower efficiency in processing the demands of the working memory task. This pattern of inefficient prefrontal cortical processing of executive cognition and working memory is thought to be a core biological aspect of schizophrenia.[19] Normal controls showed opposite relationships of performance and fMRI signal than the patients and no relationship of NAA/Cre to fMRI signal. Both patient cohorts (i.e. those who underwent the PET and the fMRI studies) had reduced NAA/Cre in the DLPFC as compared to normal values, however no correlations between reduced NAA/Cre and fMRI signal were present in the hippocampus, which also showed altered fMRI signal.[18] Therefore, although the direction of the abnormality in BOLD signal measured with fMRI and CBF measured with PET is opposite, in both cases the NAA/Cre level specifically in the DLPFC predicted the degree of abnormality in activation of the working memory network in patients with schizophrenia. This suggests that the underlying biology in prefrontal cortex monitored by the NAA measures has specific functional implications for processing of prefrontally critical information.

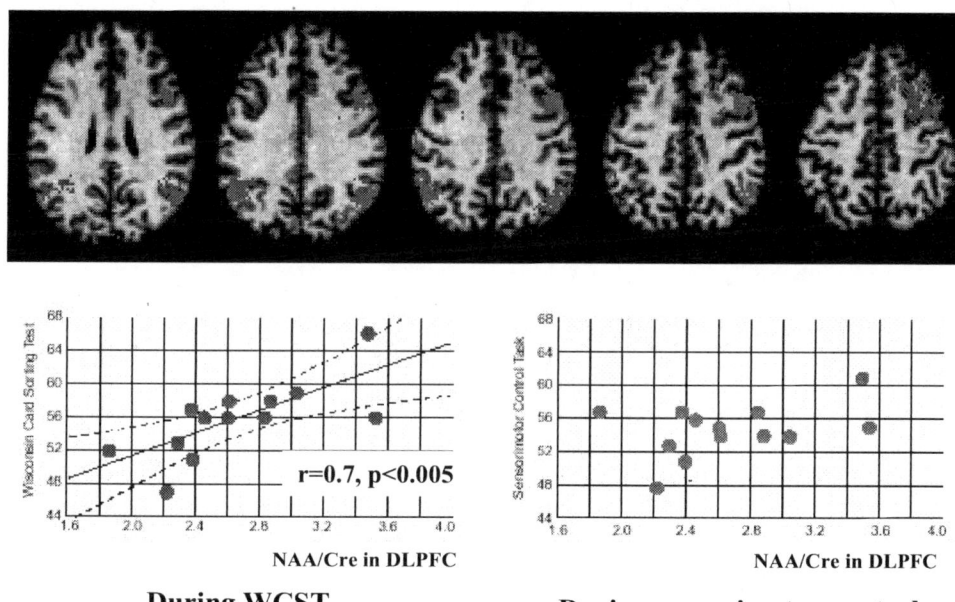

Figure 3. Relationship of NAA/Cre in the DLPFC to blood flow measured with PET. The top panel shows areas where there was a statistically significant correlation between NAA/Cre measured in the DLPFC and blood flow measures during the execution of the WCST in 13 patients with schizophrenia. Multiple areas in a network that has been associated with working memory are highlighted. The graphs below show the correlation between NAA/Cre in the DLPFC and performance on the WCST (on the left) and on a sensorimotor control task. (From Bertolino et al.[17]; see color insert between pages 364 and 365.)

NAA/Cre abnormalities in the DLPFC have also been shown to predict the degree of dopamine release in the basal ganglia as measured by the displacement of C^{11} raclopride by amphetamine in patients with schizophrenia.[20] This is an important finding because it shows that NAA/Cre ratios offer a measurement of the functional status of cortical circuitry that is relevant to dopamine transmission in the basal ganglia, the focus of much attention regarding the dopamine hypothesis of schizophrenia. These data also demonstrated for the first time in human beings evidence that prefrontal cortical function related to the activity of subcortically projecting dopamine neurons. Following the same thread of evidence, Bertolino *et al.*[21] showed that NAA/Cre in the DLPFC correlated negatively with D2 receptor availability in the basal ganglia as measured by the binding potential for IBZM (a iodinated D2 ligand). This correlation was not found in any other region explored and was interpreted to mean that a dysfunction of DLPFC (resulting in low NAA/Cre) was related to lower tonic baseline dopamine levels in the basal ganglia (resulting in higher binding potential of IBZM, a ligand that can be displaced by endogenous dopamine).

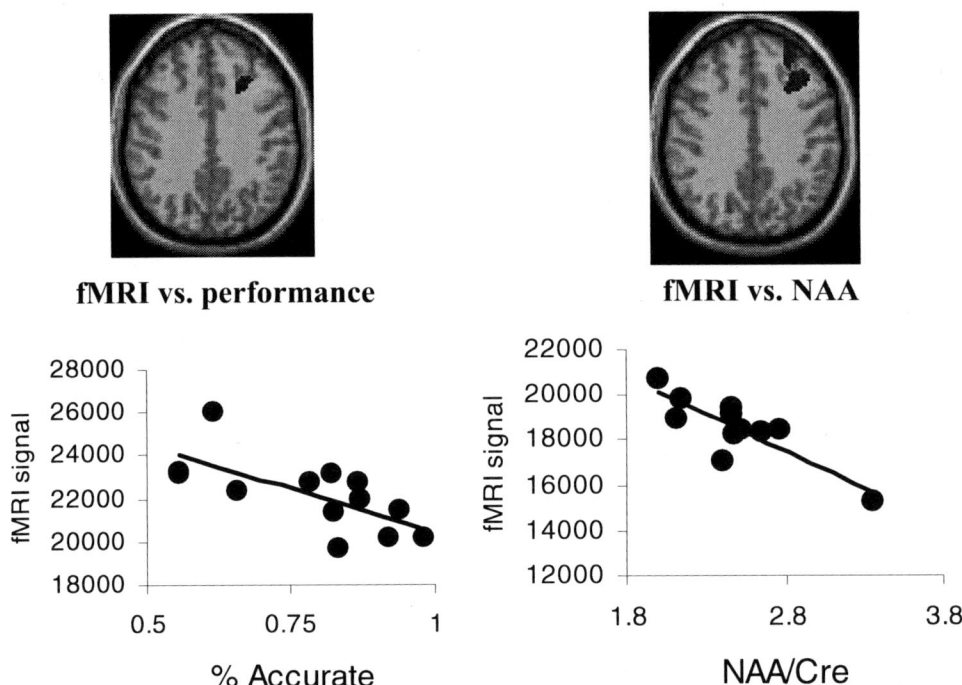

Figure 4. Relationship of NAA/Cre in the DLPFC and fMRI BOLD signal during execution of the N-Back task.The two pictures on the top show which areas of brain have a negative correlation of BOLD signal measured with fMRI during execution of the N-Back task with performance (on the left) and NAA/Cre in the DLPFC (on the right). There is overlap between the areas that correlate with performance and NAA/Cre and all belong to the DLPFC. The graphs show the distribution of the data points for the areas shown above. fMRI signal is in arbitrary units. (From Callicott *et al.*[18]; see color insert between pages 364 and 365.)

The cumulative weight of the in vivo evidence from human studies has also been strengthened by data acquired in animal models of psychosis. One such model consists of producing lesions in the bilateral ventral hippocampi of newborn rats[22] or primates.[23] After this procedure, rats developed a number of behavioral, biological and pharmacological characteristics analogous to those seen in schizophrenia, including hyperactivity in response to amphetamine or PCP only as adults. This hyperactivity responds to neuroleptics.[24] Bertolino et al.[25] demonstrated that neonatal hippocampal lesions in rats lead to reduced NAA/Cre in the DLPFC, but only after the animals reached early adult life. The NAA/Cre ratios were normal in animals prior to puberty despite the hippocampal lesion already being present. Moreover, the alterations in NAA/Cre in the DLPFC of the neonatally lesioned rats[25] were not due to cortical thinning and were accompanied by reductions in the concentration of the mRNA for the excitatory amino-acid receptor (EAAC1). EAAC1 is the glial glutamate transporter in the rat, critically responsible for regulation of synaptic glutamate levels, and its reduction could index alterations in glutamatergic function.

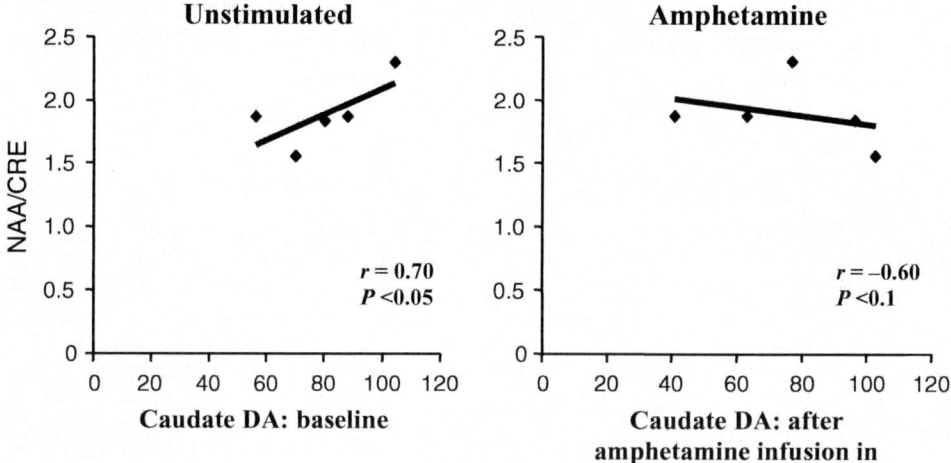

Figure 5. Relationship between DA concentration in the caudate nucleus (measured by microdialysis) and NAA/Cre in DLPFC in monkeys. (From Bertolino et al.[21])

Similarly, when young monkeys received bilateral hippocampal lesions, the NAA/Cre ratios in the DLPFC of adult animals were lower than those of control monkeys, including those with virtually identical lesions produced during adult life.[26] In this study it was only possible to assess adult animals and therefore the full extent of the developmental process could not be observed. Bertolino et al.[21] also measured striatal dopamine levels at baseline with microdialysis in these monkeys and found that it was positively correlated with NAA/Cre in DLPFC, thus confirming the suspected underlying relationship between DLPFC metabolites and dopaminergic tone in the basal ganglia of the nonhuman primate. When amphetamine was administered in the DLPFC of the monkeys (therefore mimicking a strong stimulus related response), the sign of the correlation reversed, with the animals that were lesioned neonatally releasing the greatest amount of dopamine, while having the largest reductions of NAA/Cre (Figure 5). These data, which were remarkably analogous to those observed in the amphetamine/PET experiment in patients, increase further the relevance of NAA reductions as a marker for fundamental pathophysiological alterations in schizophrenia.

The third thrust of our research has been to use NAA as a biologic phenotype to test hypotheses regarding the effect in brain of schizophrenia susceptibility genes. The first demonstration that NAA/Cre can be considered an "intermediate phenotype" for schizophrenia came from Callicott *et al.*,[13] who showed that reductions in NAA/Cre in the medial temporal lobe were also present in healthy, unaffected siblings of patients with schizophrenia, although to a lesser extent, therefore constituting a likely indication of heritability. Egan *et al.*[27,28] have further expanded this notion and linked it to specific genes. For example, brain derived neurotrophic factor (BDNF) contributes significantly to memory performance and to other measures of hippocampal structure and function. Egan *et al.*[27] showed that a functional variation in the gene, a substitution of a single aminoacid (a methionine for a valine) in the BDNF pro-protein, is associated with poorer performance on mnemonic tasks and abnormal hippocampal activation during fMRI. This variation was also associated with altered levels of NAA/Cre in the medial temporal lobe, as seen in figure 6. Normal controls, patients with schizophrenia and their siblings all showed variation associated with genotype, although in the siblings the evidence was not as clear-cut as for the other two groups. Egan *et al.*[27] also showed that this mutation causes changes in the intracellular trafficking of the protein, probably altering its secretion in response to neuronal signaling, thus providing a link between the phenomenological observation of memory deficits in schizophrenia and a biological mechanism of action.

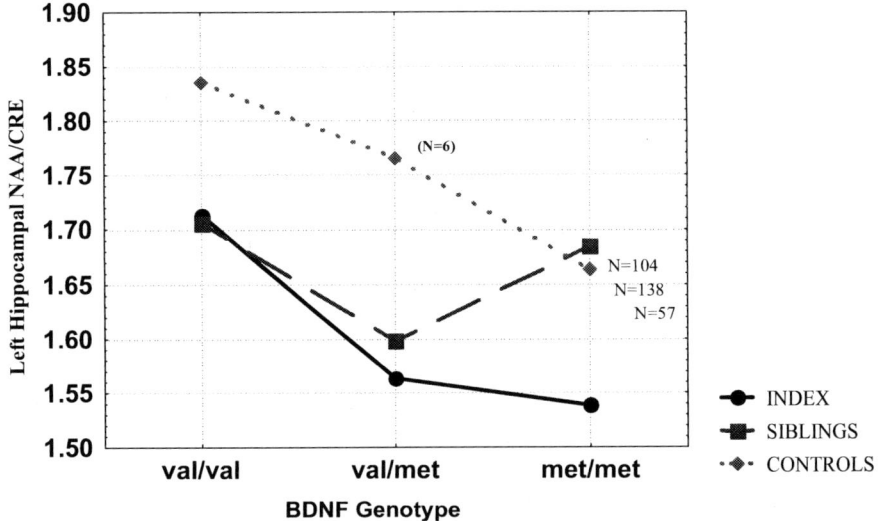

Figure 6. Relationship between BDNF genotype and left medial temporal lobe NAA/Cre. (From: Egan *et al.*[27]). The effect of genotype was $F_{(1,301)}=1.76$, $p<0.02$.

Another genetic variation that has been associated with NAA/Cre is a single nucleotide polymorphism (SNP) in intron 2 of the metabotropic glutamate receptor 3 gene (GRM3). Egan *et al.*[28] reported that the A allele at SNP rs6465084 was overtransmitted in families to offspring with schizophrenia and that it predicted important aspects of prefrontal function ranging from cognitive performance on a verbal fluency task and a memory task that require prefrontal and medial temporal processing, to cerebral activation measured with fMRI. Also MRSI provided evidence in this same direction: people carrying the A allele had lower NAA/Cre ratios in the DLPFC than G carriers. Similar effects were found across all groups studied (normal controls, patients with schizophrenia and their siblings), but, as also true for BDNF, this effect was weakest in the siblings and strongest in the normal controls (Figure 7). These results were recently confirmed by Steele *et al.*[29] using 3T methodology in a new group of normal controls, as shown in figure 8.

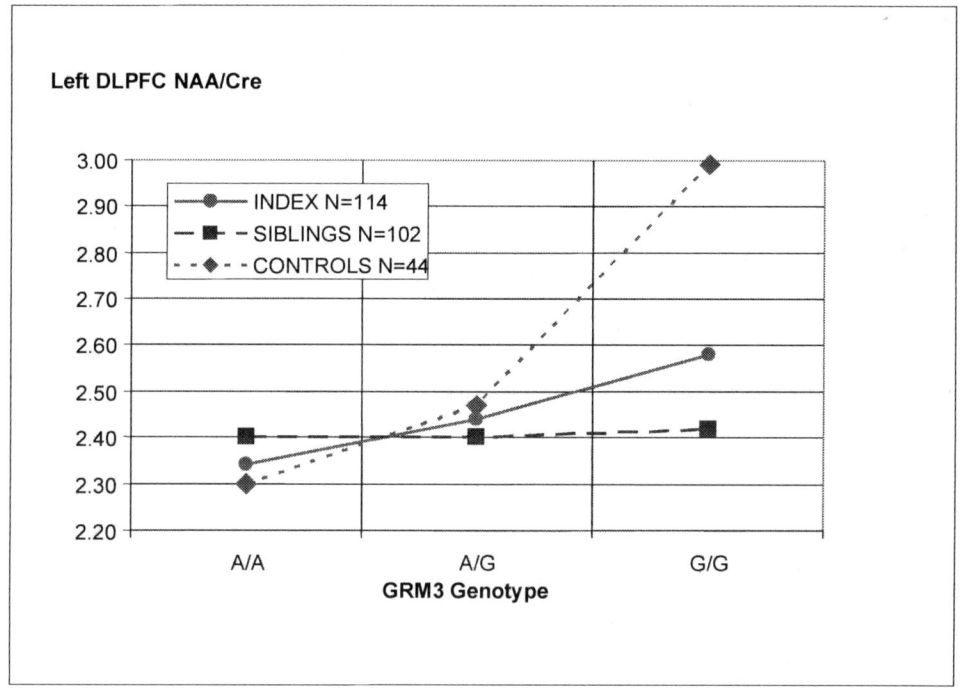

Figure 7. Effect of GRM3 genotype at rs6465084 on NAA/Cre ratios in the Left DLPFC. These data were acquired at 1.5T. (From Egan *et al.*[28]).

4. CONCLUSION

In conclusion, the work done in the Clinical Brain Disorders Branch in the last eight years has provided rich evidence demonstrating the heuristic value of measurements of NAA with MRSI. Although the functional role of this molecule is still debated (as discussed in detail in other contributions to this symposium proceedings), and likely interacts with several metabolic pathways, clearly its measurement provides information related to brain function. The work described above shows convincingly that NAA in the medial temporal lobe and possibly in the DLPFC is reduced in schizophrenia and that this reduction also predicts alterations in cerebral blood flow measured with PET and fMRI. Moreover, NAA levels are a seemingly heritable trait and covary with variations of the BDNF and GRM3 genes that may contribute to the cognitive dysfunction that constitutes one of the cardinal features of schizophrenia. The link between intermediate phenotypes at the imaging level and biologic mechanisms provided by the discovery of genes that contribute to schizophrenia risk is a critical step in understanding the pathophysiology of this illness, and measurements of NAA have helped to sort out pieces of this puzzle.

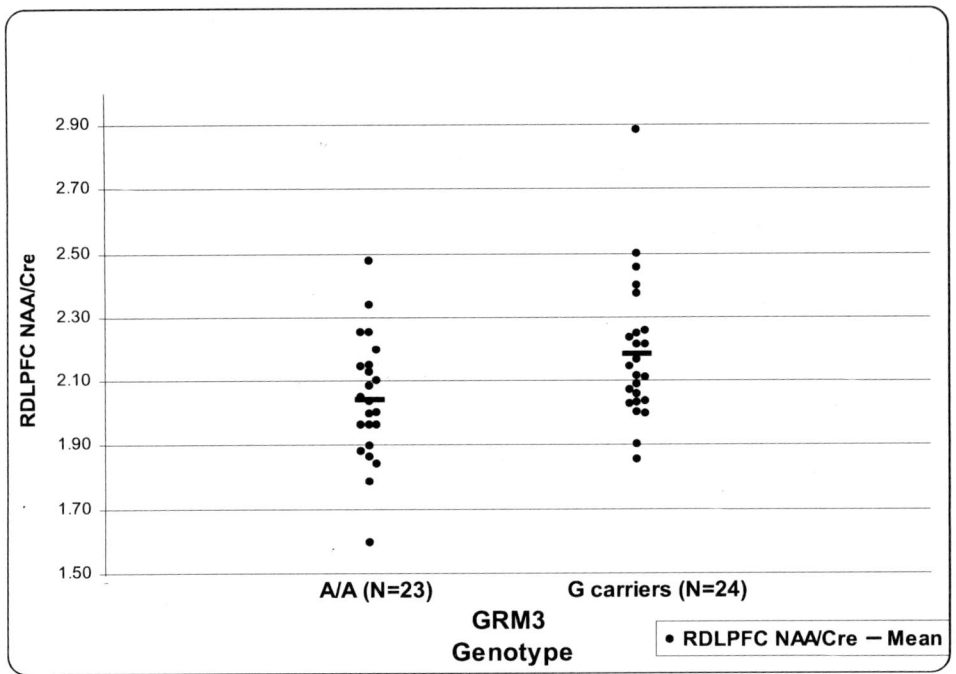

Figure 8. Effect of GRM3 genotype at at rs6465084 on NAA/Cre ratios in Right DLPFC. These data were acquired at 3T in normal controls. (From Steele *et al.*[29]).

5. QUESTION AND ANSWER SESSION

DR. DUYN: Thank you, Dr. Weinberger. We have time for a couple of questions.

DR. WEINER: It is a very beautiful story you tell. You repeatedly spoke about NAA concentrations and NAA signals, but in fact your data was almost all NAA ratios, and those ratios can change, because the creatine changes, for example. So to what extent have you tried to pursue that, to demonstrate that it is, in fact, the NAA and not some other metabolite changing?

DR. WEINBERGER: One of the things we have done in all these studies is that we have looked at NAA, not just NAA /creatine, but we have looked at NAA relative to NAA in other regions. So we have used NAA in the region of interest ratioed to NAA in other regions, which has always supported the finding.

So we don't think the data are explained by creatine variation, and in general -- in general, if we ratio to choline, the data are pretty much the same. But when we ratioed NAA in other regions, it is pretty much also the same data. So we think it is the NAA signals.

We have experimented somewhat using the water signal as a reference in the water acquired datasets, but we haven't really had this extensive a dataset to look at with the water acquired signals.

Let me just make one other point. Particularly in the genetic datasets where we have three population studies, all showing virtually the same relationships, it becomes harder to associate these changes to the denominator across all these datasets; because there are different conditions. They are obviously different methodological factors, but the absolute quantification has not been done.

DR. MATALON: Yes. I have a question regarding the NAA and other metabolites that you measured and were unaffected. I am interested, really, in metabolites that you did not measure, because NAA, to me, is a marker, and what is left to be done?

DR. WEINBERGER: Well, I mean, you know, you are absolutely correct. These are crude measures of a peak which probably is hiding other signals within it, and we would love to be able to measure NAAG, and we are hoping that with some of the newer, more sensitive methods, shorter echo times and some of the editing strategies that are available now, we will be able to resolve the NAAG signal.

We are looking with short echo time acquisitions now and less imaging based strategies at a number of other signals in the proton spectra. We have not acquired those kinds of data with these strategies, but clearly, there is a lot more biochemical information in the proton spectra than just these measurements.

You know, let me just say that, from our perspective, we actually looked at several genes in the NAA biosynthetic and degradation pathway as possible measures of these signals. We don't find associations to those genes in our datasets.

We don't really believe that we are dealing with a disorder of NAA metabolism. We believe, and these studies have all been structured around the idea, that this is not a biochemical experiment. This is an experiment using a proxy measure of something about the biology of this region of the brain.

It may well be that this is an underestimation of what the biochemical signals could tell us, but that has been basically the strategy that has been used so far.

DR. ROSS: Let me just address the issue of the heterogeneity -- a beautiful presentation, very convincing -- heterogeneity in the literature regarding NAA and schizophrenia, and you refer to a poster which is actually not displayed outside. It is carefully disguised, because we don't want to disagree with you.

DR. WEINBERGER: It has happened before, trust me.

DR. ROSS: The heterogeneity actually resides in the response of Peter Barker to my question. The standard deviation of measurements in Jeff Duyn's elegant technique is between 10 and 15 percent.

In my casting an eye on the data here, and having looked at some of your published data, the difference you are looking for in the dorsal prelateral -- however you pronounce it -- cortex is 12 percent max. So we are bound to find differences when we go to other techniques and other laboratories, and difficulties in anatomical placements.

So I don't believe there is a disagreement at that level. There does need to be some standardization, and we heard that also from Peter. But probably the take-home message is that schizophrenia does not give one diagnosable MRS data, unlike Alzheimer's Disease which we will hear from many speakers.

I think that is the only point here. So I agree thoroughly with your point. This is not a biochemical statement about a metabolic mechanism. This is a surrogate, and you demonstrate that very beautifully.

DR. WEINBERGER: I completely agree with you. These are -- at the level of contrasting groups, these are very weak signals. I mean, there is no question that the difference between schizophrenics and normals in these measures is very small, which is one of the reasons I think, in fact, there has been -- you know, there is a limit to the enthusiasm one can have about making the measure as a signature finding. It is, obviously, a finding that may have some predictive value in other aspects of the illness, but it is certainly not diagnostic, and it is certainly not dramatic.

DR. NAMBOODIRI: One comment regarding the mGluR3 and NAA. You note that NAAG is a ligand for the mGluR3. Do you think that there is something more to that than just mere correlation?

DR. WEINBERGER: That is one of the reasons we obviously looked at NAA as one of our target phenotypes for GRM3. I think, if we could get an NAAG peak out of these data, it would be extremely interesting to look at that.

My own personal guess is -- again given the limitations of resolution, the variance in the measurements, my own personal guess is we are measuring an integrative characteristic of probably the synaptic abundance and, therefore, oxidative phosphorylation related to synaptic activity of this region. This is my guess of why these measurements have any predictive value.

I don't think we are measuring something specific about NAA processing or about NAAG, GRM3 signaling. My own personal bias is -- and this is just based on the fact that these data work -- is that we are really measuring some integrative characteristic of the overall synaptic activity of this region, which indirectly is critically related to oxidative phosphorylation in mitochondria, because that is what is necessary for synaptic activity.

I think BDNF works, because it determines synaptic abundance, and I think GRM3 has to do with excitatory activity, and excitatory activity is critical for spine density, you know, synaptic abundance, and mitochondrial activity. So that is my guess about why these things work.

DR. MATALON: I just want to make a little comment to see what you think.

To me, I can say that, because I am not an MRI or MRS expert. So you can think this guy doesn't know what he is talking about. I don't mind that. I don't know.

To me, MRI and MRS are for compounds in high concentrations, macro concentrations. To say that we cannot see other things, a micro nutrient or micro components that may be extremely important -- we are neglecting them. I think this should be said right at the beginning, and what do you think about that?

DR. WEINBERGER: I mean, you know, if we can get -- I mean, I think we would like to be able to make more definitive measurements. I think it would be a great advance. There is a lot of interest in doing that.

You know, there are always limitations based on sensitivity and resolution. You know, we are taking a relatively large piece of brain which has functional differentiation and sticking it in a voxel, as if it is homogeneous. You know, there is a lot of error in these things, but I think it would be great if we could refine the biochemical resolution.

DR. PHILIPPART: You have mentioned the blood flow. I wonder if you could tell us more about what is known about blood flow, which is, after all, you know, a crucial way of measuring brain function.

DR. WEINBERGER: Well, again I think that is a whole other symposium, and there have been many. But our application of blood flow measurements was as another physiological assay of neuronal function. So the blood flow measurements I showed you were acquired during the specific cognitive processing of a working memory task and a sensory motor control task.

So presumably, these were activation data. So presumably, the response at the blood flow level of contrasting one state to another state and the regionality of those patterns reflects something of the engagement of the neuronal circuitry and the function in the circuitry.

So here, blood flow was presumably a measure that indirectly monitors neuronal activity. Actually, there have been studies now -- we have actually done studies with magnetoencephalography, looking at exactly the same task where we are not monitoring blood flow or BOLD response. We get the same patterns of activation.

DR. DUYN: Are there further questions for Dr. Weinberger? I thank all the speakers for their work here.

6. REFERENCES

1.   Weinberger DR, Laruelle M: Neurochemical and neuropharmacologic imaging in schizophrenia, in *Neuropsychopharmacology: The Fifth Generation of Progress.* Edited by Davis KL, Charney D, Coyle JT, Nemeroff C. Philadelphia, PA, Lippincott Williams & Wilkins, 2002, pp 833-855

2.   Bertolino A, Weinberger DR: Proton magnetic resonance spectroscopy in schizophrenia. *Eur J Radiol* 1999; **30**(2):132-41

3.   Marenco S, Bertolino A, Weinberger DR: *MR proton spectroscopy, in Schizophrenia: from neuroimaging to neuroscience,* vol 1. Edited by Lawrie SM, Johnstone EC, Weinberger DR. Oxford, UK, Oxford University Press, 2004, pp 73-92

4.   Keshavan MS, Stanley JA, Pettegrew JW: Magnetic resonance spectroscopy in schizophrenia: methodological issues and findings--part II. *Biol Psychiatry* 2000; **48**(5):369-80

5.   Stanley JA, Pettegrew JW, Keshavan MS: Magnetic resonance spectroscopy in schizophrenia: methodological issues and findings--part I. *Biol Psychiatry* 2000; **48**(5):357-68

6.   Duyn JH, Gillen J, Sobering G, van Zijl PC, Moonen CT: Multisection proton MR spectroscopic imaging of the brain. *Radiology* 1993; **188**(1):277-82

7.   van Der Veen JW, Weinberger DR, Tedeschi G, Frank JA, Duyn JH: Proton MR spectroscopic imaging without water suppression. *Radiology* 2000; **217**(1):296-300

8.   Weinberger DR: Cell biology of the hippocampal formation in schizophrenia. *Biol Psychiatry* 1999; **45**(4):395-402

9.   Weinberger DR, Egan MF, Bertolino A, Callicott JH, Mattay VS, Lipska BK, Berman KF, Goldberg TE: Prefrontal neurons and the genetics of schizophrenia. *Biol Psychiatry* 2001; **50**(11):825-44

10.   Bertolino A, Nawroz S, Mattay VS, Barnett AS, Duyn JH, Moonen CT, Frank JA, Tedeschi G, Weinberger DR: Regionally specific pattern of neurochemical pathology in schizophrenia as assessed by multislice proton magnetic resonance spectroscopic imaging. *Am J Psychiatry* 1996; **153**(12):1554-63

11.   Bertolino A, Callicott JH, Nawroz S, Mattay VS, Duyn JH, Tedeschi G, Frank JA, Weinberger DR: Reproducibility of proton magnetic resonance spectroscopic imaging in patients with schizophrenia. *Neuropsychopharmacology* 1998; **18**(1):1-9

12.   Bertolino A, Kumra S, Callicott JH, Mattay VS, Lestz RM, Jacobsen L, Barnett IS, Duyn JH, Frank JA, Rapoport JL, Weinberger DR: Common pattern of cortical pathology in childhood-onset and adult-onset schizophrenia as identified by proton magnetic resonance spectroscopic imaging. *Am J Psychiatry* 1998; **155**(10):1376-83

13. Callicott JH, Egan MF, Bertolino A, Mattay VS, Langheim FJ, Frank JA, Weinberger DR: Hippocampal N-acetyl aspartate in unaffected siblings of patients with schizophrenia: a possible intermediate neurobiological phenotype. *Biol Psychiatry* 1998; 44(10):941-50

14. Callicott JH, Bertolino A, Egan MF, Mattay VS, Langheim FJ, Weinberger DR: Selective relationship between prefrontal N-acetylaspartate measures and negative symptoms in schizophrenia. *Am J Psychiatry* 2000; 157(10):1646-51

15. Lewis DA, Lieberman JA: Catching up on schizophrenia: natural history and neurobiology. *Neuron* 2000; 28(2):325-34

16. Bertolino A, Callicott JH, Mattay VS, Weidenhammer KM, Rakow R, Egan MF, Weinberger DR: The effect of treatment with antipsychotic drugs on brain N-acetylaspartate measures in patients with schizophrenia. *Biol Psychiatry* 2001; 49(1):39-46

17. Bertolino A, Esposito G, Callicott JH, Mattay VS, Van Horn JD, Frank JA, Berman KF, Weinberger DR: Specific relationship between prefrontal neuronal N-acetylaspartate and activation of the working memory cortical network in schizophrenia. *Am J Psychiatry* 2000; 157(1):26-33

18. Callicott JH, Bertolino A, Mattay VS, Langheim FJ, Duyn J, Coppola R, Goldberg TE, Weinberger DR: Physiological dysfunction of the dorsolateral prefrontal cortex in schizophrenia revisited. *Cereb Cortex* 2000; 10(11):1078-92

19. Winterer G, Weinberger DR: Genes, dopamine and cortical signal-to-noise ratio in schizophrenia. *Trends Neurosci* 2004; 27(11):683-90

20. Bertolino A, Breier A, Callicott JH, Adler C, Mattay VS, Shapiro M, Frank JA, Pickar D, Weinberger DR: The relationship between dorsolateral prefrontal neuronal N-acetylaspartate and evoked release of striatal dopamine in schizophrenia. *Neuropsychopharmacology* 2000; 22(2):125-32

21. Bertolino A, Knable MB, Saunders RC, Callicott JH, Kolachana B, Mattay VS, Bachevalier J, Frank JA, Egan M, Weinberger DR: The relationship between dorsolateral prefrontal N-acetylaspartate measures and striatal dopamine activity in schizophrenia. *Biol Psychiatry* 1999; 45(6):660-7

22. Lipska BK, Jaskiw GE, Weinberger DR: Postpubertal emergence of hyperresponsiveness to stress and to amphetamine after neonatal excitotoxic hippocampal damage: a potential animal model of schizophrenia. *Neuropsychopharmacology* 1993; 9(1):67-75

23. Saunders RC, Kolachana BS, Bachevalier J, Weinberger DR: Neonatal lesions of the medial temporal lobe disrupt prefrontal cortical regulation of striatal dopamine. Nature 1998; 393(6681):169-71

24. Lipska BK, Weinberger DR: To model a psychiatric disorder in animals: schizophrenia as a reality test. *Neuropsychopharmacology* 2000; 23(3):223-39

25. Bertolino A, Roffman JL, Lipska BK, van Gelderen P, Olson A, Weinberger DR: Reduced N-acetylaspartate in prefrontal cortex of adult rats with neonatal hippocampal damage. *Cereb Cortex* 2002; 12(9):983-90

26. Bertolino A, Saunders RC, Mattay VS, Bachevalier J, Frank JA, Weinberger DR: Altered development of prefrontal neurons in rhesus monkeys with neonatal mesial temporo-limbic lesions: a proton magnetic resonance spectroscopic imaging study. *Cereb Cortex* 1997; 7(8):740-8

27. Egan MF, Kojima M, Callicott JH, Goldberg TE, Kolachana BS, Bertolino A, Zaitsev E, Gold B, Goldman D, Dean M, Lu B, Weinberger DR: The BDNF val66met polymorphism affects activity-dependent secretion of BDNF and human memory and hippocampal function. Cell 2003; 112(2):257-69

28. Egan MF, Straub RE, Goldberg TE, Yakub I, Callicott JH, Hariri AR, Mattay VS, Bertolino A, Hyde TM, Shannon-Weickert C, Akil M, Crook J, Vakkalanka RK, Balkissoon R, Gibbs RA, Kleinman JE, Weinberger DR: Variation in GRM3 affects cognition, prefrontal glutamate, and risk for schizophrenia. *Proc Natl Acad Sci U S A* 2004; 101(34):12604-9

29. Steele SU, Marenco S, Barnett AS, van der Veen JW, Egan MF, Weinberger DR: GRM3 genotype is associated with reduced N-acetyl aspartate levels in the frontal cortex: A partial replication. *Biological Psychiatry* 2004; 55:167S-167S

N-ACETYLASPARTATE AS A MARKER OF NEURONAL INJURY IN NEURODEGENERATIVE DISEASE

Norbert Schuff, Dieter J. Meyerhoff, Susanne Mueller, Linda Chao, Diana Truran Sacrey, Kenneth Laxer and Michael W. Weiner*

1. NAA IN NORMAL AGING

Considerable evidence suggests that normal aging is associated with gradual impairment of memory functioning [1]. The medial temporal lobe, especially the hippocampus, plays a central role in declarative memory processing [2]. However, magnetic resonance imaging (MRI) studies have produced controversial results concerning the age-related hippocampal volume loss, which could be due in part to the non-specificity of volume shrinkage as an indicator for neuron loss. In contrast to volume, NAA is generally considered a marker for viable neurons, because NAA reaches detectable concentrations only in neuronal tissue but not in other brain tissues, including glial cells. Using proton magnetic resonance spectroscopic imaging (^{1}H MRSI) and MRI together, we studied hippocampal metabolites and volumes in 24 healthy adults from 36 to 85 years of age. Our goals were to test whether NAA levels vary in the hippocampus as a function of normal aging and 2) to determine the relationship between hippocampal NAA and volume changes. We found NAA/Cho ratios decreased by 24% (r = -0.53, p = 0.01) and NAA/Cr ratios decreased by 26% (r = -0.61, p < 0.005) over the age range studied, while Cho/Cr remained stable, implying diminished NAA levels. In the same population, hippocampal volume shrank by 20% (r = -0.64, p < 0.05). The relationships of these measures with aging are depicted in Figure 1. Since NAA is considered a marker

---

* Magnetic Resonance Unit VA Medical Center, Department of Radiology, University of California, San Francisco, CA 94121 USA. Email: michael.weiner@radiology.ucsf.edu.

of neurons, these results provide stronger support for neuron loss in the aging hippocampus than volume measurements by MRI alone.

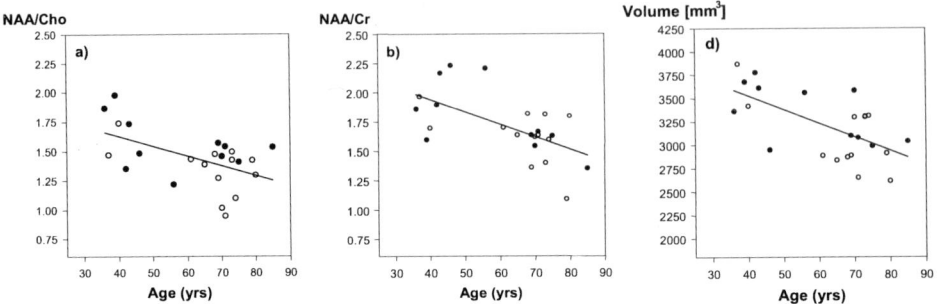

Figure 1. Age-related changes in metabolite concentrations and volume in hippocampus.

Since contributions to the NAA signal may arise from both gray and white matter tissue, it is critical to differentiate between metabolite changes of gray and white matter, and other tissue types. One approach for differentiation is the use of linear regressions to predict the relationship between metabolite intensity changes and gray/white matter variations in MRSI voxels (4,8-12). However, most previous MRSI studies that used linear regression averaged metabolite concentrations over different lobes of the brain, ignoring regional variations. In addition, tissues other than gray and white matter were ignored or not determined, such as white matter lesions, which occur frequently in the aged brain. We developed an approach for obtaining metabolite concentrations of gray, white matter, as well as of white matter lesions in different lobes of the brain using linear regression. We applied the new technique to measure NAA concentrations in the frontal and the parietal lobe in 40 normal elderly subjects (56 to 89 years, mean age 74 ± 8, 22 female, 18 male). NAA was about 15% lower in cortical gray matter and 23% lower in white matter lesions when compared to normal white matter. Cr was 11% higher in cortical gray than in white matter, and also about 15% higher in the parietal cortex compared to the frontal cortex. Cho was 28% lower in cortical gray matter than in white matter. Furthermore, NAA and Cr changes correlated with age. The results suggest that in addition to the hippocampus, age-related neuronal changes can also occur in cortical regions, while white matter regions seem to be spared.

2. NAA IN DEMENTIA

2.1 Alzheimer's Disease

The first ^{1}H MRSI study on AD from this laboratory [8] showed abnormalities of metabolite ratios of NAA, and choline (Cho) and creatine (Cr) containing compounds in white and gray matter of the centrum semiovale. Decreased levels of NAA in white matter suggest diffuse axonal loss or damage. Decreased NAA in gray matter suggests loss of neurons, while increased Cho may result from membrane breakdown products.

The second study from this laboratory [9] using similar ^{1}H MRSI methods extended the observation of metabolic abnormalities in the centrum semiovale to a larger population of AD patients and controls, and in addition, included a group of patients with subcortical ischemic vascular dementia (SIVD). While the findings from the AD and control groups were similar to the first study, different metabolite changes were noted in SIVD patients, supporting the possibility that ^{1}H MRSI may provide information to differentiate AD from SIVD. Finally, the third report from this laboratory [10] investigated the extent to which these metabolic differences between patients and controls were independent of variations in the tissue composition of the MRSI voxels (e.g. enclosed amounts of gray matter, white matter, and WM signal hyperintensities). This analysis was made possible by using information from MRI tissue segmentation coregistered with the ^{1}H MRSI data. Although the analysis revealed significant variations of the tissue composition in the regions of interest, these changes did not contribute significantly to the metabolite differences, indicating that reduced NAA/Cho and increased Cho/Cr in posterior mesial gray matter of AD were not simply an artifact of these structural variations. However, limitations of these previous studies were a relatively small number of subjects and an early MRI technology with relatively (compared to today's standards) thick slices (5 mm) and interslice gaps (0.5 mm), which compromised accuracy of tissue segmentation. Furthermore, acquisition of the ^{1}H MRSI data was performed at a relatively long spin-echo time (TE) of 272ms and further metabolite ratios rather than concentrations were reported. Subsequent MRSI studies from this lab tried to overcome these limitations.

We subsequently studied 28 AD patients and 22 healthy elderly using ^{1}H MRSI and MRI for image segmentation. ^{1}H MRSI data were aligned with MRI segmentation data to obtain volume-corrected metabolite concentrations. The results were: NAA levels were significantly reduced in frontal and posterior mesial cortex of AD, consistent with previous results from NAA ratios. Furthermore, the NAA reductions were independent from structural variations as measured by MRI, and in parietal mesial cortex correlated mildly with dementia severity. But NAA combined with MRI measures did not improve discrimination power for AD over that of MRI alone. While these results from fronto-parietal brain showed reduced NAA in AD is not an artifact of underlying structural variations, and thus may provide useful information in addition to MRI, these NAA reductions were of limited use for diagnosis of AD.

Since the hippocampus is thought to be earlier involved in AD pathology than cortical regions, we tested in another MRSI study the hypothesis that hippocampal NAA and volume used together provide greater discrimination between AD and normal elderly than does either measure alone. We used proton magnetic resonance spectroscopic imaging (^{1}H MRSI) and tissue segmented and volumetric MR images to measure atrophy corrected hippocampal NAA and volumes in 12 AD patients (mild to moderate severity) and 17 control subjects of comparable age. In AD, atrophy corrected NAA from the hippocampal region was reduced by 15.5% on the right and 16.2% on the left (both p<0.003), and hippocampal volumes were smaller by 20.1% (p < 0.003) on the right and 21.8% (p < 0.001) on the left when compared to control subjects. The NAA reductions and volume losses made independent contributions to the discrimination of AD from control subjects. When used separately, neither hippocampal NAA nor volume achieved to correctly classify AD patients better than 80%. When used together, however, the two measures correctly classified 90% of AD and 94% of control subjects. The separation between AD and controls using NAA and volume of the hippocampus together is depicted in Figure 2. In conclusion, hippocampal NAA measured by ^{1}H MRSI combined with quantitative measurements of hippocampal atrophy by MRI may improve diagnosis of AD.

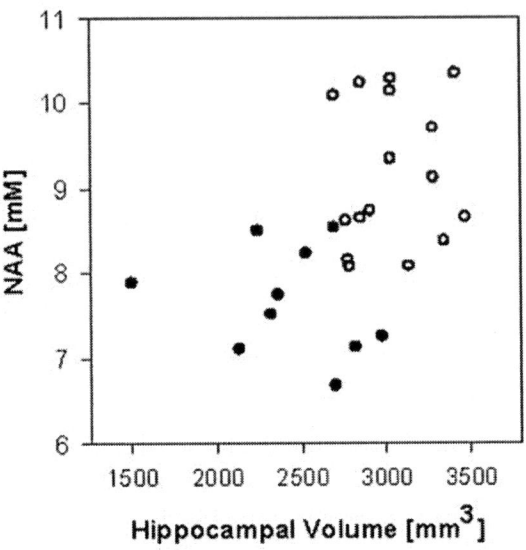

Figure 2. Comparison of hippocampal volume and NAA levels in Alzheimer's patients (filled circles) and control subjects (open circles).

2.2. Subcortial Ischemic Vascular Dementia

The contribution of subcortical ischemic vascular disease (SIVD) to cognitive impairment and dementia is poorly understood. Disruption of subcortical-cortical connections by strategically located infarctions are considered one important mechanism for cognitive impairment in SIVD.[1] MRI studies from our group found that cognitive deficits in patients with SIVD were strongly correlated with brain atrophy, while subcortical infarctions made little contributions.[2, 3] Aside from greater numbers of ischemic lesions and more extensive white matter hyperintensities (WMH), other MRI studies comparing SIVD and AD have shown atrophy in the hippocampus and entorhinal cortex in both dementias, though less prominent in SIVD than in AD.[4, 5] Furthermore, some SIVD patients without AD pathology confirmed by autopsy had reduced hippocampal and cortical gray matter volumes.[2] Therefore, MRI has limited ability to differentiate between SIVD and AD. Although we found a stereotypical regional pattern of NAA losses in AD that involved the hippocampus and parietal gray matter, but not frontal gray matter and white matter, regional differences in NAA reductions between SIVD and AD have not thoroughly been investigated before. Therefore, in a new MRSI study that included 13 SIVD patients (71 ± 8 years old), 43 AD patients of comparable age and dementia severity to SIVD, and 52 cognitively normal subjects with and without lacunes, we sought to determine the regional pattern of NAA in gray matter, white matter, and WMH in SIVD. We found that compared to controls, SIVD patients had lower NAA by 18% (p < 0.001) in frontal cortex and by 27% (p < 0.003) in parietal cortex, but no significant NAA reduction in white matter and medial temporal lobe. Compared to AD, SIVD patients had lower NAA by 13% (p < 0.02) in frontal cortex and by 20% (p < 0.002) in left parietal cortex. Cortical NAA decreased in SIVD with increasing white matter lesions (r = 0.54, p < 0.02) and number of lacunes (r=0.59, p < 0.02). In particular, thalamic lacunes were associated with greater NAA reduction in frontal cortex than lacunes outside the thalamus (p < 0.02) across groups, after adjusting for cognitive impairments, as shown in Figure 3.

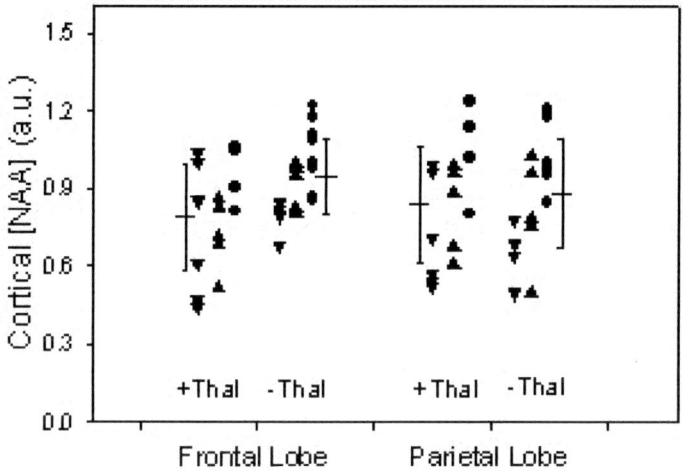

Figure 3. Thalamic lacunes associated with decreases in NAA

    The finding in SIVD that [NAA] losses in cortical regions correlated with subcortical infarction load and WMH, implies that subcortical vascular disease is responsible for cortical changes in SIVD. There are several possible explanations for this finding. First, and most likely, is that subcortical infarctions cause functional deafferentiation of the cerebral cortex, sometimes termed subcortical-cortical diaschisis. This is consistent with PET studies showing in SIVD hypometabolism and hypoperfusion in cortical regions,[6] especially in the frontal lobe.[7] In support of this view, MRS studies on animals have shown trans-synaptic decrease of NAA levels following acute deafferentation without neuronal loss.[8] It is therefore conceivable that the cortical [NAA] losses in SIVD could indicate deafferented neurons in a state of functional inactivity with a possibility for recovery rather than frank neuron loss. A second explanation is that neurons are damaged or lost via transneuronal degeneration, secondary to subcortical infarctions.[9] With the assumption that [NAA] reflects neuron density, the results imply further that a secondary degeneration causes disproportionately greater loss of neuronal than to non-neuronal cells.
    A third possibility is that cortical [NAA] reductions are due to cortical ischemia, with or without micro-infarctions in the cortex, undetectable with MRI. There are two arguments against this view: First, quite remarkably, we found no [NAA] reduction in

white matter in SIVD (or in AD). It would seem reasonable that a widespread ischemic process, which affected cortex and caused subcortical infarctions and WMH would also produce [NAA] reduction in white matter. Second, [NAA] reductions can occure preferentially in the frontal lobe, as the comparison between subjects with and without thalamic lacunes showed. In presence of a generalized ischemic process, however, one would expect that the entire cortex is affected and not the frontal region singled out. Eventually, it will be necessary to obtain autopsy information to exclude micro-infarctions and concurrent AD as potential cause of cortical [NAA] reduction in SIVD.

2.3 Reduced Medial Temporal Lobe NAA in CIND

We have previously showed that AD patients have significantly less NAA concentration in the medial temporal lobe (MTL) and parietal lobe gray matter (GM) than cognitively normal subjects.[10] This study sought to determine whether cognitively impaired but non-demented (CIND) elderly individuals who are at risk for developing dementia exhibit a similar pattern of reduced NAA in the MTL and parietal lobe GM. In addition, we also compare regional NAA patterns and hippocampal volumes in CIND patients who remained cognitively stable with those who later became demented during follow-up (mean follow-up duration: 3.6 ± 1.7 years; range: 1-7 years). Seventeen CIND patients (mean age: 75.4 ± 6.8 years), 24 AD patients (mean age: 74.8 ± 6.9 years), and 24 cognitively normal subjects (mean age: 76.0 ± 6.3 years) were studied using MRSI and MRI. There were no significant hippocampal volume differences between CIND patients and cognitively normal subjects. However, CIND subjects had 21% less MTL NAA ($p = 0.005$) than controls. Moreover, dichotomizing CIND patients revealed greater MTL NAA reductions in patients who later became demented than patients who remained cognitively stable during follow-up. Together, these results suggest that NAA reduction in the MTL can be detected in the absence of significant hippocampal atrophy and before the development of dementia. Thus, MTL NAA could potentially serve as an early marker for AD.

3. NAA IN POST TRAUMATIC STRESS DISORDER (PTSD)

Posttraumatic stress disorder (PTSD) is characterized by exposure to markedly distressing traumatic event(s), re-experiencing symptoms, emotional numbing, and increased arousal. Biological alterations include adrenergic hyperresponsiveness [11], increased thyroid activity [12], low cortisol levels, and increased negative feedback sensitivity of the hypothalamic-pituitary-adrenal (HPA) axis following low-dose dexamethasone administration [13]. In addition, magnetic resonance imaging (MRI) studies reported decreased volumes of the hippocampus in both, Vietnam combat veterans [14, 15] and noncombat trauma victims [16, 17] with PTSD. However, laterality was inconsistent across these MRI studies, with volume decreases being reported in the right, the left, and both hippocampi. In a preliminary ^{1}H MRSI study [18], we found decreased hippocampal NAA in a small number of veterans with PTSD, many of whom had been recently abusing alcohol, compared to healthy controls without a history of alcohol abuse. Another ^{1}H MRS study reported NAA reductions in medial temporal lobe structures of veteran PTSD subjects [19]. Therefore, in a new MRSI study on a new group of PTSD subjects we sought to determine if ^{1}H MRSI measurements could detect NAA changes in the hippocampus of PTSD, separate from volume changes.

Eighteen male patients with combat-related PTSD (mean age 51.2 ± 2.5 years) and 19 male control subjects (mean age 51.8 ± 3.2) were studied using MRI and Proton MR spectroscopic imaging. Both groups had no alcohol and drug abuse during the past 5 years. PTSD and control subjects had similar volumes of hippocampus and entorhinal cortex. We found NAA was significantly reduced by about 23% and creatine containing compounds were reduced between 11% and 26% bilaterally in the hippocampus of PTSD when compared to control subjects (Table 1). However, there were no significant differences in hippocampal or ERC volumes between PTSD patients without recent history of alcohol abuse and control subjects. This contrasted previous reports of hippocampal atrophy in PTSD, suggesting that alcohol abuse may have been at least in part responsible for these previous findings.

Table 1. NAA reductions in PTSD patients.

Metabolites	PTSD	Control	Difference[b]	Effect Size
Left-NAA	2.8 ± 0.8	3.7 ± 0.8	-24*	1.12
Right-NAA	2.9 ± 0.9	3.8 ± 0.7	-23*	1.12
Hippocampal tissue[c]	0.4 ± 0.7	0.4 ± 0.9	0	

[a]Concentrations in arbitrary units
[b]Difference in percent compared to controls
[c]Fraction of hippocampal to total gray and white matter tissue within a MRSI voxel
*$p < 0.01$

The finding of 23-24 % reduction of hippocampal NAA in PTSD subjects of this study, in the absence of hippocampal volume loss, were surprising for two reasons. First, the magnitude of the NAA reduction is very similar to that, which we previously reported for patients with Alzheimer's dementia [20]. Notwithstanding this similarity of NAA reductions, our PTSD patients showed no gross cognitive memory impairments. Second, hippocampal NAA reductions in Alzheimer's disease were accompanied by substantial hippocampal volume losses in the range from 20% to 40% [20]. Another explanation for NAA decrease in PTSD is impaired metabolism of neuronal processes, resulting in secondary NAA loss. Reversible NAA losses have been found in amyotrophic lateral sclerosis [21], epilepsy after surgery [22], in multiple sclerosis [23], and more recently in schizophrenic patients after treatment with antipsychotics [24, 25]. These changes have been attributed to reversible impairment of oxidative metabolism of which NAA is a product. Therefore, we cautiously interpret the finding of decreased hippocampal NAA in PTSD to reflect either neuron loss in the presence of gliosis and/or neuronal metabolic impairments. In this regard, NAA changes would be expected to be more sensitive to neuronal damage in PTSD than volume loss. We also examined the relationship of hypothalamic-pituitary-adrenal (HPA) measures and hippocampal N-Acetyl Aspartate in

11 PTSD and 11 control subjects of this study, using morning salivary cortisol samples before and after low dose dexamethasone (0.5mg) as measure of cortisol levels. We found left hippocampal NAA was strongly associated with both pre-dexamethasone cortisol levels (N= 22, r= 0.53, p= 0.013) and post dexamethasone cortisol (N=22, r= 0.63, p=0.002). After accounting for clinical symptom severity and hippocampal volume, cortisol levels accounted for 21.9% of the variance (F = 5.6, p = .004) in left hippcampal NAA and 12.6% of the variance (F = 3.2, p = .035) in right hippocampal NAA. These results show a positive relationship between cortisol levels and hippocampal NAA in subjects without hypercortisolemia. Within the range of values seen in our subjects, cortisol may have a trophic effect on the hippocampus.

4. CORRELATIONS BETWEEN NAA AND FDG-PET

The *in vivo* neuronal contribution to human cerebral metabolic rate of glucose (CMRglc), measured by [18]FDG-PET, is unknown. Since NAA is thought to reflect neuron density, evaluating how CMRglc varies as a function of NAA concentration ([NAA]) should reflect the way in which brain glucose metabolism is affected by neuron density and/or by NAA content per neuron. The CMRglc-to-[NAA] relationship could be derived for an individual subject by plotting local CMRglc against local [NAA] across that subject's brain. The CMGglc-to-[NAA] relationship might be expected to vary from subject-to-subject depending on factors such as subject cognitive status. However, there are limitations to this approach. First, while cortical gray matter is the tissue of primary interest in evaluating brain metabolic activity, PET and ^{1}H MRSI data are often expressed in terms of whole, unsegmented brain tissue, rather than as values for cortical gray matter alone. Second, even if regional data are compared, the spatial resolution of MRSI is lower than that of PET, a source of possible signal infidelity. In a study that included 19 demented, cognitively impaired, and control subjects, who had whole-brain PET data, MRI, and MRSI we aimed to establish a method for the assessment of CMRglc and [NAA] in cortical gray matter, accounting for differences between PET and ^{1}H MRSI image resolution. Furthermore, we looked for the quantitative relationship between gray matter CMRglc and [NAA] in individual cognitively normal, cognitively impaired, and demented subjects and explored whether this CMRglc-to-[NAA] relation varies with cognitive status across subjects. In 18 of 19 subjects, a significant linear regression (P < 0.05) resulted when gray matter PET was plotted against gray matter NAA, whereby gray matter PET was higher for higher gray matter NAA. A representative example of this PET-NAA correlation is shown in figure 4.

To the extent that [NAA] can be taken as a marker of neurons (5-6), the correlation with FDG-PET suggests that the metabolic activity measured by [18]FDG-PET in a sample of gray matter increases with the density of neurons present in that gray matter sample and/or with the quantity of NAA within those neurons. Furthermore, the slope of the GMCMRglc-against-GMNAA regression decreased with increasing CDR across subjects, suggesting that CMRglc per neuron is lower in cognitively impaired and demented subjects than in cognitively normal subjects. This explanation is consistent with the idea that diminished cortical metabolism is a physiologic substrate of dementia, regardless of etiology (7, for a review see 18). It also suggests that such hypometabolism may be due not simply to losses in neuron numbers in gray matter, but to an alternative or concomitant decrease in the metabolic activity per neuron of those neurons remaining. This method may be used to investigate the relationship of CMRglc to neurons in various conditions.

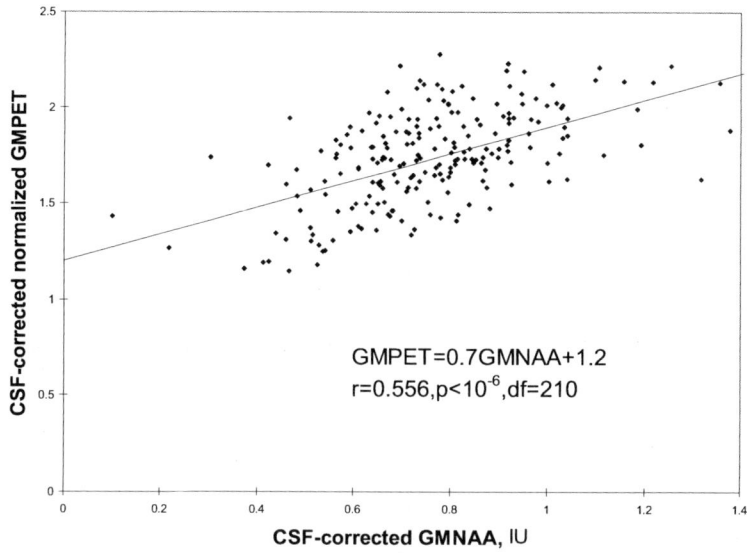

Figure 4. Correlation between FDG-PET (glucose metabolism) and NAA levels.

5. DEEP GRAY MATTER STRUCTURES IN HIV: A ^{1}H MRSI STUDY [26]

The goal was to determine the concentrations of the neuronal marker N-acetylaspartate (NAA) and of choline-containing metabolites (Cho) in the subcortical brain of HIV-seropositive patients as a function of their cognitive impairment and clinical symptoms. Pathological studies suggest that subcortical gray matter carries a heavy HIV load, and neuropsychological test results are consistent with involvement of subcortical and fronto-striatal brain systems in HIV disease. Single-volume ^{1}H MRS studies suggested neuronal preservation (i.e., unchanged NAA) and macrophage infiltration (i.e., high Cho) in subcortical brain of cognitively impaired and clinically symptomatic HIV+ individuals. Improved ^{1}H MRS methods may allow the early detection of metabolite alterations in subcortical brain of asymptomatic HIV+ individuals.

Two-dimensional ^{1}H MR spectroscopic imaging with volume preselection was performed in 30 HIV- controls and 70 HIV+ participants with varying severities of systemic disease and neuropsychological impairments.

Subcortical NAA was about 20% lower than control only in HIV+ patients with severe cognitive impairments; asymptomatic patients or those with mild cognitive impairments had normal subcortical NAA. Subcortical Cho was about 11% higher compared to controls in HIV+ patients regardless of the presence or absence of cognitive impairment or clinical symptoms. Subcortical NAA correlated with performance on a variety of neuropsychological tests but not with Center for Disease Control clinical stage.

High thalamic Cho was associated with low CD4 lymphocyte counts. The NAA findings suggest functionally significant neuronal subcortical injury only in severely cognitively impaired HIV+ patients. High subcortical Cho throughout all stages of HIV disease is consistent with early and persistent macrophage infiltration. The findings are consistent with the lack of significant subcortical neuron loss in neuropathological studies. Quantitative ^{1}H MRSI may play a role in the objective assessment of the presence, magnitude and progression of brain involvement in HIV infection.

6. BRAIN DAMAGE IN TREATED HIV-INFECTED INDIVIDUALS [27]

This was the first clinical study of the effects of HIV infection and its progression on brain metabolites using short-TE multi-slice 1H MRSI. Figure 5 shows a representative slice of a multislice ^{1}H MRSI experiment at TE=25ms from a control. Metabolite images of mI, Cho, Cr, and NAA were generated using the automatic fitting program developed in this lab. The raw (solid line) and fitted (dashed) ^{1}H MR spectrum was selected from a region in white matter. We co-registered MRSI to segmented MRI data, determined atrophy-corrected absolute metabolite concentrations in major brain regions, and analyzed by region and tissue type, using linear regression in a mixed effects model. All statistical analyses used a 2x2 ANOVA (with age as a covariate when appropriate), yielding main effects of HIV infection and heavy drinking, HIV-by-alcohol interactions and group contrasts. Statistical analyses were also done for HIV+ individuals on or off highly active antiretroviral treatment (HAART) and for CDC stages, to evaluate the contrast in HIV symptomatology. We detected effects of heavy drinking [see below or Meyerhoff et al. ACER 2004] and HIV infection, the latter reported here.

Main HIV effects in patients on HAART were observed for 7% higher mI in thalami, 6% higher Cr in temporal WM and trends to lower parietal GM NAA (-5%) and Cho. Additional data analyses including the 30% of subjects who were not on HAART, suggested that stable HAART appears to ameliorate metabolite abnormalities in HIV samples.

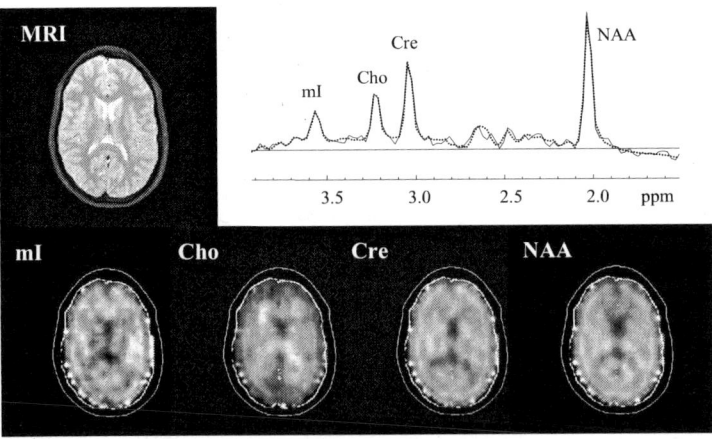

Figure 5. Metabolic imaging of myo-inositol, Choline, Creatine and N-acetylaspartate in control.

When 21 neurologically asymptomatic HIV+ participants in CDC A were compared to 35 symptomatic HIV individuals in CDC B and C, we found main effects of CDC status: symptomatic patients had higher mI in parietal and temporal WM, frontal WM, higher Cho in parietal WM and trends to higher frontal GM Cho. NAA was largely normal in these patients. Thus, the primary impact of HIV symptomatology was on WM, suggesting that inflammatory changes and perhaps myelin damage are the primary pathological events in mostly treated HIV+ samples, whereas neuronal/axonal deterioration is largely absent.

Nevertheless, lower NAA in left parietal WM and in frontal WM NAA were associated with higher viral load. CD4 did not correlate with NAA measures in this heavily treated group. Thus, these correlations support our hypotheses that lower NAA levels are associated with greater viral load.

In conclusion, the magnitude and anatomical distribution of the cross-sectional HIV effects differ from previous reports (see our previous studies above), which showed dramatic metabolic abnormalities in WM, GM and subcortical brain in pre-HAART era patients with cognitive impairments of greater severity than today. In this heavily treated population, some NAA loss was only observed in parietal GM, while WM Cho and mI increases are still significant but are not as ubiquitous a finding as in previous HIV-infected samples.

7. ^{1}H MRSI REVEALS BRAIN METABOLITE INJURY IN TREATED HIV+ PATIENTS AND IN CHRONIC HEAVY DRINKERS [28]

The above study showed that cross-sectional HIV effects were relatively subtle and it is unclear if they are premorbid or a function of HIV infection. Therefore, we studied HIV+ individuals about 2 years apart by repeated 1H MRSI to test the hypothesis of ongoing brain damage in HIV patients, who are not treated or have high viral loads despite antiretroviral therapy (HAART). Twenty-four HIV+ patients on HAART showed lower rates of frontal GM and parietal WM NAA loss than 6 HIV+ off HAART. However, even the 13 virally suppressed HIV+ (i.e., those on "successful" HAART) had NAA decreases in temporal WM at a rate of 4±6% compared to 9 viremic patients on HAART at 0±3% (p=0.09). Temporal WM Cho and lenticular Cho further decreased over time in virally suppressed patients on HAART.

Using 2x2 ANOVA in HIV+ participants on HAART (viral status x alcohol status), we found greater rates of NAA loss in suppressed patients than viremic patients in frontal GM and temporal WM, and greater increases of Cho in viremic than suppressed HIV+ patients in frontal and parietal GM, temporal WM, and thalami. This suggests that in virally suppressed subjects on HAART, neuronal damage continues while inflammatory processes are arrested.

In summary, untreated HIV infection presents with ongoing neuronal injury, in particular in parietal WM. HIV+ patients on HAART show less severe longitudinal brain metabolite damage than those off HAART, but even suppressed HIV+ patients tend to have ongoing regional NAA loss. These results demonstrate the ability of 1H MRSI to detect longitudinal metabolite changes due to HIV infection.

8. ¹H MRSI SEPARATES NEURONAL FROM GLIAL CHANGES IN ALCOHOL-RELATED BRAIN ATROPHY [29]

Eleven elderly alcoholics (61 ± 7 years), abstinent for approx. 9 months at the time of MRSI study, were compared to 9 age-matched light-drinking controls. 1H MRSI with PRESS volume preselection was used in a single slice above the ventricles. Cortical spectra were extracted from frontal and parietal gray matter regions, corrected for possible atrophy using tissue-segmented MR images and cortical metabolite ratios were compared. We found a location-by-group effect, which indicated significantly lower NAA/Cr in frontal than parietal cortex of recovering elderly alcoholics. Alcoholics with the most ventricular CSF showed the lowest NAA/Cr in frontal cortex. Together with findings of no significant tissue atrophy in these elderly alcoholics, the MRS results suggest neuronal damage or loss in frontal cortex associated with glial hyperplasia (or gliosis). This report demonstrates that: 1) that absolute metabolite measures are needed, 2) that by studying recovering alcoholics several months after cessation of drinking, tissue recovery may mask the full impact of chronic alcohol consumption on neuronal injury, and 3) that longitudinal studies in recovering alcoholics are needed to better understand the reasons for deficits in NAA and its potential recovery: Neuronal loss would not be associated with NAA recovery, whereas loss of dendritic arborization and neuronal cell body shrinkage would be reversible. Thus, this study pointed the way to further studies being performed in this lab.

9. SEPARATE AND INTERACTIVE EFFECTS OF COCAINE AND ALCOHOL [30]

In a ¹H MRSI study of the effects of concurrent cocaine and alcohol dependence on the brain, we compared non-dependent controls to subjects dependent on crack cocaine alone and to subjects dependent on both cocaine and alcohol. Our findings included lower NAA concentrations in cocaine-dependent and in cocaine and alcohol-dependent subjects, especially in dorsolateral prefrontal gray matter and in posterior parietal white matter, suggesting damage to neurons and axons in these brain regions of substance abusers. In that study, it was not possible to determine whether effects observed in subjects dependent on both cocaine and alcohol were due to cocaine abuse, to alcohol abuse, or to both. To answer that question, we examined in a follow-up study a fourth group of individuals dependent on alcohol alone and augmented the original sample by additional subjects, allowing to address the question of separate and interactive effects of chronic cocaine dependence and alcohol dependence on regional brain metabolites. We found main effects of alcohol dependence on brain atrophy and NAA concentrations. Alcohol-dependent individuals abstinent for 1-2 years had brain atrophy and about 8% lower NAA concentrations in both cortical and subcortical gray matter, but no NAA loss in white matter. NAA loss was most significant in frontal gray matter regions. There was no significant main effect of cocaine dependence on NAA concentrations, despite some regional atrophy in these cocaine-dependent individuals abstinent from cocaine for about 4 months. Cocaine-dependent individuals showed higher posterior parietal white-matter creatine concentrations than controls. In addition, no significant cocaine dependence-by-alcohol dependence interactions were found for any metabolite in any tissue or brain region. Thus, alcohol dependence, but not cocaine dependence, was associated with long-lasting NAA loss in cortical and subcortical gray matter throughout the brain, suggesting widespread neuronal injury. However, past chronic alcohol use did not aggravate any chronic cocaine-induced metabolite deficits.

10. EFFECTS OF HEAVY DRINKING

10.1 Effects of Abstinence on the Brain: Quantitative MRI and MR Spectroscopic Imaging in Chronic Alcohol Abuse [31]

Structural brain damage, especially to white matter, is well documented in chronic alcohol abuse, and there is also evidence for brain metabolic abnormalities in this condition. It is unknown, however, to what extent these structural and metabolic changes are still detectable in long-term abstinent alcoholics compared to active chronic drinkers. Therefore we compared 12 recovering alcoholics, who had been abstinent from alcohol for an average of 2 years, to 8 active heavily drinking subjects with similar alcohol use variables.

Metabolite concentrations in whole-brain and in gray matter and white matter of brain lobes did not differ significantly between the recovering alcoholics and active drinkers. However, active heavily drinking subjects had less frontal white matter than abstinent alcoholics and less gray matter in the orbital frontal pole and postcentral gyrus. However, abstinent alcoholics had smaller gray matter volumes in the anterior cingulate than active heavy drinkers.

Our cross-sectional ^{1}H MRSI measures were largely ineffective in revealing metabolic effects of abstinence on the alcohol-damaged brain. The study, however, suggests region-specific structural recovery from chronic alcohol-induced brain injury, but also region-specific long-term structural damage in abstinent alcoholics.

10.2 Brain Recovery During Abstinence from Alcohol [32]

In our ongoing studies of recovering alcoholics, we investigate the nature of brain injury in alcoholics and the potential improvements of brain metabolite concentrations and cognition during prolonged sobriety.

At one week of sobriety, NAA in alcoholics was lower by 6-19% in frontal, parietal and temporal gray matter and white matter, in the basal ganglia, the brain stem, and cerebellum, while Cho concentrations were lower by 7-13% in lobar grey and white matter regions and in the thalami. NAA deficits correlated with cognitive impairments.

Over the first month of sobriety, NAA, myo-inositol, and Cho concentrations in frontal white matter increased significantly and to a greater extent than in gray matter. Similarly, neurocognitive performance increased significantly and showed some correlation with neurochemical improvements. Seven months after cessation of drinking, abstinent alcoholics had normal concentrations of myo-inositol and Cho. Regional NAA concentrations in white and gray matter increased over 7 months of sobriety, however, they did not normalize. These long-term changes were accompanied by continued cognitive improvements in most domains except visuospatial learning and memory. Increases of Cho and myo-inositol concentrations over time are consistent with remyelination and astrocytosis. Slower increases of NAA suggest slower recovery from axonal and neuronal injury, and that NAA loss in alcoholics is not completely due to neuronal loss.

10.3 Effects of Heavy Drinking, Binge Drinking, and Family History of Alcoholism on Regional Brain Metabolites [33]

Abstinent alcoholics show regional NAA loss, primarily in frontal lobes and cerebellum. The main goals of this project were to investigate the effects of chronic active heavy drinking on NAA and other metabolites throughout the brain, and to determine if they are affected by family history (FH) of alcoholism and long-term drinking pattern.

We used quantitative MRI and multi-slice ^{1}H MRSI at a short echo-time to compare 46 chronic heavy drinkers (HD) and 52 light drinkers (LD) on regional, tissue-specific and atrophy-corrected concentrations of NAA, myo-inositol (mI), creatine- and choline-containing metabolites.

NAA in frontal white matter was 6% lower in HD than LD. NAA loss was greater in female than male heavy drinkers despite similar drinking severity and greater in FH-negative HD than FH-positive HD. FH-negative compared to FH-positive HD also had higher mI in the brainstem and tended to have lower NAA and higher mI in frontal GM. In addition, greater frontal NAA loss in HD was found as a function of age. Lower frontal white matter NAA in HD correlated with lower executive and working memory functions and with greater P3 latency.

Thus, heavy drinkers in their forties who are not in alcoholism treatment have frontal axonal injury, which is associated with lower brain function and is likely of behavioral significance. Family history of alcoholism modulates brain metabolite abnormalities. Brain injury in active heavy drinkers is less pronounced than in abstinent alcoholics and presents with a different spatial and metabolite pattern.

10.4 Magnetic Resonance Detects Brainstem Changes in Chronic, Active Heavy Drinkers [34]

Neuropathological and neuroimaging studies show cortical and subcortical volume loss in alcohol dependent individuals. The brainstem is considered critical in the development and maintenance of drug and alcohol dependence, but it has not been the focus of neuroimaging studies. Using quantitative MRI and ^{1}H MRSI, we compared the size and metabolite measures of potential cellular injury of the brainstem in 12 chronic, active heavy drinkers and 10 light drinkers. Chronic heavy drinking was associated with a significantly smaller overall brainstem volume and with significantly smaller midsagittal areas of the brainstem, midbrain, and pons. Heavy drinking was also associated with significantly lower ratios of N-acetyl-aspartate (NAA) and choline-containing metabolites (Cho) compared with creatine-containing compounds (Cr) in a region including midbrain and pons, independent of brainstem atrophy. These structural and metabolite findings are consistent with neuronal injury of the midbrain/pons of untreated chronic heavy drinkers.

11. EFFECTS OF EPILEPSY ON NAA

11.1 Identification of the Epileptogenic Focus

From the first studies in the early 1990's on, the most consistent finding in the epileptogenic focus has been a reduction of NAA. This has been first demonstrated in temporal lobe epilepsy (TLE) with evidence for hippocampal atrophy or mesial temporal sclerosis (MTS) where hippocampal NAA reductions correctly identify the epileptogenic hippocampus in up to 100%. Because early studies found a strong correlation between the degree of neuronal loss and the degree of NAA reduction, the NAA reduction in the epileptogenic hippocampus was thought to represent mainly neuronal loss [35]. However, newer evidence suggests that a substantial component of the NAA reduction is due to a not further specified, potentially reversible neuronal dysfunction in the epileptogenic tissue. This is also supported by studies in TLE without MRI evidence for MTS where histopathological studies show only mild neuronal loss despite clear hippocampal NAA reductions. However, NAA reductions in TLE without MTS are different from those found in TLE with MTS as has been demonstrated by a recent study from our laboratory. This study compared patterns of hippocampal NAA loss in 10 TLE without MTS with the patterns found in 15 TLE with MTS. The number of voxels with reduced NAA/(Cr+Cho) in the ipsilateral hippocampus was higher in TLE with MTS than in TLE without MTS (1.9 ± 1.3 vs 0.6 ± 1.3, $p = 0.02$). Furthermore, the NAA reductions in TLE without MTS were more often diffuse ($p = 0.007$) and less often concordant ($p = 0.015$) to the epileptogenic hippocampus than in TLE with MTS.

Consequently, hippocampal NAA reductions in TLE without MTS are less accurate for the identification of the epileptogenic focus but nonetheless helpful for predicting the chance of becoming seizure free after epilepsy surgery. A study in 15 TLE without MTS from our laboratory [36] found that patients who became not seizure free had lower ipsilateral hippocampal NAA/(Cr+Cho) z scores than contralateral ($p = 0.04$). Furthermore, in comparison with patients who became seizure free, patients who did not had lower ipsilateral ($p= 0.005$) and contralateral ($p=0.02$) hippocampal NAA/(Cr+Cho) z sores. Taken together, TLE patients without MTS who became seizure free had milder and less well lateralized hippocampal NAA reductions than patients who did not.

Preliminary results show, that NAA reductions may also be helpful for focus identification in patients suffering from neocortical epilepsy (NE), i.e., a form of epilepsy where identification of the seizure focus is often challenging, particularly in patients with no structural abnormality on the MRI. We studied 21 patients with NE (10 with evidence for cortical malformations on the MRI, 11 with normal MRI) and 19 age-matched controls. In controls, NAA/Cr and NAA/Cho of all voxels of a given lobe was expressed as a function of white matter content and thresholds for pathological values determined by calculating the 95% prediction intervals for NAA/Cr and NAA/Cho. Voxels with NAA/Cr or NAA/Cho below the 95% prediction interval were defined as "pathological". Z-scores were used to identify regions with a high percentage of pathological voxels. MRSI correctly identified the lobe containing the epileptogenic focus as defined by EEG in 62% of the NE patients. MRSI localization of the focus was correct in 70% of the patients with a lesion on the MRI and in 55% of the patients with normal MRI.[37]

NAA reductions are also a common finding in different types of cortical malformations. These lesions result from a disruption of the developmental processes during neuroblast proliferation and differentiation, neuroblast migration, or postmigrational cortical organization. Using a similar method as for focus localization in NE, 30% (range 0 – 78%) of all voxels in cortical malformations (8 patients with 10 malformations) were found to be metabolically abnormal. The most common abnormalities were areas with reduced NAA or increased Cho which were interspersed within metabollically normal areas. Cortical malformations are not only characterized by a disturbed tissue architecture but also by an intrinsic epileptogenicity and those NAA and Cho abnormalities probably reflect both disturbances.[38]

11.2 NAA in Brain Regions Secondarily Involved Seizure Spread

Recent studies found NAA reduction not to be restricted to the focus but also in brain areas which are involved in seizure spread. This is well known in the TLE where in up to 50% of the patients NAA reductions are not only found in the ipsilateral hippocampus but also contralaterally.[39] These contralateral NAA reductions increase to normal values after successful epilepsy surgery, but stay decreased, if surgery did not lead to seizure freedom [40]. This finding further supports the hypothesis that NAA reductions in epileptic tissue are not necessarily due to neuron loss but rather present the disturbance of the neuronal metabolism caused by epileptogenic activity. Those extrafocal NAA reductions are not restricted to the limbic system but may involve even more remote brain areas as a study done in our laboratory recently demonstrated.

In this study, we used MR spectroscopic imaging in combination with tissue segmentation in 14 TLE patients with MTS, 7 TLE patients without MTS and 12 age-matched controls. To identify voxels with abnormally low NAA, NAA/(Cr+Cho) of all voxels of a given lobe was expressed as a function of white matter content to determine the 95% prediction interval for any additional voxel of a given tissue composition. Voxels with NAA/(Cr+Cho) below the lower limit of the 95% prediction interval were defined as "pathological". Z-scores were used to identify regions with a higher percentage of pathological voxels than in controls. Additional regions with reduced NAA/(Cr+Cho) were found in the ipsilateral temporal and parietal lobes and bilaterally in insula and frontal lobes. Temporal abnormalities identified the epileptogenic focus in 70% in TLE with MTS and 83% of TLE without MTS. Extratemporal abnormalities identified the hemisphere containing the epileptogenic focus in 78% of TLE with MTS but in only 17% of TLE without MTS. Therefore, temporal and extratemporal NAA/(Cr+Cho) reductions might be helpful for focus lateralization in TLE [41]. Because volumetric studies found no evidence for tissue atrophy beyond the ipsilateral temporal lobe [42], it is reasonable that these NAA reductions probably also reflect neuronal dysfunction rather than actual neuronal loss. Furthermore, smaller, extrafocal areas with metabolic abnormalities were also found in 24% of the NE patients.[37]

12. REDUCED NAA IN ALS

Amyotrophic lateral sclerosis (ALS) is a neurodegenerative disorder that results in the loss of motor neurons both in the brain and spinal cord leading to paralysis and ultimately death. There is no definitive test for ALS and diagnoses are currently made based on clinical data. The development of a surrogate marker for disease progression would be useful for identifying individuals suffering from the early stages of ALS and for monitoring treatment effects. Previous MRSI research from our group found that the ratios NAA/Cre, NAA/Cho and NAA/Cre+Cho are reduced in probable/definite subjects as compared to controls and that NAA, Cre, and Cho decrease over time in ALS.[36] Because most previous MRS studies of ALS examined regions of interest that were sufficiently distant from the skull to avoid interference from lipids, sampling of motor cortex was often limited.[43-55] Using a Multiplanar ^{1}H MRSI technique to sample a larger region that includes brain surface, reductions of the ratio NAA/Cre+Cho, have been observed in the region of the motor cortex and corticospinal tract.[56] A quantitative analysis of these data showed that decreased NAA was responsible for the ratio changes in the motor cortex and increased Cho for ratio changes in the corticospinal tract.[57, 58].The specific aims of this project were to: 1) determine if previously reported ^{1}H MRSI differences between ALS patients and control subjects are limited to the motor cortex; and 2) determine the longitudinal metabolic changes corresponding to varying levels of diagnostic certainty. Toward this end, 21 patients with possible/suspected ALS, 24 patients with probable/definite ALS and 17 control subjects underwent multislice ^{1}H MRSI co-registered with tissue-segmented MRI to obtain concentrations of the brain metabolites NAA, Cre and Cho in the left and right motor cortex and in gray matter and white matter of non-motor regions in the brain. Of these subjects, 13 possible/suspected and 15 probable/definite patients received repeated scans (every 3 months for up to 12 months) and were studied longitudinally.

Table 2.

More Affected Hemisphere

	Control		Probable/Definite	
	Motor	Other	Motor	Other
NAA/Cre	2.16 ± 0.16	2.08 ± 0.14	2.02 ± 0.25 (-7.3%)	1.96 ± 0.17 (-5.8%)
NAA/Cho	3.60 ± 0.39	3.08 ± 0.36	3.14 ± 1.46 (-12.6%)†	2.79 ± 0.32 (-9.2%)*
NAA/(Cre+Cho)	1.34 ± 0.10	1.22 ± 0.09	1.21 ± 0.14 (-9.5%)*	1.13 ± 0.10 (-7.3%)*

Less Affected Hemisphere

	Control		Probable/Definite	
	Motor	Other	Motor	Other
NAA/Cre	2.16 ± 0.16	2.08 ± 0.14	1.92 ± 0.20 (-11.6%)‡	1.94 ± 0.17 (-6.4%)*
NAA/Cho	3.60 ± 0.39	3.08 ± 0.36	3.14 ± 0.43 (-10.8%)*	2.79 ± 0.32 (-10.3%)*
NAA/(Cre+Cho)	1.34 ± 0.10	1.22 ± 0.09	1.21 ± 0.14 (-11.5%)‡	1.13 ± 0.10 (-8.1%)*

*$p<0.05$, †$p<0.005$, ‡$p<0.0005$
Numbers in parentheses are percent change from control subject mean

In the more affected hemisphere, Reductions in the ratios, NAA/Cho and NAA/Cre+Cho were observed both within (12.6% and 9.5% respectively) and outside (9.2% and 7.3% respectively) motor cortex in probable/definite ALS (Table 2). However, these reductions were significantly greater in motor cortex ($p<0.05$ for NAA/Cho and $p<0.005$ for NAA/(Cre+Cho). Longitudinal changes in NAA were observed at three months within the motor cortex of both possible/suspected ALS patients (p<0.005) and at 9 months outside the motor cortex of probable/definite patients (p<0.005). However there was no clear pattern of progressive change over time. NAA ratios are reduced in motor cortex and outside of motor cortex in ALS, suggesting widespread neuronal injury. Longitudinal changes of NAA are not reliable, suggesting that NAA may not be a useful surrogate marker for treatment trials.

13. QUESTION AND ANSWER SESSION

DR. BARKER: We have time for one or two questions. Yes?

DR. MATALON: I find this very fascinating. You had a slide where you showed six people with lower NAA or whatever. Two of them, you showed that they, when you saw that, later on had -- One had stroke, and one had vascular abnormality.

You mean to say that you can predict that before the event that destroys cells or there was a problem with profusion that you were not aware of?

DR. WEINER: We don't know. We don't know the predictive value of NAA here, but what is interesting is that, as I showed you earlier, subjects have subcortical ischemic vascular dementia NAA reductions in their hippocampus. Therefore, NAA reductions in hippocampus are not at all specific for Alzheimer's disease. Whether the NAA reductions reflect a generalized ischemic process or whether it represents microscopic hippocampal sclerosis, it is hard to infer any specific mechanism when you see a reduction of NAA in a patient.

DR. WEINBERGER: Michael, these are really lovely studies. Let me ask a question which is inspired by your question of me, but fundamentally out of curiosity.

If you analyzed all your data by NAA-creatine ratios, how different would the data look?

DR. WEINER: We haven't done that very much because early on, we had some reviewers saying the ratio is going down, how do you know it's not because the creatine is going up? Lots of people were doing spectroscopy measurements; Brian Ross in fact started doing single voxel spectroscopy in Alzheimer's disease. It is difficult to interpret changes in ratios. A change in NAA/Creatine could be due to a change in NAA, a change of creatine, or both. One of the issues that must be dealt with concerns the effects of loss of gray matter and white matter and expansion of CSF. That is, it is important to understand the tissue composition of each MRS voxel to determine the extent to which the NAA measurements were changing independent of changes of tissue composition. To accomplish this, we developed segmentation programs and programs that co-register the spectroscopy voxels with the segmentation information, also correcting for the pulse profile, the slice offset and other instrumental parameters. This allows calculation of the absolute amount of NAA and other metabolies. We have published some work, especially in ALS where we calculated both rations and absolute values, because in some circumstances you obtain---more of this classification value (diagnostic value) by calculating ratios.

I mean, the classic example is what Brian Ross showed, is that if you just look at the NAA over creatine in Alzheimer's or look at the myo-inositol over creatine in Alzheimer's, you get one kind of classification. But if you look at the myo-inositol /NAA ratio, then you get much greater classification values.

If you are asking the question, is NAA changing? then our view has always been the best thing to do is to calculate it absolutely.

There is data that creatine does change in many circumstances, and I think we had some work we did in ALS some years ago that showed that the creatine vlaues were significantly changing in ALS patients. Those changes in creatine were giving us spurious data, making it more difficult to interpret the changes of NAA to creatine ratios.

DR. MATALON: My question is -- you didn't answer it, only partially. When you have decrease in NAA without any known reason, no dementia, no Alzheimer's, like those two you show, should we follow that with another test? profusion studies? blood supply? because some of these things may be preventable. This is really the crux of my question.

DR. WEINER: Well, first of all, what we are doing is purely research at this stage. I think it would be very difficult to start using this information to try to manage patients. The whole goal of my research is to try to identify risk factors that predict cognitive decline.

I think that NAA is a potential useful risk factor in this area. However, there is nothing you can do which prevents or slows cognitive decline in patients. There are no drugs that have been shown to slow the progression of Alzheimer's.

What we are waiting for is a drug that blocks the production of amyloid or a neuroprotective agent. Some of these are currently under clinical trials. Once one of those drugs becomes available, the importance of being able to predict future development of Alzheimer's disease becomes huge in this country, and we need to have good, robust measures that can be used widely in many MRI centers, and maybe spectroscopy will be shown to have a unique role. So we'll see.

DR. BARKER: Okay, thank you very much, Mike.

14. REFRENCES

1. Chui HC. Dementia due to subcortical ischemic vascular disease. *Clin Cornerstone* 2001;3(4):40-51.
2. Fein G, Di S, V, Tanabe J, et al. Hippocampal and cortical atrophy predict dementia in subcortical ischemic vascular disease [In Process Citation]. *Neurology* 2000;55 (11):1626-35.
3. Mungas D, Jagust WJ, Reed BR, et al. MRI Predictors of Cognition in Subcortical Ischemic Vascular Disease and Alzheimer's Disease. *Neurology* 2001;57(12):2229-35.
4. Du AT, Schuff N, Laakso MP, et al. Effects of subcortical ischemic vascular dementia and AD on entorhinal cortex and hippocampus. *Neurology* 2002;58(11):1635-41.
5. Laakso MP, Partanen K, Riekkinen P, et al. Hippocampal volumes in Alzheimer's disease, Parkinson's disease with and without dementia, and in vascular dementia: An MRI study. *Neurology* 1996;46:678-81.
6. Baron JC, D'Antona R, Pantano P, Serdaru M, Samson Y, Bousser MG. Effects of thalamic stroke on energy metabolism of the cerebral cortex. A positron tomography study in man. *Brain* 1986;109 (Pt 6):1243-59.
7. Kwan LT, Reed BR, Eberling JL, et al. Effects of subcortical cerebral infarction on cortical glucose metabolism and cognitive function. *ArchNeurology* 1999;56:809-14.
8. Rango M, Spagnoli D, Tomei G, Bamonti F, Scarlato G, Zetta L. Central Nervous System Trans-Synaptic Effects of Acute Axonal Injury: A 1H Magnetic Resonance Spectroscopy Study. MRM ed: Williams & Wilkins; 1995.

9.   Escobar A. Cerebral changes associated with senility. I. The role of transneuronal degeneration in the neocortex. *Bol Estud Med Biol* 1973;28(1):1-8.

10. Schuff N, Capizzano AA, Du AT, et al. Selective reduction of N-acetylaspartate in medial temporal and parietal lobes in AD. *Neurology* 2002;58(6):928-35.

11. Southwick SM, Paige S, Morgan CA, Bremner JD, Krystal JH, Charney DS. Neurotransmitter alterations in PTSD: catecholamines and serotonin. *Semin Clin Neuropsychiatry* 1999;4(4):242-8.

12. Wang S, Mason J, Southwick S, Johnson D, Lubin H, Charney D. Relationships between thyroid hormones and symptoms in combat-related posttraumatic stress disorder. *Psychosom Med* 1995;57(4):398-402.

13. Yehuda R, Southwick SM, Krystal JH, Bremner D, Charney DS, Mason JW. Enhanced suppression of cortisol following dexamethasone administration in posttraumatic stress disorder. *AmJPsychiatry* 1993;150(1):83-6.

14. Bremner JD, Randall P, Scott TM, et al. MRI-based measurement of hippocampal volume in patients with combat-related posttraumatic stress disorder. *Am J Psychiatry* 1995;152:973-81.

15. Gurvits TV, Shenton ME, Hokama H, et al. Magnetic resonance imaging study of hippocampal volume in chronic, combat-related post-traumatic stress disorder. *Biological Psychiatry* 1996;40:1091-9.

16. Bremner JD, Randall P, Vermetten E, et al. MRI-based measurement of hippocampal volume in posttraumatic stress disorder related to childhood physical and sexual abuse-A preliminary report. *Biological Psychiatry* 1997;41:23-32.

17. Stein MB, Koverola C, Hanna C, Torchia MG, McClarty B. Hippocampal volume in women victimized by childhood sexual abuse. *Psychol Med* 1997;27(4):951-9.

18. Schuff N, Marmar CR, Weiss DS, et al. Reduced hippocampal volume and n-acetylaspartate in post traumatic stress disorder. *The Annals of the New York Academy of Sciences* 1997; Supplement on Psychobiology of Posttraumatic Stress Disorder(821):516-20.

19. Freeman TW, Cardwell D, Karson CN, Komoroski RA. In vivo proton magnetic resonance spectroscopy of the medial temporal lobes of subjects with combat-related posttraumatic stress disorder. *MagnResonMed* 1998;40(1):66-71.

20. Schuff N, Amend D, Ezekiel F, et al. Changes of hippocampal n-acetylaspartate and volume in Alzheimer's disease: A proton MR spectroscopic imaging and MRI study. *Neurology* 1997;49:1513-21.

21. Kalra S, Cashman NR, Genge A, Arnold DL. Recovery of N-acetylaspartate in corticomotor neurons of patients with ALS after riluzole therapy. *Neuroreport* 1998;9(8):1757-61.

22. Hugg JW, Kuzniecky RI, Gilliam FG, Morawetz RB, Faught RE, Hetherington HP. Normalization of contralateral metabolic function following temporal lobectomy demonstrated by h-1 magnetic resonance spectroscopic imaging. *Ann Neurol* 1996;V40:236-9.

23. De Stefano N, Matthews PM, Arnold DL. Reversible decreases in N-acetylaspartate after acute brain injury. *Magn Reson Med* 1995;34:721-7.

24. Bertolino A, Callicott JH, Mattay VS, et al. The effect of treatment with antipsychiatric drugs on brain N-acetylaspartate measures in patients with schizophrenia. *Biological Psychiatry* 2001;49:39-46.

25. Heimberg C, Komoroski RA, Lawson WB, Cardwell D, Karson CN. Regional proton magnetic resonance spectroscopy in schizophrenia and exploration of drug effect. *Psychiatry Res* 1998;83(2):105-15.

26. Meyerhoff DJ, Weiner MW, Fein G. Deep gray matter structures in HIV infection: a proton MR spectroscopic study. AJNR Am J Neuroradiol 1996;17(5):973-8.

27. Meyerhoff DJ, Cardenas V, Studholme C, et al. Evidence for Brain Damage in Treated HIV-Infected Individuals. *Neurology* 2003;60(5):A186.

28. Meyerhoff DJ, Truran D, Flenniken D, Song E, Studholme C, Weiner MW. Longitudinal Multi-Slice Short-Te 1H MRSI Reveals On Going Brain Metabolite Injury In Treated HIV+ Patients And In Chronic Heavy Drinkers. Proc Intl Soc *Mag Reson Med* 2004;11:290.

29. Fein G, Meyerhoff DJ, Di Sclafani V, et al. 1H magnetic resonance spectroscopic imaging separates neuronal from glial changes in alcohol-related brain atrophy. Chapter in NIAAA Research Monograph No 27/Alcohol and Glial Cells 1994:227-41.

30. O'Neill J, Cardenas VA, Meyerhoff DJ. Separate and interactive effects of cocaine and alcohol dependence on brain structures and metabolites: quantitative MRI and proton MR spectroscopic imaging. *Addiction Biology* 2001;6:347-61.

31. O'Neill J, Cardenas VA, Meyerhoff DJ. Effects of abstinence on the brain: quantitative magnetic resonance imaging and magnetic resonance spectroscopic imaging in chronic alcohol abuse. *Alcohol Clin Exp Res* 2001;25(11):1673-82.

32. Gazdzinski S, Durazzo TC, Meyerhoff D. Brain Recovery During Abstinence from Alcohol: MRI, MR Spectroscopic Imaging, and Neurocognitive Studies. In: American Academy of Neurology 56th Annual Meeting; 2004; Moscone Convention Center, San Francisco, CA USA: American Academy of Neurology; 2004. p. A542.

33. Meyerhoff D, Blumenfeld R, Truran D, et al. Effects of heavy drinking, binge drinking, and family history of alcoholism on regional brain metabolites. *Alcohol Clin Exp Res* 2004;28(4):650-61.

34. , Bloomer CW, Langleben DD, Meyerhoff DJ. Magnetic resonance detects brainstem changes in chronic, active heavy drinkers. *Psychiatry Res* 2004; 132(3): 209-18

35. Duc O, Trabesinger AH, Weber OM, et al. Quanitative 1HMRS in the evaluation of mesial temporal lobe epilepsy in vivo. *Magn Reson Imaging* 1998;16:969-79.

36. Suhy J, Laxer KD, Capizzano AA, et al. 1H MRSI predicts surgical outcome in MRI-negative temporal lobe epilepsy. *Neurology* 2002;58(5):821-3.

37. Mueller S, Laxer K, Barakos J, et al. Focus identification in neocortical epilepsy with MR-spectroscopic imaging. Submitted.

38. Mueller S, Laxer K, Barakos J, et al. Metabolic characteristics causing epilepsy. Submitted.

39. Ende GR, Laxer KD, Knowlton RC, et al. Temporal lobe epilepsy: bilateral hippocampal metabolite changes revealed at proton MR spectroscopic imaging. *Radiology* 1997;202(3):809-17.

40. Cendes F, Andermann F, Dubeau F, Matthews PM, Arnold DL. Normalization of neuronal metabolic dysfunction after surgery for temporal lobe epilepsy - Evidence from proton MR spectroscopic imaging. *Neurology* 1997;49:1525-33.

41. Mueller S, Laxer K, Cashdollar N, Flenniken D, Matson G, Weiner M. Identification of Abnormal Neuronal Metabolism Outside the Seizure Focus in Temporal Lobe Epilepsy. *Epilepsia* 2004;45(4):355-66.

42. Mueller S, Suhy J, Laxer K, et al. Reduced extrahippocampal NAA in mesial temporal lobe epilepsy. *Epilepsia* 2002;43:1210-6.

43. Pioro EP, Antel JP, Cashman NR, Arnold DL. Detection of cortical neuron loss in motor neuron disease by proton magnetic resonance spectroscopic imaging in vivo. *Neurology* 1994;44(10):1933-8.

44. Jones AP, Gunawardena WJ, Coutinho CM, Gatt JA, Shaw IC, Mitchell JD. Preliminary results of proton magnetic resonance spectroscopy in motor neurone disease (amytrophic lateral sclerosis). *J Neurol Sci* 1995;129 Suppl:85-9.

45. Gredal O, Rosenbaum S, Topp S, Karlsborg M, Strange P, Werdelin L. Quantification of brain metabolites in amyotrophic lateral sclerosis by localized proton magnetic resonance spectroscopy [see comments]. *Brain* 1997;48(4):878-81.

46. Giroud M, Walker P, Bernard D, et al. Reduced brain N-acetyl-aspartate in frontal lobes suggests neuronal loss in patients with amyotrophic lateral sclerosis [published erratum appears in *Neurol Res* 1997 Aug;19(4):456]. NeurolRes 1996;18(3):241-3.

47. Cwik VA, Hanstock CC, Allen PS, Martin WR. Estimation of brainstem neuronal loss in amyotrophic lateral sclerosis with in vivo proton magnetic resonance spectroscopy. *Neurology* 1998;50(1):72-7.

48. Ellis CM, Simmons A, Andrews C, Dawson JM, Williams SC, Leigh PN. A proton magnetic resonance spectroscopic study in ALS: correlation with clinical findings. *Neurology* 1998;51(4):1104-9.

49. Block W, Karitzky J, Treaber F, et al. Proton magnetic resonance spectroscopy of the primary motor cortex in patients with motor neuron disease: subgroup analysis and follow-up measurements [see comments]. *Arch Neurol* 1998;55(7):931-6.

50. Pioro EP, Majors AW, Mitsumoto H, Nelson DR, Ng TC. 1H-MRS evidence of neurodegeneration and excess glutamate + glutamine in ALS medulla. *Neurology* 1999;53(1):71-9.

51. Bradley WG, Bowen BC, Pattany PM, Rotta F. 1H-magnetic resonance spectroscopy in amyotrophic lateral sclerosis. *J Neurol Sci* 1999;169(1-2):84-6.

52. Bowen BC, Pattany PM, Bradley WG, et al. MR imaging and localized proton spectroscopy of the precentral gyrus in amyotrophic lateral sclerosis. *Am J Neuroradiol* 2000;21(4):647-58.

53. Chan S, Shungu DC, Douglas-Akinwande A, Lange DJ, Rowland LP. Motor neuron diseases: comparison of single-voxel proton MR spectroscopy of the motor cortex with MR imaging of the brain. 1999;212(3):763-9.

54. Tarducci R, Sarchielli P, Presciutti O, et al. Study of the primary motor cortex in amyotrophic lateral sclerosis by quantitative 1HMRS. *ISMRM* 2000;1:632.

55. Petropoulos H, Mandler RN, Qualls C, et al. 1H-MRS reveals diffuse neuronal injury in Amyotrophic Lateral Sclerosis. *ISMRM* 2000;1:633.

56. Kalra S, Arnold DL, Cashman NR. Biological markers in the diagnosis and treatment of ALS. *J Neurol Sci* 1999;165 Suppl 1:S27-S32.

57. Rooney WD, Miller RG, Gelinas D, Schuff N, Maudsley AA, Weiner MW. Decreased N-acetylaspartate in motor cortex and corticospinal tract in ALS. *Neurology* 1998;50(6):1800-5.

58. Schuff N, Rooney WD, Miller RG, et al. Reanalysis of Multislice 1H MRSI in Amyotrophic Lateral Sclerosis. *Magnetic Resonanc in Medicine* 2001;45:513-6.

REGULATION OF NAA-SYNTHESIS IN THE HUMAN BRAIN *IN VIVO:* CANAVAN'S DISEASE, ALZHEIMER'S DISEASE AND SCHIZOPHRENIA

Kent Harris, Alexander Lin, Pratip Bhattacharya, Thao Tran, Willis Wong and Brian Ross[*]

1. INTRODUCTION

N-acetyl aspartate (NAA) enjoys its position as the most prominent cerebral metabolite visualized during clinical brain examinations with magnetic resonance spectroscopy because it is a 'neuronal-marker' and because its steady-state concentration in the brain is extraordinary constant. Thus normative data for [NAA] (or NAA/Cr) vary by less than 5% in the normal adult population[1]. In one rare disease, Canavan's disease, [NAA] is dramatically increased, while in a single case to date, [NAA] is essentially zero (see Burlina article, this volume). In between, there are literally dozens of brain disorders in which steady state [NAA] is significantly reduced (Table 1)[2].

The explanations for these alterations are generally to hand – absence of NAA-deacylase, destruction of neurons by hypoxia, ischemia, toxins or degenerative processes, or displacement by masses of NAA-depleted cells, as in primary glioma. However, this interpretation may be incomplete. In this paper we address another question, whether the rate of NAA-synthesis is regulated, and if so, whether in any of these well recognized disease states, altered [NAA] can be attributed to altered rates of NAA-synthesis (Figure 1). We could hypothesize for example that reduced [NAA] might be a result of reduced NAA synthesis.

[*] Magnetic Resonance Unit, Huntington Medical Research Institutes, Pasadena, CA; Rudi Schulte Research Institute, Santa Barbara CA and NARSAD.

Table 1. Some diseases that decrease NAA levels.

Regional	Unknown or Questionable
Epilepsy	Schizophrenia
Demyelination	Chronic Fatique Syndrome
Multiple sclerosis (WM)	Attention Deficit Disorder
Adrenoleukodystrophy	
Amyotrophic Lateral Sclerosis	

Focal	"Global"
Neoplasia	Hypoxia
Ischemia/Stroke	Alzheimer's disease
Multiple sclerosis (plaque)	Diabetes Mellitus
HIV lesions	Developmental delay
	Head trauma
	HIV dementia
	Mild cognitive impairment
	MELAS

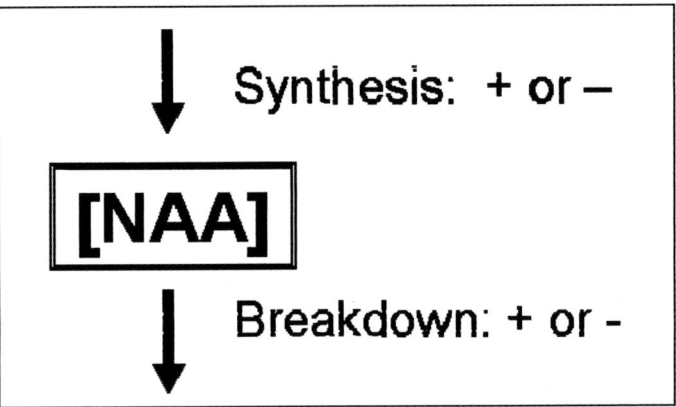

Figure 1. Mechanisms of regulating NAA concentration. As with all metabolites, there must be several (minimum of four) mechanisms which contribute to [NAA]

2. METHODS AND HUMAN SUBJECTS

[13]C Magnetic Resonance Spectroscopy: In principle, the infusion or ingestion of highly enriched ^{13}C precursors and metabolites into a patient with essentially zero (actually 1.1%) ^{13}C MR signal background, should dramatically enhance our ability to image metabolites,

determine metabolic flux rates and follow metabolic markers for the purpose of clinical diagnosis and research[3]. The first multi-purpose human MRI scanners provided all of the equipment necessary for this purpose in 1983[4], but the spectacular success of anatomical MRI resulted in 'simplification' to the exclusion of all nuclei other than proton. Broad-band clinical MR scanners essential for clinical ^{13}C MRS and MRI are more expensive to manufacture and hence confined to specialist Centers. ^{13}C MRS significantly outscores ^{1}H MRS (3 – 15 metabolites), ^{31}P MRS (5 – 7 metabolites), [23]Na, [7]Li or [19]F (1 – 5 metabolites each) in the number of human metabolites – upwards of 50 – which can be monitored.

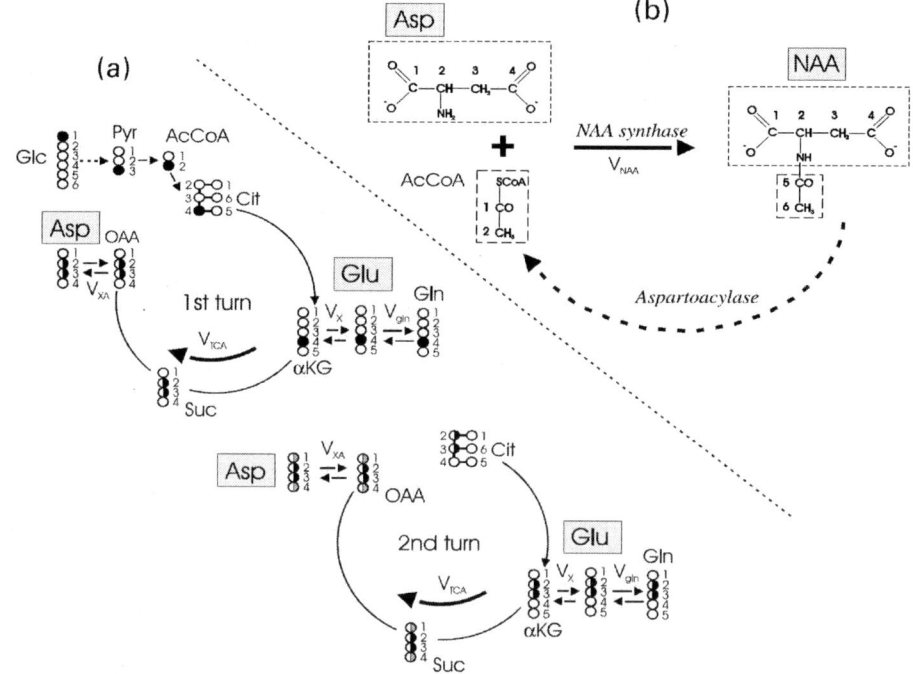

Figure 2. Modeling requires the simplest scheme which can explain the ^{13}C observations: The example given is for derivation of the *in vivo* rate of NAA-synthesis.

Nevertheless, as you would expect, for the physicians and radiologists involved in clinical MRS the actual numbers of studies (PUBMED: 1985-2005 Published MRS studies [1]H = 4000; [31]P = 400; [13]C = 40) reflect the seemingly insuperable difficulties of installing and applying proton-decoupled ^{13}C MRS to their patients. We have adapted a routine 1.5 Tesla clinical MRI scanner to permit rapid ^{13}C MRS of the brain in patients[5] and developed several clinical infusion protocols acceptable to FDA (IND 56,510) for use in infants, children and adults[6]. Using a simplified oral or intravenous infusion protocol, patients and controls received 0.23G. kg[-1] of 1-^{13}C glucose and were followed by serial proton-decoupled ^{13}C MRS for up to 180 minutes[7]. The experimental details have been published.

[13]C Data analysis: As the ^{13}C pool expands it displaces the ^{12}C pool; the extent to which this has occurred is termed fractional enrichment. A single ^{13}C MRS acquisition can be used to determine fractional enrichment of the given ^{13}C metabolite pool at a given time; more usually, multiple ^{13}C MRS acquisitions establish the rate(s) of enrichment of the critical metabolic pools. These new data are expressed as flux-rates. Several hundred peaks appear over the course of such studies so that data analysis has been automated to permit accurate identification of ^{13}C resonances, quantification of ^{13}C metabolites with reference to an internal standard, myoinositol, which is known not to become enriched during the ^{13}C glucose infusion protocol[6], and derivation of fractional enrichment over time for each metabolite of interest (JAUYANG[8]).

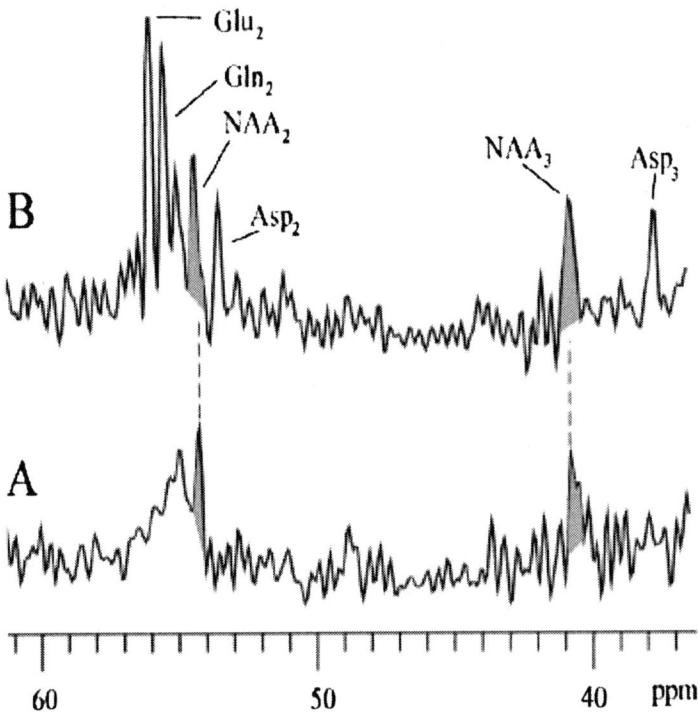

Figure 3. Expanded ^{13}C MR spectrum of Canavan's patient. (A) baseline; (B) ^{13}C enriched (120-140 min). Note labeling of Asp not detectable in the baseline and the significant labeling of NAA in carbons 2,3 (shaded).

Equation 1: $d/dt[NAA^*] = E_{Asp}V_{NAA} - E_{NAA}V_{NAA}$

Metabolic modeling of NAA-synthesis: Figure 2 explains the very simple theory behind the use of ^{13}C glucose to observed NAA-synthesis. $1\text{-}^{13}C$ glucose enters the brain and then, as pyruvate enters the mitochondrion of the neuron, oxidation through the mitochondrial enzyme pyruvate dehydrogenase (PDH) results in conversion to the 2 carbon fragment acetyl-coenzyme A (Figure 2A). After traversing the Krebs cycle (1st turn), ^{13}C atoms appear in oxaloacetate, before rejoining the Krebs cycle (2nd turn). Transaminase activity ensures that ^{13}C aspartate is formed. Incorporation of ^{13}C aspartate into specific carbon atoms of NAA also occurs within the neuronal mitochondrion, as shown (Figure 2 B). ^{13}C MRS readily distinguishes carbon 2,3 of aspartate and of NAA, so that the rate of NAA-synthesis can be derived as shown in Equation 1. For further details and validation of this method please see Moreno *et al.*[7].

NAA-synthesis was determined in the brains of 4 normal control subjects (5 examinations) and in 3 children with enzymically verified Canavan's Disease, with the informed consent of parents, the FDA (IND 56,510) and the Internal Review Board of Huntington Hospital.

3. RESULTS

3.1. NAA-Synthesis in Normal Humans

Net NAA-synthesis was readily determined in normal human subjects. As illustrated in Figure 4, ^{13}C aspartate became rapidly enriched to steady state after about 60 minutes of 1-^{13}C glucose administration. The much slower rate of appearance of ^{13}C in NAA was linear over the 3 hours of this study. Net NAA-synthesis was determined to be 9.2 nano-moles/minute/gram brain in normals, about 1% of the TCA-cycle rate.

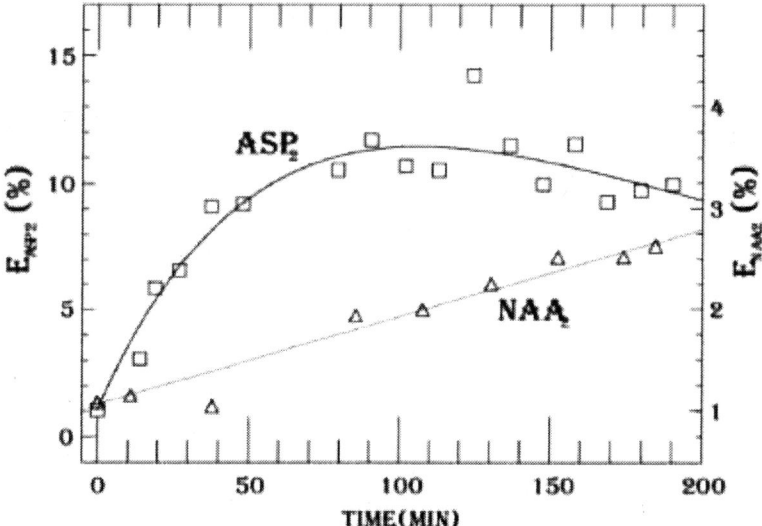

Figure 4. Demonstration of in vivo ^{13}C aspartate and NAA formation in human brain.

3.2. NAA-Synthesis in Children with Canavan's Disease

Net NAA-synthesis was readily measured in children with Canavan's Disease and was reduced by 60% (3.6 nanomoles/minute/gram brain) (Table 2). It is interesting to speculate on a mechanism for this somewhat surprising finding – whether NAA-synthetase is subject to end-product inhibition by the large accumulation of NAA, or limited by substrate supply of its precursor, aspartate? The early in vitro study in rodent brain suggested the former to be more likely, since NAA-synthetase was strongly subject to end-product inhibition[9].

Table 2. Regulation of cerebral NAA in Canavan's Disease. NAA synthesis rate is decreased, despite increased NAA concentration. Aspartate and glutamate are reduced. The hypotheses are that NAA synthetase is subject to feed-back inhibition, or that NAA synthetase is substrate limited.

	[NAA]	[Glu]	[Asp]	E_{NAA2}	V_{NAA}
Control (n=5)	9.2 mM	10 mM	2.8 mM	2.2%	9.2 nm/min/g
Canavan (n=3)	13.7 mM	5.5 mM	1.5 mM	1.4%	3.6 nm/min/g
p	<0.001	<0.001	<0.001	<0.01	<0.001

It is of great interest to enquire whether the rate of synthesis of NAAG can be determined in the human brain? Although the turnover of NAAG has been reported similar to that of glutamate in in vitro analyses of rat brain[10], the concentration of NAAG is less than 10% that of NAA. As a result, we have not observed ^{13}C NAAG or its enrichment from $1-^{13}C$ glucose in *in vivo* human brain studies to date.

3.3. Determination of NAA-Synthesis in Neurodegenerative and Other Brain Disorders

Because the assay as described requires that ^{13}C aspartate be enriched to reach steady state, the forgoing results were obtained using a high-dose $1-^{13}C$ glucose infusion protocol. When this dose is reduced, no steady state was achieved. Nevertheless, an estimate of NAA-synthesis can be obtained simply by comparing the relative enrichments of aspartate, the precursor with that of NAA, the product of the reaction[11]. Results of this assay are displayed in Figure 6 and 7. $NAA_2/Aspartate_2$ was markedly higher in all patients examined with Alzheimer's disease, but the value was not correlated with NAA/Cr (Figure 6). Results in schizophrenics (N=4) were also slightly higher than in controls. However, there was no statistical difference (Figure 7).

4. DISCUSSION

Human NAA-synthesis has been observed directly in vivo, by means of ^{13}C MRS. As expected the rate is very low, suggesting that NAA, in contrast with NAAG, is most unlikely to fulfill any function as a neurotransmitter. In human Canavan's Disease where NAA accumulates in the brain as the direct result of reduction in NAA-deacylase activity, NAA synthesis appears to be strongly inhibited. This appears to be an adaptive process, the result of either end-product inhibition or substrate limited enzyme synthesis. Nevertheless, it could be of importance as effective gene replacement therapy is achieved. Will detection of falling [NAA] be the best measure of success, or will a new equilibrium arise, where high [NAA] is maintained despite recovery from the primary enzyme defect? In neurodegenerative disease, as well as in schizophrenia where neuronal loss has been observed, it might be hypothesized that NAA synthesis would fall. The present results appear to contradict such a hypothesis. With the caveat mentioned, it appears that NAA synthesis rate may be elevated, rather than reduced in one neurodegenerative disease and one psychiatric disease. If so, there is an adaptive metabolic process which may be 'protective' against the neurodegenerative process(es).

The data is sufficiently surprising to suggest the study be repeated under the ideal conditions used to establish the rate of NAA synthesis in Canavan's Disease patients. This means a more costly and lengthier protocol with infusion of sufficient 1-^{13}C glucose to raise the fractional enrichment of aspartate and to maintain a steady state during the entire period of observation of NAA enrichment – as much as 180 minutes.

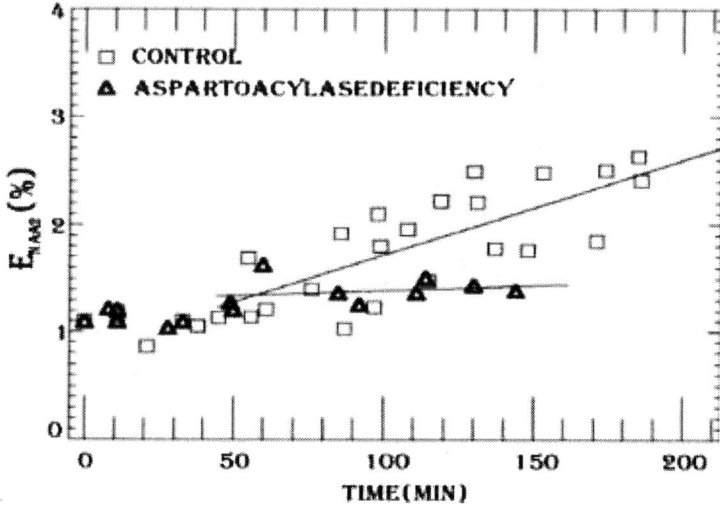

Figure 5. Impact of Canavan disease (aspartoacylase deficiency) on fractional enrichment E_{NAA2} (%).

NAA2/Asp2

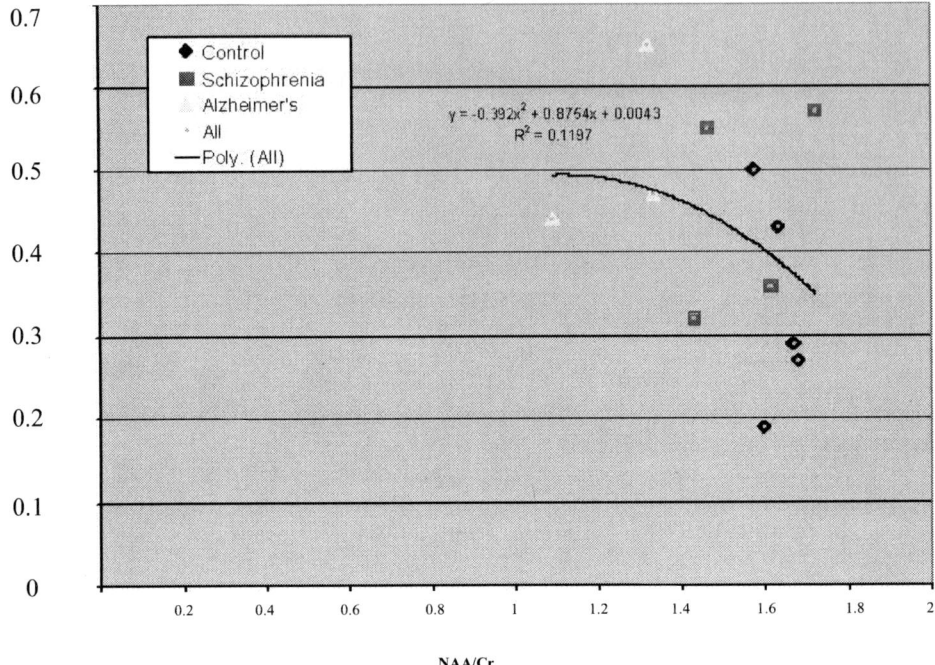

Figure 6. Impact of neurodegenerative and psychiatric deceases (schizophrenia) on cerebral NAA synthesis. Note: Apparent rate of synthesis was derived from fractional enrichment of NAA/Asp and plotted against a ^{1}H MRS measures of normal number (NAA/Cr).

5. CONCLUSION

NAA synthesis is a well regulated metabolic process in human brain. The ability to measure it directly using ^{13}C MRS offers a means of exploring pharmacological and other interventions aimed at restoring or improving neuronal function.

6. ACKNOWLEDGEMENTS

This study was supported by funds to KH and BDR from the National Alliance for Research in Schizophrenia and Depression (NARSAD), by the Boswell Fellowship awarded to PB and by Rudi Schulte Research Institute, Santa Barbara, CA (AL; TT and WW)

Synthesis per Neuron

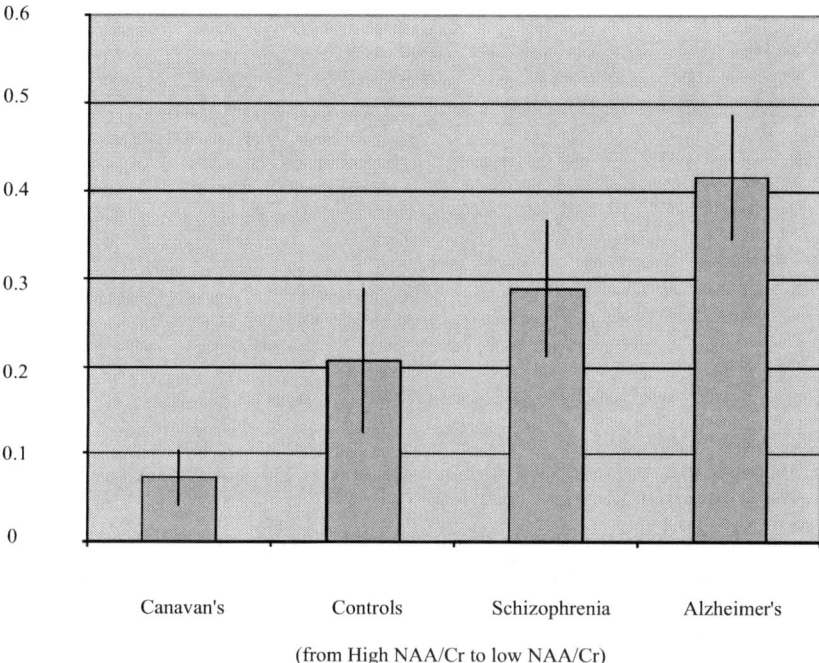

(from High NAA/Cr to low NAA/Cr)

Figure 7. Correlation of NAA Synthesis and NAA Concentration

7. QUESTION AND ANSWER SESSION

DR. GONZALEZ: Questions?

DR. WEINBERGER: Obviously, I am obligated to ask a question. It's just a spectacular series of studies, and I congratulate you.

You know, I'm sitting here thinking, now where do we exactly disagree. I could get into little intimate details about whether you have put voxels in the same place that we have with our slices.

One of the things we found and other groups have found: If you measure NAA *qua* the chemical, as if the brain were a test tube, you don't find NAA differences in schizophrenia. If you measure very discrete regions of the brain, there are some discrete regions sampled somewhat exquisitely that tend to show differences.

Now I think where we have the most fundamental difference of the phase at which we approach this signal is that you have approached this as if the pathology is in NAA, the chemical. You know, we have no evidence of that, one way or the other.

We have approached NAA very much as a proxy of the biology of the neuronal system. I think what is fascinating about your data in terms of there being a pathology of the biosynthesis or processing of NAA and that that is the pathology of the disease, which I, by the way, as in all the histories of neurochemical explanation of schizophrenia, would be very dubious of -- but nevertheless, that is a reasonable hypothesis that has to be tested in other ways. You don't need to study brain to look at glutamate or NAA signals.

There are many other tissues that could be studied if these are fundamental causative mechanisms in the disease, because the genes are the same everyplace.

Let me just ask you this question about the glutamate cycling, which is very interesting. To the extent that there is a stoichiometric relationship of this glutamate cycle and glucose utilization, is this the cause of the problem or is this itself another passenger in the fact that the tissue is less active metabolically?

DR. ROSS: Good question. The Yale Magestretti hypothesis is that the two are tightly coupled and, despite that there are many pools of glutamate, really what we are looking at is turnover rate, which is neurotransmission.

Bicarbonate, the end product, which reflects global metabolism of glucose, is not different between our small set of patients and controls. So we would say, no, this is not a reflection of metabolic disturbance, broadly speaking.

I would be much more interested again in your view. Now that could be mitochondrial metabolism at another level. We are then again looking at a surrogate, and I think this is where Dr. Gonzalez' data is so fascinating.

Clearly, we now see in real life that you can reduce the rate of NAA synthesis. You could reduce -- there must be, we would hypothesize, reduced mitochondrial cycling. How we can only reduce two terms and not the third, we are still playing with that. But I would think in the long term it will turn out to be much more focused on what the brain is doing, what the astrocytes and neurons are doing in speaking to each other, than a genetic marker of glutamate activity outside the brain, for example.

So I would still like to keep it inside the brain and relate it to function.

DR. MATALON: I heard today measurement of NAA in the entire brain tissue. Is there value for this determination under specific circumstances? Let us say, when you do gene therapy or when you do something else, maybe that is a better yardstick than a Voxel.

DR. ROSS: You mean NAA total. That is Gonen's assay, which I have never attempted to reproduce, because it is tough. So you really have to ask him directly.

In our hands, I think we only need the NAA concentration for one reason, and that is you cannot calculate fraction enrichment of a molecule pool that you don't know the size of. So, actually, I would disagree with Michael Weiner's answer to you. The ratios are just as good. We don't really need all this fancy footwork.

You do need to correct for white and gray matter, but in terms of do we need whole brain NAA levels in this context? No. Of course, we wouldn't mind having it, but it is really not important. It matters in the region in which we are measuring, and ultimately, obviously, we are going to have to measure in the dorsolateral prefrontal cortex, which you point out is very important.

DR. WEINER: Brian, did I understand you correctly when you said that you weren't actually giving us the rate of NAA synthesis, but it was the rate relative to the rate of labeling of aspartate?

DR. ROSS: In the Canavan data we are really confident that we have a single point and a single step dynamic assay in which we have collected all the data, and we make a calculation, put our hands on our hearts, and that is a rate.

In the newer data, we are trying to find a cheap and cheerful way, and it is approximately correct to say that, if you enrich an aspartate and you enrich an NAA which is made from that aspartate, then measure the ratio, the difference or the ratio will reflect NAA synthesis. But you are absolutely right. It is not anywhere near the true rate.

Nevertheless, you would have to find an explanation, for example, in Alzheimer's Disease where we have low neuron numbers, how on earth can you get more NAA incorporation. Of course, I think, while we are even standing here, we can think of ways to explain that which wouldn't necessarily mean rate. But within those limitations, these are rates.

DR. NAMBOODIRI: The inhibition of the synthetic enzyme, the purified or partially purified enzyme. The inhibitor IC50; there is an inhibition constant that is 0.5 millimolar of NAA.

DR. ROSS: 0.5 millimolar? So we have much more than that.

DR. NAMBOODIRI: -- and the same activity at 0.5 millimolar NAA. So my question is: When you have a 20-plus fold decrease in NAA, let's say, from 10 millimolar to 0.5 millimolar, how can you expect an inhibition of the NAA synthesis? You would think that NAA synthesis is already inhibited.

DR. ROSS: Yes. I think the reason we all work *in vivo* is ultimately that we know, when you break the system up, regulatory mechanisms are no longer what they were. So we are trying to mimic the *in vivo* concentration.

So all we can do is we know roughly what the Ki for the enzyme is, and it is within the physiological range. So I think it is fair to say that there's end product inhibition of this enzyme *in vivo*, as predicted by the *in vitro* experiment, but not necessarily the same Ki.

DR. GONZALEZ: Okay, thank you very much. Thank you very much, Brian. That was really wonderful.

8. REFERENCES

1. Webb PG, Sailasuta N, Kohler SJ, Raidy T, Moats RA, Hurd RE. Automated single voxel proton MRS: technical development and multisite verification. *Magn Reson Med.* Apr 1994;31(4):365-373.
2. Danielsen E, Ross BD. *Magnetic Resonance Spectroscopy Diagnosis of Neurological Diseases.* New York: Marcel-Dekker; 1999.
3. Ross BD, Lin A, Harris K, Bhattacharya P, Schweinsburg B. Clinical experience with ^{13}C MRS in vivo. *NMR Biomed.* 2003;16(6-7):358-369.
4. Bottomley PA, Hart HR, Edelstein WA, et al. NMR imaging/spectroscopy system to study both anatomy and metabolism. *Lancet.* Jul 30 1983;2(8344):273-274.
5. Bluml S, Moreno A, Hwang JH, Ross BD. 1-^{13}C glucose magnetic resonance spectroscopy of pediatric and adult brain disorders. *NMR Biomed.* Feb 2001;14(1):19-32.
6. Moreno A, Blum S, Hwang JH, Ross BD. Alternative 1- [13]C glucose infusion protocols for clinical ^{13}C MRS examinations of the brain. *Magn Reson Med.* 2001;46(1):39-48.
7. Moreno A, Bluml S, Hwang J, Ross B. Direct determination of the N-acetyl-L-aspartate synthesis rate in the human brain by ^{13}C MRS and [1-^{13}C]glucose infusion. *J Neurochem.* 2001;46(1):347-350.
8. Shic F, Lin AP, Ross BD. Automated data processing of {1H-decoupled} [13]C MR spectra acquired from human brain in vivo. *Journal of Magnetic Resonance.* 2003;162(2):259-268.
9. Truckenmuller ME, Namboodiri MA, Brownstein MJ, Neale JH. N-Acetylation of L-aspartate in the nervous system: differential distribution of a specific enzyme. *J Neurochem.* Nov 1985;45(5):1658-1662.
10. Tyson RL, Sutherland GR. Labeling of N-acetylaspartate and N-acetylaspartylglutamate in rat neocortex, hippocampus and cerebellum from [1-^{13}C]glucose. *Neurosci Lett.* Jul 31 1998;251(3):181-184.
11. Lin AP, Shic F, Enriquez C, Ross BD. Reduced glutamate neurotransmission in patients with Alzheimer's disease -- an in vivo ^{13}C magnetic resonance spectroscopy study. *Magma.* Feb 2003;16(1):29-42.

MAGNETIC RESONANCE SPECTROSCOPY FOR MONITORING NEURONAL INTEGRITY IN AMYOTROPHIC LATERAL SCLEROSIS

Sanjay Kalra and Douglas L. Arnold*

1. INTRODUCTION

ALS is the most common acquired motor neuron disease. It typically presents in the 6th decade of life, but can strike its victims anytime in adulthood. The incidence rate is 2 per 100,000. A definite cause for ALS remains unknown, although approximately 10% of cases have a familial basis. The most commonly identifiable mutation responsible for 15% of these cases is in the copper/zinc superoxide dismutase (SOD1) gene on chromosome 21. ALS is characterized by the degeneration and loss of upper motor neurons (UMNs) originating from the cerebral cortex and of lower motor neurons (LMNs) arising from the brainstem and spinal cord. Neuronal atrophy, loss and inclusions are accompanied by subcortical white matter tract degeneration and astrocytic gliosis.[1] In the brain, the burden of degeneration is most severe in the precentral gyrus followed by the postcentral gyrus, and is less severe and variably present outside this peri-rolandic zone.

Clinical manifestations consist of muscular wasting, fasciculations and severe weakness from LMN involvement. In addition to weakness, UMN degeneration produces limb spasticity, hyperreflexia and loss of fine motor control. The diagnosis of ALS requires demonstration of clinical signs of combined UMN and LMN dysfunction.[2] Due

* Sanjay Kalra, Division of Neurology, Department of Medicine, University of Alberta, 2E3.18 WMC, 8440-112 Street, Edmonton, Alberta, T6G 2B7, Canada, (780) 407-8786, fax (780) 407-1325, sanjay.kalra@ualberta.ca. Douglas L. Arnold, Montreal Neurological Institute and Hospital, Department of Neurology and Neurosurgery, McGill University, 3801 University Street, Montreal, Quebec, H3A 2B4, Canada.

to extramotor involvement cognitive impairment with frontotemporal deficits is present in upwards of 50% of patients.[3]

Depending on the regional pathological burden in the neuraxis, clinical manifestations may present in the limbs and/or bulbar territory with speech or swallowing impairment. The disease is relentlessly progressive and eventually renders its victims quadriplegic, mute, unable to swallow and completely dependent. Half of those affected die secondary to respiratory muscle weakness within 3 years of diagnosis and upwards of 90% die within 5 years of diagnosis.[4]

Despite the fact that ALS is invariably and often rapidly fatal, treatment for the most part remains symptomatic. The only disease modifying agent is riluzole which modestly slows progression and prolongs life by approximately 3-6 months without providing any functional improvement.[5] Therefore, intense effort has been directed towards discovering an effective treatment for the disease. As there is no antemortem test to positively diagnose ALS, this effort is greatly hampered by difficulties in establishing the diagnosis at an early stage and by a lack of sensitive and objective markers of disease progression. An accurate measure would aid in earlier diagnosis and allow quicker access to therapies and inclusion into research trials. A neuroprotective drug would presumably be more effective the sooner it is started. The most reliable end points in clinical trials are death and tracheostomy. Such "hard" endpoints, and others like muscle strength and disability scales (e.g. the revised Amyotrophic Lateral Sclerosis Functional Rating Scale [ALSFRS-R]), require lengthy and expensive trials with hundreds of patients per treatment arm. A more sensitive and less variable endpoint would make development of new drugs speedier and less costly. Objective, non-invasive techniques to quantitatively evaluate the targeted cell in ALS, namely the motor neuron, are lacking.

Reliable scales for evaluation of UMN involvement in particular are sorely needed. This combined with the fact that UMN signs may be absent on physical exam even in the presence of UMN pathology[6-9] has been the impetus for the application of various imaging modalities to evaluate potential biomarkers of UMN degeneration. These have been reviewed elsewhere[10] and include magnetic resonance imaging (MRI), magnetic resonance spectroscopy (MRS), diffusion weighted imaging (DWI), functional MRI, single photon computed tomography (SPECT) and positron emission tomography (PET).

The following is an overview of the application of MRS in ALS with a focus on what has been learned by imaging N-acetylaspartate (NAA).

2. TECHNICAL ASPECTS

Conventional proton MRI techniques produce exquisitely detailed images of brain structure. Since most of the "visible" protons are in water, conventional MRIs are essentially representations of the spatial distribution of water. With suppression of the MR signals arising from water, it is possible to measure the protons in metabolites that are thousands of times less abundant than water (1-10 mM for metabolites vs 70 M for water). Both single-voxel (SV) and multivoxel MR spectroscopic imaging (MRSI) techniques have been used in ALS. The former allows the acquisition of data from a single volume of brain (the volume of interest or VOI). MRSI is a powerful technique that permits analysis of individual spectra derived from brain volumes which are identified after the scan has been acquired.

In the typical long TE (135-288 ms) MRS of the brain, resonance peaks from three metabolites predominate.   The dominant peak is from the methyl groups of NAA or N-acetylaspartylglutamate (NAAG). NAA is localized exclusively to neurons and neuronal processes in the adult human brain[11-13] and provides a useful marker of neuronal integrity. Metabolites giving rise to the other peaks include membrane consti- tuents containing choline (Cho), and creatine and phosphocreatine (Cr) which are involved in energy metabolism.

The density of Cr is usually unaffected in non-acute non-destructive pathologies, including ALS.[14-17] As such it provides a convenient internal reference and metabolites such as NAA are often reported relative to Cr (i.e., as NAA/Cr). The use of this ratio largely corrects for artifacts in the data that can arise from magnetic field inhomogeneity, partial volume effects, and metabolite relaxation properties. Although absolute quantitation of metabolite concentrations is desirable, the technical challenges that must be addressed and the assumptions that are required in their estimation pose challenges to both their precision and accuracy.

3. CROSS SECTIONAL MRS STUDIES IN ALS

3.1 NAA is Decreased in Motor and Extra-Motor Regions

MRS studies in ALS have consistently demonstrated in vivo evidence of impaired neuronal integrity based on the finding of decreased NAA (or NAA/Cho, NAA/(Cho+Cr), NAA/Cr) in relevant regions of the brain, namely the motor cortex, centrum semiovale, internal capsule, and brainstem with sparing of the parietal cortices.[8,14-33] NAA/Cr is lowest in the primary motor cortex and decreased to a lesser extent in the primary sensory cortex and premotor area.[18,31,32] This is in keeping with the anatomy of the corticospinal tract which originates predominantly from these regions of the neocortex.[34]   These observations are also in accordance with the topographical pathology of ALS in which the precentral gyrus, followed by the postcentral gyrus, is most severely affected.[1] Inconsistent abnormalities in NAA between MRS studies of the frontal lobe can at least in part be explained by variable affliction of the frontal lobes in this disease.[21,26,29]

The accuracy of MRS in aiding the diagnosis of ALS has been assessed in few studies. Chan[8] found the sensitivity and specificity to be 79% and 90% respectively with a motor cortex NAA/Cr cut-off of 2.5 (2.12 standard deviations below the control mean). With a 2.5 standard deviation cut-off for NAA/Cho Pohl[35] found the sensitivity to increase with the degree of certainty in the diagnosis from 47% to 64% (specificity was not provided).

3.2 Correlations with Clinical Indices

Abnormal NAA indices have shown correlations with clinical parameters such as strength[16,36] and severity of some UMN signs.[8,14,15,18,21,22,27,29,36] Diminished frontal lobe NAA/Cr was found to correlate with cognitive dysfunction.[26] Correlations have not been demonstrated with spasticity or muscle stretch reflexes.[36] Most,[17,23,36] but not all,[16] note a relationship to the site of onset of symptoms (bulbar versus limb), but not symptoms or disease duration.[16,17,22,30,36]

Reports of associations with disease severity measures have been favourable. Correlations have been noted with the ALS Severity Scale,[36] Norris Scale,[17,26] and Jablecki Scale,[16] but not with the ALSFRS.[17] The ALSFRS bulbar subscores have been correlated with glutamate/glutamine but not NAA spectral intensities from the brainstem.[33]

The lack of uniformity with clinical correlations is likely in part due to the fact that disability scales usually are influenced by both LMN and UMN dysfunction and some elements are heavily weighted for disability secondary to LMN impairment. Weak or absent correlations with measures of UMN function, such as reflex score and spasticity scales (eg. Ashworth), is not surprising since these are imprecise and ordinal in nature.

4. LONGITUDINAL MRS STUDIES IN ALS

Longitudinal case studies and small series of patients have revealed declining NAA in most patients with ALS over a 1 to 12 month period in the motor[15,16,18,23,29,37,38] and cingulate cortices.[23] The decline is regionally specific: NAA/Cr decreased significantly more in the primary motor cortex compared to the primary sensory cortex.[39] A significant degree of inter-individual variability in the magnitude of longitudinal changes is evident.

5. OBSERVATION OF DRUG EFFECT USING MRS

Decreased NAA has often been interpreted as indicating decreased relative neuronal density either due to loss of neurons or atrophy of their processes. However, it has become clear that a low NAA can also be the consequence of neuronal metabolic dysfunction; i.e., neurons whose NAA concentration has decreased. In cell culture[40] and rodent experiments[41-43] NAA density decreased in response to a metabolic stress and then increased after removal of the stress. This "recovery" of NAA has been observed in humans in the resolving inflammatory lesions of multiple sclerosis and resolving stroke-like lesions of patients with mitochondrial encephalomyopathies.[44]

Increases in NAA have also been demonstrated in response to therapeutic interventions, including interferon treatment of multiple sclerosis,[45] cerebral revascularization for carotid stenosis,[46] surgery for temporal lobe epilepsy,[47] and zidovudine treatment of AIDS.[48] We speculate that these increases in NAA could involve two mechanisms: either increases in neuronal volume due to reversal of atrophy of the soma or dendrites or improved functioning of metabolically compromised neuronal mitochondria.[44]

We have observed a drug-induced increase in NAA/Cr as measured by MRSI in the motor cortex of ALS patients on the order of 6% approximately 3 weeks after starting the anti-glutamatergic drug riluzole.[38] As discussed above the increase in NAA/Cr is due to either an increase in neuronal volume, increased NAA concentration within neurons, or both. Neuronal volume could increase in ALS with reversal of the dendritic atrophy that is known to exist. However, it is probably more likely that the increased NAA/Cr - results from improved mitochondrial functioning, possibly associated with decreased glutamate-mediated excitotoxicity, considering the anti-glutamatergic effects of riluzole.

In a similar protocol, although with fewer patients, gabapentin and intrathecal brain derived nerve growth factor (BDNF) did not have an observable effect.[31,32] Vielhaber demonstrated a relative increase in NAA in the motor cortex of patients with ALS after one month of oral creatine supplementation.[24] Interestingly, creatine supplementation decreases the combined glutamate/glutamine (Glx) spectroscopic resonance intensity in the SOD1 mouse suggesting that it may have anti-glutamatergic properties in vivo in this animal model.[49] A drug effect on spectroscopic indices was not observed by Bowen in ALS;[15] however, the experimental paradigm was different from the aforementioned in that there was a 2 week washout period of medication before the first scan, the follow-up scans were performed at a variable interval of 2-18 weeks after starting treatment, and two drugs were tested such that each patient started either riluzole or gabapentin.

6. CONCLUSION AND FUTURE DIRECTIONS

MRS can provide important insights into the regional chemical pathology of the brain and how this chemical pathology changes in response to drug therapy. As an index of neuronal integrity, NAA holds promise as a biomarker of degeneration and a surrogate marker of therapeutic efficacy in ALS. NAA changes in the brain parallel the spatial distribution of pathologic changes and modest associations are present with clinical parameters.

However, to date, the finding of reduced motor cortex NAA lacks discriminatory power to distinguish *individual* ALS patients from healthy controls because of significant overlap between the two groups. Furthermore, longitudinal changes in NAA over the very short term are variable and relatively small and challenge the precision of the method. The increasing use of higher field magnets should substantially help in this regard. Measurement of other metabolites (e.g. glutamate) using high field techniques could supplement NAA imaging; this could increase the potential of MRS as a diagnostic instrument and as a tool to better understand pathogenesis. Additionally, technical developments are necessary to decrease scan times so that more patients can tolerate the procedure.

Further studies are required to delineate the temporal and spatial profile of the NAA response to effective drugs, to validate these responses as surrogates of clinical efficacy, and to improve the precision of the measurement. Such data will be critical in the development of NAA as a surrogate marker of therapeutic efficacy in ALS.

7. QUESTION AND ANSWER SESSION

DR. GONZALEZ: Thank you, Doug, for that very elucidating talk. Are there any questions for Professor Arnold?

DR. ROSS: I really enjoyed that talk, and by the way I had asked a question in one of my slides: In ALS, could you calculate the rate of NAA synthesis, and you have just given us the answer.

It is actually 1 nanomole per minute per gram, which is only 10 percent of the maximum rate we have shown in the human brain. So re-synthesis could be what you are showing.

DR. GONZALEZ: Doug, I have a question. The fascinating rapid reversal by therapy - - did you do any behavioral or clinical evaluation? Do these people respond at that level?

DR. ARNOLD: We don't have those data, unfortunately.

DR. WEINER: Doug, we have been following up on the glutamate story since we have been able to do studies at 4 Tesla. Unfortunately, because of shimming problems, deep in the brain stem it is difficult to get spectra at 4 Tesla. So we (i.e., Alana Kaiser in our group) have focused mostly on motor strip and corona radiata.

The major finding is that NAA is reduced in the gray matter of the cortex. Glutamate is also reduced, but glutamate is much less reduced than NAA. So, in fact, the glutamate over NAA ratio is increased.

What that means, I don't know, but that's what we found.

DR. ARNOLD: That is very interesting to hear. I don't know what it means either, because I, find it difficult to understand how changes in cytoplasmic glutamate relate to vesicular glutamate and neurotransmission.

Dr. Ross, what do you think these measurements of six millimolar glutamate, which is mostly in the cytoplasm actually tells us about glutamatergic transmission?

DR. ROSS: I think the steady state concentration, as you rightly point out, is reflecting the cytosolic neuronal concentration of glutamate, which is tenfold higher than the astrocyte, but that is not the synaptosomal pool which is responsible for neurotransmission.

So I think it gets to be very complicated. However, we definitely need the data, particularly of the kind that Michael was just telling us about. Glutamate concentrations will have some impact on the neurotransmitter rate, but they do not reflect it directly.

So I agree. It is more complicated than just the steady state.

DR. MATALON: One little question regarding riluzol, I saw you had SMA patient data on your slide. I have a patient with SMA now that is enrolled in a trial. The idea there is that glutamate leads to neuronal death. But I didn't see any data regarding that. How do we rescue neurons from glutamate toxicity?

DR. ARNOLD: Well, I think the theory is that one simply uses an antagonist such as riluzol. Right? And we have shown some evidence that neuronal integrity can be improved after riluzole treatment. Perhaps I am not understanding your question.

DR. GONZALEZ: Perhaps you can carry on this conversation at the break. We are running late. Thank you very much, Dr. Arnold.

8. REFERENCES

1. Martin,J.E. and Swash,M. The pathology of motor neuron disease. In: *Motor Neuron Disease. Biology and Management.*, edited by Leigh,P.N. and Swash,M. London: Springer-Verlag, 1995, p. 93-118.
2 . B.R. Brooks, R.G. Miller, M. Swash and T.L. Munsat, El Escorial revisited: revised criteria for the diagnosis of amyotrophic lateral sclerosis, *Amyotroph Lateral Scler Other Motor Neuron Disord* 1(5), 293-299 (2000).
3. C. Lomen-Hoerth, J. Murphy, S. Langmore, et al, Are amyotrophic lateral sclerosis patients cognitively normal? *Neurology* 60(7), 1094-1097 (2003).
4. *Amyotrophic Lateral Sclerosis*, edited by H. Mitsumoto, D. Chad, and E.P. Pioro: Philadelphia:F.A. Davis Company, 1998.
5. L. Lacomblez, G. Bensimon, P.N. Leigh, P. Guillet and V. Meininger, Dose-ranging study of riluzole in amyotrophic lateral sclerosis. Amyotrophic Lateral Sclerosis/Riluzole Study Group II, *Lancet* 347(9013), 1425-1431 (1996).
6. B. Brownell, D.R. Oppenheimer and J.T. Hughes, The central nervous system in motor neurone disease, *J Neurol Neurosurg Psychiatry* 33(3), 338-357 (1970).

7. T. Lawyer and M.G. Netsky, Amyotrophic lateral sclerosis: a clinicoanatomic study of fifty-three cases, Arch Neurol Psychiatry 69, 171-192 (1953).
8. D.K. Leung DK, A.P. Hays, K. Geysu, M.L. DelBene and L.P. Rowland, Diagnosis of ALS: clinicopathologic analysis of 76 autopsies, Neurology 52(Suppl 2), A164 (1999).
9. P.G. Ince, J. Evans, M. Knopp, et al, Corticospinal tract degeneration in the progressive muscular atrophy variant of ALS, Neurology 60(8), 1252-1258 (2003).
10. S. Kalra and D. Arnold, Neuroimaging in amyotrophic lateral sclerosis, Amyotroph Lateral Scler Other Motor Neuron Disord 4(4), 243-248 (2003).
11. M.L. Simmons, C.G. Frondoza and J.T. Coyle, Immunocytochemical localization of N-acetyl-aspartate with monoclonal antibodies, Neuroscience 45(1), 37-45 (1991).
12. J.R. Moffett, M.A. Namboodiri, C.B. Cangro and J.H. Neale, Immunohistochemical localization of N-acetylaspartate in rat brain, Neuroreport 2(3), 131-134 (1991).
13. J. Battistuta, C. Bjartmar and B.D. Trapp, Postmortem degradation of N-acetyl aspartate and N-acetyl aspartylglutamate: an HPLC analysis of different rat CNS regions, Neurochem Res 26(6), 695-702 (2001).
14. O. Gredal, S. Rosenbaum, S. Topp, et al, Quantification of brain metabolites in amyotrophic lateral sclerosis by localized proton magnetic resonance spectroscopy, Neurology 48(4), 878-881 (1997).
15. B.C. Bowen, P.M. Pattany, W.G. Bradley, et al, MR imaging and localized proton spectroscopy of the precentral gyrus in amyotrophic lateral sclerosis, AJNR Am J Neuroradiol 21(4), 647-658 (2000).
16. C. Pohl, W. Block, J. Karitzky, et al, Proton magnetic resonance spectroscopy of the motor cortex in 70 patients with amyotrophic lateral sclerosis, Arch Neurol 58(5), 729-735 (2001).
17. P. Sarchielli, G.P. Pelliccioli, R. Tarducci, et al, Magnetic resonance imaging and 1H-magnetic resonance spectroscopy in amyotrophic lateral sclerosis, Neuroradiology 43(3), 189-197 (2001).
18. E.P. Pioro, J.P. Antel, N.R. Cashman and D.L. Arnold, Detection of cortical neuron loss in motor neuron disease by proton magnetic resonance spectroscopic imaging in vivo, Neurology 44(10), 1933-1938 (1994).
19. A.P. Jones, W.J. Gunawardena, C.M. Coutinho, et al, Preliminary results of proton magnetic resonance spectroscopy in motor neurone disease (amyotrophic lateral sclerosis), J Neurol Sci 129 Suppl 85-89 (1995).
20. M. Giroud, P. Walker, D. Bernard, et al, Reduced brain N-acetyl-aspartate in frontal lobes suggests neuronal loss in patients with amyotrophic lateral sclerosis, Neurological Research 18(3), 241-243 (1996).
21. W.D. Rooney, R.G. Miller, D. Gelinas, et al, Decreased N-acetylaspartate in motor cortex and corticospinal tract in ALS, Neurology 50(6), 1800-1805 (1998).
22. W. Block, J. Karitzky, F. Traber, et al, Proton magnetic resonance spectroscopy of the primary motor cortex in patients with motor neuron disease: subgroup analysis and follow-up measurements, Arch Neurol 55(7), 931-936 (1998).
23. M.J. Strong, G.M. Grace, J.B. Orange, et al, A prospective study of cognitive impairment in ALS, Neurology 53(8), 1665-1670 (1999).
24. S. Vielhaber, J. Kaufmann, M. Kanowski, et al, Effect of creatine supplementation on metabolite levels in ALS motor cortices, Exp Neurol 172(2), 377-382 (2001).
25. N. Schuff, W.D. Rooney, R. Miller, et al, Reanalysis of multislice (1)H MRSI in amyotrophic lateral sclerosis, Magn Reson Med 45(3), 513-516 (2001).
26. K. Abe, M. Takanashi, Y. Watanabe, et al, Decrease in N-acetylaspartate/creatine ratio in the motor area and the frontal lobe in amyotrophic lateral sclerosis, Neuroradiology 43(7), 537-541 (2001).
27. W. Kenn, G. Ochs, T.A. Pabst and D. Hahn, 1H spectroscopy in patients with amyotrophic lateral sclerosis, J Neuroimaging 11(3), 293-297 (2001).
28. C.C. Hanstock, V.A. Cwik and W.R. Martin, Reduction in metabolite transverse relation times in amyotrophic lateral sclerosis, J Neurol Sci 198 37-41 (2002).
29. J. Suhy, R.G. Miller, R. Rule, et al, Early detection and longitudinal changes in amyotrophic lateral sclerosis by (1)H MRSI, Neurology 58(5), 773-779 (2002).
30. V.A. Cwik, C.C. Hanstock, P.S. Allen and W.R. Martin, Estimation of brainstem neuronal loss in amyotrophic lateral sclerosis with in vivo proton magnetic resonance spectroscopy, Neurology 50(1), 72-77 (1998).
31. S. Kalra, A. Genge and D. Arnold, A prospective, randomized, placebo-controlled evaluation of corticoneuronal response to intrathecal BDNF therapy in ALS using magnetic resonance spectroscopy: feasibility and results, Amyotroph Lateral Scler Other Motor Neuron Disord 4(1), 22-26 (2003).
32. S. Kalra, N.R. Cashman, Z. Caramanos, A. Genge and D.L. Arnold, Gabapentin therapy for amyotrophic lateral sclerosis: lack of improvement in neuronal integrity shown by MR spectroscopy, Am J Neuroradiol 24(3), 476-480 (2003).

33. E.P. Pioro, A.W. Majors, H. Mitsumoto, D.R. Nelson and T.C. Ng, 1H-MRS evidence of neurodegeneration and excess glutamate + glutamine in ALS medulla, *Neurology* **53**(1), 71-79 (1999).
34. Parent,A. *Carpenter's Human Neuroanatomy*, Baltimore:Williams and Wilkins, 1996. Ed.9 pp. 864-936.
35. C. Pohl, W. Block, F. Traber, et al, Proton magnetic resonance spectroscopy and transcranial magnetic stimulation for the detection of upper motor neuron degeneration in ALS patients, *J Neurol Sci* **190**(1-2), 21-27 (2001).
36. C.M. Ellis, A. Simmons, C. Andrews, et al, A proton magnetic resonance spectroscopic study in ALS: correlation with clinical findings, *Neurology* **51**(4), 1104-1109 (1998).
37. V.A. Cwik, C.C. Hanstock, C. Boyd, et al, Regional neuronal dysfunction in amyotrophic lateral sclerosis (ALS): in vivo measurement with proton magnetic resonance spectroscopy (MRS), Neurology 48(Suppl 2), A216 (1997).
38. S. Kalra, N.R. Cashman, A. Genge and D.L. Arnold, Recovery of N-acetylaspartate in corticomotor neurons of patients with ALS after riluzole therapy, *Neuroreport* **9**(8), 1757-1761 (1998).
39. S. Kalra, A. Genge, N.R. Cashman, and D.L. Arnold, 1H-MRSI demonstrates a regional decline in corticoneuronal integrity in amyotrophic lateral sclerosis, Amyotroph Lateral Scler Other Motor Neuron Disord 2(Suppl 2), 118-119 (2001).
40. Matthews P M, Cianfaglia L, McLaurin J, et al. *Proceedings of the Society of Magnetic Resonance and the European Society for Magnetic Resonance in Medicine and Biology* **1** 147 (1995) (Abstract)
41. B.G. Jenkins, P. Klivenyi, E. Kustermann, et al, Nonlinear decrease over time in N-acetyl aspartate levels in the absence of neuronal loss and increases in glutamine and glucose in transgenic Huntington's disease mice, *J Neurochem* **74**(5), 2108-2119 (2000).
42. C. Gasparovic, N. Arfai, N. Smid and D.M. Feeney, Decrease and recovery of N-acetylaspartate/creatine in rat brain remote from focal injury, *J Neurotrauma* **18**(3), 241-246 (2001).
43. C. Demougeot, P. Garnier, C. Mossiat, et al, N-Acetylaspartate, a marker of both cellular dysfunction and neuronal loss: its relevance to studies of acute brain injury, *J Neurochem* **77**(2), 408-415 (2001).
44. N. De Stefano, P.M. Matthews and D.L. Arnold, Reversible decreases in N-acetylaspartate after acute brain injury, *Magn Reson Med* **34**(5), 721-727 (1995).
45. S. Narayanan, N. De Stefano, G.S. Francis, et al, Axonal metabolic recovery in multiple sclerosis patients treated with interferon beta-1b, *J Neurol* **248**(11), 979-986 (2001).
46. M. Uno, S. Ueda, H. Hondo, K. Matsumoto and M. Harada, Effectiveness of revascularization surgery evaluated by proton magnetic resonance spectroscopy and single photon emission computed tomography, *Neurol. Med. Chir. (Tokyo)* **36**(8), 560-566 (1996).
47. F. Cendes, F. Andermann, F. Dubeau, P.M. Matthews and D.L. Arnold, Normalization of neuronal metabolic dysfunction after surgery for temporal lobe epilepsy. Evidence from proton MR spectroscopic imaging, *Neurology* **49**(6), 1525-1533 (1997).
48. J. Vion-Dury, F. Nicoli, A.M. Salvan, et al, Reversal of brain metabolic alterations with zidovudine detected by proton localised magnetic resonance spectroscopy, *Lancet* **345**(8941), 60-61 (1995).
49. O.A. Andreassen, B.G. Jenkins, A. Dedeoglu, et al, Increases in cortical glutamate concentrations in transgenic amyotrophic lateral sclerosis mice are attenuated by creatine supplementation, *J Neurochem* **77**(2), 383-390 (2001).

HYPOACETYLASPARTIA: CLINICAL AND BIOCHEMICAL FOLLOW-UP OF A PATIENT

Alessandro P. Burlina,[*1] B. Schmitt,[2] U. Engelke,[3] Ron A. Wevers,[3] Alberto B. Burlina[4] and Eugen Boltshauser[2]

1. INTRODUCTION

In 2001, Martin et al. reported absence of N-acetylaspartate (NAA) and N-acetylaspartylglutamate (NAAG) signals on proton magnetic resonance spectroscopy (MRS) in the brain of a 3-year-old child with "neurodevelopmental retardation and moderately delayed myelination".[1] The authors challenged provocatively the concept of NAA being a neuronal marker and its importance for the neuronal viability. Their interpretation, however, was questioned in two subsequent letters to the Editor.[2, 3] Here, we report the clinical and biochemical follow-up of the patient who suffered from a new neurometabolic disorder which likely stems from a deficiency of NAA synthesis. The implication of the "NAA-NAAG metabolic system" in this disorder is discussed.

2. CASE REPORT

This boy was adopted as a neonate, therefore his family history as well as information about the pregnancy and birth are scant. Delayed milestones became obvious in the second half of the 1st year. He was first seen for developmental assessment at 19 months. At that time a developmental quotient of 0.5 was found. Neuroimaging at 2 years 3 months and (repeated) at 3 years 6 months revealed normal MRI, except for a moderate delayed myelination, but absent NAA and NAAG at proton magnetic resonance spectroscopy. The usual metabolic work-up as well as karyotype were normal. This patient is now 8 years 6 months. His further course can be summarised as follows: he developed secondary microcephaly, actual

* [1]Department of Neuroscience, Neurological Clinic, University Hospital, Via Giustiniani 5, I-35128, Padova, ITALY, E-mail: alessandro.burlina@unipd.it; [2]Department of Neurology, University Children's Hospital, Zurich, SWITZERLAND; [3]Laboratory of Pediatrics and Neurology, University Medical Centre, Nijmegen, THE NETHERLANDS; [4]Department of Pediatrics, Metabolic Unit, University Hospital, Padova, ITALY

head circumference is 49.5 cm (-2.5 SD). With increasing age his impairment became more obvious. He finally achieved to walk unaided for short distances, but he is definitely unsteady (ataxic). He has a very short attention span, his play is explorative, stereotyped and repetitive. He does vocalise sounds but has no expressive speech. His cognitive level is estimated by medical and educational professionals to be less than 12 months. At 5 years 9 months, he had a first prolonged generalised febrile seizure. Since then, he had 11 additional episodes of non-febrile status epilepticus, often requiring hospital admission. Anticonvulsant medication proved difficult so far in view of side effects. More recently, seizure control has improved on a different anticonvulsant treatment.

The complete metabolic work-up, including plasma amino acids, urinary organic acids as well as urinary oligo- and mucopolysaccharides, did not show any abnormality. There was no evidence for a mitochondrial, organic acid, amino acid or peroxisomal disorder. Cerebrospinal fluid (CSF) study was unremarkable for cell count, glucose, protein, biogenic amines, folate, and pterines (on two separate determinations).

In summary, this boy has a profound neurological dysfunction with truncal ataxia, lack of expressive speech, behaviour abnormalities, marked cognitive impairment, epilepsy, and secondary microcephaly (for additional clinical data see Boltshauser et al. 2004).[4]

2.1. NAA and NAAG Measurements

Body fluids (CSF, serum, and urine) for NAA and NAAG measurements were obtained on two occasions: the first time at 3 years 10 months during a minor surgical intervention performed in general anaesthesia, the second time at 5 years 6 months, after sedation with midazolam. The first sample of CSF was analysed by high-resolution proton MRS and data were reported by Martin et al.[1] NAA and NAAG were not detected. The second samples (CSF and urine) were analysed either by high-resolution ^{1}H-MRS (for more details on MRS measurements see Engelke et al., 2004)[5] or by capillary electrophoresis (CE) according to Burlina et al.[6] We used a Beckman P/ACE MDQ capillary electrophoresis system which allows us to reach a sensitivity, for NAA and NAAG, 9 and 15-fold respectively higher than ^{1}H-MRS measurements.

The CE analysis of the second sample of CSF of the patient showed no detectable levels of NAA (that means < 1.14 µmol/L, which is our detection limit). NAA is, in our experience, usually undetectable also in control samples.[6] Faull and colleagues reported that lumbar CSF NAA concentrations were around 1µM in adults.[7] Furthermore, CE analysis did not detect NAAG (that means < 0.66 µmol/L, which represents our detection limit) in the CSF of our patient. The CE analysis of the urine of the patient, collected at the same time of the CSF, measured 21 and 1.9 µmol/mmol creatinine for NAA and NAAG respectively, which are in the normal range.[6] Interestingly, we could not detect NAAG in the CSF of our patient while it is generally present in micromolar range, at different age, in human CSF (see also Chapter 29, this volume).[6-8]

3. DISCUSSION

The follow-up of this patient clearly shows that hypoacetylaspartia is a severe neurological disease, with cognitive and behavior impairment, motor dysfunction, and epilepsy.

The biochemical features, besides the absence of NAA in the brain, as already reported by Martin et al.,[1] include non-detectable levels of NAAG in the CSF of the patient. The absence of NAA and NAAG support the hypothesis of a block of the enzyme L-aspartate N-acetyltransferase (ANAT) which synthesizes the formation of NAA and therefore renders NAA available for the formation of NAAG from NAA and glutamate, as recently demonstrated.[9] The study of Gehl and collaborators confirms our hypothesis that in particular clinical condition, as in Canavan disease, the proportional increase of urinary NAA and NAAG may be due to the activation of the anabolic pathway of NAAG.[10]

Surprisingly, urinary concentrations of NAA and NAAG were normal, even though at lower levels, in our patient with hypoacetylaspartia. These findings are not well understood yet, but we may hypothesize the existence of different peripheral sources (kidney, for example) for the production of small amounts of excretable NAA and NAAG.[4]

In conclusion, hypoacetylaspartia is a new disease entity, most likely an inherited neurometabolic disorder, with a moderate delayed myelination and peculiar biochemical findings: low/absent amount of NAA and NAAG. It represents the opposite biochemical condition of Canavan disease, a severe dysmyelinating disease characterized by increased concentrations of NAA and NAAG. These features suggest a possible role for the "NAA-NAAG metabolic system" in dysmyelinating disorders.

4. QUESTION AND ANSWER SESSION

DR. GONZALEZ: Thank you very much, Dr. Burlina. The paper is now open for discussion. Yes, Dr. Ledeen?

DR. LEDEEN: I would just like to ask: Was your identification of NAA and NAAG entirely based on spectroscopic data or was there additional chemical confirmation?

DR. BURLINA: We identified and measured NAA and NAAG with capillary electrophoresis, according to the specific method we developed 10 years ago. The Dutch group of researchers, who worked with us, measured NAA and NAAG with proton MRS spectroscopy, but this technique, with a high specificity, does not have enough sensitivity for low amounts of both compounds. The combination of the two technique represents one of the most powerful analytical system of measurement.

DR. TRAPP: Very lovely presentation and follow-up on this child. Certainly, I think it would be very nice if this was a case of a deficit in NAA synthesis due to a gene defect, but one question I have is: Have you checked for some major genetic defects in this child, such as deletions or duplications or chromosomal rearrangements which could not only cause changes in NAA but may be responsible for a lot of the other symptoms that this child is displaying?

DR. BURLINA: Yes. Chromosomal rearrangements were checked by the Swiss neuropediatricians. At the moment, I am not aware of other genetic investigations.

DR. BASLOW: I just wanted to comment that in the paper it is indicated that EEG was normal. So, these neurons are firing, and you can measure the activity in the brain.

DR. BURLINA: As a neurologist, I can tell you that EEG is very important, but a normal EEG doesn't necessarily mean a normal brain.

DR. BASLOW: The only point is that the NAA and NAAG system appeared not to be absolutely necessary for neuron function.

DR. BURLINA: Please, don't say that.

DR. BASLOW: Just synaptic firing. They are firing.

DR. BURLINA: Yes. Now I understand your point.

DR. BASLOW: But the information transferred usually is not useful, and that's the only point I am making.

DR. GONZALEZ: If there's no -- oh, any further comments or questions? Yes?

DR. NAMBOODIRI: You said there is a secondary microcephaly. Any relationship can you comment on?

DR. BURLINA: Yes. Again, I think the comparison with Canavan disease is interesting. Canavan disease is characterized by macrocephaly and an increase of NAA (and NAAG). What is the relationship between the size of the head and NAA/NAAG ?

I would like to see another brain MRI of our patient, to monitor the myelination process. Can we really be sure that it is only a "moderate delayed myelination"? I know there are strong difficulties to get the child again for another brain MRI, but I am confident that Prof. Boltshauser will do his best.

DR. ROSS: I seem to remember hearing the child originated in Romania. Did you tell us that?

DR. BURLINA: I know he is from an Eastern country. I don't remember which it was.

DR. ROSS: Maybe NIH should fund a magnet to be placed somewhere in that area!

DR. MEYERHOFF: What sort of therapies have been tried in this patient?

DR. BURLINA: Anticonvulsant therapy. Valproate was administered, but it didn't work because of important side effects. So, valproate was switched to phenobarbital, and seizure control started to improve.

DR. MATALON: My guess is that this patient did not have total deficiency of NAA, that synthesis was not completely blocked in the brain. The spectroscopy of the brain does not show low levels, at least not so low, enough to give some myelination. Furthermore, NAA effluxes from the brain very rapidly, as we see with Canavan disease, and, as a consequence, you get tremendous amount of NAA in urine.

In the CSF the fact that you could not measure, maybe it is beyond the level of detection of the methodology.

DR. BURLINA: Yes, I agree with you that it could be a partial deficit of NAA synthesis. That is a possibility. That is also the reason why I prefer to call the disease hypoacetylaspartia. I think there is a possibility there are effluxes of NAA from the brain, but may be, at least in this case, it would require an important blood-brain barrier damage.

Concerning our measurement method, I mean capillary electrophoresis analysis, I should say that our detection limits and our normal levels are in agreement with those reported by other groups using HPLC systems (see Ref. 7, 8). I do not think it can be a problem of level of detection. And, as I told you, NAA is usually undetectable in our control population; on the contrary, NAAG in CSF is present in micromolar range in control samples.

DR. PHILIPPART: Have you been able to get any -- studies, especially visual which are very simple.

DR. BURLINA: I do not think they were performed, I think they were not reported in the Martin's paper. We should ask Prof. Boltshauser.

DR. PHILIPPART: It is so simple, and it would give you some indication about myelination.

DR. BURLINA: Yes, you are right, it is a good suggestion. Please, do not forget, that I am a neurologist, but in this study I played the role of neurochemist. So, I did not see the patient.

DR. GONZALEZ: Thank you very much. That was really a wonderful presentation. Thank you.

5. REFERENCES

1. E. Martin, A. Capone, J. Schneider, J. Hennig, and T. Thiel, Absence of *N*-acetylaspartate in the human brain: impact on neurospectroscopy?, *Ann. Neurol.* **49**, 518-521 (2001).
2. E.V. Sullivan, E. Adalsteinsson, D.M. Spielman, R.E. Hurd, and A. Pfefferbaum, N-acetylaspartate – A marker of neuronal integrity, *Ann. Neurol.* **50**, 823 (2001).
3. D.L. Arnold, N. de Stefano, P.M. Matthews, and B. Trapp, N-acetylaspartate: usefulness as an indicator of viable neuronal tissue, *Ann. Neurol.* **50**, 823 (2001).
4. E. Boltshauser, B. Schmitt, R.A. Wevers, U. Engelke, A.B. Burlina and A.P. Burlina, Follow-up of a child with hypoacetylaspartia, *Neuropediatrics* **35**, 255-258 (2004).
5. U.F.H. Engelke, M.L.F. Liebrand-Van Sambeek, J.G.N. De Jong, J.G. Leroy, E. Morava, J.A.M. Smeitink and R.A. Wevers, N-acetylated metabolites in urine: proton nuclear magnetic resonance spectroscopic study on patients with inborn errors of metabolism, *Clin. Chem.* **50**, 58-66 (2004).
6. A.P. Burlina, V. Ferrari, P. Divry, W. Gradowska, C. Jakobs, M.J. Bennett, A.C. Sewell, C. Dionisi-Vici, and A.B. Burlina, N-acetylaspartylglutamate in Canavan disease: an adverse effector ?, *Eur. J. Pediatr.* **158**, 406-409 (1999).
7. K.F. Faull, R. Rafie, N. Pascoe, L. Marsh, and A. Pfefferbaum, N-acetyl aspartic acid (NAA) and N-acetyl aspartylglutamic acid (NAAG) in human ventricular, subarachnoid, and lumbar cerebrospinal fluid, *Neurochem. Res.* **24**, 1249-1261 (1999).
8. V. Brovia, A. Ricciardi, and L. Barbeito, N-acetyl-aspartylglutamate (NAAG) in human cerebrospinal fluid: determination by high performance liquid chromatography, and influence of biological variables, *Amino Acids* **9**, 175-184 (1995).
9. L.M. Gehl, O.H. Saab, T. Bzdega, B. Wroblewska, and J.H. Neale, Biosynthesis of NAAG by an enzyme-mediated process in rat central nervous system neurons and glia, *J. Neurochem.* **90**, 989-997 (2004).
10. A.P. Burlina, V. Ferrari, L. Bonafé, and A.B. Burlina, Clinical and biochemical follow-up of a patient with Canavan disease, *J. Inherit. Metab. Dis.* **21**(suppl. 2), 48 (1998).

CELLULAR LOCALIZATION OF NAAG

Suzannah Bliss Tieman[*]

1. INTRODUCTION

N-Acetylaspartylglutamate (NAAG), the most abundant brain peptide,[1] has been localized immunocytochemically in a variety of neurons throughout the nervous system. In many cases, NAAG's presence has been confirmed by HPLC, but these results will not be described here. NAAG is also sometimes found in glia,[2, 3] which are partially responsible for its synthesis.[4] In neurons, NAAG is seen in the cytoplasm around the nucleus, in proximal dendrites, and in axons and terminals, including synaptic vesicles. An overview of the neurons and systems in which NAAG has been found is given below. The pattern of labeled cells in subcortical structures is remarkably similar across species; in cerebral cortex however, rodents lack the pyramidal cell staining seen in cats, monkeys and humans. The overview starts with the visual system, which has been especially well studied.

2. VISUAL SYSTEM

2.1 Retina

NAAG labeling is seen in retinal ganglion cells, their dendrites in the inner plexiform layer (IPL) and their axons in the optic nerve fiber layer (ONFL) (Fig.1a). Similar patterns are seen in turtle,[5] chick,[6] rat,[7-9] cat,[10] monkey[11] and human.[12, 13] but not frog.[14, 15] Most, but not all, of the cells in the retinal ganglion cell layer are labeled.[6, 9, 10, 13, 16] Although this layer contains many "displaced" amacrine cells in addition to retinal ganglion cells,[17, 18] several lines of evidence indicate that most of the labeled cells in this layer are retinal ganglion cells and most of the unlabeled ones are amacrine cells. The labeled cells in this layer are larger, on average, than the unlabeled cells,[13] and, in

[*] Suzannah Bliss Tieman, Center for Neuroscience Research and Department of Biological Sciences, The University at Albany, State University of New York, Albany, New York, 12222, email: tieman@albany.edu.

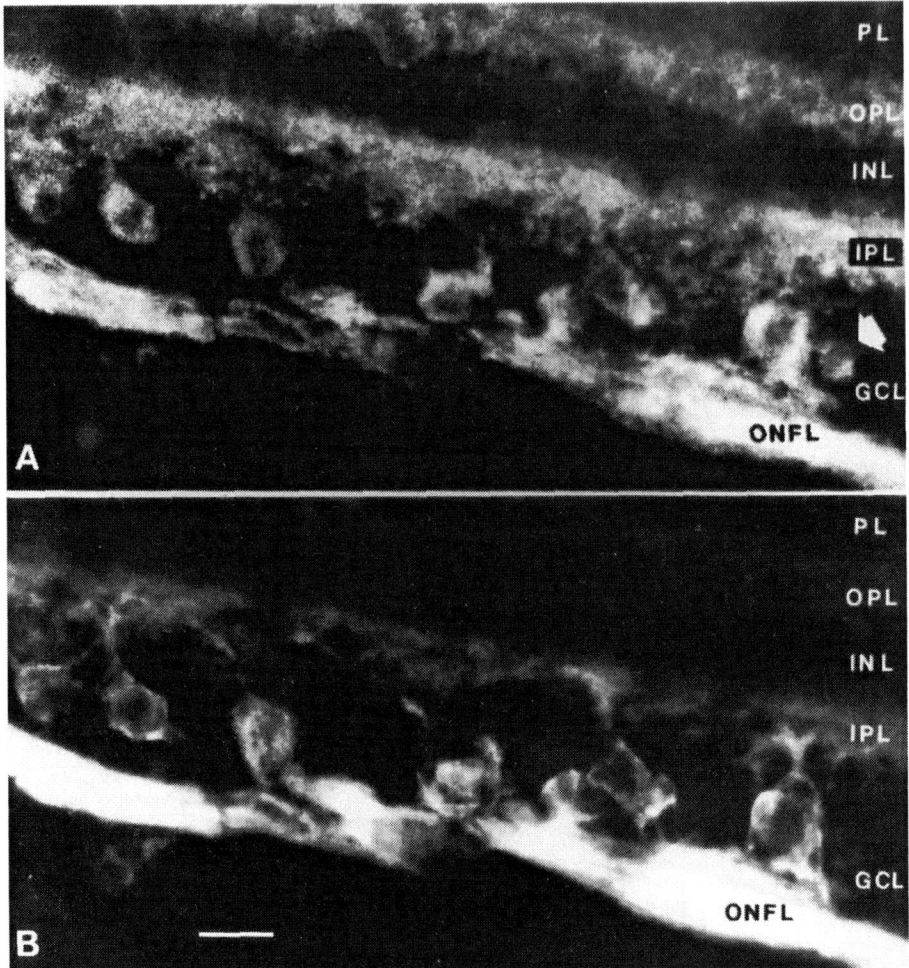

Figure 1. Co-localization of NAAG (A) and AB5 (B) in retinal sections of a cat. Numerous cells in the ganglion cell layer (GCL) are stained by both antisera. However, one small cell (arrow in A) contains NAAG, but not the AB5 antigen. This is probably a displaced amacrine cell. PL, photoreceptor layer; OPL, outer plexiform layer; INL, inner nuclear layer; IPL, inner plexiform layer; ONFL, optic nerve fiber layer. Anti-NAAG, 1:1500; AB5, 1:7500. Scale bar, 20 μm

monkey, most of those in the nasal retina disappear following section of the optic chiasm.[11] In cat, cells retrogradely labeled by DiI applied to the optic nerve also label for NAAG (Tieman, unpublished observations). In addition, NAAG co-localizes with AB5[16] (Fig. 1), a marker for retinal ganglion cells;[19] over 90% of the AB5 positive cells contain NAAG, and about 75% of the NAAG positive cells co-label for AB5. Labeled ganglion cells include small, medium and large cells, and thus, presumably, W, X, and Y cells in cat and M and P cells in monkey. Furthermore, in parafoveal retina of monkey, virtually all cells throughout the depth of the ganglion cell layer were labeled, suggesting that all major classes of ganglion cells contain NAAG (i.e., M, P, midbrain projecting, On- and

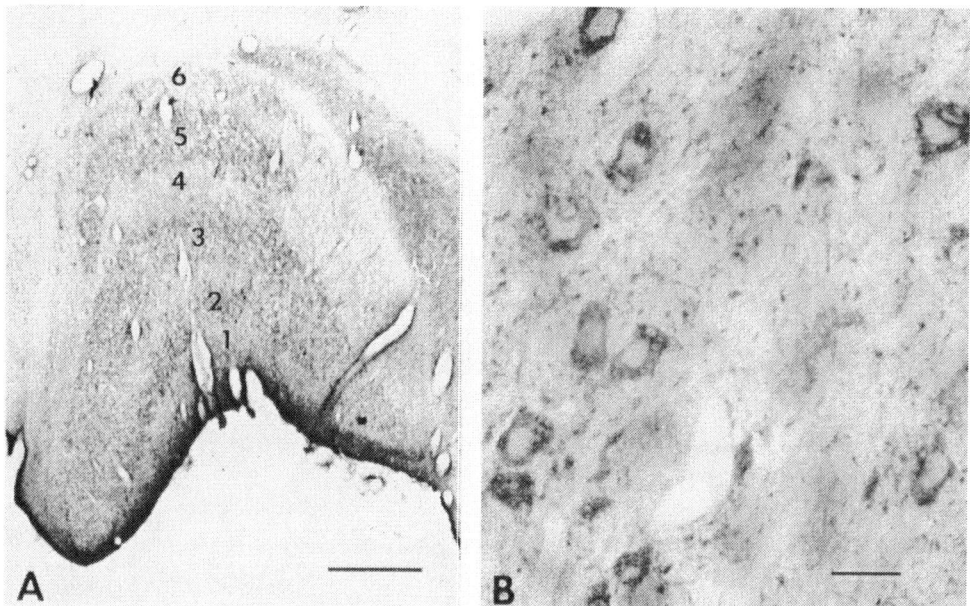

Figure 2. NAAG-like immunoreactivity in the LGN of a chiasm-sectioned monkey. A. Low-power view. Note that the staining is largely confined to the layers receiving input from the ipsilateral eye (layers 2, 3, and 5). Anti-NAAG antiserum at 1:4000. Scale bar, 1 mm. B. NAAG immunoreactivity in the neuropil and cell bodies of the LGN. Note the punctate appearance of the staining in the neuropil, which probably represents the terminal arbors of retinogeniculate axons. Antiserum at 1:4000, staining enhanced with osmium. Scale bar, 20 μm. Reprinted with permission from Tieman et al. [11]

Off-),[11] since these various types occur at different depths of the ganglion cell layer.[20] In cat, monkey and human, labeled cells include those having the morphological characteristics of ON-cells and OFF-cells (i.e., those with dendritic arborizations in sublaminae B and A of the inner plexiform layer[21]). Indeed, in parafoveal retina of monkey, the staining in the inner plexiform layer is concentrated into two bands, probably corresponding to sublaminae A and B. In addition, in chick and human, displaced ganglion cells are labeled.[6, 13] Thus, in warm blooded vertebrates and turtles, but not frogs, the overwhelming majority of retinal ganglion cells contain NAAG.

NAAG is present in ganglion cell axons and terminals. The ONFL (see above) and both the optic tract[7, 8, 22, 23] and the accessory optic tract[24] are labeled, as is the neuropil of retinal target areas such as the lateral geniculate nucleus (Fig. 2), pretectal nuclei, superior colliculus/optic tectum, suprachiasmatic nucleus and medial terminal nucleus,[6-11, 24-26] and this label decreases sharply following section of the optic nerve.[7-9, 25, 26] In monkeys, labeled neuropil is found in both magnocellular and parvicellular layers and in off-layers and o n-layers,[11] reinforcing the idea that all major classes of retinal ganglion cells contain NAAG.

NAAG is also seen in amacrine cells; labeling in amacrine cell processes probably contributes to the labeling in the IPL. In frog, with few or no labeled ganglion cells, labeling in the IPL is intense,[15] and, as in monkey parafoveal retina, is concentrated into distinct inner and outer bands. Amacrine cells are labeled in turtle,[5] frog,[15] cat,[10]

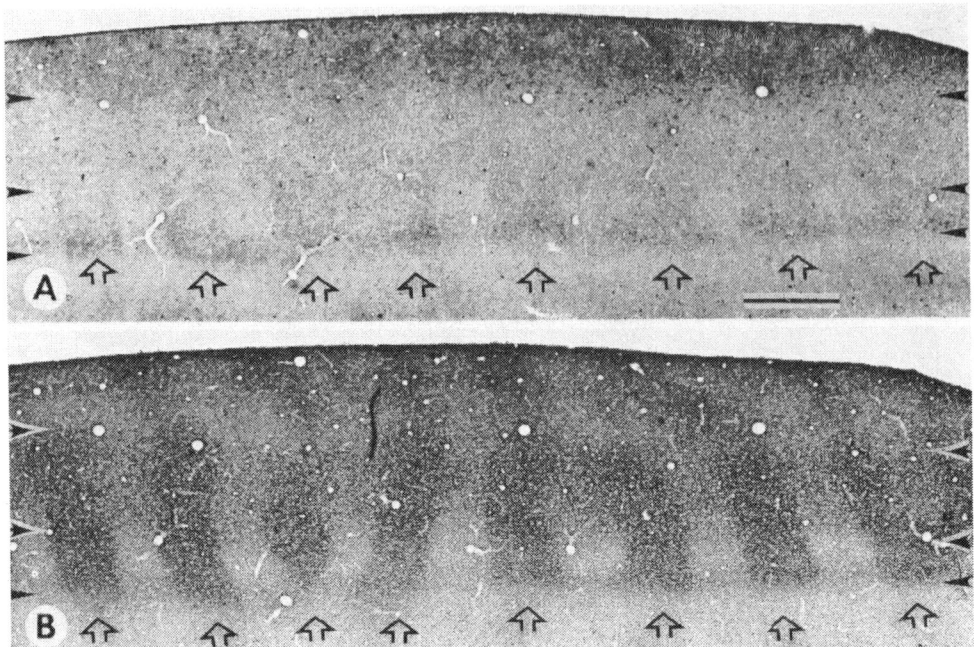

Figure 3. Patchy neuropil labeling in layer 4C of a chiasm-sectioned monkey. Shown are photographs of two adjacent tangential sections through area 17 taken near the edge of the section, where the plane of section runs through the layers at an angle. A. Section stained for NAAG, antiserum at 1:3000, osmium enhanced. B. Adjacent section reacted for cytochrome oxidase. In both A and B, the arrowheads at the side indicate the borders of layers 4C and 4Cß. The open arrows point to the ocular dominance stripes innervated by the intact, ipsilateral pathway. Scale bar, 500 μm. Reprinted with permission from Tieman et al.[11]

monkey[11] and human.[13] Furthermore, in frog, NAAG is found in the synaptic vesicles of these cells.[14] In frog, cat, monkey and human, these include displaced amacrine cells, small cells in the ganglion cell layer with thin rims of cytoplasm that fail to label with AB5 in cat[16] and survive in the nasal retina of chiasm-sectioned monkeys.[11] These are more frequent in monkey and human than in cat or frog.

Other labeled retinal cells include a few bipolar cells in the frog,[15] horizontal cells in cat[10] and monkey,[11] and interplexiform cells in monkey.[11] In general, however, these cells are less heavily labeled than the amacrine and ganglion cells.

2.2 Lateral Geniculate Nucleus

NAAG positive neurons are found in the lateral geniculate nucleus, including the medial intralaminar nucleus, of rat,[7, 9, 26-28] cat[10, 25, 29] and monkey[11] (Fig. 2). These cells are easier to see following enucleation, which decreases the neuropil labeling in the denervated layers so that the labeled cell bodies stand out. Only the relay cells, which project to visual cortex, are positive for NAAG; the interneurons, which label for glutamic acid decarboxylase (GAD), the synthesizing enzyme for GABA, are not.[9, 29]

It is likely that the axon terminals of these relay cells also contain NAAG. A distinct

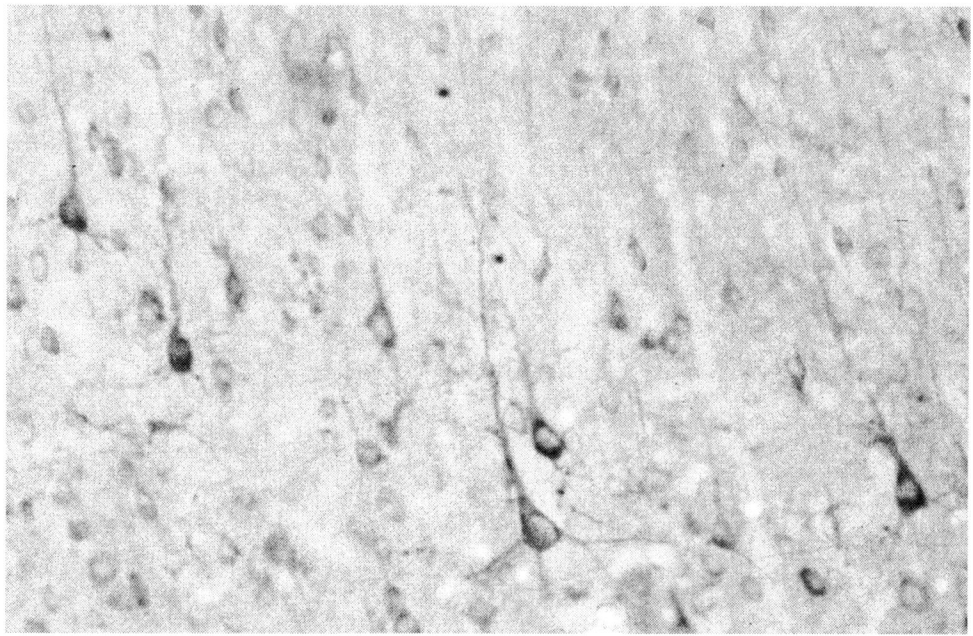

Figure 4. Both pyramidal and non-pyramidal cells are labeled for NAAG in the visual cortex of the cat. Layer 3, affinity-purified antiserum.

band of punctate label is seen in the geniculo-recipient layer of rat, monkey and human.[11, 26, 30] In rats, labeling in this layer decreases in intensity ten days following section of the optic tract.[26] Similarly, in the chiasm-sectioned monkeys, in which the cortex is deprived of input from the contralateral eye, the staining within layer 4C is patchy (Fig. 3); in tangential sections there are alternating bands of light and dark label that match the ocular dominance bands demonstrated by cytochrome oxidase histochemistry in adjacent sections.

2.3 Optic Tectum/Superior Colliculus

Most neurons in the upper, retinal receptive layers of the chick's optic tectum contain NAAG,[6] as do some of the neurons in the deeper, efferent layers. Similarly, in rat, NAAG positive neurons occur throughout the layers of the superior colliculus.[7, 9, 26] In cat and monkey, in contrast, only cells in the deeper layers of the superior colliculus contain NAAG.[10, 11]

2.4 Visual Cortex

In cat, NAAG is seen in both nonpyramidal cells and the pyramidal cells of layers 3 and 5 in areas 17, 18 and 19 of the cerebral cortex[10] (Fig. 4). In monkey, NAAG immunoreactivity is seen in some of the cell bodies in layers 2, 3A, 4B, 5 and 6 of area

17 and layers 3 and 5 of area 18. Stained cells in layer 6 of area 17 include the Meynert cells.[11] Similar results were reported for human.[30] In both cat and monkey, there are more immunoreactive neurons in areas 18 and 19 than in area 17; the layer 3 pyramidal cells of areas 18 and 19 are especially well stained. The situation appears to be different in rats: despite early reports of NAAG localization in pyramidal cells, studies with improved fixation and doubly purified antibodies found N-acetylaspartate (NAA) in pyramidal cells and NAAG mainly in interneurons, with only faint label in pyramidal cells.[23, 31] The authors suggest that reports of NAAG in rat pyramidal cells are due to cross-reactivity with NAA. In cat and monkey, however, pyramidal cell labeling could not be blocked by NAA-BSA at the highest levels tested, up to thirty times the concentration required for NAAG-BSA to block staining.[10, 11] Furthermore, even using doubly purified antibodies and enhanced fixation, in cats, monkeys and humans, NAAG antibodies label pyramidal cells.[23, 30, 32]

3. OTHER SENSORY SYSTEMS

3.1 Olfactory

NAAG is seen in the projection cells of the rat's olfactory bulb, the mitral cells;[23, 27, 28, 33, 34] this staining is punctate.[23, 28] Staining is also seen in periglomerular cells and interneurons[23] of the olfactory bulb. The same pattern is seen in the accessory olfactory bulb, but the staining is less intense.[23] NAAG is also found in mitral cell axons and terminals. The lateral olfactory tract is labeled, as is the neuropil of olfactory projection areas, such as the anterior olfactory nucleus and the pyriform cortex.[23]

3.2 Somatosensory

NAAG is found in the largest of the dorsal root ganglion cells in both frog[15] and rat,[35, 36] and there is label in the dorsal roots, as well. In the dorsal horn of the spinal cord, large-diameter spinal sensory afferents are stained for NAAG. Neuropil labeling is present in all laminae of the spinal cord except the substantia gelatinosa.[37] The absence of neuropil staining in the substantia gelatinosa is consistent with the absence of label in small neurons of the dorsal root ganglia .

Many neurons of the spinal gray are labeled, including those of Clarke's column, as are the dorsal columns.[31, 34, 37] The dorsal column nuclei are immunoreactive for NAAG,[31, 37] as are their axons in the medial lemniscus.[31] In thalamus, both cell bodies and neuropil of the ventral basal nucleus are labeled,[23] and there is a band of neuropil labeling in layer 4 of somatosensory cortex.[23]

Numerous cells of origin for mossy fibers in cerebellum, including those of the spinal grey and the lateral cervical nucleus, are moderately to strongly stained for NAAG as are the fibers of the inferior and superior cerebellar peduncles, and mossy fiber endings are among the most prominent NAAG-positive elements in the cerebellar cortex.[37] Up to 45% of neurons in the inferior olive are labeled, as are their axons (climbing fibers) in the cerebellum.[37, 38]

In somatosensory cortex, labeled neurons are mostly non-pyramidal; earlier reports of strong pyramidal cell labeling in rat somatosensory and other cortices is likely due to cross-reactivity with NAA.[23, 31]

3.3 Other Sensory Systems

Just as the larger neurons of the spinal ganglia contain NAAG (see above), so do most of the large cells of the vestibular ganglion,[37] a few cells intercalated among the axons of the eighth nerve and some axons of the eighth nerve. In addition, all divisions of the vestibular nuclei contain NAAG-labeled cell bodies and neuropil.

It is unknown whether any of the NAAG-positive axons of the eighth nerve are auditory in origin. However, the auditory thalamic nucleus, the medial geniculate nucleus (MGN), contains both cellular and neuropil labeling[28] (see also Fig. 3A of Moffett et al.[26]). One would expect, therefore, to find NAAG in neurons of the inferior colliculus, and, indeed, a micropunch radioimmunoassay study of rat brain found high levels of NAAG in cochlear nuclei, inferior colliculus and MGN.[39]

In contrast, gustatory systems may not use NAAG. The nucleus of the solitary tract, which receives taste information from the tongue, contains no NAAG-immunoreactive neurons.[31]

4. MOTOR SYSTEMS

The pyramidal cells of the motor cortex and their axons in the pyramidal tract stain for NAAG in human,[30] but not rat.[23, 31] The motor neurons of the brainstem and spinal cord are labeled for NAAG in all species examined.[23, 28, 30, 34, 36, 40]

NAAG is found in much of the rat's extrapyramidal motor system.[24, 28] Label is seen in both cells and neuropil in the globus pallidus, entopeduncular nucleus, subthalamic nucleus, and substantia nigra. In contrast, relatively few cells in the caudate-putamen and nucleus accumbens contain NAAG.

Cells of the pontine nuclei are strongly labeled,[37] as are the axons of the middle cerebellar peduncle. According to Moffett and Namboodiri,[37] some cerebellar Purkinje cells are moderately positive, but most are unstained. However, Passani et al. report intense Purkinje cell labeling in both rat and human.[30] Large, but not small cells in all divisions of the deep cerebellar nuclei are highly immunoreactive for NAAG.[37] Additionally, many neurons of the red nucleus, both parvicellular and magnocellular divisions, stain intensely for NAAG.[22, 22, 37]

Motor neurons in brainstem and spinal cord are labeled for NAAG,[28, 36, 40] as are the spinal ventral roots[36] and their terminals at the neuromuscular junction.[41]

5. OTHER REGIONS

Numerous other regions of brain are labeled. The lateral habenula is one of the most heavily labeled regions of forebrain, with staining in both neuronal cell bodies and neuropil.[23] Other regions include both neuropil and neurons in the lateral hypothalamus and paraventricular nucleus[8, 24] and the medial mammillary nucleus.[28] The neurons of both the medial septum and the diagonal band are heavily labeled.[24, 28, 42] Within the brainstem, labeled cells are found in the dorsal raphe and the locus coeruleus.[22, 40]

6. CO-LOCALIZATION

NAAG is sometimes found in neurons that are also cholinergic, nor-adrenergic, serotonergic, or GABAergic. Motoneurons have long been known to be cholinergic, and yet they, their axons, and terminals all contain NAAG (see above). Intense double staining for NAAG and ChAT is seen in both spinal and medullary motoneurons, including those of the facial, ambiguus, and trigeminal nuclei.[40] In addition, a subset of cholinergc neurons in the basal forebrain and medial septum contain NAAG.[40]

Many nor-adrenergic neurons contain NAAG. All cells of the locus coeruleus contain both NAAG and dopamine β-hydroxylase, the synthesizing enzyme for nor-adrenaline.[40] Except for the A2 cell group, which contains little or no NAAG, most of the other noradrenergic and adrenergic cell groups in the medulla label for NAAG in addition to dopamine β-hydroxylase.

NAAG co-localizes with serotonin in some cells. A few serotonergic neurons in the dorsal raphe stain for NAAG. Co-labeling is more apparent in the median raphe, although most cells are singly labeled.[40]

Some, but by no means all, GABAergic neurons contain NAAG. In cat, a few amacrine and retinal ganglion cells contain both NAAG and GABA.[43, 44] Similarly, in rat, a few cells of the inner nuclear layer (presumably amacrine cells) contain both NAAG and GAD_{67}, but most are singly labeled for NAAG.[9] In rat cerebral cortex, NAAG labeling is strongest in cells with the morphology of interneurons, suggesting likely co-localization with GAD.[23, 31] Cells of the thalamic reticular nucleus are known to be GABAergic[45] and to contain somatostatin[46] and aspartate or glutamate;[47] they also label heavily for NAAG.[9, 23] Indeed, cells of the rat's thalamic reticular nucleus and those of the cat's perigeniculate nucleus, the visual part of the thalamic reticular nucleus, contain both NAAG and GAD.[9, 29] In contrast, in LGN, GAD is found in interneurons, whereas NAAG is found in relay cells.[9, 29] In cerebellar pathways, NAAG and GAD distributions are largely complementary,[37] although since some[37] or most[30] Purkinje cells are labeled for NAAG, these should be double-labeled.

Cellular labeling for NAAG is seen in the substantia nigra,[24] whose cells are largely dopaminergic.[48] The role of this NAAG may be to modulate the release of dopamine in the striatum.[49, 50]

7. ACTIVITY-DEPENDENT REGULATION OF NAAG EXPRESSION

I examined NAAG immunoreactivity in the retina and LGN of monocularly deprived (MD) cats and compared it with that in normal cats.[51, 52] As noted above, in the LGN of the normal cat, both the neuropil and the somata of relay cells are heavily labeled. Long-term monocular deprivation decreases labeling of the somata, but not the neuropil, in the deprived layers of the LGN (Fig. 5). There is little or no loss of label in the retinal ganglion cells of the deprived eye, as might be expected, given that the labeling of their terminals in the LGN is also unchanged. There is also no change in immunoreactivity for GAD, which is found in the interneurons of the LGN. The changes

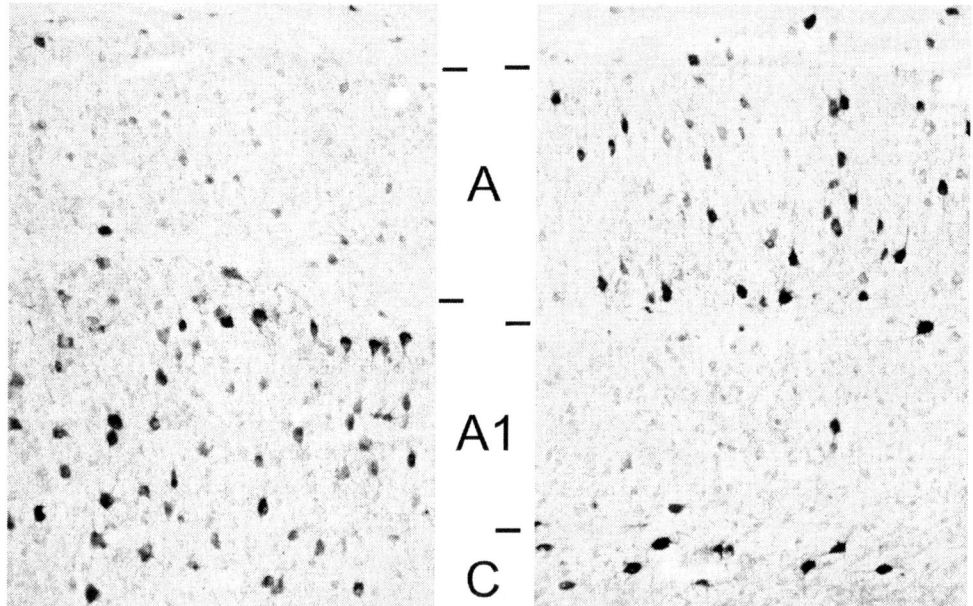

Figure 5. Monocular deprivation alters NAAG levels in the lateral geniculate nucleus of the cat. The lids of one eye were sutured together before the time of natural eye-opening, and the animal was perfused as an adult. Lamina A is deprived on the left; lamina A1 is deprived on the right. Note the dramatic difference in intensity of staining between the deprived and non-deprived laminae. Affinity-purified antiserum.

in NAAG immunoreactivity of the LGN neurons in MD cats are much more striking than changes demonstrated with immunocytochemistry for glutamate, cytochrome oxidase histochemistry or uptake and incorporation into protein of ^{3}H-leucine. Thus, the changes are unlikely to be due to an overall decrease in metabolism or protein synthesis. These results suggest that long-term monocular deprivation selectively decreases the synthesis of NAAG in geniculate neurons, or perhaps increases its hydrolysis. In contrast, 7-14 days of denervation is insufficient to reduce labeling in the cells of the cat's LGN.[25]

In the optic-chiasm sectioned monkeys, there was decreased cellular labeling for NAAG in the denervated ocular dominance columns.[11] Similarly, 10 days after optic tract section in rats, cellular labeling in the ipsilateral cortex was reduced.[26]

8. CONCLUSIONS

NAAG is found in a subset of neurons throughout the brain and spinal cord of vertebrates ranging from turtle to man. Within those neurons, NAAG is found in all parts of the cell except the nucleus. In many cases the filling of dendrites and axons is sufficient to classify the types of neuron labeled. In other cases, double-labeling paradigms have been used.

Among mammals, the pattern of localization is remarkably consistent. NAAG is found in projection neurons at all stages from periphery up to cortex of all sensory systems except perhaps gustatory. In vestibular and somatosensory systems, only large ganglion cells are labeled, whereas in visual system, all or nearly all retinal ganglion cells are labeled. Various sensory relay cells in spinal cord and medulla are labeled, including cells of dorsal horn, Clarke's column, dorsal column nuclei and vestibular nuclei. Similarly, sensory thalamic relay cells are labeled, as are the mitral cells of the olfactory bulb. Subcortical projection neurons of the motor systems are labeled, down to and including those projecting to the periphery. Only in cerebral cortex is there significant variation: in cat, monkey and man, NAAG is seen in both pyramidal and non-pyramidal cells, whereas in rat it is seen only in non-pyramidal cells.

NAAG co-localizes with a variety of classical neurotransmitters. It is seen in subsets of cells that also label for acetylcholine, GABA, dopamine, nor-adrenaline, serotonin and somatostatin.

Finally, as has been reported for a variety of other neurotransmitters, the level of NAAG expression is regulated by neural activity.

9. ACKNOWLEDGMENTS

The author's research on NAAG was supported in part by NSF grants BNS-8811039 and IBN-9212426. Drs. Joseph H. Neale of Georgetown University and David Tieman of the University at Albany collaborated on much of the research. I thank Drs. Joe Neale and John Moffett for their generous contribution of antibodies.

10. QUESTION AND ANSWER SESSION

DR. COYLE: This presentation is open for discussion.

DR. BASLOW: Just one comment about the frog. Frog brain has no measurable NAA, and so it is a different kind of a brain structure. Apparently, it does have NAAG, though. I am not sure where that is.

DR. TIEMAN: What I was hoping you would comment upon was the activity dependent expression of NAAG.

DR. MOFFETT: As it relates to what I have seen, basically, in areas of thalamus, for example, where we have extremely high levels of glutamate input -- for example, ventrobasal thalamus and lateral geniculate, which receive large amounts of glutamatergic input -- the plexus of NAAG staining is incredibly dense.

I have a sense that what we are seeing with the down-regulation of NAAG in post-synaptic neurons like that is probably due to the lack of the glutamate input. In places where glutamate input is high, even if there are no glutamatergic cell bodies there, if the terminals are glutamatergic, then the level of NAAG staining in either interneurons or principal neurons in the area, and the expression of NAAG in synaptic terminals in the neuropil is very high.

DR. TIEMAN: Except in LGN, where the interneurons are not labeled.

DR. MOFFETT: Yes. Whereas in other places they are, like in neocortex.

PARTICIPANT: I will ask a question about your activities. Which -- I mean, the up-regulation of NAAG -- we have heard Dr. Ross, I think, talking about ways that that can be regulated. It is either due to a decrease in its breakdown or an increase in its synthesis. Do you have any inference from your -- or anyone else here -- of what might be happening?

DR. TIEMAN: I would be happy to hear from anyone else on this.

DR. LIEBERMAN: I will comment about it. To some extent, when you do stimulate, you do get an increase in synthesis, in some cases both NAAG and NAA, so there is some activity-dependent regulation.

DR. COYLE: I might mention that both John and our laboratories have shown loss of NAAG release with the optic nerve transection, both radiolabeled as well as endogenous NAAG, confirming not only the localization but also the functional significance of this process. Any other questions?

Thank you very much Dr. Tieman.

11. REFERENCES

1. J. H. Neale, T. Bzdega, and B. Wroblewska. *N*-acetylaspartylglutamate: The most abundant peptide neurotransmitter in the mammalian central nervous system. *J. Neurochem.* **75**, 443-452 (2000).
2. M. Cassidy and J. H. Neale. Localization and transport of *N*-acetylaspartylglutamate in cells of whole murine brain in primary culture. *J. Neurochem.* **60**, 1631-1638 (1993).
3. L. Passani, S. Elkabes, and J. T. Coyle. Evidence for the presence of *N*-acetylaspartylglutamate in cultured oligodendrocytes and LPS activated microglia. *Brain Res.* **794**, 143-145 (1998).
4. L. M. Gehl, O. H. Saab, T. Bzdega, B. Wroblewska, and J. H. Neale. Biosynthesis of NAAG by an enzyme-mediated process in rat central nervous system neurons and glia. *J. Neurochem.* **90**, 989-997 (2004).
5. W. D. Eldred, J. T. Coyle, and C. A. Joneckis. Colocalization of NAA and NAAG in ganglion cells in the turtle retina. *Invest. Ophthalmol. Vis. Sci.* **33 (Suppl.)**, 1029 (1992).
6. L. C. Williamson, D. A. Eagles, M. J. Brady, J. R. Moffett, M. A. A. Namboodiri, and J. H. Neale. Localization and synaptic release of *N*-acetylaspartylglutamate in the chick retina and optic tectum. *Eur. J. Neurosci.* **3**, 441-451 (1991).
7. K. J. Anderson, M. A. Borja, C. W. Cotman, J. R. Moffett, M. A. A. Namboodiri, and J. H. Neale. *N*-acetylaspartylglutamate identified in the rat retinal ganglion cells and their projections in the brain. *Brain Res.* **411**, 172-177 (1987).
8. J. R. Moffett, L. Williamson, M. Palkovits, and M. A. A. Namboodiri. *N*-acetylaspartylglutamate: A transmitter candidate for the retinohypothalamic tract. *Proc. Natl. Acad. Sci. USA* **87**, 8065-8069 (1990).
9. J. R. Moffett. Reductions in N-acetylaspartylglutamate and the 67 kDa form of glutamic acid decarboxylase immunoreactivities in the visual system of albino and pigmented rats after optic nerve transections. *J. Comp. Neurol.* **458**, 221-239 (2003).
10. S. B. Tieman, C. B. Cangro, and J. H. Neale. *N*-Acetylaspartylglutamate immunoreactivity in neurons of the cat's visual system. *Brain Res.* **420**, 188-193 (1987).
11. S. B. Tieman, J. H. Neale, and D. G. Tieman. *N*-acetylaspartylglutamate immunoreactivity in neurons of the monkey's visual system. *J. Comp. Neurol.* **313**, 45-64 (1991).
12. S. B. Tieman, K. Butler, and J. H. Neale. *N*-acetylaspartylglutamate: a neuropeptide in human visual system. *J. Amer. Med. Assoc.* **259**, 2020 (1988).
13. S. B. Tieman and D. G. Tieman. *N*-Acetylaspartylglutamate immunoreactivity in human retina. *Vision Res.* **36**, 941-947 (1996).
14. L. C. Williamson and J. H. Neale. Ultrastructural localization of *N*-acetylaspartylglutamate in vesicles of retinal neurons. *Brain Res.* **456**, 375-381 (1988).
15. M. M. Kowalski, M. Cassidy, M. A. A. Namboodiri, and J. H. Neale. Cellular localization of *N*-acetylaspartylglutamate in amphibian retina and spinal sensory ganglia. *Brain Res.* **406**, 397-401 (1987).
16. S. B. Tieman and K. R. Fry. Do all retinal ganglion cells contain NAAG? *Soc. Neurosci. Abstr.* **15**, 1208 (1989).
17. A. Hughes and E. Wieniawa-Narkiewicz. A newly identified population of presumptive microneurones in the cat retinal ganglion cell layer. *Nature* **284**, 468-470 (1980).
18. A. Hughes. Population magnitudes and distribution of the major modal classes of cat retinal ganglion cell as estimated from HRP filling and a systematic survey of the soma diameter spectra for classical neurones. *J. Comp. Neurol.* **197**, 303-339 (1981).
19. K. R. Fry and D. M. K. Lam. Specific labelling of ganglion cells in the cat retina by a monoclonal antibody. *Vision Res.* **26**, 373-382 (1986).

20. V. H. Perry and L. C. L. Silveira. Functional lamination in the ganglion cell layer of the macaque's retina. *Neuroscience* **25**, 217-223 (1988).

21. R. Nelson, E. V. Famiglietti, Jr., and H. Kolb. Intracellular staining reveals different levels of stratification for on- and off-center ganglion cells in cat retina. *J. Neurophysiol.* **41**, 472-483 (1978).

22. A. S. Guarda, M. B. Robinson, L. Ory-Lavollée, G. L. Forloni, R. D. Blakely, and J. T. Coyle. Quantitation of N-acetyl-aspartyl-glutamate in microdissected brain nuclei and peripheral tissues: findings with a novel liquid phase radioimmunoassay. *Mol. Brain Res.* **3**, 223-232 (1988).

23. J. R. Moffett and M. A. A. Namboodiri. Differential distribution of N-acetylaspartylglutamate and N-acetylaspartate immunoreactivities in rat forebrain. *J. Neurocytol.* **24**, 409-433 (1995).

24. J. R. Moffett, M. Cassidy, and M. A. A. Namboodiri. Selective distribution of N-acetylaspartylglutamate immunoreactivity in the extrapyramidal system of the rat. *Brain Res.* **494**, 255-266 (1989).

25. S. B. Tieman, J. R. Moffett, and S. M. Irtenkauf. Effect of eye removal on N-acetylaspartylglutamate immunoreactivity in retinal targets of the cat. *Brain Res.* **562**, 318-322 (1991).

26. J. R. Moffett, L. C. Williamson, J. H. Neale, M. Palkovits, and M. A. A. Namboodiri. Effect of optic nerve transection on N-acetylaspartylglutamate immunoreactivity in the primary and accessory optic projection systems in the rat. *Brain Res.* **538**, 86-94 (1991).

27. K. J. Anderson, D. T. Monaghan, C. B. Cangro, M. A. A. Namboodiri, J. H. Neale, and C. W. Cotman. Localization of N -acetylaspartylglutamate-like immunoreactivity in selected areas of the rat brain. *Neurosci. Lett.* **72**, 14-20 (1986).

28. G. Tsai, B. S. Slusher, L. Sim, and J. T. Coyle. Immunocytochemical distribution of N-acetylaspartylglutamate in the rat forebrain and glutamatergic pathways. *J. Chem. Neuroanat.* **6**, 277-292 (1993).

29. L.-C. S. Xing and S. B. Tieman. Relay cells, not interneurons, of cat's lateral geniculate nucleus contain N-acetylaspartylglutamate. *J. Comp. Neurol.* **330**, 272-285 (1993).

30. L. A. Passani, J. P. G. Vonsattel, and J. T. Coyle. Distribution of N-acetylaspartylglutamate immunoreactivity in human brain and its alteration in neurodegenerative disease. *Brain Res.* **772**, 9-22 (1997).

31. J. R. Moffett, M. A. A. Namboodiri, and J. H. Neale. Enhanced carbodiimide fixation for immunohistochemistry: Application to the comparative distributions of N-acetylaspartylglutamate and N-acetylaspartate immunoreactivities in rat brain. *J. Histochem. Cytochem.* **41**, 559-570 (1993).

32. J. R. Moffett and M. A. A. Namboodiri. in *The First International Symposium on N-Acetylaspartate.* edited by M.A.A.Namboodiri and J.R.Moffett, (Springer, 2006).

33. R. D. Blakely, L. Ory-Lavollée, R. Grzanna, K. J. Koller, and J. T. Coyle. Selective immunocytochemical staining of mitral cells in rat olfactory bulb with affinity purified antibodies against N-acetyl-aspartyl-glutamate. *Brain Res.* **402**, 373-378 (1987).

34. C. G. Frondoza, S. Logan, G. Forloni, and J. T. Coyle. Production and characterization of monoclonal antibodies to N-acetyl-aspartyl-glutamate. *J. Histochem. Cytochem.* **38**, 493-502 (1990).

35. C. B. Cangro, M. A. A. Namboodiri, L. A. Sklar, A. Corigliano-Murphy, and J. H. Neale. Immunohistochemistry and biosynthesis of N-acetylaspartylglutamate in spinal sensory ganglia. *J. Neurochem.* **49**, 1579-1588 (1987).

36. L. Ory-Lavollée, R. D. Blakely, and J. T. Coyle. Neurochemical and immunocytochemical studies on the distribution of N-acetyl-aspartylglutamate and N-acetyl-aspartate in rat spinal cord and some peripheral tissues. *J. Neurochem.* **48**, 895-899 (1987).

37. J. R. Moffett, M. Palkovits, M. A. A. Namboodiri, and J. H. Neale. Comparative distribution of N-acetylaspartylglutamate and GAD_{67} in the cerebellum and precerebellar nuclei of the rat utilizing enhanced carbodiimide fixation and immunohistochemistry. *J. Comp. Neurol.* **347**, 598-618 (1994).

38. W. M. Renno, J. H. Lee, and A. J. Beitz. Light and electron microscopic immunohistochemical localization of N-acetylaspartylglutamate (NAAG) in the olivocerebellar pathway of the rat. *syn* **26**, 140-154 (1997).

39. S. Fuhrman, M. Palkovits, M. Cassidy, and J. H. Neale. The regional distribution of N-acetylaspartylglutamate (NAAG) and peptidase activity against NAAG in the rat nervous system. *J. Neurochem.* **62**, 275-281 (1994).

40. G. Forloni, R. Grzanna, R. D. Blakely, and J. T. Coyle. Co-localization of N-acetylaspartylglutamate in central cholinergic, noradrenergic and serotonergic neurons. *syn* **1**, 455-460 (1987).

41. U. V. Berger, R. E. Carter, and J. T. Coyle. The immunocytochemical localization of N-acetylaspartyl glutamate, its hydrolysing enzyme NAALADase, and the NMDAR-1 receptor at a vertebrate neuromuscular junction. *Neuroscience* **64**, 847-850 (1995).

42. M. C. Senut, F. De Bilbao, and Y. Lamour. Medial septal neurons containing N-acetyl-aspartyl-glutamate-like immunoreactivity project to the hippocampal formation in the rat. *Neurosci. Lett.* **113**, 12-16 (1990).

43. L. Shelton, S. B. Tieman, and K. R. Fry. GABAergic ganglion cells in cat retina. *Soc. Neurosci. Abstr.* **16**,

1217 (1990).
44. L. A. Shelton GABA in the Retinal Ganglion Cells of the Cat. M.S. Thesis, State University of New York at Albany, 1-31 (1991).
45. C. R. Houser, J. E. Vaughn, R. P. Barber, and E. Roberts. GABA neurons are the major cell type of the nucleus reticularis thalami. *Brain Res.* **200**, 341-345 (1980).
46. W. H. Oertel, A. M. Graybiel, E. Mugnaini, R. P. Elde, and D. E. Schmechel. Coexistence of glutamic acid decarboxylase- and somatostatin-like immunoreactivity in neurons of the feline nucleus reticularis thalami. *J. Neurosci.* **3**, 1322-1332 (1983).
47. A. Gonzalo-Ruiz, J. M. Sanz, and A. R. Lieberman. Immunohistochemical studies of localization and co-localization of glutamate, aspartate and GABA in the anterior thalamic nuclei, retrosplenial granular cortex, thalamic reticular nucleus and mammillary nuclei of the rat. *J. Chem. Neuroanat.* **12**, 77-84 (1996).
48. N. E. Anden, A. Dahlström, K. Fuxe, and K. Larsson. Further evidence for the presence of nigro-neostriatal dopamine neurons in the rat. *Am. J. Anat.* **116**, 329-333 (1965).
49. J. K. Weatherspoon, A. R. Frank, and L. L. Werling. Neurotensin, N-acetyl-aspartyl-glutamate and β-endorphin modulate [³H]dopamine release from guinea pig nucleus accumbens, prefrontal cortex and caudate putamen. *Neuropeptides* **30**, 497-505 (1996).
50. T. Galli, G. Godeheu, F. Artaud, J. M. Desce, A. Pittaluga, L. Barbeito, J. Glowinski, and A. Chéramy. Specific role of N-acetyl-aspartyl-glutamate in the *in vivo* regulation of dopamine release from dendrites and nerve terminals of nigrostriatal dopaminergic neurons in the cat. *Neuroscience* **42**, 19-28 (1991).
51. S. B. Tieman and S. M. LePage. Monocular deprivation alters N-acetylaspartylglutamate immunoreactivity in the lateral geniculate nucleus of the cat. *Invest. Ophthalmol. Vis. Sci.* **29 (Suppl.)**, 32 (1988).
52. S. B. Tieman. Morphological changes in the geniculocortical pathway associated with monocular deprivation. *Ann. NY Acad. Sci.* **627**, 212-230 (1991).

SYNTHESIS OF *N*-ACETYLASPARTYL-GLUTAMATE (NAAG) AND *N*-ACETYL-ASPARTATE (NAA) IN AXONS AND GLIA OF THE CRAYFISH MEDIAL GIANT NERVE FIBER

Edward M. Lieberman, Mohit Achreja and Albert K. Urazaev[*]

1. INTRODUCTION

The biosynthesis of n-acetylaspartylglutamate (NAAG) along with its precursor, N-acetylaspartate (NAA), has been of interest since they were first shown to be present in nerve tissue in relatively high concentration approximately 4 decades ago[1,2]. In 1973 Reichelt and Kvamme[3] demonstrated that biosynthesis of NAAG occurred in homogenates of mouse brain with an absolute dependence on ATP and NAA as a primary substrate. In addition, inhibitors of cytoplasmic and mitochondrial protein synthesis were without effect suggesting that an enzyme-mediated rather than post-translational protein processing was responsible for NAAG biosynthesis. This work has since been confirmed in spinal sensory ganglion[4] and more recently in rat spinal cord [5].

NAAG, considered to have appeared late in evolution[6], has now been shown to be present in axons and glia of crayfish medial giant nerve fibers along with its glial membrane-associated degradative enzyme, Glutamate Carboxypeptidase II (GCPII) and the glial group II metabotropic glutamate receptor, for which it is a specific agonist[7-11]. The metabotropic glutamate receptor is responsible for regulating glial membrane potential, nitric oxide production[12] and regulation of GCPII. There is also developing evidence that the mGLUR$_{II}$ may participate in nervous system volume regulation and perineural K$^+$ homeostasis [13-15].

In this report we review recently published and ongoing investigations from our laboratory that show that there is a significant presence of NAAG and NAA in both axoplasm and glia of the medial giant nerve fiber of the crayfish and that their synthesis occurs in the presence of a single substrate – glutamate. Intermediates of NAAG synthesis, e.g. aspartate and NAA, are also effective substrates and glutamine can be rapidly deaminated by glutaminase present in axons and glia of the crayfish nerve fiber[16]

[*] E.M. Lieberman and M. Achreja, Dept. of Physiology, The Brody School of Medicine of East Carolina University, Greenville, NC 27834; A.K. Urazaev, Present address: Dept. of Biological Sciences, Purdue University, West Lafayette, IN 47907

forming glutamate, allowing it to enter into the NAAG biosynthetic pathway.

2. METHODS

2.1 Biological samples

Desheathed crayfish whole nerve cord or the medial giant nerve fibers isolated from nerve cords were used in most experiments. The details of the procedure have been described previously[8-11]. Cannulation of the medial giant axon with a glass micropipette was used to withdraw axoplasm for analysis of NAAG and NAA allowing for characterization and differentiation of biosynthesis occurring in axons and glia[17]. The axon is then perfused with physiological saline to remove the remaining axoplasm while leaving an intact glial sheath that can be analyzed for its content of NAAG and NAA.

2.2 Intact Nerve

Biosynthesis of NAAG and NAA in intact nerve cord or giant nerve fibers was monitored using incorporation of radiolabel from the precursors glutamine, glutamate, aspartate and NAA. Following incubation, the tissue was rinsed in several changes of normal saline to remove extracellular radiolabel. The rinsed tissues were homogenized using freeze-thaw then sonicated to free radiolabeled products of amino acid substrates from the tissue. In the earlier experiments, the homogenate was centrifuged and the supernatant taken for analysis by cation-exchange HPLC for its content of NAAG, GLU, ASP and GLN. Identification of NAAG was made by acid hydrolysis and amino acid analysis of the products[9]. In more recent experiments a TSK-GEL SAX column is being used for HPLC analysis of the content of NAAG and NAA in isolated axoplasm.

2.3 Native-Isolated and Dialyzed-Isolated Axoplasm

In ongoing investigations, native-isolated or isolated-dialyzed axoplasm are being using for the investigation of substrate and cofactor requirements for NAAG biosynthesis. These procedures remove the uncertainty in the characterization of NAAG and NAA biosynthesis imposed by the use of homogenates of a heterogeneous multicellular tissue. Axoplasm, removed by a glass microcannula, was expelled into a small polyethylene cup filled with light mineral oil where it formed spherical droplets whose diameter was optically measured and volume calculated. Additions were made directly to these droplets for incubation in place[18]. Following incubation, the droplets were removed, and processed for analysis of metabolic products of amino acid substrates by HPLC[9].

Dialysis of axoplasm was accomplished by drawing up axoplasm droplets from the oil-filled cups into Spectropore hollow tube dialysis membranes (13 kD cutoff) using oil to isolate the axoplasm from solutions at either end of the fiber. The fiber, containing the axoplasm was placed in an artificial axoplasm saline containing in mM: 140 K_2SO_4; 15 NaCl; 2.4 $MgCl_2$ and 20 Tris at pH 7.4. Sucrose was added to equalize dialysis solution osmolarity to that of axoplasm. Pharmacological agents can also be added to the dialysis solution as required. Axoplasm was dialyzed for 3-5 hours before use.

3. RESULTS

3.1 NAAG is Found in Both Axons and Glia of the Crayfish Medial Giant Nerve Fiber

Lieberman et al[7] were the first to report that NAAG appeared to be a product of glutamate metabolism in crayfish nerve cord. This finding has since been confirmed[9-11] and illustrated in Fig. 1. Following a 2 hour incubation of the medial giant nerve fiber with bath-applied radiolabeled glutamate, axons (right panel) and glia (left panel) contained significant amounts of radiolabeled NAAG, glutamate and glutamine with lesser amounts of GABA and aspartate (not shown). Stimulation of the nerve fiber caused a 30% reduction of NAAG in glia and a total depletion of NAAG in the axon. The radiolabeled NAAG that appeared in the superfusate following stimulation was increased by 300% over non-stimulated controls and represented 49% of the total recovered radiolabel efflux. Glutamate had the next largest efflux at approximately 20% and was unchanged by stimulation. The remaining radiolabel was distributed between aspartate, glutamine and GABA. Total radiolabel recovered in these four glutamate metabolites was greater than 95%.

NAAG content of axons and glia, normalized to 1 µl of cytoplasm, suggested that glia synthesize and accumulate NAAG an order of magnitude greater than axons. When the volume of the axon, relative to glia is considered (axoplasm represents 95% of the nerve fiber cytoplasm) the increase in radiolabel in axonal NAAG content gives approximately the same rate of synthesis as that seen in glia.

Radiolabeled glutamate directly injected into the axoplasm of medial giant axons appeared in axonal NAAG at 3x the quantity in glia. Stimulation of the nerve fiber

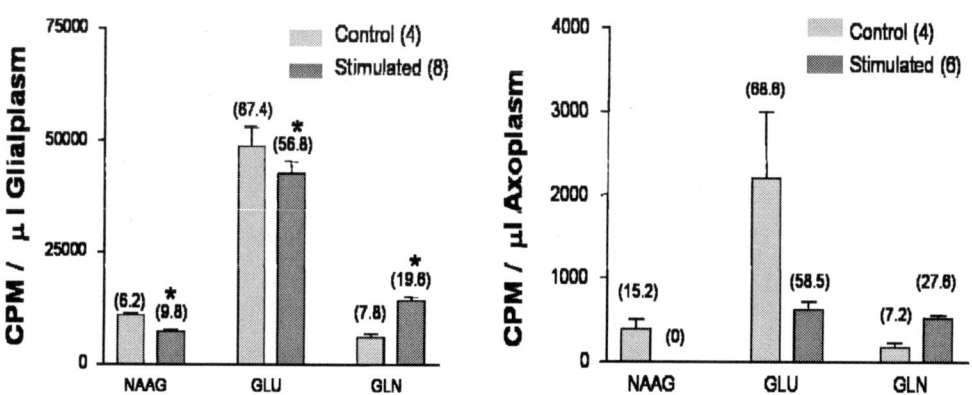

Figure 1. Effect of stimulation on the content and distribution of radiolabeled products of glutamate metabolism in axons and glia following 2 h incubation with [14C]-glutamate. Modified from data of Urazaev et al[9]. Means ± 1 SEM shown in this figure were derived from the totality of samples analyzed, the statistical analysis was performed only on the 4 paired samples. Numbers in parentheses represent the percent of the sum of the radiolabeled compounds recovered in samples of axonal or glial cytoplasm averaged for each experimental pair. * Indicates statistical significance between stimulated and unstimulated nerve fibers at $p \leq 0.05$.

depleted NAAG from the axon but not the glia, in contrast to that seen with nerve fibers bath incubated with radiolabel. In these same experiments it was also shown that nerve fibers incubated with radiolabeled NAAG and the NAAG peptidase inhibitors, β-NAAG or quisqualate, the superfusate concentration of glutamate was reduced by about 60% of the control suggesting that NAAG peptidase, with its hydrolytic site in the extracellular space, was present.

In a continuing series of experiments[10], it was found that stimulation of nerve in the presence of radiolabeled NAAG dramatically increased the amount of NAAG, glutamate and glutamine in axons and their associated glia (Fig.2). These results and those described above suggest that the glial-associated NAAG peptidase was activated by stimulation to hydrolyze extracellular NAAG. The glutamate released can be taken up by glia and recycled into the NAAG, glutamate and glutamine pools of both axons and glia. In the absence of exogenous NAAG, endogenous NAAG released by stimulation is also hydrolyzed. The glutamate released can be detected as an increased NMDA receptor activation that potentiates the NAAG activated metabotropic glutamate receptor-initiated hyperpolarization response of glia (Fig.3). These results clearly point to the fact that axons and their associated glia utilize glutamate to synthesize NAAG and that NAAG is an axon-to-glia signaling agent that acts specifically on a glial group II metabotropic receptor to activate a number of physiological responses of the glia just now being identified[9-11,19,20]. These include, but not limited to, activation of signaling cascades that regulate glial membrane potential, extracellular K^+ homeostasis and glial volume regulation and the activation of GCPII to terminate the NAAG signal. Glutamate, the

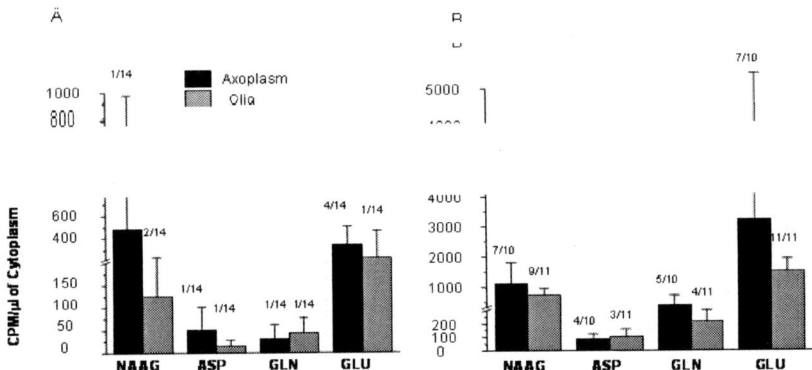

Figure 2: Effect of stimulation on accumulation and cellular distribution of radiolabeled metabolites of NAAG in medial giant nerve fibers. **Panel A:** The resting nerve cord was incubated with [³H]-NAAG for 30 min in the absence of 2-PMPA, a potent GCPII inhibitor. The number above each bar indicates the fraction of 14 axons and glia that had detectable radiolabel in NAAG, ASP, GLN and GLU. Mean radiolabel content is for all 14 fibers (samples in which there were no detectable radioactivity were given a value of zero). **Panel B:** Nerve fibers were incubated for 29 min incubation with [³H]-NAAG and then stimulated at 50 Hz during the 30th minute of incubation. The number of samples containing detectable amounts of radiolabeled NAAG, GLU and GLN increased dramatically for the 10-11 nerve fibers and the mean radiolabel content increased by a factor of 3 to 6. Data are presented as mean ± 1 SEM. The axonal and glial contents of all radiolabeled substances with stimulation (panel B) were significantly greater than those of the unstimulated controls (panel A) at p ≤ 0.001. Reprinted by permission from Urazaev et al[10].

hydrolysis product of NAAG, primarily acts on glial NMDA receptors participating in a number of cellular activities including glial membrane potential regulation, increased intracellular Ca^{2+} and NO production.

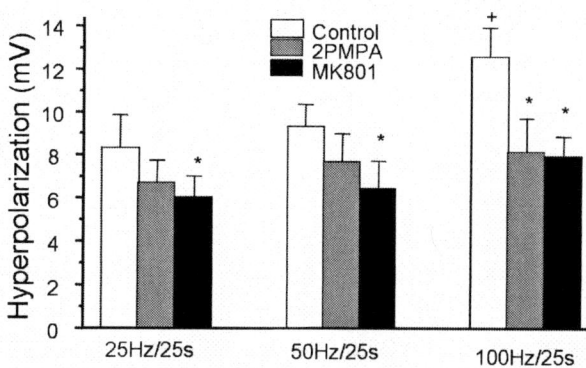

Figure 3: The effect of 2PMPA and MK 801 on the amplitude of the stimulation-induced glial cell hyperpolarization. The data of this figure demonstrates that the amplitude of glial cell hyperpolarization increased proportionally with increasing stimulation frequency between 25 and 100Hz. The GCP II inhibitor, 2-PMPA (0.1 μM), or the NMDA antagonist, MK801 (10 μM) blocked a portion of the hyperpolarization to nearly the same extent at each stimulation frequency. Data are presented as mean $\pm$ SEM. Asterisks indicate a statistically significant difference from the control at $p \leq 0.05$ (n=16-20). Reprinted by permission from Urazaev et al.[10]

3.2 NAAG Synthesis in Axoplasm Isolated from the Medial Giant Axon of the Crayfish

Synthesis of NAAG and NAA (Figs. 4, 5) was examined in axoplasm isolated from the axon by microcannulae as described in the METHODS. Radiolabeled glutamate, used as the substrate for NAAG synthesis, was added to isolated axoplasm in distilled H_2O or in an incubation medium (IM) containing in mM; .25 NAA; 0.1 cAMP; 0.5 ATP; 50 KCL; 10 $MgCl_2$; 30 Tri.HCl at pH 7.0^3. The volume of IM or glutamate in water as a fraction of the axoplasm volume varied from 1% to 25% depending on the protocol. Incubation of the axoplasm with the IM was 2-3 hours at room temperature. Following incubation, the axoplasm droplets were diluted in 100μl of water containing non-labeled GLU (100nM), NAA (100nM) and NAAG (50nM) to facilitate identification of NAA and NAAG peaks with A TSK-GEL SAX coulmn HPLC system. The individual peaks were collected and analyzed for radioactivity.

In the experiments illustrated in Fig. 4, radiolabeled glutamate was added only to track synthesis and represented approximately 10^{-9} M. This amount was insignificant with respect to the endogenous glutamate in native isolated axoplasm of approximately $(10^{-2}$ M). When radiolabeled glutamate in water was added to axoplasm, in the absence of cofactors or other substrates, little NAAG was synthesized (filled triangles) even up to a volume fraction of 0.25. On the other hand, with incubation medium (IM) (filled squares) there was a steep slope of increasing NAAG synthesis to approximately 100 pmoles/μl.hr at volume fraction of 0.17. In a direct comparison between glutamate in water and IM solutions at a volume fraction of 0.1, synthesis of NAAG was 6.4$\pm$3.7 pmole/μl.hour (n=4) compared to 56.8$\pm$4.2 pmole/μl.hour, respectively. These results suggest that NAA and/or one or more of the cofactors in IM are necessary for the efficient synthesis of NAAG.

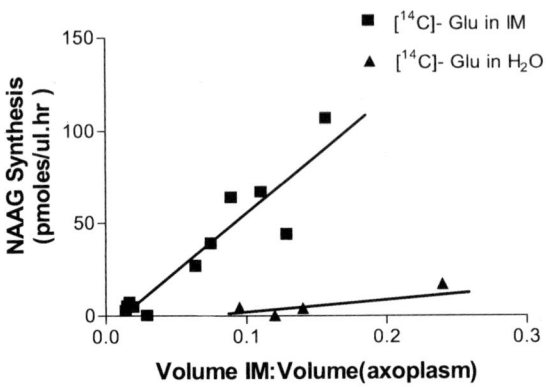

Figure 4. The effects of increasing radiolabeled glutamate, NAA and cofactors on NAAG synthesis in isolated axoplasm.

For the experiments illustrated in Fig. 4 our initial assumption was that there would be sufficient substrate in native axoplasm to support synthesis using [^{14}C]-glutamate in water as the only additive. As seen that is clearly not the case. Basing the protocol on the above described experiments, NAA was the first to be examined as a substrate for NAAG synthesis. These experiments were performed with a constant volume fraction of IM solution, radiolabeled glutamate to track synthesis and varying amounts of NAA. The results are shown in Fig. 5.

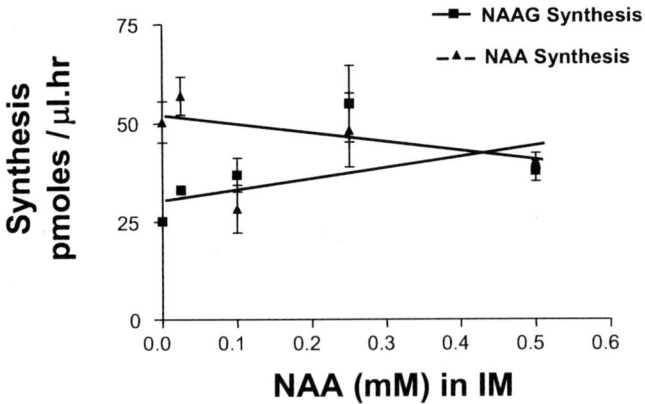

Figure 5: The effect of NAA on the synthesis of NAAG and NAA in native-isolated axoplasm. Axoplasmic samples were incubated for 2 hrs with constant volume fraction of IM containing varying amounts of non-radiolabeled NAA and radiolabeled GLU to track synthesis. Calculation of synthesis was based on an average GLU concentration in axoplasm of 10^{-2} moles/L. Values are given as the mean $\pm$ 1 SEM.

There appears to be sufficient NAA in native axoplasm to support NAAG synthesis without its addition to IM. Statistically, neither regression line slope is different from zero. Both NAAG and NAA contained radiolabel suggesting that glutamate was converted to aspartate which was subsequently acetylated to form NAA to which glutamate was added forming NAAG. HPLC column effluents containing NAA and NAAG were hydrolyzed in 6N HCl to their amino acid components which were analyzed by cation exchange HPLC. NAA contained radiolabeled aspartate and NAAG contained both radiolabeled glutamate and aspartate. Free radiolabeled aspartate was also found in the non-hydrolyzed amino acid fraction and is consistent with our earlier published experiments showing that radiolabeled aspartate was found in nerve fibers incubated with radiolabeled glutamate. These results suggest that the only necessary substrate for NAA and NAAG synthesis in axoplasm is glutamate and the experiments with radiolabeled glutamate in water lacked a cofactor, most likely cAMP - not substrate.

3.3 Preferred Substrates for NAAG Biosynthesis in Native-Isolated Axoplasm.

In an attempt to determine the most effective substrates for NAAG biosynthesis in freshly isolated axoplasm the experiments described in section 3.2 were repeated keeping the volume fraction of IM to axoplasm constant. Only the radiolabeled species was altered. As previously described, the amount of radiolabeled substrate was small compared to the amount of these substrates in native axoplasm. Glutamate appeared to be the most effective substrate for NAAG synthesis. This determination was based on the percentage of radiolabel recovered in NAAG compared to the total recovered in axoplasm. The order of effectiveness was GLU (32%) > ASP (14%) > NAA (12%) > GLUC (8%). Radiolabeled acetate did not appear to contribute to NAAG synthesis.

These results should only be taken as a preliminary indicator of substrate effectiveness since the content of each of these substrates in axoplasm is not precisely known and determination of the specific activity of the substrate pool and the amount of radiolabel of a specific substrate that should appear in NAAG, relative to the other substrates, is not possible. Nonetheless, GLU is known to be approximately 10mM in crayfish axoplasm while the other substances are approximately equal to or less than GLU. These results suggest it is fair to say that GLU, at the least, is a highly effective substrate for NAAG biosynthesis.

3.4 NAAG Biosynthesis in Dialyzed-Isolated Axoplasm

As demonstrated above, there appears to be sufficient glutamate, aspartate and NAA in native (undialyzed) axoplasm to support synthesis in the absence of these substrates in the incubation medium. Consistent with the above results, preincubation of intact nerve fiber with 10mM amino-oxyacetic acid (AOAA), a glutamate-oxaloacetate transaminase inhibitor, reduced NAA synthesis by about 40% but had no discernible effect on NAAG synthesis in isolated axoplasm (Table 1).

To provide a system in which substrate concentrations can be controlled and specific activities determined, all endogenous substrates present in axoplasm were removed by dialysis against a simple salt solution approximating that found intracellularly (artificial axoplasm). Synthesis of NAAG and NAA was reconstituted in dialyzed axoplasm by replacing those substrates and cofactors thought to support synthesis. The IM added to axoplasm provided final incubation concentrations of 5mM glutamate as the sole substrate, radiolabeled glutamate to follow synthesis and the cofactors, acetyl-CoenzymeA (5nM), c-AMP (13 μM) and ATP (65μM). A comparison of NAAG and NAA synthesis under various experimental conditions in isolated and dialyzed-isolated axoplasm are shown in Table 1. NAAG and NAA were synthesized to about the same extent in dialyzed as in native axoplasm and both molecules contained radiolabel. It is also demonstrated by the results that a significant NAAG and NAA synthesis occurred in the absence of acetyl-CoA in the IM added to dialyzed axoplasm. NAAG synthesis was reduced approximately 60% and NAA by 77%. These results suggest that CoA probably is complexed with acetyltransferase and not readily dialyzable. More significantly these results suggest that glutamate may serve as a substrate for the citric acid cycle providing acetate to produce acetyl-CoA and oxidative phosporylation to provide ATP. The experiments demonstrating that the removal of ATP reduced NAAG synthesis by only approximately 40% and NAA by 50% is consistent with the hypothesis that glutamate is a substrate for intermediary metabolism.

Table 1. A comparison of NAAG and NAA synthesis in native-isolated and dialyzed-isolated axoplasm.

	Isolated Axoplasm	Dialyzed Axoplasm
	NAAG / NAA pmole/µl.hr $\pm$ SEM	NAAG / NAA pmole/µl.hr $\pm$ SEM
IM 0.5 mM NA	38.0$\pm$5.3 / 40.5$\pm$4.1 (4)[1]	
IM - 0 NAA	25.2$\pm$1.9 / 50.4$\pm$1.6 (5)	43.8$\pm$13.2 / 34.8 $\pm$4.3 (5)
AOAA	26.8$\pm$2.6 / 29.1$\pm$3.6 (10)	
IM − 0 AcCoA		17.6 $\pm$ 3.7 / 7.9 $\pm$ 2.1 (5)
IM − 0 ATP		26.0 $\pm$ 2.6 / 15.5 $\pm$ 8.0 (3)

[1] Number of experiments. All values given as mean $\pm$ 1 SEM

4. DISCUSSION

It would not be unfair to question the relevance of our investigations on axons and glia of the medial giant nerve fibers of the crayfish, and in particular, the synthesis, degradation and recycling of NAAG and NAA to mammalian neuron/glia systems.

In our view the functional and pharmacological properties of the crayfish glial cell appear to be a composite of the properties of cells found in mammalian systems including receptors; ion, substrate and neurotransmitter transporters; intracellular transduction pathways; neurotransmitter and ion channel characteristics of Schwann cells, oligodendrocytes, astrocytes and microglia. Table 2 compares some of the properties of mammalian and crayfish glia that illustrate this concept. Because functional axon-glia interactions have been and are continuing to be defined in the crayfish nerve fiber, this model system represents an opportunity to investigate glial and axonal functional properties in an intact, interactive system in which contributions of each cell in the system can be differentiated.

Although it would not be appropriate to assume functional properties of the crayfish glial cell directly reflect the functional properties of a given mammalian glial cell, many of the fundamental processes appear to be conserved. We believe investigations in definable model systems such as the crayfish medial giant nerve fiber may provide guidance in developing the protocols that elucidate functional properties of the mammalian axon/neuron-glial systems, not unlike studies in mammalian neuron-glia cell culture models that serve a similar purpose.   The difference between the crayfish axon-glia preparation and mammalian cell culture is that the former is a "real world" working system in an anatomically and functionally appropriate relationship.

Table 2. A comparison of functional components of mammalian and crayfish axons, neurons and glia with regard to NAAG.

	Mammalian		Crayfish	
	Neurons (perikarya)	Astrocytes/ Microglia	Axons	glia
NAAG (NAA)	+[6,21,22]	+[6,21,22]	+ [7-11]	+[7-11]
GCPII	-	+[6,22,23]	-	+[10]
NMDAR	+[24,25]	+[25]	-	+[19,20]
mGLUR$_{II}$	interneurons/axons[26]	+[27-30]	+[34]	+[21]
NOS	+[30,31]	nNOS/iNOS[32,33]	+[35,36]	+[12]

The early work from our laboratory established that NAAG was present in both axons and glia of the medial giant nerve fiber of the crayfish. Since NAAG was not taken up into either axons or glia as an intact molecule the only viable conclusion was that it was synthesized in both cells[9-11]. These investigations also demonstrated that exogenous and endogenous NAAG are hydrolyzed by a nerve activity-dependent peptidase that has pharmacological characteristics of Glutamate Carboxypeptidase II[9, 10]. Although we did not specifically address the question of the presence of NAA in axons and glia in these earlier investigations, it is clear that NAA must also be present to synthesize NAAG. As demonstrated in the investigation of isolated axoplasm presented here, NAA is found in axoplasm in about equal amounts to NAAG. As yet, we have not determined the amount of NAA in crayfish glia.

It became clear from these early studies that attempts to elucidate the metabolic pathway(s) for synthesis and identification of the preferred substrates in the intact nerve-glial preparation would not be easy due to the difficulty in determining the absolute concentrations of substrates in axoplasm and glial cytoplasm without sophisticated techniques such as Mass Spectrometry. In addition, highly developed metabolic interactions between the giant axon and its associated glial cells would further obscure the cells of origin of synthesized and accumulated products and intermediates. We found no evidence that the axon membrane has functional glutamate transporters thus the glutamate that eventually appears in axons is most probably due to a glutamate – glutamine shuttle between glia and axons such that glutamate generated by the hydrolysis of NAAG was transported into glia where it was utilized for metabolism, synthesis of NAAG and conversion to glutamine[16]. Glutamine is transported across the adaxonal space to be taken up by the axon and converted into glutamate for NAAG synthesis as well as utilized by intermediary metabolism as an energy source in the axon[11].

To better understand the synthesis of NAAG we began a series of experiments using axoplasm removed from the axon. The results of these experiments demonstrated that the metabolic machinery required for NAAG synthesis was present in native-isolated axoplasm and that most of the substrates and cofactors required were also present. Nonetheless, these experiments demonstrated that the addition of radiolabeled glutamate in water, in the absence of ATP and cAMP, was not sufficient to support NAAG synthesis. It was also demonstrated that glutamate as the sole substrate with cofactors included was sufficient to support synthesis. Both aspartate and glutamate released by acid hydrolysis of NAAG and NAA synthesized with radiolabeled glutamate were radiolabeled supporting the conclusion that glutamate is a sufficient substrate for synthesis of aspartate and acetate needed for NAAG and NAA biosynthesis.

When experiments were performed with axoplasm dialyzed to remove substrates and cofactors, only acetyl-CoA appeared to have a major effect, reducing NAAG and NAA synthesis by 60 and 77%, respectively. An ATP regeneration system is apparently present in dialyzed axoplasm that can use glutamate to fuel oxidative phosphorylation. To further define the necessary substrates, intermediates and cofactors supporting NAAG biosynthesis, dialysis of axoplasm alone will not provide the necessary conditions. The TCA Cycle and ATP regeneration will also need to be inhibited.

4.1 Summary and Conclusions

Three major conclusions can be proposed as a result of these studies: 1. The enzymes and cofactors for generation of the intermediates required for NAAG synthesis and the NAAG synthase appear to be present in native-isolated and dialyzed axoplasm; 2. As a substrate, glutamate is sufficient for NAA and NAAG synthesis and is utilized by the TCA cycle and oxidative phosphorylation to generate all intermediates, most of the cofactors and ATP needed for NAA and NAAG biosynthesis and 3. The crayfish giant nerve fiber appears to be an advantageous model for elucidation of the mechanism of NAAG and NAA synthesis in both axons and glia and potentially for the identification and characterization of the NAAG synthase.

4.2 Future Experiments

Experiments planned for the immediate future will utilize native axoplasm to determine the role of cAMP in synthesis. Based on where the block occurs in the pathway for NAAG biosynthesis, in the absence of cAMP, defines its role. For example, if neither NAA nor NAAG are synthesized then cAMP is required for either the glutamate-oxaloacetate transamination to produce aspartate or for the N-acetyltransferase that forms NAA from aspartate. In either case the appearance - or not - of radiolabeled aspartate will provide the necessary information.

Dialyzed axoplasm will be used to define which substrates are effective in supporting synthesis and where ATP is necessary in the synthetic pathway. For these experiments it will be necessary for metabolism of glutamate to the intermediates used in NAA and NAAG synthesis to be inhibited. This can be accomplished by adding metabolic inhibitors directly to the dialysis solution while removing all endogenous substrates.

6. QUESTION AND ANSWER SESSION

DR. COYLE: This paper is open for discussion. Well, I will make a comment. Congratulations. As I said in my introductory comments, since the mid-1980s we have been trying to periodically characterize synthesis of NAAG, and we tried the Reichle strategy in rat brain extracts and could not get that to work with dialysis, putting in all sorts of other additives.

We then tried strategies that are cocktails that are used for glutathione synthesis, and again we haven't been successful. So I am very excited about the leads you have in this model system, because I think it will be very important for characterizing that enzyme or family of enzymes responsible for NAAG synthesis.

DR. NAMBOODIRI: I had one question on dialysis. When you dialyze it, you don't lose all the activity. That is to say, it only decreases by 50 percent.

You would expect when you dialyze, you remove all the ATP, everything. So that unless you add them back, you should not have any activity.

DR. LIEBERMAN: I expect that we removed all of the free ATP during the dialysis while most of the enzymes associated with NAAG and NAA synthesis and those of the citric and glycolytic cycles are present and functional in the dialyzed axoplasm. When substrates like glutamate and cofactors other than ATP were added back to the dialyzed axoplasm enough ATP could be synthesized to fuel NAA and NAAG synthesis. The conclusion we came to is that the regeneration system is functional if it has substrate to use. We will have a better idea of the extent that this is the case when we do the dialysis in the presence of iodoacetic acid and cyanide to knockout ATP regeneration. This should knockdown NAA and NAAG synthesis, which we predict will be restored when we add back ATP.

DR. COYLE: John.

DR. MOFFETT: I think it is interesting that in our neuroblastoma cell system we also found that dibuteryl cyclic AMP increased the level of NAAG synthesis significantly.

DR. LIEBERMAN: There is almost no NAA and NAAG synthesis when we add glutamate to either native or dialyzed axoplasm in the absence of cyclic AMP, other co-factors and substrates. It is clear something was missing.

DR. TRAPP: Just to keep on that theme, I think there is accumulating evidence now that in oligodendrocyte-axon interactions that cyclic AMP is a major mediator of that interaction that is playing a role in stabilizing the cytoskeleton of the axon, probably along with other activities which you have just described here. But it is coming to be a key molecule.

DR. LIEBERMAN: Yes. A best guess at this point is that cyclic AMP is necessary for phosphorylation of one or more of the enzymes to make the synthetic pathway functional.

PARTICIPANT: Is there any evidence for systematic modification of NAAG synthesis by phosphorylation or cyclic AMP?

DR. LIEBERMAN: No one has isolated and characterized the synthetase as of yet. Due to its "simplicity", as compared to the intact cell or whole cell homogenates, isolated and dialyzed axoplasm represents a system in which the possibility for finding the enzyme is improved.

PARTICIPANT: What has been the problem of identifying it?

DR. COYLE: What's been the problem?

DR. LIEBERMAN: I can't answer that question other than to say that, from a metabolic standpoint, intact cells or tissues or their homogenates are very complicated. As I have suggested and the results of our experiments seem to support, the simplest system that still supports NAA and NAAG synthesis opens opportunities to find and characterize the NAAG synthetase.

DR. COYLE: We have done a fair amount of enzymology over the last 30 years, and we cannot get synthesis in broken cell preparations. I think Dr. Namboodiri maybe may have gotten a handle on it. We will hear, but this is the first reliable description I am aware of. John?

DR. MOFFETT: Well, we do have a new neuroblastoma cell line, and we are able to get very good NAAG synthesis. One of the regulators we found was dibutyryl cyclic AMP

DR. COYLE: But that is in an intact system. Right.

DR. MOFFETT: It is intact. You can get a small blip of NAAG synthesis in the model when you homogenize the cells, and we are hoping that we will be able to increase specific activity of the enzyme with this preparation.

7. REFERENCES

1. A. Curatelo, P. D'Archangelo, A. Lino, and A. Brancati, Distribution of N-acetyl-aspartic and N-acetyl-aspartyl-glutamic acids in nervous tissue. J. Neurochem. 12: 339-342 (1965)
2. E. Miyamoto, Y. Kakimoto, and I. Sano, Identification of N-acetyl-alpha-aspartylglutamic acid in the bovine brain. J. Neurochem. 13(10):999-1003 (1966)
3. K.L. Reichelt and E. Kvamme, Histamine-dependent formation of N-acetyl-aspartyl peptides in mouse brain. J. Neurochem 21(4):849-859 (1973)
4. C.B. Cangro, M.A. Namboodiri, L.A. Sklar, A. Corigliano-Murphy, and J.H. Neale, Immuno-histochemistry and biosynthesis of N-acetylaspartylglutamate in spinal sensory ganglia. J. Neurochem. 49(5):1579-88 (1987)
5. L.M. Gehl, O.H. Saab, T. Bzdega, B. Wroblewska and J.H. Neale, Biosynthesis of NAAG by an enzyme-mediated process in rat central nervous system neurons and glia. J. Neurochem, 90(4):989-997 (2004)
6. J.H. Neale, T. Bzdega and B. Wroblewska, N-acetylaspartylglutamate: the most abundant peptide neurotransmitter in the mammalian central nervous system. J. Neurochem. 75: 443-452 (2000)
7. E.M. Lieberman, A. Urazaev, R.M. Grossfeld, and J. Christian, NAAG and NAALadase are present in the crayfish medial giant nerve fiber: Implication for axon-Schwann cell signaling. J. Neurochem. Suppl. 72: S56 (1999)
8   L.S. Kane, J.G. Buttram, A.K. Urazaev, and E.M. Lieberman, Uptake and metabolism of glutamate at non-synaptic regions of crayfish central nerve fibers: Implications for axon-glia signaling. Neuroscience 97: (3) 601-609 (2000)
9. A.K. Urazaev,. R.M., Grossfeld, P.L. Fletcher, H.Speno, B.Gafurov, J.G. Buttram Jr. and E.M. Lieberman, Synthesis and release of N-acetylaspartylglutamate (NAAG) by crayfish nerve fibers: Implications for axon-glia signaling. Neuroscience 106(1):237-247 (2001a)
10. A. Kh. Urazaev, J.G. Butram Jr., J.P. Deen, B. Sh. Gafurov, B.S.,Slusher, R. M. Grossfeld, and E. M. Lieberman, Mechanisms of clearance of released N-acetylaspartylglutamate in crayfish nerve fibers: Implications for axon-glia signaling. Neuroscience 107(4): 697-703 (2001b)
11. J.G. Buttram, Jr., J.A. Engler, A.Kh. Urazaev, R.M. Grossfeld, and E.M. Lieberman, Uptake and metabolism of glutamine at non-synaptic regions of crayfish central nerve fibers: Implications for axon-glia signaling. Comp. Biochem. Physiol. B 133:2 09-220 (2002)
12. R.A. Khairova, V. Harrell, M.T. Powers, A.K. Urazaev, E.M. Lieberman, The effect of endogenous and exogenous nitric oxide on action potential generation and propagation in crayfish medial giant nerve fibers: implications for axon-glia signaling. Program# 730.9 2004 Abstract Viewer and Itinerary Planner Washington DC, Society for Neuroscience 2004 online.
13. D.G. Brunder and E.M. Lieberman, Studies of axon-glial cell interactions and periaxonal K^+ homeostasis: I. The influence of Na^+, K^+, Cl^- and cholinergic agents on the membrane potential of the adaxonal glia of the crayfish medial giant axon. Neuroscience 25(3):951-959 (1988)
14. S. Hassan, and E.M. Lieberman, Studies of axon-glial cell interactions and periaxonal K^+ homeostasis II. The effect of axonal stimulation, cholinergic agents and transport inhibitors on the resistance in series with the axon membrane. Neuroscience 25(3): 961-969 (1988)
15. E.M. Lieberman and S.Hassan, Studies of axon-glial cell interactions and perineural K^+ homeostasis III. The effect of anisosmotic media and potassium on the relationship between the resistance in series with the axon membrane and glial cell volume. Neuroscience 25: 971-981 (1988)
16. E. McKinnon, P.T. Hargittai, R.M. Grossfeld, and E.M. Lieberman, Glutamine cycle enzymes in the crayfish giant nerve fiber: Implications for axon-to-glia signaling. Glia 14(3): 198-208 (1995)
17. B.G. Wallin, The relation between external potassium concentration, membrane potential and internal ion concentrations in crayfish axons, Acta Physiol Scand. 70(3):431-448 (1967)

18. E.M.Lieberman. Enzymatic determination of subpicomole quantities of phospho-L-arginine in nerve cytoplasm. Anal Biochem. Oct 15;134(2):413-23 (1983)
19. B. Gafurov, A.K. Urazaev, R.M. Grossfeld, and E.M. Lieberman, Evidence that N-acetylaspartylglutamate (NAAG) is the probable mediator of axon-to-glia signaling in the crayfish medial giant axon. Neuroscience. 106(1): 227-235 (2001)
20. B. Gafurov, A.K. Urazaev, R.M. Grossfeld, and E.M. Lieberman, Mechanism of NMDA receptor contribution to axon-to-glia signaling in the crayfish medial giant nerve fiber. Glia 38(1): 80-86 (2002)
21. L.A. Passani, J.P. Vonsattel and J.T. Coyle, Distribution of N-acetylaspartylglutamate immunoreactivity in human brain and its alteration in neurodegenerative disease. Brain Res.;772(1-2):9-22 (1997)
22. M. Cassidy and J.H. Neale, Localization and transport of N-acetylaspartylglutamate in cells of whole murine brain in primary culture. J.Neurochem. 60 (5):1631-1637 (1993)
23. U.V. Berger, R. Luthi-Carter, L.A. Passani, S. Elkabes, I. Black, C. Konradi and J.T. Coyle. Gluatamate carboxypeptidase is expressed by astrocytes in the adult rat nervous system; J. Comp. Neurol. 415(1):52-64 (1999)
24. F. Conti, P. Barbaresi, M. Melone, and A.Ducati et al. Neuronal and glial localization of NR1 and NR2A/2B subunits of the NMDA receptor in the human cerebral cortex. Cerebral Cortex 9(2):110-120 (1999)
25. F. Conti, S. DeBiasi, A. Minelli, M. Melone. Expression of NR1 and NR2A/B subunits of the NMDA receptor in cortical astrocytes. Glia; 17(3): 254-8 (1996)
26. A. Pisani, P. Bonsi, M.V. Catania, R. Giuffrida, M. Morari, M. Marti, D. Centonze, G. Bernardi, A.E. Kingston and P. Calabresi. Metabotropic glutamate group 2 receptors modulate synaptic inputs and calcium signals in striatal cholinergic interneurons. J. Neurosci. 22 (14): 6176-85 (2002)
27. R.S. Petralia, Y.X. Wang, A.S. Niedzielski and R.J. Wenthold. The metabotropic glutamate receptors, mGluR2 and mGluR3, show unique postsynaptic, presynaptic and glial localizations. Neuroscience 71(4):949-76 (1996)
28. B. Wroblewska M.R. Santi and J.H. Neale. N-acetylaspartylglutamate activates cyclic AMP-coupled metabotropic glutamate receptors in cerebellar astrocytes. Glia 24:172-179 (1998)
29. Y. Tanabe, A. Nomura, M. Masu, R. Shigemoto, N. Mizuno, S..Nakanishi. Signal transduction, pharmacological properties, and expression patterns of two rat metabotropic glutamate receptors, mGluR3 and mGluR4. J.Neurosci. 13(4):1372-8 (1993)
30. D.S. Bredt, P.M. Hwang, S.H. Snyder. Localization of nitric oxide synthase indicating a neural role for nitric oxide. Nature 347(6295):768-70 (1990)
31. K. Iwase K. Iyama, K. Akagi, S. Yano, K. Fukunaga, E. Miyamoto, M. Mori and M. Takiguchi. Precise distribution of neuronal nitric oxide synthase mRNA in the rat brain revealed by non-radioisotopic in situ hybridization. Brain Research, Molecular Brain Research 53(1-2):1-12 (1998)
32. M. Oka, M. Wada, A. Yamamoto, Y. Itoh and T. Fujita. Functional expression of constitutive nitric oxide synthases regulated by voltage-gated Na+ and Ca2+ channels in cultured human astrocytes. Glia 46(1):53-62 (2004)
33. A.E. Wiencken and V.A. Casagrande. Endothelial nitric oxide synthetase (eNOS) in astrocytes: another source of nitric oxide in neocortex. Glia 26(4):280-901(1999)
34. M. Schramm and J. Dudel. Metabotropic glutamate autoreceptors on nerve terminals of crayfish muscle depress or facilitate release. Neurosci Lett. 234(1):31-4. (1997)
35. K. Johansson and M. Carlberg. NADPH-diaphorase histochemistry and nitric oxide synthase activity in deutocerebrum of the crayfish, Pacifastacus leniusculus (Crustacea, Decapoda).Brain Res.; 649(1-2):36-42 (1994)
36. H. Schuppe, H. Aonuma and P.L. Newland. Distribution of NADPH-diaphorase-positive ascending interneurones in the crayfish terminal abdominal ganglion. Cell Tissue Res. 305(1):135-46 (2001)

NAAG AS A NEUROTRANSMITTER

Barbara Wroblewska[*]

1. INTRODUCTION

N-acetylaspartylglutamate (NAAG) was discovered in the central nervous system of mammals in 1965 (Curotalo et al., 1965). Since then, several groups have described the localization of NAAG in the brain (Fuhrman et al., 1994; Shave et al., 2001) and spinal cord (Moffett and Namboodiri 1995; Moffett et al., 1990; Renno et al., 1997; Slusher et al., 1992; Williamson et al., 1991) as well as the presence of NAAG in the nervous systems of different vertebrate (Anderson et al., 1986, 1987; Cangro et al., 1987; Tieman et al., 1991) and invertebrate species (Gafurov et al., 2001). A considerable amount of data indicates that NAAG may function as a neurotransmitter (co-transmitter) or a neuromodulator in the central nervous system.

Several excellent reviews summarize progress in this field (Neale et al., 2000; Baslow 2000; Coyle 1997), although "the naaging question of the function of N-acetylaspartylglutamate" (Coyle 1997) still has not been fully answered. Recently, more interest has been focused on the role of NAAG in pathologies of the central nervous system. Barbara Slusher and her group from Gilford Pharmaceuticals discovered that inhibition of the enzyme hydrolyzing NAAG increases endogenous levels of NAAG, decreases release of glutamate, and reduces ischemic brain injury in rats (Slusher et al., 1999). They synthesized a series of NAAG peptidase inhibitors derived from (phosphonomethyl)-pantanedioic acid (PMPA), and used the inhibitors to study the effects of increase in endogenous NAAG on brain ischemia, neurotoxicity and neuropathic pain (Rojas et al., 2003, Carpenter et al., 2003; Slusher et al., 1999; Thomas et al., 2001).

Our approach to understanding the function of endogenous NAAG was based on the possibility of manipulating levels of extracellular levels of NAAG after the synaptic release. To investigate the effects of increased levels of endogenous NAAG on the neurotransmission we used mice with a knockout of the NAAG peptidase gene.

---

[*] Department of Biology, Georgetown University, Washington D.C., USA; wroblewb@georgetown.edu.

   The enzyme hydrolyzing NAAG (Riveros and Orrego 1984) has been characterized by Robinson and collaborators (Robinson et al., 1987) as a metallo-exopeptidase with the affinity for acidic peptides. Activity of this enzyme has been found on retinal cells, mixed cortical cultures, and glial cells (Williamson et al., 1991, Cassidy and Neale 1993). Protein was purified (Slusher et al., 1990), partially sequenced and identified (Carter et al., 1996) as a homolog of prostate specific membrane antigen (PSMA) (Israeli et al., 1993). Further, cDNA for the peptidase has been cloned from rat brain (Bzdega et al., 1997, Luthi-Carter et al., 1998) and is known as glutamate carboxypeptidase type II (GCPII). The gene for GCPII is expressed predominantly/exclusively in the glial cells (Berger et al., 1995). In collaboration with Warren Heston from Cleveland Clinic we prepared Folh1 knockout mice. Mice harboring a disruption of the gene for GCPII/Folh1 were generated by inserting into the genome a targeting cassette in which the intron-exon boundary sequence of exon 1 and 2 were removed and stop codons were inserted into both exons (Bacich et al., 2002). Although these mice did not express mRNA and protein encoding for NAAG peptidase (GCPII) they developed normally, and expressed normal neurological responses and mating activity. Several behavioral tests did not show any abnormalities (Bacich et al., 2002). We did not observe any significant changes in the overall levels of NAAG and glutamate in the brains of knockout mice as measured by HPLC. However, in various tissues in the knockout mice (e.g. brain, kidney, and spinal cord) NAAG-hydrolyzing activity was not completely abolished. The remaining activity of the membranes collected from KO mice was low (about 10-20% of the wild type), but consistently expressed. Using ESTs from the murine database we cloned the second peptidase which acted to hydrolyze NAAG (Bzdega et al., 2004). cDNA for this peptidase, named carboxypeptidase III (GCPIII), is about 70% identical to mouse GCPII, and 85% identical to its human homolog. Both peptidases had a very similar activity profiles (pH dependency and metal ions requirements). Interestingly, GCPIII is expressed in both neuronal and glial cells in culture and throughout several brain regions (Bzdega et al., 2004). There is no difference in the expression of GCPIII (as measured by the real time quantitative PCR) between wild type and GCPII- knockout mice.

   Presently, mice that do not express both peptidases (GCPII and GCPIII) are not available; therefore we used peptidase inhibitors to investigate the role of endogenous NAAG in the neurotransmission in the central nervous system. In collaboration with Alan Kozikowski we synthesized a series of peptidase inhibitors based on the chemical structure of NAAG, but lacking the peptide bond (Kozikowski et al., 2001, 2004; Nan et al., 2000; Rong et al., 2002). We measured NAAG hydrolyzing activity of these inhibitors in the cells stably expressing GCPII and GCPIII (Bzdega et al., 2004). We also investigated the effects of these inhibitors on several of glutamatergic receptors, using cell lines stably expressing metabotropic receptors. The activity of peptidase inhibitors on other neurotransmitter receptors was screened in NIMH Psychoactive Drug Screen. Several of the most effective peptidases inhibitors were used in the pain studies and in the schizophrenia model in rats.

2. PAIN MODELS

   In collaboration with Tatsuo Yamamoto from Chiba University (Yamamoto et al., 2004; Yamamoto et al., 2001) we examined the effects of peptidase inhibitors in two different pain models: inflammatory pain (measurements of flinching responses after injection of dorsal surface of the right paw with 5% formalin) and neuropathic pain (measurements of pain responses using von Frey filaments after partial ligation of right sciatic nerve). Inhibitors of

NAAG peptidases (which blocked the activity of both GCPII and GCPIII – named ZJ11, ZJ46 and ZJ 17) reduced pain responses in both pain models studied. Application of peptidase inhibitors also significantly decreased fos-like immunoreactivity in lamine I and II of the spinal cord suggesting, that peptidase inhibitors affected pain imputs from primary afferents. Because studied peptidase inhibitors did not show any significant interactions with receptors studied we hypothesized that beneficial effects of NAAG peptidases inhibitors in the pain models were due to the increased levels of endogenous NAAG and decreased levels of glutamate. Because we had shown earlier that NAAG is a selective agonist of mGluR3 metabotropic glutamate receptors (Wroblewska et al., 1997) animals were treated with group II antagonist (LY341495). The antagonist significantly blocked the beneficial effects of NAAG peptidase inhibitors (Yamamoto et al., 2004). There was no effect of either of the studied peptidase inhibitors on acute pain perception (hot plate tests and mechanical nociceptive pain).

3. SCHIZOPHRENIA

It has been shown by Bita Moghaddam (Moghaddam and Adams 1998, Cartmell et al., 1999, 2000) that stimulation of group II metabotropic glutamate receptors (using mGluR2/3 agonist, LY354740) attenuated the disruptive effects of phencyclidine (PCP) on working memory, stereotypy, locomotion and cortical glutamate release in rats. Phencyclidine (blocker of NMDA receptor) elicits positive and negative schizophrenia-like symptoms in animals (Zukin and Javitt 1989).

Schizophrenia is one of the most common debilitating neurological disorders, since about 1% of the world population exhibits symptoms of schizophrenia. About 60% of schizophrenia sufferers live in poverty and about 5% end up homeless (Javitt and Coyle 2004). Cognitive deficiencies, emotional withdrawal, lack of social skills, agitation, paranoia, and hallucinations are common symptoms of schizophrenia. Due to the complexity of this disease, understanding of the development and the progression of schizophrenia still remains unknown.

For many years hyperactivity of dopamine receptors was perceived as a primary cause of schizophrenia (dopamine theory). Dopamine receptor antagonists (D2 antagonists) are the only known effective treatments for schizophrenic psychosis. However, the discovery that phencyclidine and ketamine cause symptoms which resemble schizophrenia (Javitt 1991; Zukin and Javitt 1989) and that studies on postmortem brains (Konradi and Heckers 2003) indicate that hypofunction (Tsai and Coyle, 2002) of glutamatergic neurotransmission is involved in the development and manifestation of schizophrenia (glutamate theory). Interestingly, even a single dose of phencyclidine elicits schizophrenic symptoms in animals (Zukin and Javitt 1989), whereas other treatments that produce schizophrenia-like symptoms require chronic application.

Group II metabotropic glutamate receptors (mGluR2 and mGluR3) are localized on the presynaptic terminal and regulate the release of glutamate from the synapse (Ohishi et al., 1993). NMDA antagonists, like phencyclidine and ketamine (Adams and Moghaddam 1998), increase release of glutamate in the prefrontal cortex, while the agonists of mGluR2/3 reverse the effects of PCP through the decrease in presynaptic release of glutamate. Group II metabotropic glutamate receptor agonists (e.g. LY341495) stimulate both pre- and postsynaptic receptors. Although the role of postsynaptic mGluR2/3 receptors is not clear yet, recent data shows that stimulation of postsynaptic mGluR2/3 enhances NMDA receptor currents in pyramidal neurons of rat prefrontal cortex (Tyszkiewicz et al., 2004).

In our studies, instead of using exogenous agonists of group II metabotropic glutamate receptors, we used peptidase inhibitors (ZJ43) presuming that inhibition of peptidases (ZJ43 – Ki of 0.8 nM and 23 nM for GCPII and GCPIII, respectively) increases levels of NAAG (an endogenous, selective agonist of mGluR3 receptor, (Wroblewska et al., 1997) and increased NAAG activates mGluR3 receptors. Following the application of PCP rats displayed series of stereotypic behaviors like, including stereotypic mouth movements, falling while walking, walking in circles, head bobbing, head sideways movement, and tremors (Olszewski et al., 2004). Some of these behaviors were significantly decreased in the presence of peptidase inhibitor (stereotypic mouth movements, falling while walking, walking in circles, head bobbing), while others were not affected. When animals were co-injected with metabotropic glutamate receptor antagonist (LY341495) some of these beneficial effects were abolished (Olszewski et al., 2004). Application of LY341495 alone exacerbated some of the PCP-induced motor activity. This may suggest that in the PCP-induced motor activity blocking the interaction between endogenous NAAG and mGluR3 receptor increases effects of PCP.

It has been shown that NAAG activates presynaptic mGluR3 receptors and participates in the regulation of the release of neurotransmitters (Berent-Spillson et al., 2004; Garrido Sanabria et al., 2004; Slusher et al., 1999; Zhao et al., 2001). Since NAAG is colocalized with several neurotransmitters (Renno et al., 1997) the effects on the presynaptic release maybe very important in both physiological and pathological conditions.

However, there is also mGluR3 receptor present on the glial cells, astrocytes (Wroblewska et al., 1998), giant axon myelinizing cells (Urazaev et al., 2001) Schwann cells (Berger et al., 1995) and microglia (Taylor et al., 2002, Taylor et al., 2005). Very little is known about the role of this receptor in the physiology of NAAG.

Our understanding of the consequences of the increase in NAAG levels in the central nervous system and the interactions of NAAG with its receptors increased considerably in the recent years. Nevertheless, we still don't know enough about the synthesis of NAAG, and we have no information about transport of NAAG in the brain. More research is this field is needed, especially with regard to recent discoveries on the possible importance of endogenous NAAG in the physiology and pathology of the brain.

4. QUESTION AND ANSWER SESSION

DR. COYLE: We have time for a couple of questions.

DR. WEINBERGER: I was confused about one thing. Is this the same compound that Moghaddam used?

DR. WROBLEWSKA: It is either the same or it is very similar. They have several--

DR. WEINBERGER: You actually found something different than she found?

DR. WROBLEWSKA: No. She didn't use the antagonist. She used the agonist of the group II receptors.

DR. WEINBERGER: Right.

DR. WROBLEWSKA: And actually, in her paper she suggested that NAAG could work as an endogenous agonist of the Group 2 receptors. So it was the reason we tried this. We used the antagonist to block the effects of NAAG.

PARTICIPANT: Just a brief comment on your PCP results. A lot of those behaviors look very dopaminergic and maybe 10 milligrams per kilogram is rather high to get selective glutamatergic effects, and I am sure, if you went down to 1 1/2 or 2.5 milligrams--

DR. WROBLEWSKA: We did. I mean, we tried -- we tried five -- I think 3, 5 and 10, and the response reached a plateau at 10. So that is why we used the higher dose. You are absolutely right. With mice we used a much lower dose per kilogram.

PARTICIPANT: Absolutely worth looking at something like cognition which is certainly affected at low concentrations.

DR. WROBLEWSKA: Yes. Thank you.

PARTICIPANT: Just a brief question. I just wanted to make sure I got it right.

Did you say that blocking NAAG peptidase did not have an effect on acute pain and only on the longitudinal?

DR. WROBLEWSKA: Yes. It didn't. The same thing was observed in the knock-out mice. When we didn't have the carboxypeptidase activity in mice, there was no significant difference in the response on the hotplate, and tail flick. We didn't look at anything else.

DR. NAMBOODIRI: So when you inhibit the knock-out element, you don't see any increase in NAAG. Right?

DR. WROBLEWSKA: No.

DR. NAMBOODIRI: Earlier studies from Coyle's group, done mostly by Barbara Slusher, have shown that when you block the enzyme with inhibitors, the NAAG will go up. That is the whole idea that this enzyme is involved in the degradation of NAAG. Right? But you haven't seen any change in the NAAG levels. So how do you explain that?

DR. COYLE: I'm confused. In our early studies, we showed that if you inject *in vivo* NAAG inhibitors in exogenous NAAG, that delayed the degradation of NAAG. We have published a lot of papers, so I may lose track, but I don't remember that we published a paper showing increases in steady state levels of endogenous NAAG.

DR. WROBLEWSKA: Barbara Slusher had shown with the inhibitors in the myocardial model that she did have increased levels of NAAG. Remember the Nature magazine paper?

DR. COYLE: Right. That was with stroke, and that was also *in vivo* dialysis.

DR. WROBLEWSKA: Right.

DR. COYLE: Right. That was not -- I am unaware of showing any change in NAAG levels steady state with inhibitors in the tissue.

DR. WROBLEWSKA: Well, I think that there are changes in NAAG levels with the knock-outs after treatment with enzyme inhibitors. The problem is that we looked at whole brain homogenates, and you know, there was no overall change, no overall significant difference. I think that, if we looked at the area where the enzyme is, or where NAAG is released, or in areas with mGluR3 receptors specifically, we would see a change. We don't really have the means to do this currently.

DR. COYLE: Dan?

DR. WEINBERGER: Joe, maybe you can both speak to this, because if you try to map out these effects, they seem to get very complicated, to me. Maybe you have some thoughts about how to sort of understand these mechanisms.

So Moghaddam's theory was that the PCP effect, which clearly is blocking the NMDA channel, was not because of that effect, but because of presynaptic increased glutamate release. That was her model, and that presumably your effect with the peptidase inhibitor is because you are increasing NAAG binding or NAAG availability at the presynaptic GRM3 site, which would block glutamate release. But you should also be modulating the postsynaptic NMDA site, and you should also be stimulating GRM3 receptors in astrocytes.

Do you have thoughts about how this works? I mean, the basic model that the effect is because of increases in presynaptic release is a very sticky model. It would argue that, when you give neuroleptics -- the same analogous model would be the reason you get Parkinson's is because there is too much presynaptic dopamine, which is clearly not the case.

So can you maybe try to map out how you think this is working a little bit?

DR. WROBLEWSKA: I will present what I think about it. It wasn't our idea to increase endogenous NAAG in the synapse per se, because I think that with all the complications with the effects of NAAG on NMDA receptors with the fact that, if you have the peptidase inhibition, you also not only have more NAAG present, but you have less glutamate released.

This has been shown by Barbara Slusher that glutamate really goes down. So you have so many mixed signals toward NMDA with both presynaptic mGluR3 and postsynaptic NMDA receptors that I don't think we are able right now to say how this is working. If you use NAAG and inhibit presynaptic mGluR3, you get similar effects, like decreasing the levels of glutamate. Right? Or whatever transmitter is present in the brain area or in the synapse where NAAG is colocalized.

We know that NAAG is colocalized with many different transmitters, and on the presynaptic level it can affect the release of these transmitters. For me the most interesting effects of NAAG are on the postsynaptic or extrasynaptic mGluR3.

DR. COYLE: Let me give you my spin on this thing, and I can't explain everything either.

The first thing is that these dissociative anesthetics act as a primary site of blocking NMDA receptors. So you've got to get to the first cause.

The second thing I would say is that the doses used, especially in humans, are doses that would be hitting a discrete subpopulation of NMDA receptors, because the doses used in humans to reproduce the syndromic features of schizophrenia are not anaesthetic doses. If they were anesthetics, they would mimic coma, not schizophrenia.

As a matter of fact, these subjects have a normal mini-mental status examination, which is a monitor for delirium and dementia. So it is a discrete subpopulation. We believe that there is evidence to support the notion that the NMDA receptors on the GABAnergic interneurones in these cortical limbic regions may be the ones that are most sensitive.

This could explain, in addition, the repeated findings of reduction in GABAnergic markers. That can be demonstrated in animals that get these dissociative anesthetics.

Then consistent with Rita Moghaddam's results, this would result in a disinhibition of glutamatergic output and, downstream, increase in dopamine. The behaviors that are described are behaviors that are seen with excessive dopaminergic release.

So my feeling is that, if you want to look at negative cognitive impairments, you want to look at doses that do not result in this profound behavioral disruption, and that those doses are the ones that are associated with disruption in cortical function.

Now where mGluR3 fits in this scenario, I think it can fit in many different spots, depending upon which neuronal system's function are being disrupted. Here, if you give a drug that reduces glutamate release, you are going to reduce the impact of disinhibited glutamatergic output, and I think the behavioral responses are consistent with that.

DR. WROBLEWSKA: We had shown that mGluR3 activation reduces GABA release. There are papers on norepinephrine, and I think there was a paper on the possible effects of NAAG on the presynaptic inhibition of dopamine release.

DR. COYLE: I would like to thank the speakers for excellent presentations, and the audience.

5. REFERENCES

Adams B. and Moghaddam B. (1998) Corticolimbic dopamine neurotransmission is temporally dissociated from the cognitive and locomotor effects of phencyclidine. *J Neurosci* **18**, 5545-54.

Anderson K. J., Borja M. A., Cotman C. W., Moffett J. R., Namboodiri M. A., and Neale J. H. (1987) N-acetylaspartylglutamate identified in the rat retinal ganglion cells and their projections in the brain. *Brain Res* **411**, 172-7.

Anderson K. J., Monaghan D. T., Cangro C. B., Namboodiri M. A., Neale J. H., and Cotman C. W. (1986) Localization of N-acetylaspartylglutamate-like immunoreactivity in selected areas of the rat brain. *Neurosci Lett* **72**, 14-20.

Bacich D. J., Ramadan E., O'Keefe D. S., Bukhari N., Wegorzewska I., Ojeifo O., Olszewski R., Wrenn C. C., Bzdega T., Wroblewska B., Heston W. D. W., and Neale J. H. (2002) Deletion of the glutamate carboxypeptidase II gene in mice reveals a second enzyme activity that hydrolyzes N-acetylaspartylglutamate. *J Neurochem* **83**, 20-29.

Baslow M. H. (2000) Functions of N-Acetyl-L-Aspartate and N-Acetyl-L-Aspartylglutamate in the Vertebrate Brain: Role in Glial Cell-Specific Signaling. *J Neurochem* **75**, 453-459.

Berent-Spillson, A., Robinson, A.M., Golovoy, D., Slusher, B., Rojas, C., and Russell, J. W. (2004) Protection against glucose-induced neuronal death by NAAG and GCP II inhibition is regulated by mGluR3. *J Neurochem* **89**, 90-99.

Berger U. V., Carter R. E., McKee M., and Coyle J. T. (1995) N-acetylated alpha-linked acidic dipeptidase is expressed by non- myelinating Schwann cells in the peripheral nervous system. *J Neurocytol* **24**, 99-109.

Bzdega T., Crowe S. L., Ramadan E. R., Sciarretta K. H., Olszewski R. T., Ojeifo O. A., Rafalski V. A., Wroblewska B., and Neale J. H. (2004) The cloning and characterization of a second brain enzyme with NAAG peptidase activity. *J Neurochem* **89**, 627-35.

Bzdega T., Turi T., Wroblewska B., She D., Chung H. S., Kim H., and Neale J. H. (1997) Molecular cloning of a peptidase against N-acetylaspartylglutamate from a rat hippocampal cDNA library. *J Neurochem* **69**, 2270-7.

Cangro C. B., Namboodiri M. A., Sklar L. A., Corigliano-Murphy A., and Neale J. H. (1987) Immunohistochemistry and biosynthesis of N-acetylaspartylglutamate in spinal sensory ganglia. *J Neurochem* **49**, 1579-88.

Carpenter K. J., Sen S., Matthews E. A., Flatters S. L., Wozniak K. M., Slusher B. S., and Dickenson A. H. (2003) Effects of GCP-II inhibition on responses of dorsal horn neurones after inflammation and neuropathy: an electrophysiological study in the rat. *Neuropeptides* **37**, 298-306.

Carter R. E., Feldman A. R., and Coyle J. T. (1996) Prostate-specific membrane antigen is a hydrolase with substrate and pharmacologic characteristics of a neuropeptidase. *Proc Natl Acad Sci U S A* **93**, 749-53.

Cartmell J., Monn J. A., and Schoepp D. D. (1999) The metabotropic glutamate 2/3 receptor agonists LY354740 and LY379268 selectively attenuate phencyclidine versus d-amphetamine motor behaviors in rats. *J Pharmacol Exp Ther* **291**, 161-70.

Cartmell J., Monn J. A., and Schoepp D. D. (2000) Tolerance to the motor impairment, but not to the reversal of PCP- induced motor activities by oral administration of the mGlu2/3 receptor agonist, LY379268. *Naunyn Schmiedebergs Arch Pharmacol* **361**, 39-46.

Cassidy M. and Neale J. H. (1993) N-acetylaspartylglutamate catabolism is achieved by an enzyme on the cell surface of neurons and glia. *Neuropeptides* **24**, 271-8.

Coyle J. T. (1997b) The nagging question of the function of N-acetylaspartylglutamate. Neurobiol Dis 4, 231-8.

Curatolo A., D Arcangelo P, Lino A., and Bracanti A. (1965) Distribution of N-acetylaspartic and N-acetylaspartylglutamic acids in nervous tissue. *J Neurochem* **12**, 339-42.

Fuhrman S., Palkovits M., Cassidy M., and Neale J. H. (1994) The regional distribution of N-acetylaspartylglutamate (NAAG) and peptidase activity against NAAG in the rat nervous system. *J Neurochem* **62**, 275-81.

Gafurov B., Urazaev A. K., Grossfeld R. M., and Lieberman E. M. (2001) N-acetylaspartylglutamate (NAAG) is the probable mediator of axon-to- glia signaling in the crayfish medial giant nerve fiber. *Neuroscience* **106**, 227-35.

Garrido Sanabria, Emilio R., Wozniak, Krystyna M., Slusher, Barbara S., and Keller, Asaf. (2004) GCP II (NAALADase) inhibition suppresses mossy fiber-CA3 synaptic transmission by a presynaptic mechanism. *J Neurophysiol* **91**: 182-193.

Israeli R. S., Powell C. T., Fair W. R., and Heston W. D. (1993) Molecular cloning of a complementary DNA encoding a prostate-specific membrane antigen. *Cancer Res* **53**, 227-30.

Javitt D. C. and Coyle J. T. (2004) Decoding schizophrenia. *Sci Am* **290**, 48-55.

Javitt DC, and Zukin SR. (1991) Recent advances in the phencyclidine model of schizophrenia. *Am J Psychiatry* **148**, 1301-8.

Konradi C. and Heckers S. (2003) Molecular aspects of glutamate dysregulation: implications for schizophrenia and its treatment. *Pharmacol Ther* **97**, 153-79.

Kozikowski A. P., Nan F., Conti P., Zhang J., Ramadan E., Bzdega T., Wroblewska B., Neale J. H., Pshenichkin S., and Wroblewski J. T. (2001) Design of remarkably simple, yet potent urea-based inhibitors of glutamate carboxypeptidase II (NAALADase). *J Med Chem* **44**, 298-301.

Kozikowski A. P., Zhang J., Nan F., Petukhov P. A., Grajkowska E., Wroblewski J. T., Yamamoto T., Bzdega T., Wroblewska B., and Neale J. H. (2004) Synthesis of Urea-Based Inhibitors as Active Site Probes of Glutamate Carboxypeptidase II: Efficacy as Analgesic Agents. *J Med Chem* **47**, 1729-1738.

Luthi-Carter R., Berger U. V., Barczak A. K., Enna M., and Coyle J. T. (1998) Isolation and expression of a rat brain cDNA encoding glutamate carboxypeptidase II. *Proc Natl Acad Sci U S A* **95**, 3215-20.

Moffett J. R. and Namboodiri M. A. (1995) Differential distribution of N-acetylaspartylglutamate and N-acetylaspartate immunoreactivities in rat forebrain. *J Neurocytol* **24**, 409-33.

Moffett J. R., Williamson L., Palkovits M., and Namboodiri M. A. (1990) N-acetylaspartylglutamate: a transmitter candidate for the retinohypothalamic tract. *Proc Natl Acad Sci U S A* **87**, 8065-9.

Moghaddam B. and Adams B. W. (1998) Reversal of phencyclidine effects by group II metabotropic glutamate receptor agonist in rats. *Science* **281**, 1349-1352.

Nan F., Bzdega T., Pshenichkin S., Wroblewski J. T., Wroblewska B., Neale J. H., and Kozikowski A. P. (2000) Dual function glutamate-related ligands: discovery of a novel, potent inhibitor of glutamate carboxypeptidase II possessing mGluR3 agonist activity. *J Med Chem* **43**, 772-4.

Neale J. H., Bzdega T., and Wroblewska B. (2000) N-Acetylaspartylglutamate: The Most Abundant Peptide Neurotransmitter in the Mammalian Central Nervous System. *J Neurochem* **75**, 443-452.

Ohishi H., Shigemoto R., Nakanishi S., and Mizuno N. (1993) Distribution of the messenger RNA for a metabotropic glutamate receptor (mGluR3) in the rat brain - an *in situ* hybridization study. *J Comp Neurol* **335**, 252-266.

Olszewski R. T., Bukhari N., Zhou J., Kozikowski A. P., Wroblewski J. T., Shamimi-Noori S., Wroblewska B., Bzdega T., Vicini S., Barton F. B., and Neale J. H. (2004) NAAG peptidase inhibition reduces locomotor activity and some stereotypes in the PCP model of schizophrenia via group II mGluR. *J Neurochem* **89**, 876-85.

Renno W. M., Lee J. H., and Beitz A. J. (1997) Light and electron microscopic immunohistochemical localization of N- acetylaspartylglutamate (NAAG) in the olivocerebellar pathway of the rat. *Synapse* **26**, 140-54.

Riveros N. and Orrego F. (1984) A study of possible excitatory effects of N-acetylaspartylglutamate in different in vivo and in vitro brain preparations. *Brain Res* **299**, 393-5.

Robinson M. B., Blakely R. D., Couto R., and Coyle J. T. (1987) Hydrolysis of the brain dipeptide N-acetyl-L-aspartyl-L-glutamate. Identification and characterization of a novel N-acetylated alpha- linked acidic dipeptidase activity from rat brain. *J Biol Chem* **262**, 14498-506.

Rojas C., Thomas A. G., Majer P., Tsukamoto T., Lu X. M., Vornov J. J., Wozniak K. M., and Slusher B. S. (2003) Glutamate carboxypeptidase II inhibition as a novel therapeutic target. *Adv Exp Med Biol* **524**, 205-13.

Rong S. B., Zhang J., Neale J. H., Wroblewski J. T., Wang S., and Kozikowski A. P. (2002) Molecular modeling of the interactions of glutamate carboxypeptidase II with its potent NAAG-based inhibitors. *J Med Chem* **45**, 4140-52.

Shave E., Pliss L., Lawrence M.L., FitzGibbon T., Stastny F., Balcar V.J. (2001) Regional distribution and pharmacological characteristics of [3H]N-acetyl-aspartyl-glutamate (NAAG) binding sites in rat brain. *Neurochem International* **38**, 53-62

Slusher B. S., Robinson M. B., Tsai G., Simmons M. L., Richards S. S., and Coyle J. T. (1990) Rat brain N-acetylated alpha-linked acidic dipeptidase activity. Purification and immunologic characterization. *J Biol Chem* **265**, 21297-301.

Slusher B. S., Tsai G., Yoo G., and Coyle J. T. (1992) Immunocytochemical localization of the N-acetyl-aspartyl-glutamate (NAAG) hydrolyzing enzyme N-acetylated alpha-linked acidic dipeptidase (NAALADase). *J Comp Neurol* **315**, 217-29.

Slusher B. S., Vornov J. J., Thomas A. G., Hurn P. D., Harukuni I., Bhardwaj A., Traystman R. J., Robinson M. B., Britton P., Lu X. C., Tortella F. C., Wozniak K. M., Yudkoff M., Potter B. M., and Jackson P. F. (1999) Selective inhibition of NAALADase, which converts NAAG to glutamate, reduces ischemic brain injury. *Nat Med* **5**, 1396-402.

Taylor, D. L., Diemel, L. T., Cuzner, M. L., and Pocock, J. M. (2002) Activation of group II metabotropic glutamate receptors underlies microglial reactivity and neurotoxicity following stimulation with chromogranin A, a peptide up-regulated in Alzheimer's disease. *J Neurochem* **82**, 1179-1191.

Taylor, D. L., Jones, F., Kubota, E. S. F. Chen S., and Pocock, J.M.(2005) Stimulation of Microglial Metabotropic Glutamate Receptor mGlu2 Triggers Tumor Necrosis Factor {alpha}-Induced Neurotoxicity in Concert with Microglial-Derived Fas Ligand. *J. Neurosci.* **25**, 2952-2964.

Thomas A. G., Olkowski J. L., and Slusher B. S. (2001) Neuroprotection afforded by NAAG and NAALADase inhibition requires glial cells and metabotropic glutamate receptor activation. *Europ J Pharmacol* **426**, 35-38.

Tieman S. B., Neale J. H., and Tieman D. G. (1991) N-acetylaspartylglutamate immunoreactivity in neurons of the monkey's visual pathway. J Comp Neurol 313, 45-64.

Tsai G. and Coyle J.T. (2002) Glutamatergic mechanisms in schizophrenia. *Annu Rev Pharmacol Toxicol.* **42**, 165-179.

Tyszkiewicz J. P., Gu Z., Wang X., Cai X., and Yan Z. (2004) Group II metabotropic glutamate receptors enhance NMDA receptor currents via a protein kinase C-dependent mechanism in pyramidal neurones of rat prefrontal cortex. *J Physiol* **554**, 765-77.

Urazaev, A. K., Buttram, J. G., Deen, J. P., Gafurov, B. S., Slusher, B. S., Grossfeld, R. M., Lieberman, E. M. (2001) Mechanisms for clearance of released N-acetylaspartylglutamate in crayfish nerve fibers: Implications for axon-glia signaling *Neurosci* **107**, 697-703.

Williamson L. C., Eagles D. A., Brady M. J., Moffett J. R., Namboodiri M. A., and Neale J. H. (1991) Localization and Synaptic Release of N-acetylaspartylglutamate in the Chick Retina and Optic Tectum. *Eur J Neurosci* **3**, 441-451.

Wroblewska B., Santi M. R., and Neale J. H. (1998) N-acetylaspartylglutamate activates cyclic AMP-coupled metabotropic glutamate receptors in cerebellar astrocytes. *Glia* **24**, 172-9.

Wroblewska B., Wroblewski J. T., Pshenichkin S., Surin A., Sullivan S. E., and Neale J. H. (1997) N-acetylaspartylglutamate selectively activates mGluR3 receptors in transfected cells. *J Neurochem* **69**, 174-81.

Yamamoto T., Hirasawa S., Wroblewska B., Grajkowska E., Zhou J., Kozikowski A., Wroblewski J., and Neale J. H. (2004) Antinociceptive effects of N-acetylaspartylglutamate (NAAG) peptidase inhibitors ZJ-11, ZJ-17 and ZJ-43 in the rat formalin test and in the rat neuropathic pain model. *Eur J Neurosci* **20**, 483-94.

Yamamoto T., Nozaki-Taguchi N., Sakashita Y., and Inagaki T. (2001) Inhibition of spinal N-acetylated-alpha-linked acidic dipeptidase produces an antinociceptive effect in the rat formalin test. *Neuroscience* **102**, 473-9.

Yamamoto T., Hirasawa S., Wroblewska B., Grajkowska E., Zhou J., Kozikowski A., Wroblewski J., and Neale J. H. (2004) Antinociceptive effects of N-acetylaspartylglutamate (NAAG) peptidase inhibitors ZJ-11, ZJ-17 and ZJ-43 in the rat formalin test and in the rat neuropathic pain model. *Eur J Neurosci* **20**, 483-494.

Zhao J., Ramadan E., Cappiello M., Wroblewska B., Bzdega T., and Neale J. H. (2001) NAAG inhibits KCl-induced [(3)H]-GABA release via mGluR3, cAMP, PKA and L-type calcium conductance. *Eur J Neurosci* **13**, 340-6

Zukin S. R. and Javitt D. C. (1989) Mechanisms of phencyclidine (PCP)-n-methyl-d-aspartate (NMDA) receptor interaction: implications for drug abuse research. *NIDA Res Monogr* **95**, 247-54.

GLUTAMATE CARBOXYPEPTIDASE II (NAALADase) INHIBITION AS A NOVEL THERAPEUTIC STRATEGY

Ajit G. Thomas, Krystyna M. Wozniak, Takashi Tsukamoto, David Calvin, Ying Wu, Camilo Rojas, James Vornov and Barbara S. Slusher[*]

1. INTRODUCTION

The dipeptide *N*-acetylaspartylglutamate (NAAG) was first identified in bovine brain in 1965.[1] NAAG is present in high millimolar concentrations in mammalian brain and peripheral nervous system.[2] Its role was largely unknown until the early 1980s when the concept of its potential function as a neurotransmitter was postulated.[3] As neurotransmitters generally undergo inactivation via re-uptake or by enzymatic degradation, the enzyme for the catabolism of NAAG was sought, identified, purified and characterized from the brains and kidneys of rodents.[4] Glutamate carboxypeptidase (GCPII) is a zinc peptidase that hydrolyzes the neuropeptide NAAG to glutamate (G) and *N*-acetyl aspartate (NAA) (Fig. 1).

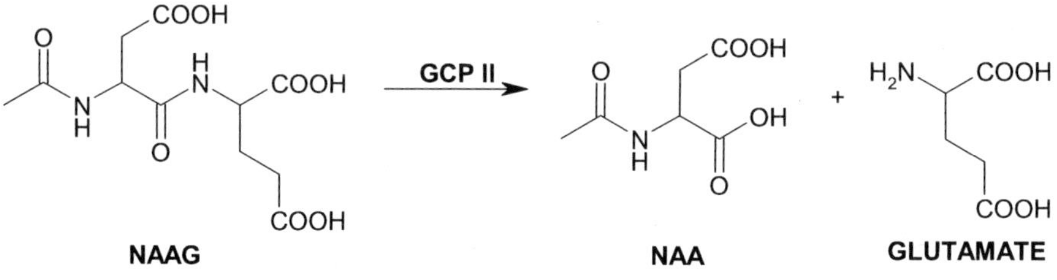

Figure 1. Hydrolysis of NAAG catalyzed by GCP II.

Rat GCP II was cloned in 1996 and found to be homologous to human prostate-specific membrane antigen (PSMA).[5] GCP II has been found in new vasculature of

[*] Guilford Pharmaceuticals Inc., 6611 Tributary Street, Baltimore, Maryland 21224 USA; phone, 410-631-6804, email, tsukamotot@guilfordpahrm.com

several solid tumors.[6] Additionally, GCP II catalyzes the hydrolysis of folate polyglutamate to folate and several molecules of glutamate in the membrane brush border of the small intestine.[7] It is localized on the plasma membrane of glial cells and positioned with its catalytic region facing the synapse.[8] Depending on localization and function, GCP II has been referred to as NAALADase (*N*-acetylated-α-linked acidic dipeptidase) when studying NAAG hydrolysis in the brain[9], as PSMA when studying the role of the enzyme in prostate cancer[10] or as folate hydrolase when focusing on the potential function of this enzyme in human nutrition.[7] However, the preferred official name for the enzyme is GCP II (EC 3.4.17.21)

Excessive glutamate has been implicated in a variety of neurodegenerative disorders including stroke, ALS, and chronic pain. Conventional therapies have focused on blockade of post-synaptic glutamate receptors with small molecules. To this end, several glutamate receptors have been evaluated and exploited as therapeutic targets for neurological disorders associated with excess glutamate toxicity. Of these, the NMDA receptors have received the most attention culminating in several antagonists in clinical trials.[11] However, this class of compounds has been historically associated with major side effects thought to be associated with blockade of normal physiological neurotransmission, such as learning and memory.[11]

An alternative therapeutic approach to blocking postsynaptic glutamate receptors would be upstream reduction of presynaptic glutamate. If NAAG were functioning as a storage form of glutamate, then inhibition of GCP II would, in theory, prevent the release of excess glutamate and thereby be neuroprotective. In addition, NAAG has been shown to have other functions including acting as an mgluR3 agonist and a partial antagonist of postsynaptic NMDA receptors.[12] Activation of mGluR3 by NAAG has been shown to inhibit glutamate release[13] and increase transforming growth factor β (TGFβ) release, both of which have been shown to provide neuroprotection.[14] Consequently, elevation of NAAG would be beneficial in its own right via its interaction at the mGluR3. Therefore, inhibition of GCP II should be neuroprotective via a dual mechanism involving both a direct reduction of glutamate and an elevation of NAAG.

2. 2-PMPA AND 2-MPPA ARE POTENT AND SPECIFIC GCP II INHIBITORS

Research at Guilford has focused on the potential utility of GCP II inhibitors wherein excess glutamate neurotransmission has been implicated. 2-(Phosphonomethyl) pentanedioic acid (2-PMPA) is a potent, competitive GCP II inhibitor with a K_i value of 0.2 nM[15; 16] (Fig 2). It exhibits a fast association rate and a slow dissociation rate.[17] The high potency of 2-PMPA can be attributed to the strong chelation of the phosphonate group to an active site zinc atom as well as the interaction of the glutamate moiety (pentanedioic) portion of the inhibitor with the glutamate recognition site of GCPII.[18] 2-PMPA seems to be quite specific for GCP II, i.e., no significant activities were observed at 10 µM (more than 10,000-fold higher than the K_i for NAALADase inhibition) in over 100 different receptor and enzyme assays, including glutamate receptors and transporters.[19]

The poor pharmacokinetic profile of 2-PMPA prompted the search for other small molecule GCP II inhibitors. Using 2-PMPA as template, and extending structure-activity relationship studies to other zinc binding groups, 2-MPPA was identified as another potent, selective and competitive GCP II inhibitor with a K_i value of 30 nM.[20] 2-MPPA has an enhanced oral bioavailabilty when compared to 2-PMPA.

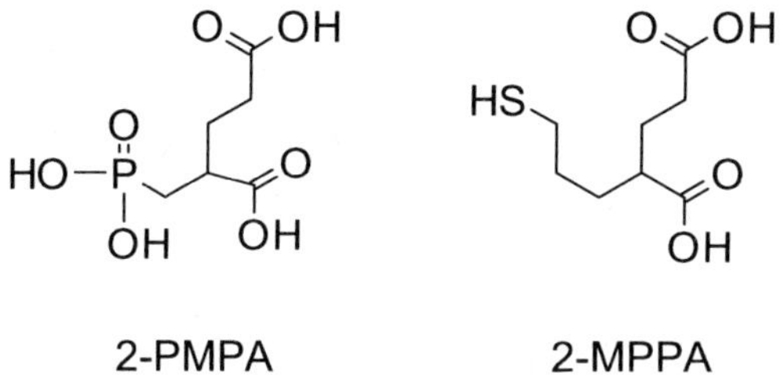

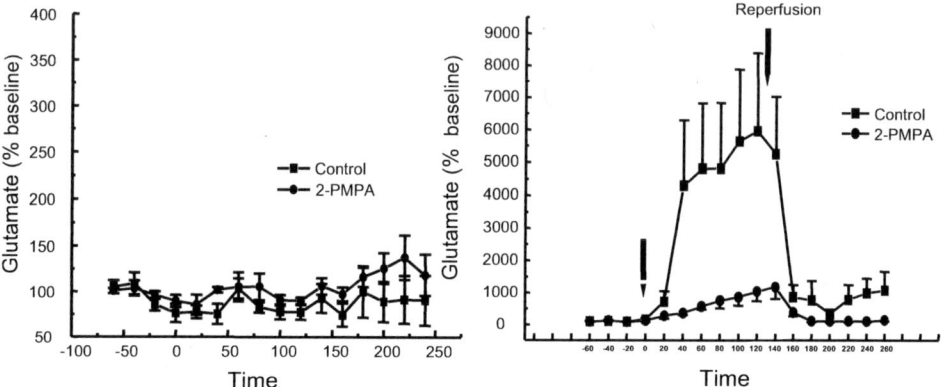

2-PMPA   2-MPPA

Figure 2. 2-(Phosphonomethyl)pentanedioic acid (2-PMPA) and 2-(3-mercaptopropyl) pentanedioic acid (2-MPPA).

3. GCP II INHIBITION IS A NOVEL MECHANISM FOR REDUCTION OF EXCITOTOXIC GLUTAMATE

The roles of NAAG and GCP II in excitotoxic neurotransmission were investigated upon the identification of these potent and selective inhibitors of GCP II. Rats implanted with microdialysis probes were treated with either 2-PMPA or vehicle and subjected to middle cerebral artery occlusion (MCAO). Dialysates were collected up to 4 hours after occlusion and analysed for glutamate. While 2-PMPA had no significant effect on extracellular glutamate in normal, non-ischemic rats (left panel, Fig. 3), 2-PMPA signifi-

Figure 3. 2-PMPA reduces excitotoxic extracellular glutamate levels after MCAO but has no effect on basal glutamate levels.

cantly attenuated the ischemia-induced increase in extracellular glutamate seen in vehicle-treated rats (right panel, Fig. 3: 6,000% rise reduced to 1,200% ($p < 0.05$). The fact that 2-PMPA selectively attenuated the ischemia-induced rise in glutamate suggests a potential role for GCP II inhibitors in excitotoxic mechanisms.[19]

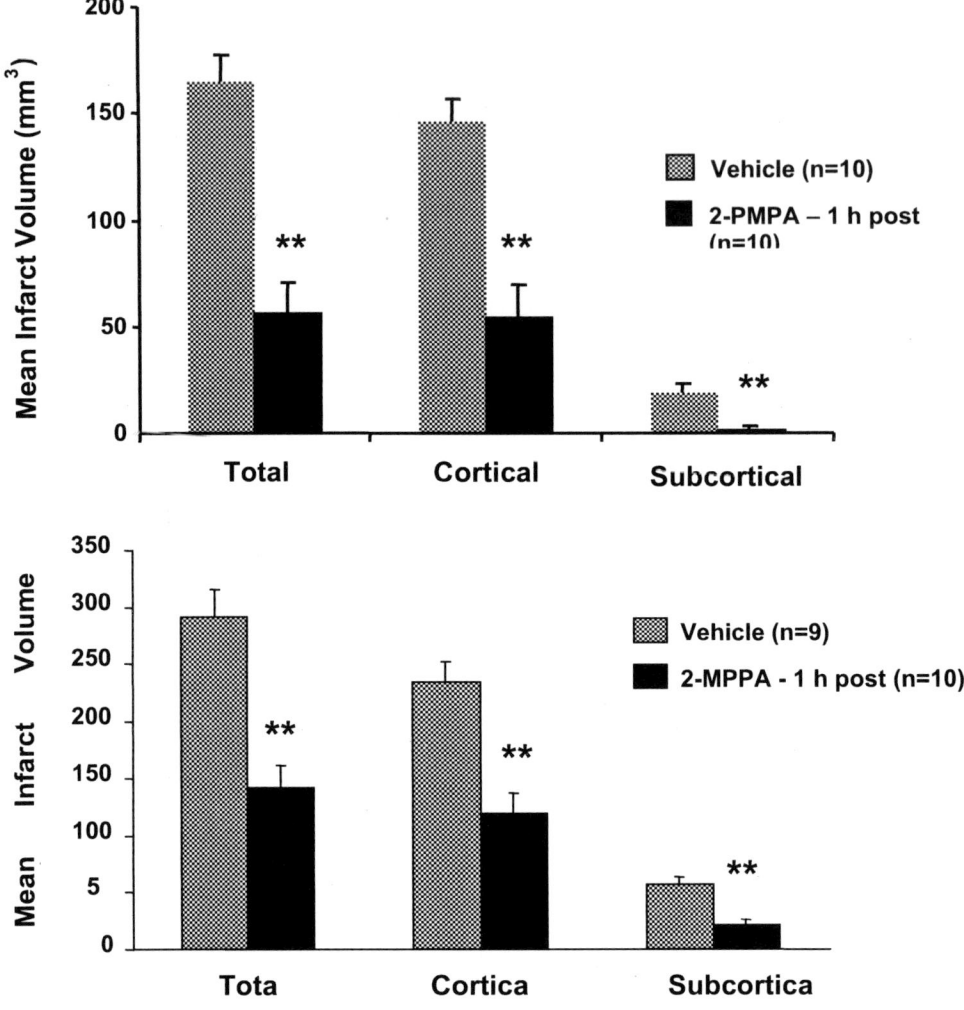

Figure 4. GCP II inhibitors reduce brain injury following MCAO. (A), 2-PMPA; (B), 2-MPPA (**p < 0.01 for both compounds).

4. GCP II INHIBITORS PROVIDE NEUROPROTECTION

The efficacy of GCP II inhibition by 2-PMPA (100 mg/kg, i.p.) was assessed in the middle cerebral artery occlusion (MCAO) model of ischemic stroke. Focal ischemia was induced in rats by MCAO for 2 hours, followed by 22 hours of reperfusion.[19] The rats were then sacrificed and their brains removed and evaluated by TTC (2,3,5-triphenyltetrazolium chloride) staining to determine brain injury volumes. 2-PMPA, when dosed at 1 hour following the beginning of the ischemia, significantly reduced the total brain injury volume, with predominant and significant effects in the cortical hemisphere[19] (Fig. 4A). Similar efficacy was observed with 2-MPPA administered at 2 hours post onset of MCAO (30 mg/kg, i.v.) (Fig. 4B).

5. GCP II INHIBITORS PREVENT NEURODEGENERATION

To examine the potential usefulness of GCP II inhibitors in familial amyotrophic lateral sclerosis (FALS), 2-MPPA (30 mg/kg/day po) was administered to SOD G93A mice daily from 37-39 days of age. Clinical signs were measured weekly. In mice treated with 2-MPPA there were statistically significant delays in the onset of all of the clinical end-points measured except for shaking of limbs (table below).   There was also a significant prolongation in the median survival - from 190 days in the vehicle treated group to 219 days in the 2-MPPA treated group, as illustrated by the Kaplan-Meier survival curve[21] (Fig. 5). The results support the use of GCP II inhibitors as potential neuroprotective agents against neurodegenerative disorders such as ALS.

Clinical Symptoms	Delay in Onset (days)	p value
Crossing of limbs	17	0.017
Dragging of Limbs	23	0.012
Gait	17	0.037
Righting Reflex	20	0.032
Shaking of limbs	7	0.173

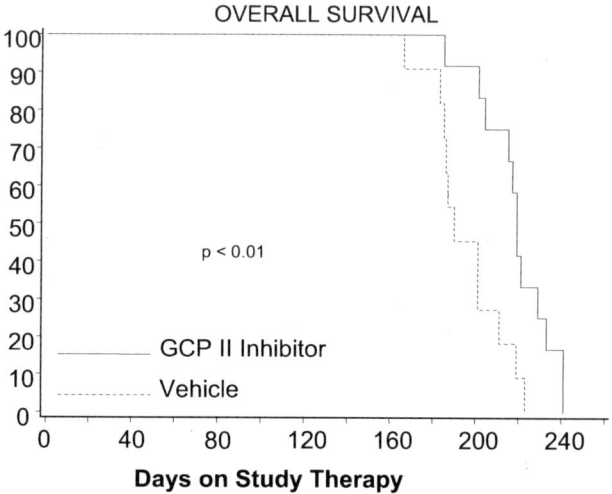

Figure 5. 2-MPPA, a prototype GCP-II inhibitor, significantly prolongs FALS transgenic mice survival (delta = 29 days, a 15% increase in the mean life span, p= 0.0059)[21].

6. GCP II INHIBITORS PROVIDE ANALGESIA

GCP II inhibitors were also examined as potential analgesics for neuropathic pain and peripheral neuropathy using a chronic constrictive injury (CCI) model.[22] Briefly, one sciatic nerve was exposed by blunt dissection proximal to nerve trifurcation and four ligatures loosely tied. The other side was sham operated. After twelve days, thermal pain threshold (withdrawal latency) was assessed by means of the plantar test.[23] The ligated and non-ligated hind limbs of the CCI rats were tested and a difference score for each animal was determined by subtracting the mean withdrawal latency of the non-ligated (sham-operated) leg from the mean withdrawal latency of the ligated leg.[23] Negative values indicate a relative hyperalgesia on the operated side as compared to the sham side. The unoperated animals (untreated control) showed no difference between right and left leg withdrawal latencies. Vehicle-treated animals remained hyperalgesic over the period of testing as indicated by the negative withdrawal latency difference over the entire course of the study. 2-PMPA significantly attenuated the CCI-induced hyperalgesia beginning at 11 days of treatment and continuing through the end of the study on day 21 (Fig. 6).

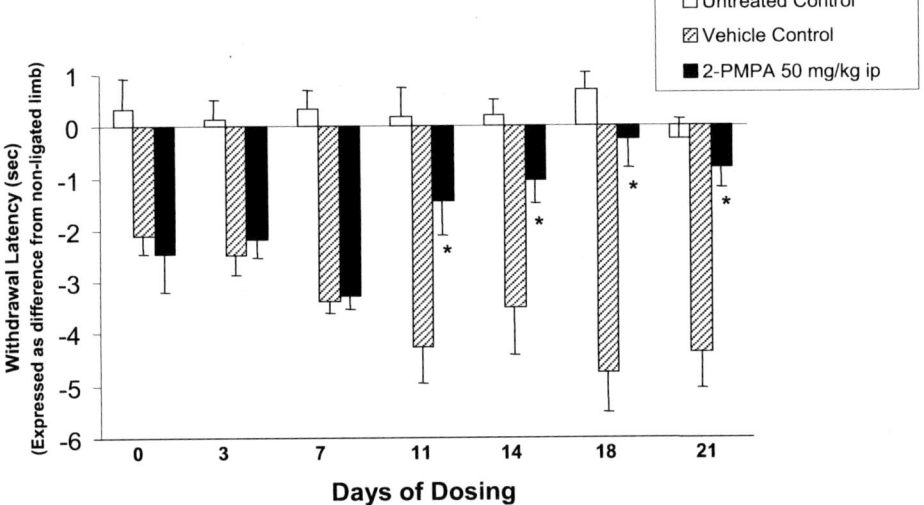

Figure 6. 2-PMPA reduces neuropathic pain in CCI.

2-MPPA was also evaluated in the same model and was found to significantly reduce thermal hyperalgesia (Fig. 7). The antinociceptive effect was significant from day 8 of treatment and persisted throughout the duration of the study. 2-MPPA has the major advantage over 2-PMPA in that it induced efficacy at a lower dose consistent with its pharmacokinetic profile.

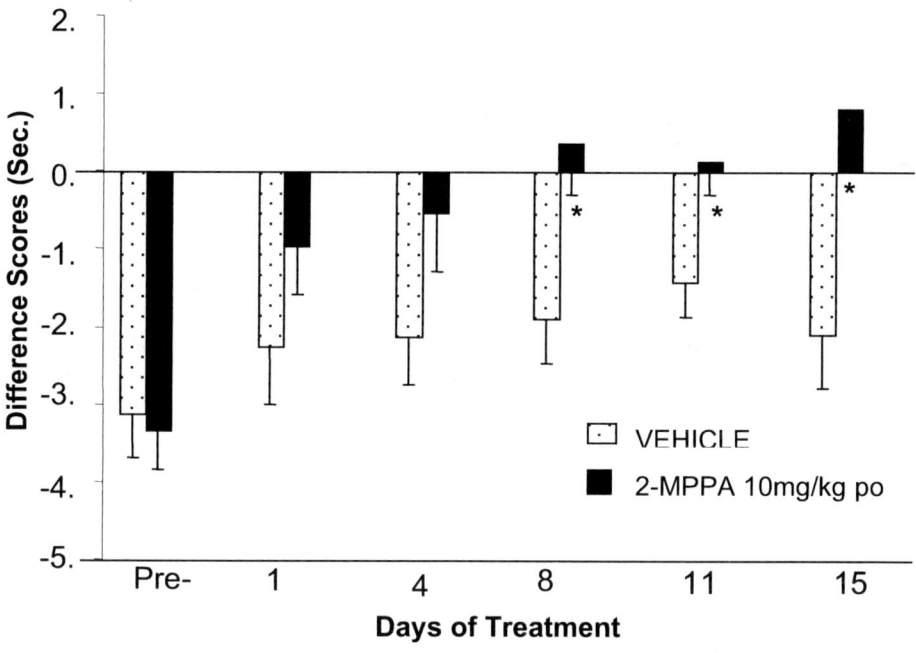

Figure 7. 2-MPPA reduces neuropathic pain in CCI

7. GCP II INHIBITORS IN THE CLINIC

Based on the preclinical efficacy of GCP II inhibitors in models of stroke, ALS and peripheral neuropathy, a potent and selective GCP II inhibitor was chosen to be administered to 77 individuals in three Phase I clinical trials. In the first trial, doses of up to 1500 mg were administered to volunteers. Oral bioavailability of the drug, particularly in the fasted state, was very good. Plasma levels were achieved that were above those needed to produce effects in animal models of diabetic neuropathy and neuropathic pain. The compound was safe and well tolerated at all doses without any CNS effect (EEG, visual tracking, coordination, etc). Gastrointestinal complaints constituted the most common category of adverse event, with dyspepsia being most commonly reported.

In further trials, daily doses of 375 mg and 750 mg were administered for 14 consecutive days to healthy volunteers and to patients with diabetes. In normal volunteers, the safety and tolerability profile was similar to placebo at doses up to 750 mg in the fasted state. In diabetic patients, complaints of mild hypoglycemia were reported at the highest dose, but were not associated with documented laboratory changes. In neither group were there any apparent effects on glutamate sensitive parameters.

8. SUMMARY

GCP II inhibition decreases extracellular excitotoxic glutamate and increases extracellular NAAG, both of which provide neuroprotection. We have demonstrated with our potent and selective GCP II inhibitors efficacy in models of stroke, ALS and neuropathic pain. GCP II inhibition may have significant potential benefits over existing glutamate-based neuroprotection strategies. The upstream mechanism seems selective for excitotoxic induced glutamate release, as GCP II inhibitors in normal animals induced no change in basal glutamate. This suggestion has recently been corroborated by Lieberman and coworkers[24] who found that both NAAG release and increase in GCP II activity appear to be induced by electrical stimulation in crayfish nerve fibers and that subsequent NAAG hydrolysis to glutamate contributes, at least in part, to subsequent NMDA receptor activation. Interestingly, even at relatively high doses of compounds, GCP II inhibition did not appear to be associated with learning/memory deficits in animals. Additionally, quantitative neurophysiological testing data and visual analog scales for 'psychedelic effects' in Phase I single dose and repeat dose studies showed GCP II inhibition to be safe and well tolerated by both healthy volunteers and diabetic patients.

GCP II inhibition may represent a novel glutamate regulating strategy devoid of the side effects that have hampered the development of postsynaptic glutamate receptor antagonists.

9. QUESTION AND ANSWER SESSION

DR. MEYERHOFF: Questions? Go ahead.

DR. COYLE: Yes?

PARTICIPANT: That was really very interesting stuff, especially with regard to ALS and stroke. I have interest in both areas. I am a clinician.

You administered -- well, first PMPA and then 2-PMPA in stroke and ALS, but in pre-symptomatic stages. So in stroke, I think you gave it before occlusion.

DR. TSUKAMOTO: Right.

PARTICIPANT: And in ALS you gave it at one month, and there really isn't any pathology at that point to that mouse. I think it comes at 90 days. So have you tried further to see if there are effects when the mice become symptomatic or post-occlusion in stroke?

DR. TSUKAMOTO: No, we haven't. One reason is, as you can see, each experiment takes almost a year; it is a very long, intensive study. So what we are interested in is how this drug worked compared to the existing drug. So we used the protocol they used for Rilusol, and we haven't done any modification in that protocol at this point. Yes.

PARTICIPANT: Just now we heard about two different types of GCP, II and III. Then we talked about it had to be knock-out mice, in which GCPII has been knocked out, but there is no symptomatic phenotype. So how specific is your inhibitor for GCPII versus GCPIII?

DR. TSUKAMOTO: I am not familiar with GCPIII, but what I think is --

DR. WROBLEWSKA: It appears not GCPIII, we checked 2-PMPA --

DR. TSUKAMOTO: Right. What I think is that our inhibitor probably has a good chance to inhibit both GCPII and GCPIII.

DR. WROBLEWSKA: 2-PMPA does it, for sure.

DR. TSUKAMOTO: Right. So in our case, we are working at the glutamate receptors II or III. Probably GCPII inhibitors will shut them both down. Yes?

PARTICIPANT: But you are calling GCPII NAALADase? Does this term refer to GCPII and III?

DR. TSUKAMOTO: Dr. Coyle?

DR. COYLE: Actually, you know, as we got into the 21st Century, I think that term was developed simply as an acronym for the action of the enzyme, and I think GCPII and GCPIII are the appropriate names now, or if you look at the gene, at least for GCPII it is also known as folate hydrolase I, FOH-I.

PARTICIPANT: But if you use this inhibitor, and if it is not specific for II or III, and if it works, what does it do to the folate, because the folate conjugates. It is very critical for the absorption of folic acid in the intestine.

DR. TSUKAMOTO: Yes.

PARTICIPANT: So if you do give this drug to the patients, are you risking the patient's health?

DR. TSUKAMOTO: That is a very good question. We had a collaboration with a researcher at California Davis -- what is his name at University of California?

DR. COYLE: Dr. Halstead.

DR. TSUKAMOTO: Yes, Halstead.

DR. COYLE: Halstead. He was the researcher we worked with.

DR. TSUKAMOTO: Yes. He actually was independently studying this folate hydrase, which turns out to be, as we have been discussing, the same enzyme. So we gave him our inhibitors and see effect of that on the folate and the methionine in metabolism. It didn't show any changes in that study.

DR. COYLE: Actually, the American diet is so fortified with folate, which does not have a polyglutamate tail on it, so that if you eat a normal diet, this would have absolutely no effect on your folate level.

PARTICIPANT: But the individuals who depend on dietary folate, which is predominantly polyglutamate, the inhibitor is going to have an effect.

DR. COYLE: Right. Right.

DR. MADHAVARAO: Would you like to comment on: Since the GCPII knock-out mice did not actually have any abnormalities, is there any actual role for this enzyme there? You know, it assumes that it is the glutamate released from NAAG is actually inducing all this toxicity or mediating the pain. So the pain related effects in the absence of such enzyme in the GCPII knock-out mice, would you not expect that such GCPII mediated pain responses should not be there? So would you like to comment on that?

DR. TSUKAMOTO: I'm not sure if I get your question right. But what Guilford has been focused the last couple of years is really -- we are really focused on reducing glutamate now, but after attending the conference and seeing all the effects that NAAG could have--

I am not really sure that the effect we see in efficacy is something to do with increasing NAAG or decreasing the glutamate. What we care most is that it works.

DR. NAMBOODIRI: I have two questions. Regarding the mechanics of action here, is it acting by preventing the formation of glutamate or is it preventing the release of glutamate by increasing the NAAG level?

DR. TSUKAMOTO: That is again the same type of question. We don't really know yet, but what microdialysis studies tell us is as far as stroke is concerned, we are able to suppress the extra concentration in the glutamate by treating with the GCPII inhibitors.

Now what I cannot prove is that all this glutamate really comes from NAAG hydrolysis. We don't really know that yet, but at least we are able to reduce glutamate, and it correlates with efficacy in our stroke studies. Yes.

DR. NAMBOODIRI: So it is generating glutamate from NAAG?

DR. TSUKAMOTO: Yes, but I cannot say that all the increases you see in stroke are all coming from NAAG hydrolysis by this enzyme? No. We don't have any proof yet.

DR. LIEBERMAN: One is a comment, and it goes to his question. That would be: It would be very interesting to see in the cerebral ligature model, the stroke model, of whether or not the knock-out mouse would give you the same results as the peptidase inhibition.

DR. TSUKAMOTO: Yes. We have done that.

DR. LIEBERMAN: Oh.

DR. TSUKAMOTO: Well, actually, I looked at those, but I am a chemist. Larry, can you comment on this?

DR. WILLIAMS: Using the Hudson knock-out that Barbara Wroblewska described, there is a new part that is smaller --

DR. LIEBERMAN: The second question: What is your estimate of why you are getting hydrolysis of NAAG after some sort of an insult, or injury, versus you don't get hydrolysis under resting conditions? Have you infused NAAG into these areas which are showing the damage versus those that have no damage? Is it a substrate activated breakdown or is it an enzyme regulation effect?

DR. TSUKAMOTO: Right. So you are talking about what triggers NAAG peptidase activity in under various conditions? I don't think I am the right person to answer that question, but we have a couple of --

DR. WROBLEWSKA: The peptidase works when it is activated by hydrolyzing NAAG, and the actions of NAAG are enhanced when the peptidase is inhibited.

DR. LIEBERMAN: If peptidase action blocks the mGluR by degrading NAAG -- it may be the NAAG that is there is more potent when the enzyme is inhibited.

DR. TSUKAMOTO: And one other factor that we cannot underestimate is the concentration of phosphate, of course, because this enzyme is very sensitive to phosphate concentration, and easily inhibited if you have more than 10 micromolar phosphate. So phosphate might be also another regulator of this enzyme.

DR. COYLE: One perhaps unappreciated fact: When you see that the rise in glutamate gets blocked with the PMPA, it kind of knocks your socks off, and you say, wait a second, what happened to glutamate release?

I should point out that, at least in the rat, the concentration of glutamate is about 3 millimolar, but if ten percent of it -- only ten percent of it is synaptic glutamate, that means it is about 200 micromolar, 300 micromolar -- concentration of NAAG in that region is 200 to 300 micromolar, which means that the NAAG contributes at least equally to the total extra-cellular burden of glutamate, when it is released and hydrolyzed.

9. REFERENCES

1.   Curatolo, A., D'Arcangelo, P., Lino, A. & Brancati, A., 1965, Distribution of N-Acetyl-Aspartic and N-Acetyl-Aspartyl-Glutamic Acids in Nervous Tissue, *J Neurochem.* **12:** 339-42.
2.   Neale, J. H., Bzdega, T. & Wroblewska, B., 2000, N-Acetylaspartylglutamate: the most abundant peptide neurotransmitter in the mammalian central nervous system, *J Neurochem.* **75:** 443-52.
3.   Coyle, J. T., Stauch-Slusher, B., Tsai, G., Rothstein, J. D., Meyerhoff, J. L., Simmons, M., Blakely, R. D. (1991). In *Excitatory Amino Acids*, Vol. 69. Raven Press, Ltd.
4.   Robinson, M. B., Blakely, R. D., Couto, R. & Coyle, J. T., 1987, Hydrolysis of the brain dipeptide N-acetyl-L-aspartyl-L-glutamate. Identification and characterization of a novel N-acetylated alpha-linked acidic dipeptidase activity from rat brain, *J Biol Chem.* **262:** 14498-506.

5.   Carter, R. E., Feldman, A. R. & Coyle, J. T., 1996, Prostate-specific membrane antigen is a hydrolase with substrate and pharmacologic characteristics of a neuropeptidase, *Proc Natl Acad Sci U S A.* **93:** 749-53.

6.   Chang, S. S., O'Keefe, D. S., Bacich, D. J., Reuter, V. E., Heston, W. D. & Gaudin, P. B., 1999, Prostate-specific membrane antigen is produced in tumor-associated neovasculature, *Clin Cancer Res.* **5:** 2674-81.

7.   Heston, W. D., 1997, Characterization and glutamyl preferring carboxypeptidase function of prostate specific membrane antigen: a novel folate hydrolase, *Urology.* **49:** 104-12.

8.   Cassidy, M. & Neale, J. H., 1993, N-acetylaspartylglutamate catabolism is achieved by an enzyme on the cell surface of neurons and glia, *Neuropeptides.* **24:** 271-8.

9.   Slusher, B. S., Robinson, M. B., Tsai, G., Simmons, M. L., Richards, S. S. & Coyle, J. T., 1990, Rat brain N-acetylated alpha-linked acidic dipeptidase activity. Purification and immunologic characterization, *J Biol Chem.* **265:** 21297-301.

10. Pinto, J. T., Suffoletto, B. P., Berzin, T. M., Qiao, C. H., Lin, S., Tong, W. P., May, F., Mukherjee, B. & Heston, W. D., 1996, Prostate-specific membrane antigen: a novel folate hydrolase in human prostatic carcinoma cells, *Clin Cancer Res.* **2:** 1445-51.

11. Doble, A., 1999, The role of excitotoxicity in neurodegenerative disease: implications for therapy, *Pharmacol Ther.* **81:** 163-221.

12. Wroblewska, B., Wroblewski, J. T., Pshenichkin, S., Surin, A., Sullivan, S. E. & Neale, J. H., 1997, N-acetylaspartylglutamate selectively activates mGluR3 receptors in transfected cells, *J Neurochem.* **69:** 174-81.

13. Xi, Z. X., Baker, D. A., Shen, H., Carson, D. S. & Kalivas, P. W., 2002, Group II metabotropic glutamate receptors modulate extracellular glutamate in the nucleus accumbens, *J Pharmacol Exp Ther.* **300:** 162-71.

14. Bruno, V., Battaglia, G., Copani, A., Giffard, R. G., Raciti, G., Raffaele, R., Shinozaki, H. & Nicoletti, F., 1995, Activation of class II or III metabotropic glutamate receptors protects cultured cortical neurons against excitotoxic degeneration, *Eur J Neurosci.* **7:** 1906-13.

15. Jackson, P. F., Cole, D. C., Slusher, B. S., Stetz, S. L., Ross, L. E., Donzanti, B. A. & Trainor, D. A., 1996, Design, synthesis, and biological activity of a potent inhibitor of the neuropeptidase N-acetylated alpha-linked acidic dipeptidase, *J Med Chem.* **39:** 619-22.

16. Rojas, C., Frazier, S. T., Flanary, J. & Slusher, B. S., 2002, Kinetics and inhibition of glutamate carboxypeptidase II using a microplate assay, *Anal Biochem.* **310:** 50-4.

17. Tiffany, C. W., Cai, N. S., Rojas, C. & Slusher, B. S., 2001, Binding of the glutamate carboxypeptidase II (NAALADase) inhibitor 2-PMPA to rat brain membranes, *Eur J Pharmacol.* **427:** 91-6.

18. Jackson, P. F. & Slusher, B. S., 2001, Design of NAALADase inhibitors: a novel neuroprotective strategy, *Curr Med Chem.* **8:** 949-57.

19. Slusher, B. S., Vornov, J. J., Thomas, A. G., Hurn, P. D., Harukuni, I., Bhardwaj, A., Traystman, R. J., Robinson, M. B., Britton, P., Lu, X. C., Tortella, F. C., Wozniak, K. M., Yudkoff, M., Potter, B. M. & Jackson, P. F., 1999, Selective inhibition of NAALADase, which converts NAAG to glutamate, reduces ischemic brain injury, *Nat Med.* **5:** 1396-402.

20. Majer, P., Jackson, P. F., Delahanty, G., Grella, B. S., Ko, Y. S., Li, W., Liu, Q., Maclin, K. M., Polakova, J., Shaffer, K. A., Stoermer, D., Vitharana, D., Wang, E. Y., Zakrzewski, A., Rojas, C., Slusher, B. S., Wozniak, K. M., Burak, E., Limsakun, T. & Tsukamoto, T., 2003, Synthesis and biological evaluation of thiol-based inhibitors of glutamate carboxypeptidase II: discovery of an orally active GCP II inhibitor, *J Med Chem.* **46:** 1989-96.

21. Ghadge, G. D., Slusher, B. S., Bodner, A., Canto, M. D., Wozniak, K., Thomas, A. G., Rojas, C., Tsukamoto, T., Majer, P., Miller, R. J., Monti, A. L. & Roos, R. P., 2003, Glutamate carboxypeptidase II inhibition protects motor neurons from death in familial amyotrophic lateral sclerosis models, *Proc Natl Acad Sci U S A.* **100:** 9554-9.

22. Bennett, G. J. & Xie, Y. K., 1988, A peripheral mononeuropathy in rat that produces disorders of pain sensation like those seen in man, *Pain.* **33:** 87-107.

23. Hargreaves, K., Dubner, R., Brown, F., Flores, C. & Joris, J., 1988, A new and sensitive method for measuring thermal nociception in cutaneous hyperalgesia, *Pain.* **32:** 77-88.

24. Urazaev, A. K., Buttram, J. G., Jr., Deen, J. P., Gafurov, B. S., Slusher, B. S., Grossfeld, R. M. & Lieberman, E. M., 2001, Mechanisms for clearance of released N-acetylaspartylglutamate in crayfish nerve fibers: implications for axon-glia signaling, *Neuroscience.* **107:** 697-703.

N-ACETYLASPARTYLGLUTAMATE (NAAG) IN SPINAL CORD INJURY AND DISEASE

James L. Meyerhoff, Debra L. Yourick, Barbara S. Slusher and
Joseph B. Long[*]

1. INTRODUCTION

This review will focus on *N*-acetylaspartylglutamate (NAAG) and its catalytic enzyme in motor systems of the spinal cord during injury and disease. NAAG has been studied in the central nervous systems (CNS) of a number of species. It is a dipeptide that fulfills most of the criteria for classification as a neurotransmitter.[1,2] Notably, it is localized to neurons,[3,4] in synaptic vesicles,[5,6] and is released in a calcium-dependent manner upon neuronal depolarization.[6,7,8] NAAG has been reported to act as a mixed agonist/antagonist at NMDA receptors,[9-13] as an agonist at mGluR3 group II metabotropic glutamate (Glu) receptors[14-17] and to be a potential source of synaptic Glu.[18]

In an early study, NAAG levels were measured by gas chromatography in tissue samples from regions of the central nervous system (CNS) from rat, rabbit and guinea pig, by Miyake *et al.*, (1981)[19] who reported levels to be higher in the spinal cord than in any other region. An increasing rostro-caudal gradient of NAAG was likewise reported in an HPLC analysis of CNS regions in the rat.[20] Levels of NAAG in this study were

---

[*] Contact author: James L. Meyerhoff , Division of Neuroscience, Walter Reed Army Institute of Research, 503 Robert Grant Avenue, Silver Spring, MD 20910-7500, Telephone: 301-319-9871, Fax: 301-319-9706, email: james.meyerhoff@na.amedd.army.mil; Debra L. Yourick, Associate Director for Research, Marketing and Policy Development, Walter Reed Army Institute of Research, Silver Spring, MD 20910-7500; Barbara S. Slusher, Guilford Pharmaceuticals, Inc., 611 Tributary Street, Baltimore, MD 21224; Joseph B. Long, Division of Military Casualty Research, Walter Reed Army Institute of Research, Silver Spring, MD 20910-7500. Research was conducted in compliance with the Animal Welfare Act and other federal statutes and regulations relating to animals and experiments involving animals and adheres to principles stated in the Guide for the Care and Use of Laboratory Animals, NRC Publication, 1996 edition. The views of the author(s) do not purport to reflect the position of the Department of the Army or the Department of Defense, (para 4-3, AR 360-5).

confirmed as being highest in the spinal cord. In a subsequent study comparing NAAG concentrations in dorsal vs. ventral halves of cervical, thoracic and lumbar spinal cord, concentrations were found to be higher in the ventral half at all three spinal levels, and were higher in the thoracic than at cervical or lumbar levels.[21] NAAG levels were found to be particularly high in the ventral horn,[22] where an association with motoneurons has been noted.[23,24] A more recent study surveying a large number of CNS regions has confirmed the rostro-caudal gradient, reporting levels as low as 2.4 nmol/mg protein in the median eminence of the hypothalamus, and less than 5 nmol/mgprotein in the cerebral cortices, compared with levels as high as 64 nmol/mg protein in the thoracic spinal cord, where high concentrations were noted in motoneuron cell bodies, as well as in ascending sensory tracts.[25]

The possibility of the presence in brain of a peptidase targeting Glu-containing dipeptides was suggested in 1984 by Riveros and Orrego,[26] and by Blakely *et al.* in 1986.[27] The enzyme was then fully characterized in 1987.[18] NAAG is hydrolyzed by an extracellular metalloproteinase, glutamate carboxypeptidase (GCPII; EC 3.4.17.21) to liberate Glu and N-Acetylaspartate (NAA). GCP II has been previously known variously as N-acetyl-alpha-linked acidic dipeptidase (NAALADase), NAAG-peptidase and NAAG-hydrolyzing enzyme. The peptidase activity was demonstrated by subcellular fractionation to be enriched in synaptic plasma membranes.[28] GCP II is expressed by astrocytes in the central nervous system and is localized predominantly on the extracellular face of plasma membranes.[14,29]

Using a radioenzymatic assay, the enzyme activity was shown to be largely restricted to the central nervous system, with the exception of the kidney. Small amounts of activity were found in the sciatic nerve and the adrenal gland. Moderate amounts of activity were demonstrated in the cervical spinal cord, but a rostro-caudal gradient for the peptidase similar to that for the peptide was not found.[28] Distribution of GCP II in the CNS has been studied immunocytochemically in the rat, with the enzyme found to be localized in the neuropil of the amygdala, caudate putamen, central gray, dorsal raphe, globus pallidus, hippocampus, hypothalamus, locus coeruleus, medial and lateral geniculate, olfactory bulb, periaqueductal gray, solitary nucleus, spinal trigeminal nucleus, substantia nigra, superior colliculus, and thalamus.[30] Activity was seen in neuropil, but not cytoplasm. Regional CNS analysis of GCP II activity was assessed using regional microdissection of 58 brain and spinal cord regions and radioenzymatic assay.[25] Peptidase activity ranged from 148 pmol Glu produced/mg protein/minute in the superior colliculus to 54 pmol/mgp/min in the median eminence. Activity was also high in the inferior colliculus, the reticular formation and the mammillary bodies. The ratio of the dipeptide concentration to peptidase activity was > 0.3 in thoracic spinal cord, as well as in the ventrolateral medulla and the reticular formation. Using enzyme assays and immunoblotting, GCP II has also been localized in the peripheral nervous system, including the sciatic nerve, phrenic nerve, cervical dorsal root ganglion and superior cervical ganglion of the rat.[31] In summary, although the levels of peptide in the spinal cord are the highest in the CNS, the peptidase distribution does not follow the same rostro-caudal gradient as does the peptide.

2. PATHOLOGICAL CHANGES IN NAAG AND GCP II IN SPINAL MOTO-NEURON INJURY AND DISEASE

In 1984 Koller *et al.*[32] reported reductions in NAAG levels caudal to spinal cord transactions. Given this finding, along with the known association of NAAG with motoneurons[21], its co-localization with cholinergic motoneurons[33] and the report of marked reduction in choline acetyltransferase activity in the ventral horn of the spinal cord in postmortem studies of patients with amyotrophic lateral sclerosis [ALS],[34] it was logical to study levels of NAAG in conditions known to affect motoneurons. NAAG was reported to be significantly decreased in the spinal cord (32%) and in the cerebral cortex (43%) of male mice with hereditary myodystrophy manifesting an associated hindlimb paralysis.[23] Levels of NAA, Glu and aspartate (Asp) had previously been shown to be decreased in the brains of these mice.[35]

Several studies have been carried out in spinal cord tissue obtained at autopsy from patients that had died of ALS. The comparison groups were autopsy specimens taken from patients that had died of non-neurological disease or Parkinson's disease. Both NAAG and NAA levels were found to be reduced in the cervical spinal cords of patients with ALS.[36] Levels of NAAG and NAA were demonstrated to be lower than normal by 60% and 40% respectively, in the ventral horns of cervical spinal cords from patients with ALS.[37] In a more extensive postmortem regional CNS study of patients with ALS, NAAG levels were found to be decreased in both the ventral and the dorsal horns, as well as in the dorsal, lateral and ventral columns of the spinal cord.[38] NAA levels were similarly decreased in both the ventral and the dorsal horns, as well as in the dorsal column. Levels of Glu and Asp were also both decreased in the ventral horn. Conversely, GCPII activity was significantly increased in the spinal cord, but only in the ventral column.

Abnormalities analogous to those reported in spinal cord were found in postmortem brain tissue, as well, but were limited to motor cortex.[38] Levels of NAAG were significantly reduced in gray matter from motor cortex. Both Glu and Asp were reduced in gray as well as white matter in motor cortex. Once again, the alteration of the peptidase activity was in a direction opposite to that of the peptide, as it was increased in both gray and white matter in motor cortex.

Other studies have focused on cerebrospinal fluid obtained by lumbar puncture from patients diagnosed with ALS or from patients with other neurological diseases. In contrast to the reductions in levels of NAAG, NAA, Glu, and Asp in CNS tissues, levels of these compounds were increased in cerebrospinal fluid of ALS patients compared to patients with other neurological diseases.[37] A subsequent study has replicated the finding of increased csf levels of Glu, but did not find increased levels of Asp.[39] A decrease in Glu transport has also been reported in autopsy samples of spinal cord, motor cortex and somatosensory cortex from patients with ALS.[40]

Hereditary canine spinal muscular atrophy (HCSMA) has been viewed as a potential animal model for ALS. In dogs with HCSMA, levels of NAAG, Glu and Asp were significantly reduced in cervical spinal cords of pups with the homozygous form of the disease.[41] As in the studies of humans with motor neuron disease, activity of GCPII was increased in the spinal cords. In contrast to findings in postmortem cerebral cortical tissue from patients with ALS, changes in peptidase activity were not found in the motor cortices of dogs with HCSMA.

In summary, decreases in spinal cord levels of NAAG have been reported in mouse and canine models of motor neuron disease, as well as in postmortem tissue from patients with ALS. These changes, as well as alterations of Glu and Asp levels, suggest that disruptions of metabolism of NAAG and excitatory amino acids may be a hallmark of ALS. While these changes could be contributory to pathogenesis, they could also be merely reflective of underlying pathology.

3. THERAPEUTIC STUDIES WITH GCP II INHIBITORS IN RODENT MODELS

Degenerative diseases or injuries to the central nervous system (CNS) are believed to result in elevations of extracellular concentrations of the excitatory amino acid neurotransmitter Glu which, through excessive stimulation of excitatory amino acid receptors, including NMDARs, can evoke a cascade of disruption of intracellular processes inducing cell dysfunction and death.[42-44] Although a number of pharmacological agents that selectively antagonize the action of Glu at its receptors have been shown to reduce injury in experimental models of cerebral ischemia and trauma, none have yet been effective in clinical trials, due primarily to unacceptable side effects.[45] Consequently, to some extent the focus for development of therapeutic agents has shifted away from NMDA receptor antagonists and has been expanded to include alternative means to interrupt this Glu-mediated excitotoxic cascade.

Because Glu is a product of hydrolysis of NAAG via GCP II activity, NAAG is a potentially important source of synaptic Glu under pathophysiological conditions, and inhibition of GCP II has been considered a promising alternative neuroprotective strategy. Because of its reciprocal influence on the synaptic concentrations of NAAG and Glu, GCP II has recently been identified as a potentially important therapeutic target to pharmacologically counter the Glu-mediated excitotoxic cascade associated with stroke or traumatic injury.[46] In particular, in addition to reducing pathophysiologically deleterious concentrations of extracellular Glu, it has been envisioned that by promoting the extracellular accumulation of NAAG, GCP II inhibitors can elicit neuroprotective mechanisms that are mediated through both neuronal and glial group II metabotropic Glu receptors, since NAAG is a relatively potent and selective agonist at these receptors,[15,17,47] which have been shown to mediate neuroprotection through presynaptic inhibition of Glu release and through induction of TGF-β synthesis and release.[17,48,49]

2-phosphonyl-methyl pentanedioc acid (2-PMPA), a potent inhibitor of GCP II, has been shown to protect against acute neuronal injury caused by middle cerebral artery occlusion (MCAo) in vivo[46]; and also to protect against injury by metabolic inhibition, hypoxia, or excitatory amino acid exposure in vitro.[47,50-52] In the MCAo study, reductions in extracellular Glu and elevations in extracellular NAAG concentrations were reported to occur after administration of 2-PMPA.[43]

Inhibitors of GCP II have also been shown to be therapeutically effective when administered chronically in a rodent model of hereditary motoneuron disease. Ten percent of patients with ALS suffer from a familial subtype (FALS) of the disease, and of those 20% are caused by mutations of copper/zinc superoxide dismutase type 1 (SOD1). Glu toxicity has been implicated in ALS, including the familial type. Rodents carrying mutant SOD1 have motoneuron degeneration similar to patients with ALS. GCP II inhibitors have been shown to be protective *in vitro* in cultured neurons transfected with mutant SOD1, as well as in *vivo* in G93A FALS transgenic mice carrying the SOD1 mutation[53]. Chronic treatment with 2-MPPA [2-(3-mercaptopropyl)pentanedioc acid], an orally-adminstered analog of 2-PMPA, increased longevity and delayed the onset of 4 out of 5 rated neurological motor symptoms in the G93A FALS transgenic mice. In cultured rat motoneurons transduced to express mutant SOD1, treatment with 2-PMPA was neuroprotective. Interestingly, however, the protective effects of 2-PMPA were not diminished by co-administration of either the group II mGluR antagonist EGLU [(2S)-alpha-ethylglutamic acid] or an antibody against TGF-beta. The authors concluded that in that sytem, the neuroprotective effects of 2-PMPA were mediated neither via TGF-beta nor through the effects of NAAG on mGluR3 receptors.

2-PMPA has been reported to be highly selective for GCP II, and shown not to interact with a range of receptors,[46] and discussion continues on the latter point.[2,54,55]

Repeated administration of 2-PMPA was also shown to be effective in enhancing recovery in an *in vivo* model of peripheral nerve injury. Four weeks of daily oral administration of a GCP II activity inhibitor was reported to improve nerve morphology, physiology and endurance in a walking test after sciatic nerve crush injury.[56] Thus, evidence is accumulating that inhibition of catabolism of NAAG may have significant potential as a therapy for neuronal injury as well as for neurodegenerative disorders.

4. EVALUATION OF INHIBITORS OF GCP II IN ACUTE SPINAL CORD INJURY IN THE RAT

Spinal intrathecal injection of the peptide Dynorphin A has been shown to reliably induce ischemia, neuronal injury and persistent flaccid hindlimb paralysis.[57-63] Concentrations of the excitatory amino acid neurotransmitters Glu and Asp are significantly increased in lumbar cerebrospinal fluid in the rat, shortly after the onset of dynorphin A-induced hindlimb paralysis.[62] In addition, a variety of competitive and noncompetitive inhibitors of the NMDA receptor complex have been shown to significantly improve recovery of hindlimb motor function following this insult.[62-66] In light of these findings suggesting excitoxic mechanisms in this model, we chose to use it to assess GCP II inhibition as a means to ameliorate excitoxic injury in the rat spinal cord.

Anesthetized male Sprague-Dawley rats (250-300 g) were injected intrathecally with dynorphin A or vehicle, between the lumbar L4-L5 vertebrae according to a published method.[62,63] Injections were delivered in a total volume of 22 µl [containing the dynorphin A (20 nmoles), 2-PMPA (1-4 umoles) or vehicle]. 2-PMPA was provided by Guilford Pharmaceuticals (Baltimore, MD). Locomotor function was independently evaluated by blinded observers at 2 h and again at 24 h following dynorphin A injection, using a 5 point ordinal scale. Neurological scores were assigned as follows: 4 = normal motor function; 3 = mild paraparesis, with the ability to support weight and walk with impairment; 2 = paraparesis, making walking movements without supporting weight; 1 = severe paraparesis, in which rats could make limited hindlimb movement, but not walking movement; 0 = flaccid paralysis, with complete absence of any hindlimb movement.

For amino acid measurements, CSF (approximately 70 µl) was collected on ice for approximately 10 min beginning 15 min after spinal subarachnoid injections of dynorphin A, acidified and frozen until assayed by method of Robinson (1993).[67] After euthanasia and formalin infusion, a section of lumbosacral spinal cord corresponding to L5-L6, was sectioned (6 μm) and stained with cresyl violet. Neurons with a cell dimension ≥ 25 μm and a visible nucleus were counted by an observer blinded as to treatment condition.

Dynophin A reliably caused flaccid hindlimb paralysis persisting beyond 24 h and significant neuronal injury throughout the lumbosacral cord. When coadministered with dynorphin A, 2-PMPA caused significant dose-dependent improvements in motor scores by 2 and 24 h postinjection. Higher doses of 2-PMPA alone (> 4 μmoles) also caused neurological deficits. On the basis of their assigned neurological scores (0-4), rats were categorized as ambulatory (3 or 4) or nonambulatory (0,1,or 2), at 24 hr postinjection. 80% of the rats co-treated with 4 μmoles of 2-PMPA could walk when tested 24 h post-injury, in contrast to 17% of the rats co-treated with saline vehicle.[68]

To rule out the possibility that protection against dynorphin A-induced spinal cord injury might result from direct interactions of 2-PMPA and dynorphin A, we also evaluated the effects of 2-PMPA in rats receiving spinal subarachnoid injections of somatostatin, a structurally unrelated peptide that has also been shown to cause hindlimb paralysis, vasospasm and ischemic spinal cord injury in rats.[59,61] 2-PMPA also significantly improved recovery of hindlimb motor function in rats acutely paralyzed after spinal subarachnoid injection of 25 nmoles of somatostatin.

The protective effects of 2-PMPA were also clearly evident by histopathological assessment of spinal cords removed from rats 72 h after dynorphin A injection. In contrast to the necrosis, hemorrhage, and cellular infiltration typically seen in the spinal cords of dynorphin A-injured rats co-treated with saline vehicle, significant sparing of neurons in both dorsal and ventral horns were observed in rats co-treated with 4 μmoles of 2-PMPA. The spongy rarefaction of white matter that was characteristically seen in dynorphin A-injured cords was also less evident in the cords of rats co-treated with 2-PMPA. Cell counts in the ventrolateral portions of the L5-L6 anterior horn quantitatively revealed the deleterious and salutary effects of dynorphin A and 2-PMPA, respectively. Rats paralyzed by dynorphin A had a significant loss of large neuronal cell bodies in this region, whereas the dynorphin A-injected rats co-treated with PMPA had a dose-related preservation of these cells that in all likelihood are α-motoneurons that are directly involved in locomotion.[68,69]

In lumbar CSF samples taken following dynorphin A injection, the concentrations of Glu were significantly increased by approximately 3 fold, to 12.3 μM. When coinjected with dynorphin A, 2-PMPA (which by itself did not alter CSF concentrations of this or other amino acids) the dynorphin A-induced elevations in Glu were significantly reduced.[69]

As has been previously shown with ischemic injuries to the brain[46,47] and with hypoxic or metabolic injuries to neurons in culture,[46,51,52] 2-PMPA exhibited dose-dependent neuroprotective actions against the spinal cord injuries and persistent neurological deficits caused by spinal subarachnoid injection of dynorphin A. Significant motor recovery was evident about 2 h after dynorphin A injection in rats cotreated with 2-PMPA, suggesting that the protective effects of this compound resulted from the interruption of secondary pathophysiological events rather than from a direct interruption of or interference with the initial pharmacological actions of dynorphin A that trigger an ischemic insult to the lumbosacral cord.

2-PMPA eliminated the dynorphin A-induced increases in Glu in CSF samples taken from the lumbar injection site, as would be expected if the hydrolysis of NAAG serves as a primary source of the elevated extracellular Glu seen under these pathophysiological conditions. 2-PMPA did not elicit a change in CSF Glu concentrations in rats not injected with dynorphin A, indicating that the effects of the drug on extracellular Glu were restricted to ameliorating the disrupted tissue milieu associated with injury. The diminished elevations in Glu concentrations might have also resulted from alternative actions of 2-PMPA that served to lessen the severity of the dynorphin A-induced injury and in turn reduce the associated release and extracellular accumulation of Glu. Although csf levels of NAAG were not measured in the present study, in a reported study using the rat MCAo model, reductions in extracellular Glu along with elevations in extracellular NAAG concentrations followed treatment with 2-PMPA,[46] as might be predicted with GCP II as the site of drug action.

Beta-NAAG is a non-hydrolyzable analog of NAAG reported to protect cultured neurons against both hypoxia and NMDA-induced injury.[70] We have also found that Beta-NAAG was highly protective *in vivo* against spinal injury induced by Dynorphin-A.[71] Although it has been suggested that NAAG may be neuroprotective via agonist effects at mGluR3 receptors, it has been reported that beta-NAAG is an antagonist at mGluR3 receptors.[72] Accordingly, the neuroprotective effects of beta-NAAG might be mediated via reduction of synaptic availability of Glu, or direct or indirect antagonism at the NMDA receptor.

As noted, the concentrations of NAAG in the spinal cord are the highest in the central nervous system.[19,32,73] This suggests that a neuroprotective intervention targeted at enhancing NAAG actions might be particularly well-suited as a therapeutic means of suppressing Glu-mediated excitotoxicity associated with spinal cord injuries or other neuropathological conditions.

Consistent with these expectations, we observed the selective GCP II inhibitor 2-PMPA to significantly reduce post-injury elevation of extracellular Glu and to greatly improve recovery from dynorphin A-induced ischemic spinal cord injury. These data reinforce and extend earlier indications pointing to the potential importance of NAAG and GCP II under pathophysiological circumstances associated with excitotoxic actions of extracellular Glu.

5. DISCUSSION

In addition to the therapeutic benefits derived from reductions in extracellular Glu, the neuroprotection resulting from GCP II inhibition can also arise from NAAG acting at several different excitatory amino acid receptors. NAAG has been shown to function as a partial agonist/antagonist at NMDA receptors.[9-13] As a partial agonist (with reduced efficacy), NAAG would act as an NMDA receptor antagonist under conditions where synaptic concentrations of Glu are at or near saturation. On the other hand, Losi et al.[74] reported a lack of antagonism of NAAG at synaptic or extrasynaptic NMDA receptors in cultured cerebellar granule cells. Very recently however, Bergeron et al.[75] demonstrated that NAAG reduces NMDA receptor current in CA1 pyramidal neurons in slices from rat hippocampus and in dissociated neurons. The antagonism of NAAG at the NMDAR was shown to be overcome by glycine, suggesting that NAAG and glycine might compete at the glycine site on the NMDAR. Furthermore, it was suggested that elevated glycine levels in the culture media might account for the absence of NAAG antagonism at the NMDAR in some reports of *in vitro* studies.

NAAG has also been demonstrated to act as an agonist at group II metabotropic Glu receptors, with greatest selectivity being seen for group II mGluR3 receptors, which are found on both neuronal and glial cells.[15,16,76] Group II metabotropic Glu receptors have been proposed to elicit neuroprotective actions through inactivation of voltage-sensitive calcium channels,[77] inhibition of cAMP formation,[78] and presynaptic reduction of Glu release.[79] In addition, the importance of glial mGluR3 receptors to the neuroprotective effect of NAAG and 2-PMPA in certain experimental situations has recently been recognized and has been shown to involve the synthesis and release of TGF-β.[47,49,52]

In summary, inhibition of GCP II can be expected to reduce levels of Glu and to increase NAAG levels. Possibly, following inhibition of GCP II, NAAG could be neuroprotective via a variety of actions (e.g. via NMDAR, mGluR3 and/or TGF-β mediated mechanisms), and different pathways might predominate in a particular model system. Collectively, this expanding body of data points to potential actions of NAAG contributing to the neuroprotective effects of 2-PMPA.

6. CONCLUSIONS

As previously proposed,[46] a broad spectrum "upstream" approach to counter glutamate excitotoxicity through GCP II inhibition offers several advantages over therapies targeted directly at blockade of receptors. Such approaches may provide an attractive new strategy for development of drugs for spinal cord injury and disease.

7. ACKNOWLEDGEMENTS

The authors wish to thank Dr. Michael B. Robinson for his review of this paper and for his helpful comments. They also thank Ms. Carma Rogers for excellent assistance with preparation of this manuscript.

8. QUESTIONS AND ANSWER SESSION

DR. MEYERHOFF: All right. Thank you very much. I am pleased to share my time with my esteemed colleague from Walter Reed, Dr. Debra Yourick, who is going to describe some of our in vitro experiments, and then I will describe some in vivo experiments on spinal injury.

While Dr. Yourick is setting up, I wanted to comment on one of Dr. Namboodiri's questions to the previous speaker, with respect to mechanism. The last speaker cited an experiment by Gadge *et al.*, using a rodent ALS variant that is an SOD transgene.

In neurons cultured from that species, there was neuro protection achieved with 2PMPA. Dr. Slusher was the co-author on that paper by Gadge. They tested the mechanism by administering the mGluR3 antagonist ethylglutamate, eGlu, and that did not reverse the PMPA amelioration and increase survival of these neurons in culture with the SOD defect.

They also applied antibodies to transforming growth factor Beta, and that also did not reverse the ameliorative effect of 2PMPA, which is very interesting in terms of the ongoing discussions about whether the primary therapeutic effect is at the presynaptic mGluR3 or whether it is on glutamate levels or whether it is at the NMDA receptor, and it may be that these things are somewhat model specific. But I just wanted to mention that, because if Dr. Slusher had been able to be here, I imagine she might have brought it up in terms of mechanisms, a very interesting aspect of that experiment.

DR. WEINBERGER: Has there actually been any pharmacologic characterization of NAAG as a partial agonist *per se*? Has it been actually characterized as a partial agonist?

DR. YOURICK: Yes, but not a partial agonist.

DR. COYLE: I think we have to be careful. I mean, we have just been working like for five years on this Schaffer collateral CA1 synapse, and done it five ways from Sunday. At the concentrations used at that synapse, it is a noncompetitive glycine reversible antagonist.

Now that is not seen at the granule cells in the cerebellum, and in the whatchamacallit, in the hippocampus -- no, not CA1, the area you all looked. But that is a gyrus. LTD in the dentate gyrus, the mGluR3 effect predominates. But, actually, the work done in 1986 on primary cultures of motor neurons -- there it was shown to have a partial agonist effect or a weak agonist effect.

One problem with tissue cultures in trying to tease this out is the influence of high levels of glycine in the culture medium, if it works by a glycine reversible effect. So it may be very dependent on the NR2 subunit and other factors in terms of how it behaves at different sites.

DR. LIEBERMAN: That was my question, is whether or not it had been characterized as the NMDA receptor there versus the NMDA receptors in other places where you don't see any NAAG effects on that receptor.

DR. WROBLEWSKA: There was some work done on the NMDA receptors -- but in our hands, NAAG is -- it is not an antagonist NMDA receptor of the granular type -- but it has some other effects in very high concentrations.

DR. COYLE: Right. The CA1 Schaffer collateral synapse, as far as we can see, has no influence of mGluR3 receptors.

DR. WROBLEWSKA: I have a question about the ALS. I mean, protective effect in the ALS are 2-PMPA receptor. Right?

DR. MEYERHOFF: If you mean in the in vitro model by Gadge, *et al.*, on which Barbara Slusher was a co-author, they just suggested it wasn't acting through the presynaptic mGluR3. But that leaves the possibility that the drug could have simply reduced synaptic glutamate availability through inhibition of the enzyme or that it could have had an effect through the NMDA receptor. But the paper, as I recall, doesn't really address what the mechanism is, except to say what it doesn't appear to be. That was just in the 2-PMPA paper that was done in cultured rat neurons that had the transgene.

9. REFERENCES

1. Coyle, J.T., The nagging question of the function of N-acetylaspartylglutamate. *Neurobiol. of Dis.* **4**, 231-238 (1997).
2. Neale, J.H., Bzdega, T., Wroblewska, B., N-Acetylaspartylglutamate: The most abundant peptide neurotransmitter in the mammalian central nervous system. *J. Neurochem.* **75**, 443-452, (2000).
3. Moffett, J.R., Namboodiri, M.A.A., Cangro, C.A. and Neale, J.H. Immunohistochemical localization of N-acetylaspartate in rat brain. *Neuroreport* **2**, 131-134, (1991).

4.   Anderson, K.J., Monaghan, D.T., Cangro, C.B., Namboodiri, M.A.A., Neale, J.H. and Cotman, C.W., Localization of N-acetylaspartylglutamate-like immunoreactivity in selected areas of the rat brain. *Neuroscience Letters* **72**:14-20, (1986).
5.   Williamson, L.C., Neale, J.H., Ultrastructural localization of N-acetylaspartylglutamate in synaptic vesicles of retinal neurons. *Brain Res.* **456**, 375-381, (1988a).
6.   Williamson, L.C., Neale, J.H. Calcium-dependent release of N-acetylaspartylglutamate from retinal neurons upon depolarization. *Brain Res.* **475**, 151-155, (1988b).
7.   Tsai, G., Forloni, G., Robinson, M.D., Stauch, B.L., Coyle, J.T., Calcium-dependent evoked release of N-[³H]acetylaspartylglutamate from the optic pathway. *J. Neurochem.* **51**, 1956-1959, (1988).
8.   Tsai, G., Stauch, B.L., Vornov, J.J., Deshpande, J.K. and Coyle, J.T., Selecive release of N-acetyl-aspartylglutamate from rat optic nerve terminals in vivo. *Brain Res.* **518**, 313-316, (1990).
9.   Sekiguchi, M., Okamoto, K., Sakai, Y., Low-concentration N-acetylaspartylglutamate suppresses the climbing fibre response of Purkinje cells in guinea pig cerebellar slices and the responses to excitatory amino acids of *Xenopus laevis* oocytes injected with cerebellar mRNA. *Brain Res.* **482**, 87-96, (1989).
10. Burlina, A.P., Skaper, S.D., Rosaria Mazza, M., Ferrari, V., Leon, A., Burlina, A.B. N-acetylaspartylglutamate selectively inhibits neuronal responses to N-methyl-D-aspartic acid *in vitro. J. Neurochem.* **63**, 1174-1177, 1994.
11. Puttfarcken, P.F., Handen, J.S., Montgomery, D.T., Coyle, J.T., Werling, L.S., 1993. N-acetyl-aspartylglutamate modulation of N-methyl-D-aspartate-stimulated [³H]norepinephrine release from rat hippocampal slices. *J. Pharmacol. Exp. Ther.* **266**, 796-803.
12. Valivullah, H.M., Lancaster, J., Sweetnam, P.M., Neale, J.H., Interactions between N-acetylaspartylglutamate and AMPA, kainate and NMDA binding sites. *J. Neurochem.* **63**, 1714-1719, (1994).
13. Koenig, M.L., Rothbard, P.M., DeCoster, M.A., Meyerhoff, J.L., N-Acetyl-aspartyl-glutamate (NAAG) elicits rapid increases in intraneuronal Ca²⁺ *in vitro. Neuroreport* **5**, 1063-1068, (1994).
14. Cassidy, M. and Neale, J. N-acetylaspartylglutamate catabolism is achieved by an enzyme on the cell surface of neurons and glia. *Neuropeptides.* **24**, 271-278, 1993.
15. Wroblewska, B., Wroblewski, J.T., Saab, O.H., Neale, J.H., N-Acetylaspartylglutamate inhibits forskolin-stimulated cyclic AMP levels via a metabotropic glutamate receptor in cultured cerebellar granule cells. *J. Neurochem.* **61**, 943-948, (1993).
16. Wroblewska, B., Wroblewski, J.T., Pshenichkin, S., Surin, A., Sullivan, S.E., Neale, J.H., N-acetylaspartylglutamate selectively activates mGluR3 receptors in transfected cells. *J. Neurochem.* **69**, 174-181, (1997).
17. Bruno, V., Wroblewska, B., Wroblewski, J.T., Fiore, L., Nicoletti, F., Neuroprotective activity of N-acetylaspartylglutamate in culture cortical cells. *Neurosci.* **85**, 751-757, (1998a).
18. Robinson, M.B., Blakely, R.D., Couto, R., Coyle, J.T., Hydrolysis of the brain dipeptide N-acetyl-L-aspartyl-L-glutamate. Identification and characterization of a novel N-acetylated alpha-linked acidic dipeptidase activity from rat brain. *J. Biol. Chem.* **262**, 14498-14506, (1987).
19. Miyake, M., Kakimoto, Y., Sorimachi, M., A gas chromatographic method for the determination of N-acetyl-L-aspartic acid. N-acetyl-α-aspartylglutamic acid, and α-citryl-L-glutamic acid and their distributions in the brain and other organs of various species of animals. *J. Neurochem.* **36**, 804-810, (1981).
20. Koller, K.J., Zaczek, R., and Coyle, J.T., N-acetyl-aspartyl-glutamate: regional levels in rat brain and the effects of brain lesions as determined by a new HPLC method. *J. Neurochem.* **43**, 1136-1142, (1984).
21. Ory-Lavolee, L., Blakely, R.D. and Coyle, J.T. Neurochemical and immunocytochemical studies on the distribution of N-acetylaspartyl glutamate and N-acetylaspartate in rat spinal cord and some peripheral nervous tissues. *J. Neurochem.* **48**, 895-899, (1987).
22. Guarda, A.S.; Robinson, M.B.; Ory-Lavollee, L.; Lorloni, G.L.; Blakely, R.D.; and Coyle, J.T., Molecular *Brain Res.*, **3**:223-232, (1998).
23. Blakely, R.D., Ory-Lavolee, L. and Coyle, J.T. Specific alterations in the levels of N-acetylaspartyl-glutamate in the nervous system of the dystrophic mouse. *Neuroscience Letters.* **79**:223-228, (1987).
24. Frondoza, C.G., Logan, S., Forloni, L. and Coyle, J.T., Production and characterization of monoclonal antibodies to N-acetylaspartylglutamate. *J. Histochem. Cytochem.* **38**, 493-502, (1990).
25. Fuhrman, S.; Palkovits, M.; Cassidy, M.; and Neal J.H. The regional distribution of N-acetylaspartyl-glutamate (NAAG) and peptidase activity against NAAG in the rat nervous system. *J. Neurochem.* **62**, 275-281 (1994).
26. Riveros, N. and Orrego, F., A study of possible excitatory effects of N-acetyl-aspartylglutamate in different in vivo and in vitro brain preparations. *Brain Res.* **299**, 393-395, (1984).

27. Blakely, R.D., Ory-Lavolee, L., Thompson, R.C. and Coyle, J.T., Synaptosomal transport of radiolabel from N-Acetyl-Aspartyl [H³] glutamate suggests a mechanism of inactivation of an excitatory neuropeptide. *J. Neurochem.* **47**, 1013-1019, (1986).

28. Blakely, R.D., Robinson, M.B., Thompson, R.C. and Coyle, J.T., Hydrolysis of the brain dipeptide N-acetylaspartylglutamate: subcellular and regional distribution, ontogeny and the effect of lesions on N-acetylated-alpha-linked acidic dipeptidase activity. *J. Neurochem.* **50**, 1200-1209, (1988).

29. Berger, U.V., Luthi-Carter, R., Passani, L.,A., Elkabes, S., Black, I., Konradi, C. and Coyle, J.T., Glutamate carboxypeptidase II is expressed by astrocytes in the adult rat nervous system. *J. Comp. Neurol.* **415**, 52-64, (1999).

30. Slusher, B., Tsai, G., Yoo, G., and Coyle, J.T. Immunocytochemical localization of the N-acetyl-aspartyl-glutamate (NAAG) hydrolyzing enzyme N-acetylated alpha-linked acidic dipeptidase (NAALADase). *J. Comp. Neurol.* **315**, 217-229, (1992).

31. Berger, U., Carter, R., McKee, M., and Coyle, J.T., N-acetylated alpa-linked acidic dipeptidase is expressed by non-myelinating Schwann cells in the peripheral nervous system. *J. Neurocytol.* **24**:99-109, (1995).

32. Koller, K.J., Coyle, J.T., Ontogenesis of N-acetyl-aspartate and N-acetyl-aspartylglutamate in rat brain. *Dev. Brain Res.* **15**, 137-140, (1984).

33. Forloni, G., Grzanna, R., Blakely, R.D. and Coyle, J.T., Co-localization of N-acetylaspartyl glutamate in central cholinergic, noradrenergic and serotoninergic neurons. *Synapse* **1**:444-460, (1987).

34. Gilberg, P., Aquilonius, S.; Eckernas, S.; Lundqvist, G.; and Winblad B. Choline acetyltransferase and substance P-like immuno-reactivity in the human spinal cord: changes in amyotrophic lateral sclerosis. *Brain Res.,* **250**:394-397 (1982).

35. Marcucci, F., Colombo, L., De Ponte, G. and Mussini, E., Decrease in N-acetylaspartic acid in brain of myodystrophic mouse. *J. Neurochem.* **43**:1484-1486, (1984).

36. Constantakis, E. and Plaitakis, A. N-Acetylaspartate and N-Acetylaspartylglutamate are altered in the spinal cord in amyotropihic lateral sclerosis. *Ann. Neurol.* (abstract) **24**:478, 1988.

37. Rothstein, J.D. MD, PhD; Tsai, G, MD; Kuncl, R.W. MD, PhD.; Clawson, L., RN, BSN; Cornblath, D.R. MD; Drachman, D.B., MD; Pestronk, A., MD; Stauch, B.L., BS; and Coyle, J.T., Abnormal excitatory amino acid metabolism in amyotrophic lateral sclerosis. *Ann Neurol,* **28**:18-25, (1990).

38. Tsai, G.; Stauch-Slusher, B.; Sim, L.; Hedreen, J. C.; Rothstein, J.D.; Kuncl, R.; and Coyle, J.T. Reductions in acidic amino acids and N-acetylaspartylglutamate in amyotrophic lateral sclerosis CNS. *Brain Res.,* **556**: 151-156, (1991).

39. Shaw, P.J; Forrest, V.; Ince, P.G.; Richardson, J.P.; and Wastell, H.J. CSF and Plasma Amino Acid Levels in Motor Neuron Disease: Elevation of CSF Glutamate in a Subset of Patients. *Neurodegeneration,* **4**: 209-216, (1995).

40. Rothstein, J.D. MD and Ph.D.; Martin, L.L. Ph.D.; and Kungl, R.W., MD, Ph.D. Decreased Glutamate Transport by the Brain and Spinal Cord in Amyotrophic Lateral Sclerosis. *N. Engl. J. Med.,* **326**: 1464-8, (1992)

41. Tsai, G.; Cork, L.C.; Slusher, B.S.; Price, D.; and Coyle, J.T. Abnormal acidic amino acids and N-acetylaspartylglutamate in hereditary canine motoneuron disease. *Brain Res.,* **629**, 305-309, (1993).

42. Choi, D.W. Glutamate toxicity and diseases of the nervous system. *Neuron.* **1**, 623-634, (1988).

43. McIntosh, T.K., Smith, D.H., Meaney, D.F., Kotapka, M.J., Gennarelli, T.A., Graham, D.I., Neuropathological sequelae of traumatic brain injury: relationship to neurochemical and biochemical mechanisms. *Lab. Invest.* **74**, 315-342, (1996).

44. McIntosh, T.K., Juhler, M., Wieloch, T., Novelpharmacologic strategies in the treatment of experimental traumatic brain injury: *J. Neurotrauma* **15**, 731-769, (1998).

45. Olney, J.W., Labruyere, J. and Price, M.T., Pathological changes induced in cerebrocortical neurons by phencyclidine and related drugs. *Science* **244**, 1360-1362, (1989).

46. Slusher, B.S., Vornov, J.J., Thomas, A.G., Hurn, P.D., Harakuni, I., Bhardwaj, A., Traystman, R.J., Robinson, M.B., Britton, P., Lu, X., Tortella, F.C., Wozniak, K.M., Yudkoff, M., Potter, B.M., Jackson, P.F., Selective inhibition of NAALADase, which converts NAAG to glutamate, reduces ischemic brain injury. *Nat. Med.* **5**, 1396-1402, (1999).

47. Thomas, A.G., Olkowski, J.L., Slusher, B.S., Neuroprotection afforded by NAAG and NAALADase inhibition requires glial cells and metabolic glutamate receptor activation. *Eur. J. Pharmacol.* **426**, 23-26, (2001a).

48. Bruno, V., Sureda, F.X., Storto, M., Casabona, G., Caruso, A., Knopfel, T., Kuhn, R., Nicoletti, F., The neuroprotective activity of group-II metabotropic glutamate receptors requires new protein synthesis and involves a glial-neuronal signaling. *J. Neurosci.* **17**, 1891-1897, (1997).

49. Bruno, V., Battaglia, G., Casabona, G., Copani, A., Caciagli, F., Nicoletti, F., Neuroprotection by glial metabotropic glutamate receptors is mediated by transforming growth factor-beta. *J. Neurosci.* **18**, 9594-9600, (1998b).

50. Hacker, H.D., Yourick, D.L., Koenig, M.L., Slusher, B.S. and Meyerhoff, J.L., Neuroprotection in rabbit retina with N-acetylaspartylglutamate and 2-phosphonyl-methyl pentanedioic acid. In (Eds. B.Stuck & M. Belkin) *Laser and noncoherent light ocular effects: epidemiology, prevention and treatment III. Proceedings of Ophthalmic Technologies IX.* **3591**:422-429, (1999).

51. Tortella, F.C., Lin, Y., Ved, H., Slusher, B.S., Dave, J.R. 2000. Neuroprotection produced by the NAALADase inhibitor 2-PMPA in rat cerebellar neurons. *Eur. J. Pharmacol.* 402, 31-37.

52. Thomas, A.G., Liu, W., Olkowski, J.L., Tang, Z., Lin, Q., Lu, X.-C.M., Slusher, B.S. Neuroprotection mediated by glutamate carboxypeptidase II (NAALADase) inhibition requires TGF-β. *Eur. J. Pharmacol.* **430**, 33-40, (2001b).

53. Ghadge, G., Slusher, B.S., Bodner, A., Del Canto, M., Wozniak, K., Thomas, A., Rojas, C., Tsukamoto, T., Majer, P., Miller, R., Monti, A. and Roos, R. Glutamate carboxypeptidase II inhibition protects motor neurons from death in familial amyotrophic lateral sclerosis models. *PNAS* **100**(16): 9554-9559, (2003).

54. Nan, F., Bzdega, T., Pshenichkin, S., Wroblewski, J.T., Wroblewska, B., Neale, J.H. Dual function glutamate-related ligands: discovery of a novel, potent inhibitor of glutamate carboxypeptidase II possessing mGluR3 agonist activity. *J. Med. Chem.* **43**, 772-774, (2000).

55. B. Slusher (personal communication).

56. Wozniak, K., Callizot, N., Poindron, P. and Slusher, B. NAALADase inhibition enhances behavioral and morphological recovery following sciatic nerve crush in mice. *Neurosci. Abs.,* **26**, 111, (2000).

57. Faden, A.I., Jacobs, T.P., Dynorphin-related peptides cause motor dysfunction in the rat through a non-opiate mechanism. *Br. J. Pharmacol.* **81**, 271-276, 1984.

58. Long, J.B., R.C. Kinney, D.S. Malcolm, Graeber, G.S., Holaday, J.W., Intrathecal dynorphin A (1-13) and dynorphin A (3-13) reduce rat spinal cord blood flow by non-opioid mechanisms. *Brain Res.* **436**, 374-379, (1987).

59. Long, J.B., J.M. Petras, W.C. Mobley, Holaday, J.W., Neurological dysfunction following intrathecal injection of dynorphin A (1-13) in the rat: II. non-opioid mechanisms mediate loss of motor, sensory, and autonomic function. *J. Pharmacol. Exp. Ther.* **246**, 1167-1174, (1988)

60. Long, J.B., Spinal subarachnoid injection of somatostatin causes neurological deficits and neuronal injury in rats. *Eur. J. Pharmacol.* **149**, 287-296, (1988).

61. Long, J.B., Rigamonti, D.D., Dosaka, K., Kraimer, J.M., Martinez-Arizala, A., Somatostatin causes vasoconstriction, reduces blood flow and increases vascular permeability in the rat central nervous system. *J. Pharmacol. Exp. Ther.* **260**, 1425-1432, (1992).

62. Long, J.B., D.D. Rigamonti, M.A. Oleshansky, C.P. Wingfield, Martinez-Arizala, A., Dynorphin A-induced rat spinal cord injury: evidence for excitatory amino acid involvement in a pharmacological model of ischemic spinal cord injury. *J. Pharmacol. Exp. Ther.* **269**, 358-366, (1994a).

63. Long, J.B., Skolnick, P.P., 1-aminocyclopropanecarboxylic acid protects against dynorphin A-induced spinal injury. *Eur. J. Pharmacol.* **261**, 295-301, (1994b).

64. Bakshi, R., Faden, A.I., Competitive and non-competitive NMDA antagonists limit dynorphin A-induced rat hindlimb paralysis. *Brain Res.* **507**, 1-5, (1990a).

65. Bakshi, R., Faden, A.I., Blockade of the glycine modulatory site of the NMDA receptors modifies dynorphin-induced behavioral effects. *Neurosci. Lett.* **110**, 113-117, (1990b).

66. Caudle, R.M. Isaac, L., A novel interaction between dynorphin (1-13) and an N-methyl-D-aspartate site. *Brain Res.* **443**, 329-332, (1988).

67. Robinson M.B., Djali S., and Buchhalter J. R., Inhibition of glutamate uptake with L-trans-pyrrolidine-2,4-dicarboxylate potentiates glutamate toxicity in primary hippocampal cultures. *J. Neurochem.* **61**, 2099-2103, (1993).

68. Long, J.B. Meyerhoff J.L. and Slusher B.S., NAALADase inhibition protects against dynorphin A-induced ischemic spinal cord injury in rats. *Neurosci. Abs.* **23**:2301, (1997).

69. Long, J.B., Yourick, D.L., Slusher, B.S., Robinson, M.B., and Meyerhoff, J.L. Inhibition of glutamate carboxypeptidase II (NAALADase) protects against dynorphin A-induced ischemic spinal cord injury in rats. *Eur. J. Pharmacol.* (in press, 2005).

70. Yourick, D.L., Koenig, M.L., Durden, A.V. and Long, J.B. N-Acetylaspartylglutamate and beta-NAAG protect against injury induced by NMDA and hypoxia in primary spinal cord cultures. *Brain Res.* 99156-64, (2003).

71. Long, J.B., Yourick, D.L and Meyerhoff, J.L., N-Acetylaspartylglutamate (NAAG) protects against dynorphin A-induced ischemic spinal cord injury in rats. *Neurosci. Abs.* **24**:463, (1998).

72. Lea, P.M., Wroblewska, B., Sarvey, J.M. and Neale, J.H. beta-NAAG rescues LTP from blockade by NAAG in rat dentate gyrus via the type 3 metabotropic glutamate receptor. *J. Neurophysiol.* **85,**1097-1106, (2001).

73. Koller KJ, Coyle JT. The characterization of the specific binding of [3H]-N-acetylaspartylglutamate to rat brain membranes, *J Neurosci.* **5**(11), 2882-8, (1985).

74. Losi, G., Vicini, S. and Neale, J. NAAG fails to antagonize synaptic and extrasynaptic NMDA receptors in cerebellar granule cells. Neuropharmacology. 46:490-496, 2004.

75. Bergeron, R., Coyle, J.T., Tsai, G. and Greene, R.W. NAAG reduces NMDA receptor current in CA1 hippocampal pyramidal neurons of acute slices and dissociated neurons. *Neuropsycholopharm.* 1-10, (2004).

76. Wroblewska, B., Santi, M.R., Neale, J.H., N-Acetylaspartylglutamate activates cyclic-AMP coupled metabotropic glutamate receptors in cerebellar cells. *Glia* **24**, 172-180, (1998).

77. Schoepp, D.D., Conn, P.J., Metabotropic glutamate receptors in brain function and pathology. *Trends Pharmacol. Sci.* **14**, 13-20, (1993).

78. Buisson, A. Choi, D.W., The inhibitory mGluR agonist, s-4-carboxy-3-hydroxy-phenylglycine selectively attenuates NMDA neurotoxicity and oxygen-glucose deprivation-induced neuronal death. *Neuropharmacol.* **34**, 1081-1087, (1995).

79. Sanchez-Prieto, J., Budd, D.C., Herrero, I., Vazquez, E., Nicholls, D.G., Presynaptic receptors and the control of glutamate exocytosis. *Trends Neurosci.* **19**, 235-239, (1996).

N-ACETYLASPARTYLGLUTAMATE (NAAG) IN PELIZAEUS-MERZBACHER DISEASE

Alessandro P. Burlina, Vanni Ferrari, Alberto B. Burlina, Mario Ermani, Odile Boespflug-Tanguy, Enrico Bertini, and the Clinical European Network on Brain Dysmyelinating Disease (ENBDD)[*]

1. INTRODUCTION

Pelizaeus-Merzbacher disease (PMD) is a rare X-linked leukodystrophy caused by mutations in the proteolipid protein 1 gene (*PLP1*). Neuropathological and MRI studies show diffuse and symmetrical hypomyelination in brain (see for review Nave and Boespflug-Tanguy,[1] Hudson and colleagues[2]). *PLP1* maps to the long arm of the chromosome at the Xq22 region and encodes the two major myelin proteins, PLP and its isoform DM20. All types of mutation are encountered, but duplications are the most frequent causative mutations.[2] Clinical features of PMD include impairment of psychomotor development, nystagmus, ataxia, dystonia, and spasticity.[2]

Despite of the tremendous progress in understanding the molecular basis of PMD, knowledge about the neurochemical mechanisms underlying the dysmyelination process remains incomplete. Evidence for neuronal dysfunction, a significant and widespread decreased of brain *N*-acetylaspartate (NAA), has been reported in patients with known PLP gene mutations.[3] Recently, Takanashi *et al.* (2002)[4] reported increased concentrations of total NAA in the brain of five patients with PLP duplications using

[*] Alessandro P. Burlina, Mario Ermani, Department of Neuroscience, Neurological Clinic, University Hospital, Via Giustiniani 5, I-35128, Padova, E-mail: alessandro.burlina@unipd.it, Vanni Ferrari, Department of Veterinary Medicine, University of Padova, ITALY; Alberto B. Burlina, Department of Pediatrics, Metabolic Unit, University Hospital, University of Padova, ITALY; Odile Boespflug-Tanguy, INSERM U 384, University of Clermont-Ferrand, Clermont-Ferrand, FRANCE; Enrico Bertini, Unit of Molecular Medicine, Department of Laboratory Medicine, Bambino Gesù Hospital, Rome, ITALY; Clinical ENBDD (European Network on Brain Dysmyelinating Disease): M. Baethmann, J. Gärtner, F. Hanefeld, A. Kohlschutter, J.-M. Lopez-Terradas, S. Peudenier, J.M. Prats-Vinas, D. Rodriguez, R. Surtees, G. Uziel, L.Vallee, and T. Voit.

proton magnetic resonance spectroscopy (MRS).[4] Therefore, *in vivo* MRS data suggest an involvement of the NAA biochemical pathway in the pathophysiology of PMD.

N-acetylaspartylglutamate (NAAG) is one of the most abundant dipeptide in the brain and has a neuronal localization.[5] However, in vivo proton MRS demonstrated that NAAG concentrations in human brain are higher in white matter (1.5-2.7 mM) than those found in gray matter (0.6-1.5 mM).[6] Rat cell cultures have shown that NAAG can be synthesized from NAA and glutamate in astrocytes,[7] and is catabolized to NAA and glutamate by two NAAG peptidases, also known as glutamate carboxypeptidase II and III (GCPII and GCPIII).[8,9] Our study shows that N-acetylaspartylglutamate is increased in the CSF of PMD patients.

2. PATIENTS AND METHODS

We analysed 32 CSF samples from patients with clinically defined diagnosis of PMD, with a confirmed *PLP* mutation and without confirmed mutations (PMLD, Pelizaeus-Merzbacher like Disease). Control CSF samples were collected after consent from 31 hospitalized patients, aged 1-10 years, free from dysmyelinating or metabolic disorders, in whom examination of CSF was indicated for other reasons. CSF was also collected from the following groups of patients:

- ULD group: patients with unclassified leukodystrophies (ULD);
- OLD group: patients with other leukodystrophies (OLD), including one patient with metachromatic leukodystrophy (MLD), 5 patients with vanishing white matter disease (VWMD), one patient with Alexander disease; one patient with phenylketonuria (PKU) and secondary demyelination;
- CD group: patients with Canavan disease (CD).

The amount of NAAG and NAA in the CSF was measured using two capillary electrophoresis systems: Beckman P/ACE 2100 and a Beckman Coulter P/ACE MDQ, according to Burlina *et al.*[10] Our detection limits for NAA and NAAG were 1.14 μmol/L and 0.66 μmol/L respectively, at a signal to noise ratio of approximately 10.

2.1. Statistical Analysis

Data were analyzed by Kruskall-Wallis test and Mann-Whitney U test. The probability level accepted for significance was $p < 0.05$.

3. RESULTS

High concentrations of NAAG were detected in 29 CSF samples out of 32 samples of patients with a clinical presentation of PMD. NAA was elevated in three CSF samples.

Figure 1 shows the NAAG concentrations in CSF of PMD patients compared with controls (CTRL), OLD, CD, and ULD groups. The mean NAAG concentration in the CSF of PMD patients was 43.45 ± 40.63 (median value = 32.99, range 0-203.58). It differed significantly from the control group ($p < 0.00001$), from the OLD group ($p < 0.00001$), the ULD group ($p < 0.00009$), and the CD patients ($p < 0.008$) (see table 1).

NAAG was normal in CSF of all patients, except three, with unclassified leukodystrofies. NAAG was also normal in the CSF of the patients with Alexander disease, VWMD, MLD, and PKU. Conversely, NAAG was elevated in the CSF of the CD patients together with elevation of NAA because of the aspartoacylase deficiency, as we already reported[10]. NAA was in the normal range in the CSF samples of ULD patients, except four cases.

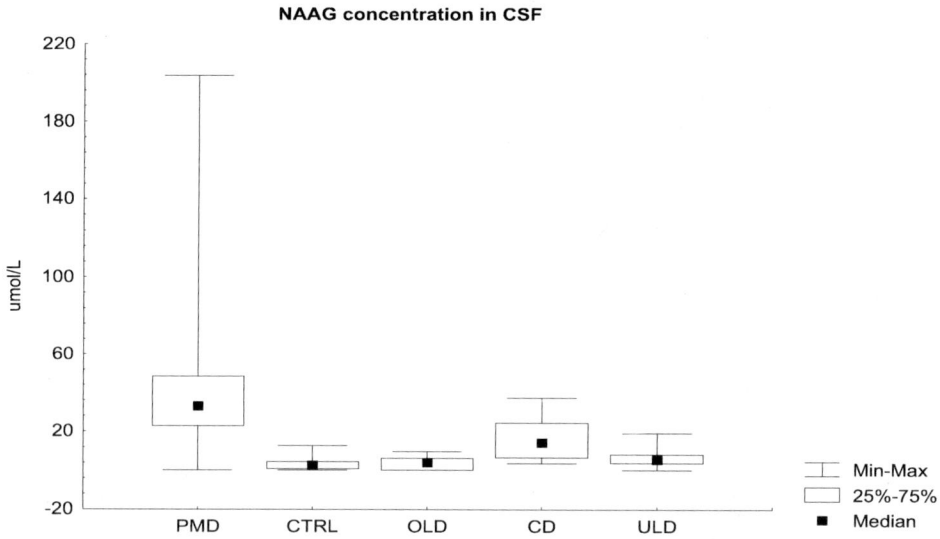

Figure 1. The box plot graph shows the concentration of N-acetylaspartylglutamate (NAAG), expressed in μmol/L, in the CSF of all patients studied. PMD group (including PMLD patients), n=32; CTRL group, n=31; OLD group, n=8; CD group, n=8; ULD group, n=32. PMD = Pelizaeus-Merzbacher disease; PMLD = Pelizaeus-Merzbacher like disease; CTRL = Controls; OLD = Other Leukodystrophies; CD = Canavan disease; ULD = Unclassified Leukodystrophies.

Table 1. NAAG concentrations in the CSF of each group of patients. Mean ± standard deviation, median value, and range are represented

	Mean ± SD	Median	Range (min – max)
PMD (n=32)	43.45 ± 40.63	32.99	0.0 – 203.58
CTRL (n=31)	3.42 ± 3.31	2.59	0.0 – 12.68
OLD (n=8)	3.73 ± 3.61	3.93	0.0 – 9.65
CD (n=8)	16.37 ± 12.02	14.16	3.55 – 37.30
ULD (n=32)	6.44 ± 4.41	5.42	0.0 – 19.03

PMD group includes PMLD patients. Values are expressed in μmol/L.
PMD = Pelizaeus-Merzbacher disease; PMLD Pelizaeus-Merzbacher like disease;
CTRL = Controls; OLD = Other Leukodystrophies; CD = Canavan disease;
ULD = Unclassified Leukodystrophies.

4. DISCUSSION

We detected high concentrations of NAAG in the CSF of 22 patients with PMD and mutations in the *PLP1* gene. Furthermore, we found high levels of NAAG in the CSF of 7 patients with a dysmyelinating disorder with a clinical presentation of PMD, but without mutations in the *PLP1* gene (PMLD). Conversely, the CSF NAA was in the normal range in PMD and PMLD patients (except for two PMD samples and one PMLD sample with increased NAA content). These results confirm our preliminary data (Burlina *et al.* J. Inherit. Metab. Dis. 2000; 23 (suppl 1: 213). Moreover, NAAG was not increased in other leukodystrophies that were analyzed such as VWMD, Alexander disease and other leukodystrophy of known origin.

We demonstrated that NAAG is significantly and consistently increased in CSF of patients with genetically confirmed PLP disorders but also in patients with an undefined genetically dysmyelinating disorder (PMLD). Thus we have evidence that increased NAAG is a common marker in dysmyelinating leukodystrophies with or without confirmed PLP mutations.

An involvement of the NAA/NAAG cycle has been described only in few disorders. Indeed, in 1990 Rothstein reported elevated levels of both NAA and NAAG in the CSF of patients with amyotrophic lateral sclerosis.[11] High concentrations of NAA and NAAG have been detected in urines of CD patients and in the CSF of one patient with Canavan disease, a leukodystrophy due to aspartoacylase deficiency leading to accumulation of NAA.[10] We suppose that in Canavan disease the accumulation of NAA lowers the threshold of NAAG synthesis and may inhibit NAAG hydrolysis, leading to a paralleled increased of NAAG and later on a decrease of both NAA and NAAG in the long-term of the disease, as we have shown.[12] Recently, Wolf and co-workers described two unrelated girls with severe brain hypomyelination and increased levels of NAAG in the CSF.[13]

NAAG is the most abundant dipeptide neurotransmitter in the mammalian CNS.[14] It exhibits mixed agonist/antagonist properties at the *N*-Methyl-D-Aspartate (NMDA) receptor. Indeed, NAAG renders NMDA less effective in promoting cerebellar granule cells survival during differentiation, and also raises the threshold of NMDA toxicity to the differentiated granule cells.[15] Moreover, NAAG acts as an agonist at metabotropic glutamate receptors group 2 (mGluR3).[14]

Recently, it has been shown that metabotropic glutamate receptors (mGluRs) are involved in regulating oligodendrocyte excitotoxicity. Deng and co-workers demonstrated that group 1 mGluRs mitigates excitotoxicity damage to oligodendrocyte precursor cells.[16] Oligodendrocyte precursor cells share with neurons a high vulnerability to glutamatergic excitotoxicity. Furthermore, a secondary role for NAAG could be the release of glutamate after NAAG hydrolysis.

But NAAG is a "complex" molecule with multiple actions. NAAG can act as a neurotoxin, and also as a neuroprotective agent, depending on the activity of its peptidase.[17] Ghadge *et al.* have demonstrated that the inhibition of GCPII, the dipeptidase which hydrolyzes NAAG to NAA and glutamate, prevented motor neurons death in a cell culture model of amyotrophic lateral sclerosis.[18] The authors suggested that increasing NAAG levels, by blocking NAAG hydrolysis may be neuroprotective by activating mGluR3 receptors.[18]

It has also been shown that in PMD patients axonal dysfunction and/or injury can be detected by proton MRS and in the mouse model of PMD axonal swelling and degeneration are present.[3,19] Therefore, the increase levels of NAAG in CSF may be expression of an augmented release of *N*-acetylaspartylglutamate after neuronal damage.

Because of its prominent signal, always detectable in human healthy brain by in vivo proton MRS, NAA is generally considered an important marker of viable functions in neurons. Indeed, many neurodegenerative disorders with either loss of neural cells or neuronal dysfunction exhibit a decreased signal of NAA in the proton MR spectra.[20] In three PMD patients we detected elevated levels of NAA in the CSF. Therefore, we cannot confirm that NAA was increased in all our PMD patients and this does not fully correlate with the increased concentrations of total NAA in the brain of five patients with PLP duplications, as detected by proton magnetic resonance spectroscopy.[4] The involvement of the metabolic pathway of NAA and NAAG has been recently confirmed by the report of a new severe neurological disorder. Indeed, Martin and colleagues (2001) reported a child with neurodevelopmental retardation and moderately delayed myelination. In vivo proton MRS of the brain did not detect an NAA signal (hypoacetylaspartia).[21] We analysed the CSF of the patient and NAAG was not detectable, indicating that in the case of hypoacetylaspartia both NAA and NAAG can be decreased (see Burlina, Schmitt et al., this volume).[22]

In conclusion, our findings show that NAAG may be a useful biochemical marker in the diagnosis of Pelizaeus-Merzbacher disease and may help to understand the myelination process.

5. ACKNOWLEDGEMENTS:

This study was supported by a grant from the European Union's Biomed 2 Programme (PL 95 1405), European Network on Brain Dys-myelinating Disease: Diagnostic, Molecular and Therapeutic Approach (ENBDD).

6. QUESTION AND ANSWER SESSION

DR. COYLE: Dr. Burlina's paper is open for comment. Yes?

DR. ROSS: So what is your working hypothesis for the elevation in NAAG in Palizaeus-Merzbacher Disease?

DR. BURLINA: I am starting to think that NAAG is really a complex molecule, because we heard from Dr. Wroblewska and also from the other speakers that it has a neuroprotective effect.

It, to me, does seem that is not the only effect. As I have shown, NAAG is high in the CSF of PMD patients, and from our preliminary data it appears that NAAG levels correlate to the severity of disease.

At the moment, I don't have an exact explanation for the elevation of NAAG in the CSF of PMD patients. I know that the functions of PLP are not completely understood. Maybe there is an interaction between PLP and the hydrolytic NAAG enzymes. In some way, NAAG seems to be involved in the dysmyelination process.

A glutamatergic action can be hypothesized in the case of PMD pathogenesis, and, as is well known, developing oligodendrocytes are highly vulnerable to excitotoxicity. Therefore, the presence of elevated levels of NAAG could be relevant for the disease.

I think we need to work more on the animal model of PMD, the jimpy mouse. Interestingly, we detected increased levels of NAAG in the urine of jimpy mice. The results are preliminary, but very encouraging.

DR. COYLE: Regarding the possibility of a second international NAA meeting, I can only speak for myself and not for the entire NAAG and NAA community, but Padua sounds like a fantastic place to have a follow-up second international meeting. Thank you. Danny?

DR. WEINBERGER: I just had a sort of an additional clinical question. Any relationship between NAAG levels in CSF or urine and age of onset, clinical manifestations, anything like that?

DR. BURLINA: As I told you before, it seems that we have some relationship with NAAG in the CSF and the clinical severity of the disease, no clear relationship with the age has appeared yet. Concerning urine data in PMD patients, we detected an increased amount of NAAG, but not in all patients; anyway, analyses are still in progress. I just remind you that few years ago we reported an increased amount of NAAG also in the urine of Canavan patients, in which, of course, NAA is elevated too.

I should mention that when I received the samples I didn't know the genetic result, or the clinical diagnosis, except that it was a leukodystrophy. Therefore, I didn't know if I had a PMD CSF sample, or another kind of leukodystrophy (i.e. Alexander disease, vanishing white matter disease, etc.). In some cases, the NAAG elevation strongly suggested a diagnosis of PMD before the genetic analysis of PLP mutations could be completed.

DR. COYLE: Thank you very much.

7. REFERENCES

1. K.A. Nave and O. Boespflug-Tanguy, X-linked developmental defects of myelination: from mouse mutants to human genetic diseases, *Neuroscientist* **2**, 33-43, (1996).
2. L.D. Hudson, J.Y. Garbern and J.A. Kamholz, Pelizaeus-Merzbacher disease, in: *Myelin Biology and Disorders*, vol. II, edited by R.A. Lazzarini (Elsevier, London), pp. 867-885, (2004)
3. S. Bonavita, R. Schiffmann, D.F. Moore, K. Frei, B. Choi, N. Patronas, A. Virta, O. Boespflug-Tanguy, and G. Tedeschi, Evidence for neuroaxonal injury in patients with proteolipid protein gene mutations, *Neurology* **56**, 785-788 (2001).
4. J. Takanashi, K. Inoue, M. Tomita, A. Kurihara, F. Morita, H. Ikehira, S. Tanada, E. Yoshitome, and Y. Kohno, Brain N-acetylaspartate is elevated in Pelizaeus-Merzbacher disease with *PLP1* duplication, *Neurology* **58**, 237-241 (2002).
5. J.T. Coyle, The nagging question of the function of N-acetylaspartylglutamate. *Neurobiol. Dis.* **4**, 231-238, (1997).
6. P.J.W. Pouwels and J. Frahm, Differential distribution of NAA and NAAG in human brain as determined by quantitative localized proton MRS, *NMR Biomed.* **10**, 73-78, (1997).
7. L.M. Gehl, O.H. Saab, Y. Bzdega, B. Wroblewska, and J.H. Neale, Biosynthesis of NAAG by an enzyme-mediated process in rat central nervous system neurons and glia, *J. Neurochem.* **90**, 989-997, (2004).
8. B. Stauch Slusher, M.B. Robinson, G. Tsai, M.L. Simmons, S.S. Richards, and JT Coyle. Rat brain N-acetylated α-linked acidic dipeptidase activity. Purification and immunological characterization, *J. Biol. Chem.* **265**, 21297-21301, (1990).
9. T. Bzdega, S.L. Crowe, E.R. Ramadan, K.H. Sciarretta, R.T. Olszewski, O.A. Ojeifo, V.A. Rafalski, B. Wroblewska, and J.H. Neale, The cloning and characterization of a second brain enzyme with NAAG peptidase activity, *J. Neurochem.* **89**, 627-635, (2004).
10. A.P. Burlina, V. Ferrari, P. Divry, W. Gradowska, C. Jakobs, M.J. Bennett, A.C. Sewell, C. Dionisi-Vici, and A.B. Burlina, N-acetylaspartylglutamate in Canavan disease: an adverse effector? *Eur. J. Pediatr.* **158**, 406-409, (1999).
11. J.D. Rothstein, G. Tsai, R.W. Kuncl RW, L. Clawson, D.R. Cornblath, D.B. Drachman, A. Pestronk, B.L. Stauch, and J.T. Coyle, Abnormal excitatory amino acid metabolism in amyotrophic lateral sclerosis, *Ann. Neurol* . **28**, 18-25, (1990).
12. A.P. Burlina, V. Ferrari, L. Bonafé, and A.B. Burlina, Clinical and biochemical follow-up of a patient with Canavan disease, *J. Inherit. Metab. Dis.* **21**(suppl. 2), 48, (1998).

13. N.I. Wolf, M.A.A.P. Willemsen, U.F. Engelke, M.S. van der Knaap, P.J.W. Pouwels, I. Harting, J. Zschocke, E.A. Sistermans, D. Rating, and R.A. Wevers, Severe hypomyelination associated with increased levels of N-acetylaspartylglutamate in CSF, *Neurology* **62**, 1503-1508, (2004).

14. J.H. Neale, T. Bzdega, and B. Wroblewska. N-acetylaspartylglutamate: the most abundant peptide neurotransmitter in the mammalian central nervous system, *J. Neurochem.* **75**, 443-452, (2000).

15. A.P. Burlina, S.D. Skaper, M.R. Mazza, V. Ferrari, A. Leon, and A.B. Burlina, N-acetylaspartylglutamate selectively inhibits neuronal responses to N-methyl-D-aspartic acid in vitro. *J. Neurochem.* **63**, 1174-1177, (1994).

16. W. Deng, H. Wang, P.A. Rosenberg, J.J. Volpe, and F.E. Jensen, Role of metabotropic glutamate receptors in oligodendrocyte excitotoxicity and oxidative stress, *Proc. Natl. Acad. Sci.* **101**, 7751-7756, (2004).

17. A.G. Thomas, J.J. Vornov, J.L. Olkowski, A.T. Merion and B.S. Slusher, N-acetylated α-linked acidic dipeptidase converts N-acetylaspartylglutamate from a neuroprotectant to a neurotoxin. *J. Pharmacol. Exp. Therap.* **295**, 16-22, (2000).

18. G.D. Ghadge, B.S. Slusher, A. Bodner, M. Dal Canto, K. Wozniak, A.G. Thomas, C. Rojas, T. Tsukamoto, P. Majer, R.J. Miller, A.L. Monti and R.P. Roos, Glutamate carboxypeptidase II inhibition protects motor neurons from death in familial amyotrophic lateral sclerosis, *Proc. Natl. Acad. Sci.* **100**, 9554-9559, (2004).

19. I. Griffiths, M. Klugmann, T. Anderson , D. Yool, C. Thomson, M.H. Schwab, A. Schneider, F. Zimmermann, M. McCulloch, N. Nadon, and K.A. Nave, Axonal swellings and degeneration in mice lacking the major proteolipid of myelin, *Science* **280**, 1610-1613, (1998).

20. A.P. Burlina, T. Aureli, F. Bracco, F. Conti, and L. Battistin, MR spectroscopy: a powerful tool for investigating brain function and neurological diseases, *Neurochem. Res.* **25**, 1365-1372, (2000).

21. E. Martin, A. Capone, J. Schneider, J. Hennig, and T. Thiel, Absence of N-acetylaspartate in the human brain: impact on neurospectroscopy? *Ann. Neurol.* **49**, 518-521, (2001).

22. E. Boltshauser, B. Schmitt, R.A. Wevers, U. Engelke, A.B. Burlina, and A.P. Burlina, Follow-up of a child with hypoacetylaspartia. *Neuropediatrics* **35**, 255-258, (2004).

CONCLUDING REMARKS

DR. COYLE: Thank you. It is hard to believe, but in two days we are only 12 minutes late. So we are going to skip the break and go to Danny Weinberger.

I can't think of a more difficult task than summing up a meeting. It requires incredible intellectual dexterity, and I can't think of a better person to do this than Dan Weinberger, who is going to provide the concluding remarks.

DR. WEINBERGER: The concluding remarks are going to be very brief. But first I want to make a few thank-yous. Firstly, I think we all -- Joe said it earlier today, but we all owe Dr. Namboodiri an enormous debt of gratitude. He has brought together perhaps 50 percent of the people who have toiled over NAA for some time, and I think we all owe you an enormous debt of gratitude for being so persistent on this. Give him a round of applause.

(Applause.)

I am not sure that people actually even appreciate how much work this was, because the NIH did not make it easy for Namboo to get this done.

The other person who I have to acknowledge and thank is Shari Thompson, the Administrator for our program, who also carried an enormous burden on getting this done. So, Shari, thank you for persisting. We appreciate it enormously.

So let me just make a few final comments. This has been an enormously informative and educational two days; it has been 50 years since the discovery of NAA to have the first international conference. Padua in two years sounds like a great idea, though I am not completely convinced we are going to be ready in two years, but whenever the number two international symposium is, that would be a great venue. I hope it is not 50 years until the second one, and I hope we can see between now and the next international symposium at least as many publications on this subject as appeared in the first 50 years. We should be able to equal that, I think.

One of the things that is very clear is that we still haven't nailed the molecular, cellular, and neurobiology of NAA and NAAG. The life cycle of NAA still seems to me quite unclear. We seem pretty confident that it is synthesized in neurons. We seem pretty confident that it is hydrolyzed in oligodendrocytes, but is it only hydrolyzed there? Is it possibly hydrolyzed extracellularly? Why is it being taken up by astrocytes, if it is not metabolized by astrocytes? I think there is still some uncertainty that the astrocyte carrier transport mechanisms really exist *in vivo*, and how this is done. It is not clear how it is getting into oligodendrocytes and how it is getting back into neurons.

What is also very mysterious about NAA are the cellular processes that contribute to variations in NAA concentrations. Obviously, NAA is a very abundant molecule which has made it easy to measure, but this also has probably made some people less enthusiastic about its biological relevance, because there seems to be a huge buffering of NAA concentrations.

So what are the molecular processes that regulate NAA concentrations in cells? I guess this comes down to one of the issues in using NAA as a measure of cellular metabolic function and dysfunction, which is: Are there just diverse molecular processes that affect NAA concentrations, and there could be many, many different pathways to changes in NAA levels, or is ultimately NAA monitoring some final common functional pathway that is affected in conditions where NAA concentrations vary?

There have been a number of proposals of what this might be – for example, oxidative phosphorylation, mitochondrial cycling, tissue glutamate cycling, and ultmately synaptic activity and maybe the abundance of neuronal processes.

This general question seems to reduce to whether the many situations where NAA concentrations are affected in tissue, whether this be in animal tissue, or in human tissue, are the changes in NAA concentration because of some selective metabolic derangement in NAA metabolism or NAA trafficking across neuronal or glial cells, or is this some more general integrative manifestation of changes in mitochondrial cycling and phosphorylation that ultimately is changed because -- I mean, I think in Canavan's Disease there is obviously a derangement of the metabolic pathway of NAA, but in all these other conditions where NAA is implicated, is there something specific or is this really just a sort of nonspecific manifestation of what is ultimately the neuropathological change of all neurological disorders, which is a change in the synaptic activity because of a change in excitatory inputs to spines and changes in synaptic expression, etcetera?

There are many issues in the *in vivo* measurements in living human beings of NAA concentrations. It is clear that the technology for making these measurements has improved dramatically, but there is still a long way to go on this. I think a lot of the issues that come up around NAA measurements in human brain *in vivo* still are unresolved about improving signal to noise, improving resolution, improving sensitivity, imaging time.

So there is a tremendous amount of work to be done, still at the basic cellular level of understanding NAA trafficking and variations in its levels, and then this gets even more complicated at the clinical level. But at the end of the day, it seems impossible to dismiss the conclusion that measuring NAA does reflect something of the biological state of the tissue, and that this does have some predictive clinical value in a number of neurological conditions.

The interesting thing to me about NAAG is that it seems much more approachable, because it is fundamentally a ligand. While there are all the same problems in understanding the metabolic cycle of NAAG and, obviously, the relationship of NAAG to NAA and glutamate processing is complicated -- this does have ligand characteristics at the GRM3, in the genetic parlance mGluR3, receptor and also the NMDA receptor. So this would seem to make it much more approachable, and it obviously has value in some of these neuroprotective model systems. But again, it seems also that there are many, many difficult questions to answer about its biological role *in vivo*.

One of the other things to emerge from this conference is that we have new tools, particularly at the genetic level, to begin to explore these questions, because maybe one of the ways to look at NAAG in the future is not only to try to answer some of these unresolved questions about the life cycle of NAAG and how it is processed and how it contributes presynaptically, in glia, and postsynaptically, to glutamate synaptic activity, and also possibly to heteroreceptor functions of mGluR3, but also to manipulate genetically some of the signaling systems that NAAG speaks to, particularly GRM3. There will be ways not just to knock it out, but to manipulate the efficacy of GRM3 signaling and to look at maybe how

NAAG will affect that.

For my money, this is a thriving field, complicated but thriving, and I think that at the second international symposium we will see some real progress here beyond the enormous progress that has been made in the last few years.

I look forward to whatever is left of this discussion. Thank you.

(Applause.)

DR. LIEBERMAN: What is interesting about it: If, in fact, it is hydrolyzed by acylase in the extracellular space, that aspartate is a wonderful NMDA receptor agonist. So it may have some role there as well.

DR. NAMBOODIRI: Any other comments?

DR. COYLE: Thank you, Namboo. Thank you all.

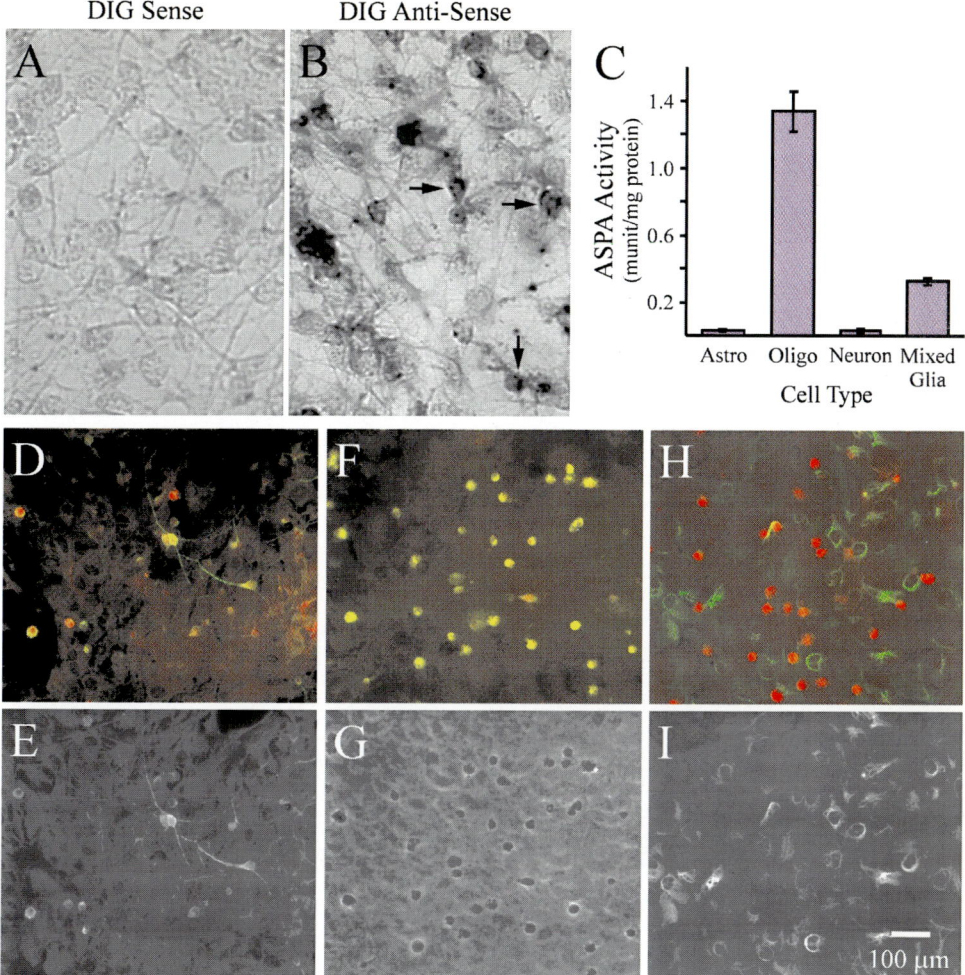

Figure 1. Characterization of primary cultures of neural cells from newborn rat cortex for expression of ASPA mRNA, enzyme activity and co-immunostaining of ASPA protein with neural cells markers. The in situ detection of ASPA mRNA in oligodendrocytes cells in cultures mixed glia probed with sense (A) and anti-sense (B) ASPA digoxigenin labeled probes. The oligodendrocyte cells show dark staining (arrows) on the bed layer of unstained astrocytes. Optimal ASPA enzyme activity (mU/mg protein) can be seen in culture of oligodendrocytes (C) and about 29% ASPA activity can also be seen in mixed glial cultures. The pure cultures of astrocytes and cortical neurons exhibit insignificant level of ASPA activity. Immunostaining of mixed glial cultures from normal rat brain shows a co-localization of ASPA (red) with oligodendrocyte cell specific markers A_2B_5 (D) and CNP (F), both green, or exhibiting a yellow staining for co-expression. The immunostaining of ASPA with GFAP (green), an astrocyte cell specific marker shows a discrete localization of the two markers (H). (epifluorescence images D, F, H; Phase image E, G, I).

1

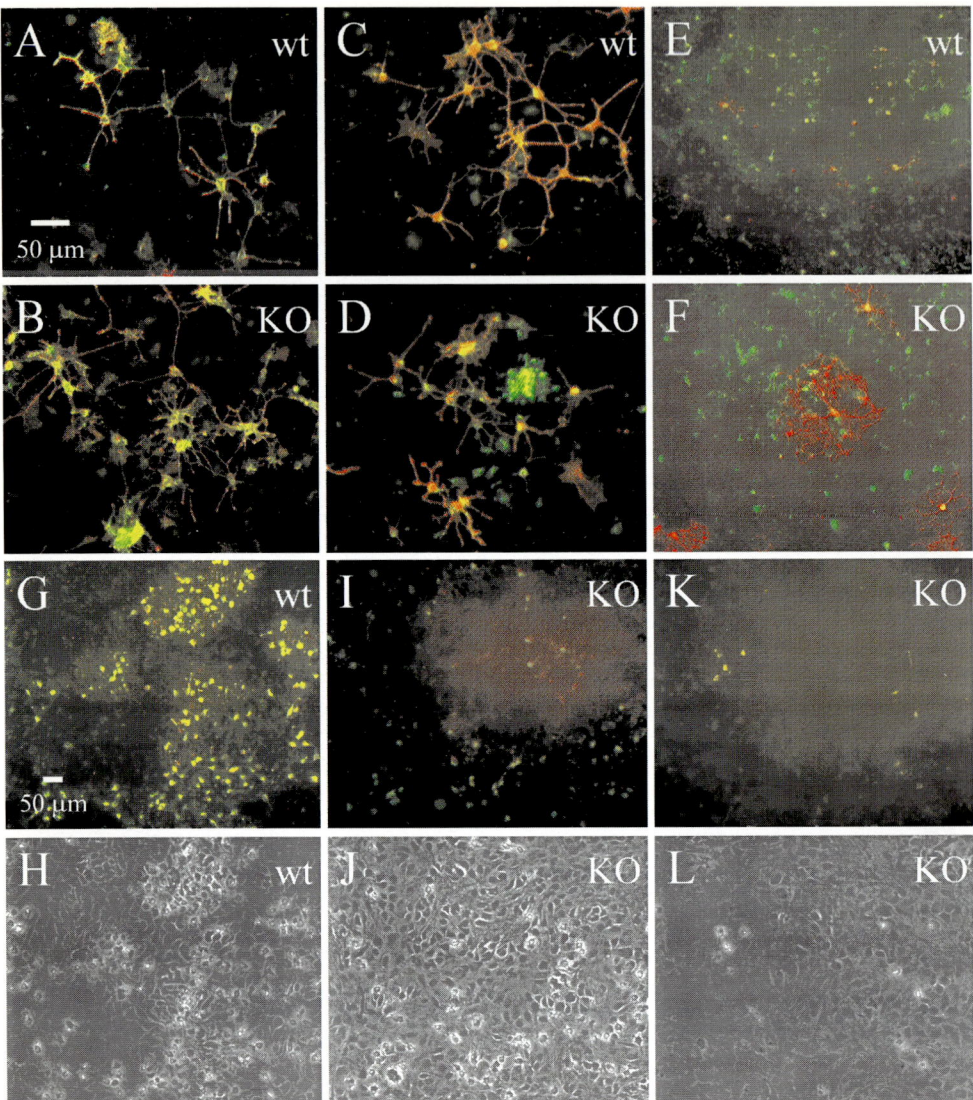

Figure 2. Immunocytochemistry of mixed glia cultures from ASPA wt and KO mice cerebral cortex. The immunostaining of markers for progenitor and mature oligodendrocytes were carried out in 6 and 15d-old cultures. The immunostaining for marker NG2 (green) and GD3 (Red; A and B), A_2B_5 (red) and GPDH (green; C and D), and O4 (red; E and F) showed the status of oligodendrocytes, while GFAP (green; E and F), the marker for astrocyte shows no co-localization with oligodendrocyte marker O4. A delay in MBP (green) and PLP (red) expressing oligodendrocyte cells was detected in 15day old ASPA KO mice cortical cultures (G, I and K). The double immunostaining of MBP astrocyte, showed a discrete staining for ASPA (red) and GFAP (green) in mixed glial cultures (Fig. 1G and panel H as phase image).

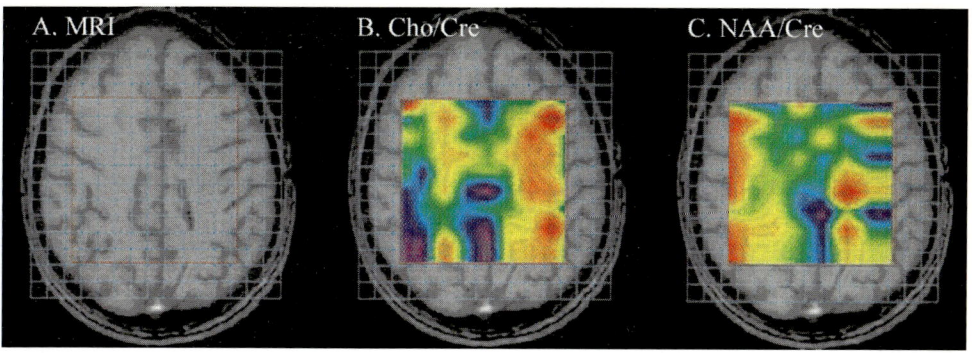

Figure 2. Spectroscopic Imaging (SI) of child with TBI maximally affecting the left cerebral hemisphere.

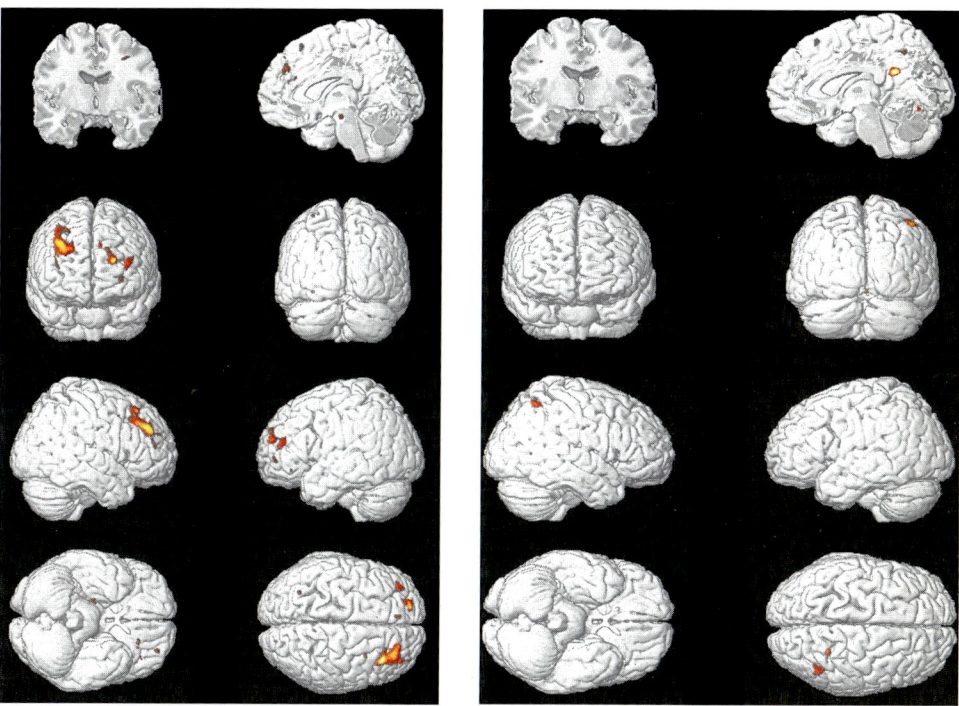

Figure 3 (left panel). Voxel Based Morphometry (VBM) analysis of GM cluster volumes covarying with posterior left hemisphere white matter NAA concentration.

Figure 4 (right panel). Voxel Based Morphometry (VBM) analysis of GM cluster volumes covarying with left hemisphere frontal white matter NAA concentration.

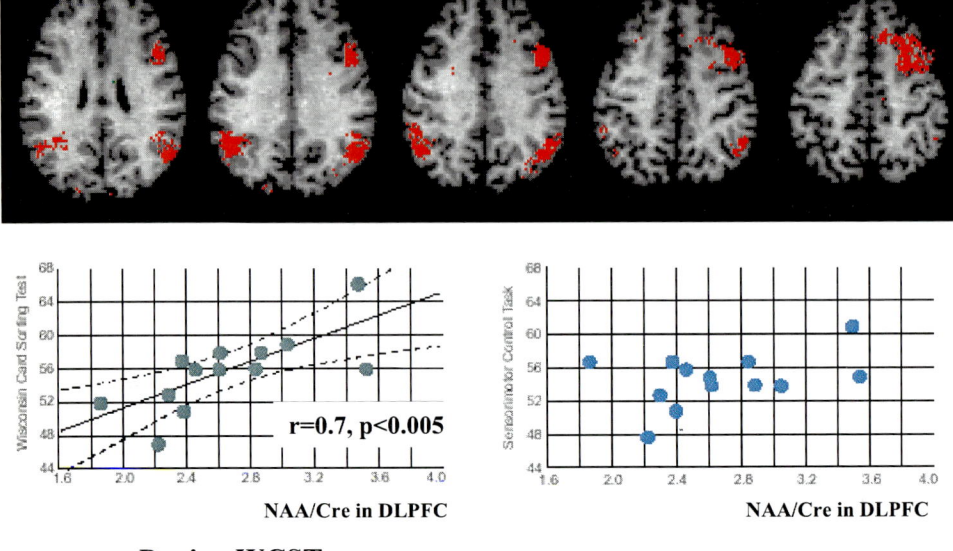

During WCST   **During sensorimotor control**

Figure 3. Relationship of NAA/Cre in the DLPFC to blood flow measured with PET. The top panel shows areas where there was a statistically significant correlation between NAA/Cre measured in the DLPFC and blood flow measures during the execution of the WCST in 13 patients with schizophrenia. Multiple areas in a network that has been associated with working memory are highlighted. The graphs below show the correlation between NAA/Cre in the DLPFC and performance on the WCST (on the left) and on a sensorimotor control task. (From Bertolino et al.[17]).

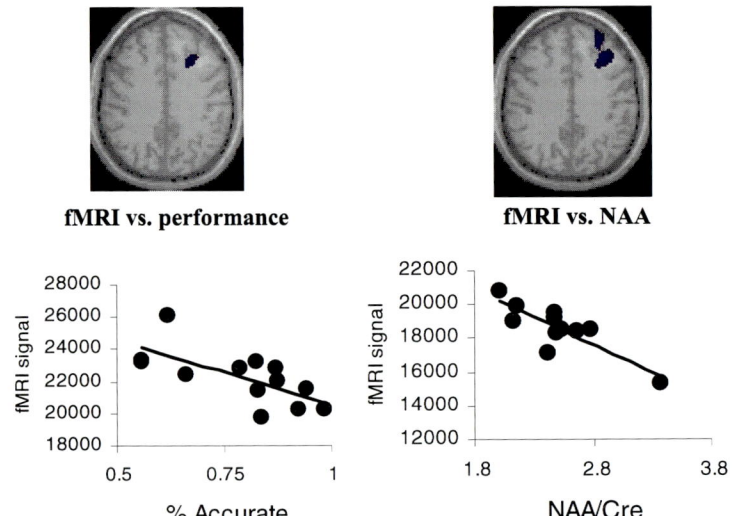

Figure 4. Relationship of NAA/Cre in the DLPFC and fMRI BOLD signal during execution of the N-Back task.The two pictures on the top show which areas of brain have a negative correlation of BOLD signal measured with fMRI during execution of the N-Back task with performance (on the left) and NAA/Cre in the DLPFC (on the right). There is overlap between the areas that correlate with performance and NAA/Cre and all belong to the DLPFC. The graphs show the distribution of the data points for the areas shown above. fMRI signal is in arbitrary units. (From Callicott et al.[18]).

INDEX